Lecture Notes in Bioengineering

Lecture Notes in Bioengineering (LNBE) publishes the latest developments in bioengineering. It covers a wide range of topics, including (but not limited to):

- Bio-inspired Technology & Biomimetics
- Biosensors
- Bionanomaterials
- Biomedical Instrumentation
- Biological Signal Processing
- Medical Robotics and Assistive Technology
- Computational Medicine, Computational Pharmacology and Computational Biology
- Personalized Medicine
- Data Analysis in Bioengineering
- Neuroengineering
- Bioengineering Ethics

Original research reported in proceedings and edited books are at the core of LNBE. Monographs presenting cutting-edge findings, new perspectives on classical fields or reviewing the state-of-the art in a certain subfield of bioengineering may exceptionally be considered for publication. Alternatively, they may be redirected to more specific book series. The series' target audience includes advanced level students, researchers, and industry professionals working at the forefront of their fields.

Indexed by SCOPUS, INSPEC, zbMATH, SCImago.

Mohd Hasnun Arif Hassan · Raihana Sharir · Syed Faris Syed Omar · Hiroyuki Nunome · Weerawat Limroongreungrat
Editors

Proceedings of the 3rd International Conference on Innovation and Technology in Sports

ICITS 2024, 13–15 December, Bukit Jalil, Malaysia

Springer

Editors
Mohd Hasnun Arif Hassan
Sports Technology Innovation Centre (STIC)
Universiti Malaysia Pahang Al-Sultan Abdullah
Pekan, Malaysia

Syed Faris Syed Omar
National Sports Institute of Malaysia
Kuala Lumpur, Malaysia

Weerawat Limroongreungrat
Sports Science and Technology
Mahidol University
Nonthaburi, Thailand

Raihana Sharir
Sports Science and Recreation
Universiti Teknologi MARA
Shah Alam, Malaysia

Hiroyuki Nunome
Sports and Health Science
Fukuoka University
Fukuoka, Japan

ISSN 2195-271X ISSN 2195-2728 (electronic)
Lecture Notes in Bioengineering
ISBN 978-981-95-0803-7 ISBN 978-981-95-0804-4 (eBook)
https://doi.org/10.1007/978-981-95-0804-4

This Springer imprint is published by the registered company Springer Nature Singapore Pte Ltd.
The registered company address is: 152 Beach Road, #21-01/04 Gateway East, Singapore 189721, Singapore

Preface

It is with great pleasure that we present the proceedings of the 3rd International Conference on Innovation and Technology in Sports (ICITS) and the 10th Asian Society of Sports Biomechanics Conference (ASSB), held from December 13 to 15, 2024, at the National Sports Institute of Malaysia, Bukit Jalil, Kuala Lumpur. This joint conference brought together researchers, practitioners, and industry experts from around the world to discuss and explore the latest advancements in sports technology and biomechanics.

The theme of this year's conference, *Empowering Tomorrow's Champions Through Cutting-Edge Sports Technology and Biomechanics*, underscores the crucial role of innovation in shaping the future of sports. As the demands of modern sports evolve, so too must the tools, techniques, and scientific approaches used to enhance athletic performance, improve safety, and refine training methodologies. This conference provided a platform for the exchange of knowledge and collaboration among experts dedicated to pushing the boundaries of sports science.

The papers compiled in this proceedings volume cover a broad spectrum of topics that are at the forefront of contemporary sports research. Key areas of discussion include:

- **Sports Instrumentation**: Advancements in state-of-the-art tools and devices revolutionizing sports performance analysis.
- **Sports Materials**: Innovations in material engineering contributing to enhanced safety and efficiency in sports equipment.
- **Sports Biomechanics**: Insights into human movement analysis to optimize athletic performance.
- **Sports Equipment**: The latest breakthroughs in the design and functionality of sports gear.
- **Design and Simulation in Sports**: The use of virtual environments and computational models to drive sports excellence.
- **Data Analytics in Sports**: Leveraging big data and artificial intelligence to refine athlete performance and strategy.

- **Performance and Motion Analysis**: Advanced methodologies for assessing and improving movement efficiency.
- **Coaching Innovation**: Integration of new coaching techniques and technologies to support athlete development.
- **Sports Medicine**: Exploring medical advancements aimed at improving athlete health and longevity.
- **Sports Trauma and Injury Prevention**: Strategies to mitigate and rehabilitate sports-related injuries.
- **Foot, Footwear, and Gait Biomechanics**: Investigating the role of footwear and gait mechanics in athletic performance and injury prevention.

This collection of research represents the collective efforts of distinguished scholars and professionals who are committed to driving the future of sports science. We extend our deepest gratitude to all authors for their valuable contributions, as well as to the reviewers whose meticulous evaluations have ensured the high quality of these proceedings. Additionally, we would like to express our sincere appreciation to the organizing committee, sponsors, and all participants whose support and engagement made this conference a success.

We hope that this compilation of cutting-edge research will serve as an invaluable resource for academics, practitioners, and industry leaders alike, inspiring further innovation in the realm of sports technology and biomechanics.

Pekan, Malaysia — Mohd Hasnun Arif Hassan
Shah Alam, Malaysia — Raihana Sharir
Kuala Lumpur, Malaysia — Syed Faris Syed Omar
Fukuoka, Japan — Hiroyuki Nunome
Nonthaburi, Thailand — Weerawat Limroongreungrat

Contents

Biomechanical Responses Following a 90 min Ball-Oriented Soccer Simulation ... 1
Muhammad Hamdan, Raihana Sharir, Wee Kian Yeo, and Raja Mohammed Firhad Raja Azidin

Ball Speed Assessment of a New Sepaktakraw Ball Launcher ... 11
Burhan Basyiruddin, Zulfakri Mohamad, Mohad Anizu Mohd Nor, and Shariman Ismadi Ismail

Kinematics Tracking Performance of In-Match Football Activities Using Smartphone-Based Radar Gun ... 21
Siti Nur Aqilah Rusli and Khairul Imran Sainan

Optimization of Retinoic Acid Loaded Nanosponge Formulation for Active Individuals: Impact of Drying Method and Homogenization Speeds ... 35
Sharifah Shakirah binti Syed Omar and Hazrina Ab Hadi

Optimizing Athletic Abilities: Exploring the Effects of Complex Training on Speed, Change of Direction and Explosive Power in Youth Soccer Players ... 45
Muhammad Iftiqar Ahmad Termizi, Raja Mohammed Firhad Raja Azidin, and Raihana Sharir

The Effect of Physical Activity Brain Breaks Among University Students ... 69
Ngien Siong Chin, Boon Sian Teo, and Matheus Yan Yu Chin

Correlation of Strength Characteristics with Running Economy in Well-Trained Male Distance Runners ... 79
Tham Thaiyanont, Thareerat Sairaya, Paweena Tanapin, Purinat Haowhan, Nattawut Meepol, Benthiwa Surasartpisal, Panupong Patarachao, Phitchapa Konthasing, Taspol Keerasomboon, and Parunchaya Jamkrajang

EMG Monitoring of Motorcyclists' Muscle Activation During Two Different Riding Postures 95
Wan Zaim Hazim Wan Hilmi, Nur Muhamad Farhan Mohd Johan, Mohammad Azzeim Mat Jusoh, and Helmi Rashid

The Association of *ACE I/D* and *ACTN3 R577X* Polymorphisms with Athlete Status, Playing Status and Vertical Jump Performance in Malaysian Footballers 111
Nur Fatihah Saidi, Chee Hian Tan, Raja Mohammed Firhad Raja Azidin, Choo Yee Yu, and Geik Yong Ang

Match Performance Variables of Thai Football Players in the First Division and the Second Division Teams 121
Sirawut Pukaror, Arom Treeraj, Sarit Suwanmana, and Weerawat Limroongreungrat

Biomechanical Effects of Golf Club Length on Address Position Among Collegiate Golfers 127
Muhamad Safiq Saiful Annur, Hosni Hasan, Shariman Ismadi Ismail, and Muhamad Noor Mohamed

Optical Tracking System Review: Results of Validity and Reliability Studies on Prescribed Circuits in Football 139
Khairul Imran Sainan, Ahmad Khushairy Makhtar, and Zulkifli Mohamed

Evaluating the Impact of Acceptance and Commitment Therapy Combined with Structured Exercise (ACT-EX) on Weight Loss in Overweight and Obese Young Adults: A Randomized Controlled Trial 155
Patricia Pawa Pitil and Siti Raudzah Ghazali

The Need for Software in Assisting Coaches in Johor with Generating Periodization 171
Mohamed Rafi M. S. Solaiman, Asha Hasnimy Binti Mohd Hashim, Adjah Naqkiah Mazlan, and Nur Fatihah Raimi

Does the Dynamics Strength Index Associate with the Swimming Performance in Age-Group Swimmers? 179
Radomyos Matjiur, Siwakorn Akkaneekool, Sakchai Pitakwong, Weerawat Limroongreungrat, Rodrigo Zacca, Karla de Jesus, and Phornpot Chainok

The Use of Technology Integration Software to Enhance Level of Movement Skills Based on Simpson's Psychomotor Domain 189
Adjah Naqkiah Mazlan, Jamalludin Harun, Zainal Abidin Zainuddin, and Nur Fatihah Raimi

A Meta-Analysis of Drone Technology in Sports 199
Fiderica Jeply, Diyana Zulaika binti Abdul Ghani, and Ahmad Nabil bin Md. Nasir

Enhancing Volunteer Management Using Preference Matching Algorithm 209
Suhizaz Sudin, Norfadila Mahrom, Priyadharishini Thurairaju, and Ahmad Faizal Annuar

EMG Analysis of Selected Lower Leg Muscles During Plantar Flexion in Various Sagittal Foot Positions 219
B. B. Shyamily, K. H. Shreaya, and Gerald Santhosh Dsouza

Comparative Analysis of Heel Contact Force and Electromyographic Activity Among Different Nordic Hamstring Exercises 239
Tibeth Sukantameth, Jutharat Promson, Nawarat janpolsaen, Prisana phonhan, Sirinya Audonphon, Radomyos Matjiur, Siwakorn Akkaneekool, Rodrigo Zacca, and Phornpot Chainok

Sex Comparison of ACL Injury Risk Using a Clinic-Based Algorithm Among Volleyball Athletes UiTM 249
Raihana Sharir, Muhammad Hidayat Jamaludin, Ezhan Iqbal Zulkiflie, and Radin Rafeeuddin

Effectiveness of Foam Rolling in Reducing Delayed Onset Muscle Soreness (DOMS) 255
Nur Wahida Amalin Muhammad Helmi Kannan, Nur Fatihah Raimi, Adjah Naqkiah Mazlan, and Ayu Suzailiana Muhamad

Effects of Hand–Eye Coordination Training on Accuracy and Reaction Time Among E-Sports Players 267
Mohamad Aiman Hakimi Mamat, Nur Fatihah Raimi, and Adjah Naqkiah Mazlan

Correlation Between Leg Muscle Power, Flexibility and Shooting Precision in Blind Football: A Study of Malaysian Athletes 277
Sohaila Ishak and Hosni Hasan

A Study on the Muscle Performance of Drivers During Long Versus Short-Cornered Course Using Electromyography (EMG) 289
Muhammad Izzdihar Mahadzir, Mohammad Azzeim Mat Jusoh, Ahmad Khushairy Makhtar, and Sukarnur Che Abdullah

Early Phase Development of Training Efficiency Application for Seni Silat Cekak Malaysia Using EEG Signal 299
Aznor Hanah Abdul Halim, Rohani Farook, Fakhrul Mukmin Mansor, M. F. Mohd Fairuz, Siti Azura Abuzar, A. R. Mohamad Rizal, and Suhizaz Sudin

Real-Time Monitoring and Alert System for Hikers Using IoT Technology 311
Muhammad Shazwan Baderol, Siti Azura Abuzar, Dayang Khadijah Hamzah, Aznor Hanah Abdul Halim, Rohani Farook, and Suhizaz Sudin

Evaluating the Hydrodynamic Performance of Macon, Big, and Fat Blades During the Drive Phase in Rowing 319
Ab Aziz Mohd Yusof, Hazim Sharudin, Haszeme Abu Kasim, Ainaa Maya Munirah Ismail, Ahmad Naufal Adnan, and Kamariah Md Isa

A Systematic Review of Position-Specific for Anthropometric in Basketball Elite Athlete 333
Gao Yili, Hadafi Fitri Mohd Latip, and Mohd Shahrizal Sunar

Comparative Study of Anthropometry and Body Composition of University Basketball Players in China and Malaysia 345
Gao Yili, Hadafi Fitri Mohd Latip, and Mohd Shahrizal Sunar

A Low-Cost Goal Line Technology (GLT) Solution for Grassroots Football: Performance Analysis of Ball Detection across Variable Speeds 357
Ilmi Mohd Ariffin, Zulkifli Mohamed, Mohd Hanif Mohd Ramli, Mohd Rozaiman Aziz, and Azlan Aziz

The Relationship between Dynamic Balance and Soccer Skill Performance among Female Soccer Players 367
Nurul Nabilah Azizul, Nur Hafizatul Alya Yahya, Nur Syazliana Suhaimi, Aimie Shyzleena Che Suhaimi, and Siti Azilah Atan

Investigation on Golfer's Stability Towards the Swing Profile 381
Atiqullah Huzaifah, Muhammad Firdaus Mohd Nor, Mohd Nadzeri Omar, Nasrul Hadi Johari, Mohd Hasnun Arif Hassan, Mahadzir Ishak, and Zuazizi Aziz

Development of Low-cost LED-Driven Boxing Machine 393
Muhammad Muhaimim Mahmud, Muhammad Haniff Gusrial, Nur Aqilah Othman, and Mohd Hasnun Arif Hassan

A 3-Dimensional Finite Element Modelling of Insole for Flatfoot Children 407
Mohd Firdaus Mahamad Ali and Nadzifah Yaakub

Data Analytics in Sport Science—Current Applications and Opportunities in an Unexplored Frontiers 419
Muhammad Nur Farhan Saniman, M. H. Muhammad Sidik, Ahmad Shahir Jamaludin, Mohd Amir Shahlan Mohd Aspar, Devin Babu, Abdul Nasir, and Norain Binti Abdullah

Current Trend in Preferred Deadlift Stance Among Elite Powerlifters 435
Fareedzul Hareez Mohd Pudzi, Mohd Hanifa Sariman, Raja Mohammed Firhad Raja Azidin, Raihana Sharir, and Radin Rafeeuddin

Optimizing Stack Height in Running Footwear: Engineering Insights for Enhancing Biomechanical Performance 461
Abdul Nasir, Norain Binti Abdullah, Devin Babu, Muhammad Haziq, Mohd Amir Shahlan Mohd Aspar, and Muhammad Nur Farhan Saniman

Wearable Gait Analysis in Neurological and Musculoskeletal Disorders: From Monitoring to Prevention 471
Mohd Nizar Mhd Razali, Nurul Hasya Md Kamil, Muhamad Ridzuan Radin Muhamad Amin, Dewan Muhammad Nuruzzaman, and Ahmad Shahir Jamaludin

Leveraging Virtual Reality to Optimize Mental Imagery and Skill Acquisition for Elite Athletes: A Review 481
Ainur Munira Rosli, Mohd Nizar Mhd Razali, Muhamad Rusydi Muhamad Yasin, Nurul Hasya Md Kamil, and Ahmad Shahir Jamaludin

Data-Driven Strategies for Injury Prevention in Sports: Integrating Biomechanics and Wearable Analytics 493
Amirul Hakim Sufian, Mohd Nizar Mhd Razali, Nurul Hasya Md Kamil, Ahmad Redza Ahmad Mokhtar, and Ahmad Shahir Jamaludin

AI-Driven Performance Metrics: Enhancing Athlete Monitoring and Injury Prevention 503
Nurul Hasya Md Kamil, Mohd Nizar Mhd Razali, Syukran Hakim Norazman, Mohd Arifpin Mansor, and Ahmad Shahir Jamaludin

Biomechanical Profiling: Review on Real-Time Data Integration and Predictive Models for Performance Optimization 513
Noorazliza Sulaiman, Ahmad Shahir Jamaludin, Mohd Nizar Mhd Razali, Muhamad Ridzuan Radin Muhamad Amin, and Abdul Nasir

Flexible Anti-Stabbing Fabrics in Martial Arts Gear: Design, Material Properties, and Performance Evaluation 523
Ahmad Shahir Jamaludin, Nurul Hasya Md Kamil, Khairul Ikmal Azmi, Mohamad Rusydi Mohamad Yasin, and Mohd Nizar Mhd Razali

Simulating Reality: High-Fidelity Simulators and Its Impact on Cognitive and Physical Performance of an Athlete 531
Ahmad Shahir Jamaludin, Mohamad Rusydi Mohamad Yasin, Abdul Nasir, Mohd Arifpin Mansor, and Mohd Nizar Mhd Razali

Development of a Sensor-Based Tool for Assisting Beginners in Table Tennis: A Case Study on Forehand Grip Consistency 539
Syukran Hakim Norazam, Muhamad Ridzuan Radin Muhamad Amin, Mohd Amir Shahlan Mohd Aspar, Muhammad Nur Farhan Saniman, and Abdul Nasir Abd Ghafar

A Study on Muscle Activities During Stoop Lifting Movement 553
Nurhidayah Omar, Munawwarah Solihah Muhammad Isa, Mohammad Shahril Salim, Saidatul Ardeenawati Awang, and Suhizaz Sudin

The Association Between Pain Characteristics, Psychological Aspects, Functional Disability and Postural Sway in Individuals with Nonspecific Chronic Low Back Pain: The Preliminary Study 559
Rahmat Adnan, Sarah Syafieqah Anuwar, Dzulhafiz Zikrie Dzolkapli, Siti Nurul Ezzah Che Man, Hammad Haziq Ismail, Muhammad Shaleh Mazuki, Motohide Arayama, Norasrudin Sulaiman, Hashbullah Ismail, Mazuin Mohd Razalli, and Shariman Ismadi Ismail

Biomechanical Responses Following a 90 min Ball-Oriented Soccer Simulation

Muhammad Hamdan, Raihana Sharir, Wee Kian Yeo, and Raja Mohammed Firhad Raja Azidin

Abstract

Background It is believed that biomechanical changes occur in the lower limbs because of physiological and neuromuscular loading accumulated over an extended duration and compromise the integrity of the Anterior Cruciate Ligament (ACL). Generally, joint kinematics and ACL loading were assessed in previous studies via a selection of unilateral or bilateral landing of dynamic tasks following a set of soccer running profile. However, very few studies have fully incorporated the ball-handling component into their investigations. The aim of this study was to investigate the lower limb biomechanical changes following a Ball-Oriented Soccer Simulation (BOSS) during unilateral and bilateral landing tasks.

Methods Seventeen (n = 17) male participants volunteered for this study. Participants were required to complete the BOSS. At selected points in the study, participants were required to perform three successful trials of bilateral landing tasks, and unilateral landing tasks to assess biomechanical properties of both the right and left

M. Hamdan (✉)
Faculty of Sports Science and Recreation, Universiti Teknologi MARA, Pahang Branch, Jengka Campus, Jengka, Malaysia
e-mail: muhammadhamdan@uitm.edu.my

R. Sharir · R. M. F. Raja Azidin
Faculty of Sports Science and Recreation, Universiti Teknologi MARA, Selangor Branch, Shah Alam Campus, Shah Alam, Malaysia

W. K. Yeo
National Sports Institute, Bukit Jalil, Malaysia

R. Sharir
Research Institute for Sport and Exercise Sciences, Liverpool John Moores University, Liverpool, UK

R. M. F. Raja Azidin
Selangor Football Club, Shah Alam, Malaysia

Universitas Negeri Malang, Malang, Indonesia

M. H. A. Hassan et al. (eds.), *Proceedings of the 3rd International Conference on Innovation and Technology in Sports*, Lecture Notes in Bioengineering,
https://doi.org/10.1007/978-981-95-0804-4_1

lower limbs during landing. Knee and hip extension angles at initial contact and peak knee abduction moments were of interest in this study.

Results Mixed-models ANOVA indicate no significant difference across all knee and hip angles throughout the simulation. Peak knee abduction moments were significantly higher in the left knee during bilateral landing throughout the simulation ($p < 0.05$), however, the situation was somewhat reversed during unilateral landings.

Discussion Findings may indicate compensation strategies during bilateral landing that may mask injury risk status. Further investigation is warranted to seek deeper understanding of locomotor control and limb dominance on biomechanical injury risk following prolonged, repetitive exertions in soccer.

Keywords Biomechanics · Anterior cruciate ligament (ACL) · Soccer simulation · Landing mechanics · Knee abduction moment

1 Introduction

Anterior cruciate ligament (ACL) injuries have been a topic of extensive research. In soccer, the lower limb biomechanics is an important aspect in the study of ACL injury risk, prevention, as well as rehabilitation. ACL injuries are generally non-contact in nature and have been observed to occur at a more likelihood during the later stages of each half of a match from biomechanical impairments such as muscular strength reductions, increased knee laxity and kinematic alterations. Previous biomechanical studies into non-contact ACL injuries have been studying the influence of soccer-specific exertions on the lower limb joint kinematics and the loading of the ACL using treadmill or overground running simulations matching running activity profiles of actual match-play (Greig 2009; Hamdan et al. 2018a, b; Hamdan et al. 2018a, b; O'Connor et al. 2009; Raja Azidin et al. 2015; Sanna and O'Connor 2008). It is believed that biomechanical changes occur in the lower limbs because of physiological and neuromuscular loading accumulated over an extended duration and compromise the integrity of the ACL. Generally, joint kinematics and ACL loading were assessed in these studies via a selection of unilateral or bilateral landing of dynamic tasks.

However, very few studies have fully incorporated the ball-handling component into their investigations. Fine, coordinative tasks such as dribbling a ball or passing and shooting as well as repeated jump-landing movements such as during ball heading in soccer are seldom recreated up to scale in simulated soccer match-play and improve ecological validity to existing findings in literature. In a match-play setting, counter movement jumps performed during a ball heading and dribbling or passing a ball are open chained and may be influenced by a variety of conditions occurring on-field. Thus, the utility of ball-handling tasks needs to be considered in biomechanical studies on ACL loading during soccer match-play. Recently, a study by Hamdan et al. (2020) introduced a soccer match-play simulation that incorporates these ball-handling tasks up to the scale of their utility profiles in actual matches. In their study, eccentric muscle strength was studied, but no kinematic assessment was performed with respect to ACL injury risk status following the ball-handling activities.

The aim of this study was to investigate the lower limb biomechanical changes following a Ball Oriented Soccer Simulation (BOSS) during unilateral and bilateral landing tasks.

2 Materials and Methods

A priori power calculation was conducted to determine an appropriate sample size to establish differences between limbs using G*Power (G*power v 3.1.9.2, Universität Kiel, Germany). To achieve an approximate of 80% statistical power with an alpha (α) level of 0.05, 8 samples were required per group (limb). Seventeen (n = 17) male, healthy and recreationally active participants volunteered for this study (age: 25.4 ± 4 years old; body mass: 65.7 ± 9.6 kg; height: 1.7 ± 0.05 m). All participants trained for at least two sessions per week, with each session lasting for at least an hour. Participants were questioned on their history of injuries, and none had admitted to any previous lower limb injury within six months prior to the testing. Written informed consent was obtained from the participants, and the study was performed in accordance with the ethics committee guidelines from Universiti Teknologi MARA (Ref: 600-IRMI (5/1/6)) and the National Sport Institute, Malaysia (ISNRE: RE/A/004/2020–006/2019).

The study took form of a single-group, repeated measures design where participants completed a 90-min BOSS protocol (Hamdan et al. 2020). Further details regarding the BOSS has been described elsewhere (Hamdan et al. 2020; Hamdan and Raja Azidin 2022). Participants were reminded to avoid strenuous exercise, caffeine, and alcohol for 24 h prior to testing and were invited to attend a familiarization session at least three days before the actual session. During the BOSS protocol a 15-min passive rest was implemented between minutes 45 and 60 to represent half-time in soccer matches. At selected time points throughout the protocol, participants completed three trials of drop vertical jump task from a 0.3 m elevated platform placed at 50% of the participants' height from the designated landing area and 3 trials of a single leg hop task 75% of the participants' height from the designated landing area. To limit inter-trial variability, a successful trial was defined as a landing where participants maintained their balance at least 2 s after landing. Heart rate and rating of perceived exertion were monitored continuously at 5-min intervals throughout the protocol.

Twenty-four (24) optoelectronic cameras (OQUS 3, Qualisys AB, Gothenburg, Sweden) and two 0.9 × 0.6 m embedded force plates (Kistler, Winterthur, Switzerland) sampling at 240 Hz were used to collect motion and ground reaction force data in three-dimensional (3D) space. These cameras recorded the trajectories of 44 spherical reflective markers positioned according to the LJMU kinematic model

(Vanrenterghem et al. 2010) on the participants to generate biomechanical markers of ACL injury risk. Functional hip and knee joint centres were calculated for maximum error reduction of anatomical landmark placement over these joints (Robinson and Vanrenterghem 2012) using Visual3D (C-motion, Germantown, MD, USA). An inverse kinematic model was utilized to reduce errors deriving from surface movement and soft tissue artifacts. Knee joint moments were calculated using inverse dynamics.

Marker trajectories and force data were low-pass filtered at 18 Hz (Kristianslund et al. 2012) prior to inverse dynamics computations. Peak knee abduction moments were derived from the frontal plane knee moments during weight acceptance phases in both unilateral and bilateral landings as defined in Dempsey et al. (2007). For both unilateral and bilateral landings, initial contact was defined as the time instance when ground reaction force exceeds 20 N. During this instance, sagittal plane knee and hip joint angles were calculated.

Knee and hip extension angles at initial contact (IC) as well as peak knee abduction moments were treated as dependent variables with time and limb side as independent variables. A 2 (side: left, right) $\times$ 8 (time: 0 min, 15 min, 30 min, 45 min, 60 min, 75 min, 90 min, 105 min) repeated measures ANOVA was employed to identify significant differences with α set at 0.05 using the statistical software package SPSS (Version 28, IBM SPSS Inc., USA). The assumption of equality of variance was tested using Mauchly's test of sphericity. If there was a violation of the assumption, the Greenhouse–Geisser epsilon was referred to. The Greenhouse–Geisser correction values was used if the epsilon was <0.75, and the Hyunh-Feldt correction was used if otherwise (Girden 1992). To reduce the probability of Type I errors, post-hoc analyses were conducted using Bonferroni procedures.

3 Results

No significant main effect of time was observed in both unilateral and bilateral landings for all dependent variables ($p > 0.05$). However, there was an observable change of in knee extension angles at initial contact. The knees were more extended towards the later stages of the BOSS protocol by up to 17% in both unilateral and bilateral landing tasks. Hip extension angles during bilateral landings were 3–5% more extended compared to before the commencement of the BOSS. Interestingly, the knee joint moments in bilateral landings were significantly different between the left and right limbs from the start of the study ($p < 0.001$). The left knees endured $2 - 3$ times the abduction moments acting on the right knees. On the other hand, the during unilateral landings. Peak knee abduction moments were generally 1–2 times higher in the right knees compared to the left knees, especially at the end of both halves of the BOSS (min 45: $p = 0.035$; min 105: $p = 0.018$). A graphical representation of the study findings can be found in Fig. 1.

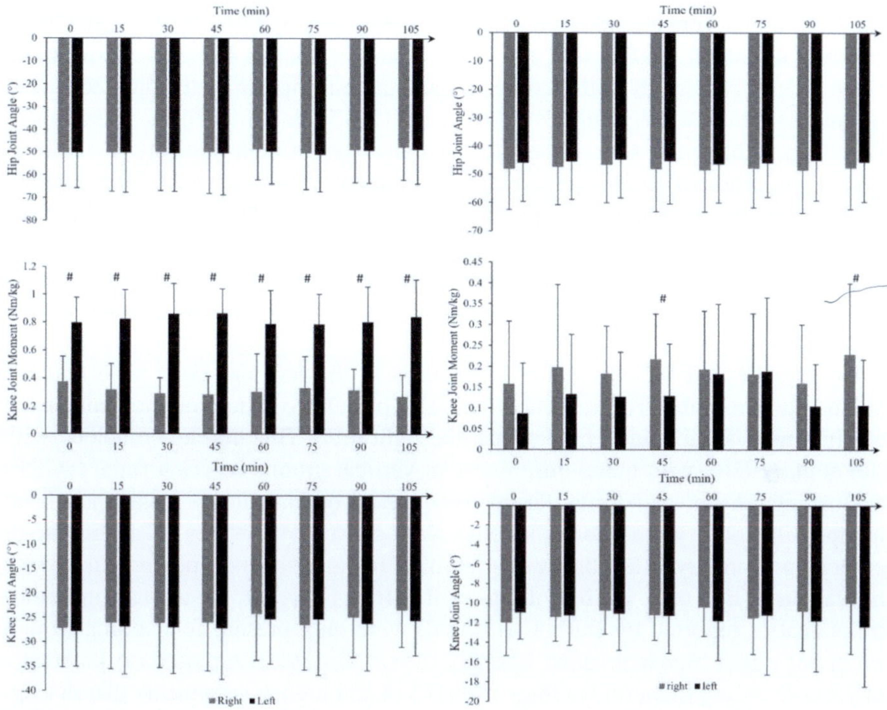

Fig. 1 Hip angles (top row), knee angles (middle row), and knee moments (bottom row) during bilateral (left column) and unilateral (right column) landings. # Represents significant differences between left and right limbs at their corresponding time point

4 Discussion

4.1 Landing Biomechanics

Increasing knee extension angles (up to 16.9%) were present throughout unilateral and bilateral landings throughout the BOSS. However, primary findings in the repeated measures ANOVA with post-hoc analyses revealed no differences at any one point in time compared to the beginning of the protocol (0 min), indicating that the changes were not significant. The increased knee extension angle (~2°) matched observations in existing literature (Bossuyt et al. 2015). Previous studies by Raja Azidin et al. (2015) and Bossuyt et al. (2015) have reported increases in knee extension angles during initial contact of unilateral landing tasks. They argued that the biomechanical changes may be resultant kinematics caused by reductions in muscle strength in the lower limbs. Specifically, reductions in eccentric hamstring contractions would indicate a compromised ability to counteract the torque from concentric quadriceps contractions during a knee extension motion (Hashemi et al. 2011). Several factors may explain the discrepancies between the current study findings and the findings from previous studies. Firstly, the study by Raja Azidin et al. (2015) utilized a unilateral landing task at a high entrance velocity (45° side-cutting task), which includes a change-of-direction component, thus increasing the biomechanical demand of the task. Secondly, Bossuyt et al. (2015) investigated the biomechanical changes following soccer-specific exertions using a high intensity, short duration fatigue protocol on female participants as opposed to an aerobic, long duration soccer simulation on male participants in the current study. The differences between the male and female skeletal structure (i.e., lumbar curve, hip width, femur angle) as well as the predominant energy systems utilized in the match-play simulations (aerobic versus anaerobic) may influence the outcome differences between the two studies.

Despite the lack of observable differences in the left and right knee and hip kinematics during the drop vertical jump tasks, peak knee abduction moments in the left limb were significantly higher than the right limb. This finding coincided with Lim et al. (2020), who noted differences in vertical ground reaction force (vGRF) and loading rate between limbs during plyometric drop jump tasks. It was speculated that the bilateral asymmetries in kinetics were caused by the step-off technique or temporal asymmetry due to the trailing limb having insufficient time to catch up with the lead limb (Ball et al. 2010; Collings et al. 2019). However, bilateral asymmetries in kinematics reported by Ball et al. (2010) were only present for landing from a 0.2 m but not at higher heights, whereas Lim et al. (2020) reported no kinematic differences during bilateral landings from 0.3 m and higher, suggesting that in step-off landings from >0.3 m, the trail limb may have ample time to catch up to the lead limb and improve bilateral symmetry in kinematics. This argument could also be proven in the bilateral symmetry in knee kinematics during bilateral landings in this study. Lim et al. (2020) suggested that it may seem that the limb that was tasked with greater load had taken a functional lead during the task, and that athletes may gradually overload the limbs with functional lead limb and become detrimental to the athletes' injury risk status. As demonstrated in this study, the functional lead limb during bilateral landings may contradict the participants' dominant/preferred limb, thus may present confusion when considering the speculation that limb preference may be the cause of bilateral kinetic asymmetries in drop jumps by De Britto et al. (2015).

An interesting point to note in this observation is that the peak knee abduction moments during the single leg hop trials did not reveal any significant differences between the left and right limbs apart from at the end of the first half (45 min) and at the end of the second half (105 min), however, the loads were notably higher in the right compared to the left limbs. This situation may seem similar to bilateral landings observations, however, is reversed in limb sides. In this particular case, the speculation on the role of limb preference by De Britto et al. (2015) seems easier to comprehend, as the full derivative of the knee abduction moment is loaded onto one knee at a time, and unilateral landings are utilized more in soccer (i.e., sprinting, striding, bounding, and cutting). Upon seeing the reversal of knee joint loading across the unilateral and bilateral landing tasks in this study, another speculation comes into the narration. Perhaps, bilateral landings may allow the opportunity to implement compensation strategies to 'protect' or 'mask' the dominant limb from vulnerability to injuries. Unfortunately, this does not completely explain the difference in landing kinetics across the left and right knees during bilateral landings in this study. This may be concerning as the jumps and landings in soccer ball headings may cause alterations in the jump performance and biomechanics (Ford et al. 2005) as knee joint moments are observed to increase significantly in jump performances involving an overhead target. This may cause the landing mechanics and strategy to be compromised, thus increasing vulnerability towards ACL injury risk. Further research into between-limb asymmetries should be warranted to improve understanding on the role of limb dominance on the joint kinetics during unilateral and bilateral landings.

4.2 Practical Implications

The findings in this study highlights a more extended knee orientation at IC during landings as soccer-specific exertions accumulate overtime. This may implicate a need to utilize intervention methods to offset the changes due to fatigue such as training for neuromuscular control in a fatigued state or incorporating injury prevention exercises into training regimes. Apart from that movement kinematics, rigorous, in-depth kinetic assessments of various dynamic tasks should also be considered in clinical screening for injuries and return-to-play assessments in sports with bilateral symmetry included as an objective in these assessments. Furthermore, according to Lim et al. (2020), since kinetic asymmetries are impossible to be observed visually, perhaps bio-feedback systems specialised in monitoring landing forces and may be considered during interventions for correcting the uneven loading in the lower limbs.

4.3 Limitations

This study does not come with no limitations, and therefore some of the findings should be interpreted with caution. Previous studies into non-contact ACL injuries have been pursuing for more aggressive interrogation of the ACL loading by using higher velocity movements with multidirectional components withing the biomechanical trials such as the triple hop tasks and the side cutting tasks. These tests were less feasible for the laboratory layout in this study, thus were not conducted. Thus, the utility of the drop vertical jump and single leg hop tasks in this study may reveal a smaller picture of the actual changes occurring in the lower limbs due to fatigue. In order to improve external validity of the tests, lesser experimental control was applied (Collings et al. 2019) to allow participants perform the tasks to the best of their ability, such as freedom of footwear (court shoes) models and landing strategy (toe-to-heel or heel-to-toe) employment.

5 Conclusions

Notable increased knee extension could be observed over time following the BOSS protocol. This indicates increased biomechanical injury risk with fatigue development form soccer-specific exertion. There was also a notable difference in peak knee abduction moments between the left and right limbs during the bilateral landing trials, which was reversed in the unilateral landing trials, suggesting that there may be an influence of limb dominance landing strategy during bilateral landing that may be detrimental to the risk of injury among soccer players. Further investigation should be warranted to address the bilateral differences in joint kinetics despite the symmetries in joint kinematics of a bilateral landing task.

Acknowledgements The authors would like to extend their appreciation to the National Sports Institute, Bukit Jalil, Malaysia, for providing the facilities, equipment, and software needed for conducting the study.

References

Ball NB, Stock CG, Scurr JC (2010) Bilateral contact ground reaction forces and contact times during plyometric drop jumping. J Strength & Cond Res 24(10):2762–2769

Bossuyt FM, García-Pinillos F, Raja Azidin R, Vanrenterghem J, Robinson MA (2015) The utility of a high-intensity exercise protocol to prospectively assess ACL injury risk. Int J Sports Med 95(02):125–133

Collings TJ, Gorman AD, Stuelcken MC, Mellifont DB, Sayers MGL (2019) Exploring the justifications for selecting a drop landing task to assess injury biomechanics: a narrative review and analysis of landings performed by female netball players [Review]. Sports Med 49(3):385–395. https://doi.org/10.1007/s40279-018-01045-x

De Britto MA, Franco PS, Pappas E, Carpes FP (2015) Kinetic asymmetries between forward and drop jump landing tasks [Article]. Rev Bras Cineantropometria E Desempenho HumO 17(6):662–671. https://doi.org/10.5007/1980-0037.2015v17n6p661

Dempsey AR, Lloyd DG, Elliott BC, Steele JR, Munro BJ, Russo KA (2007) The effect of technique change on knee loads during sidestep cutting. J Med Sci Sport Exerc 39(10):1765

Ford KR, Myer GD, Smith RL, Byrnes RN, Dopirak SE, Hewett TE (2005) Use of an overhead goal alters vertical jump performance and biomechanics. J Strength Cond Res 19(2):394–399

Girden ER (1992) ANOVA: repeated measures. SAGE

Greig M (2009) The influence of soccer-specific activity on the kinematics of an agility sprint. Eur J Sport Sci 9(1):23–33

Hamdan M, Raja Azidin RMF (2022) Lab-controlled soccer: a review of soccer match-play simulations. Malays J Sport Sci Recreat (MJSSR) 18(2):255–273

Hamdan M, Mohd Noh SN, Hasan H, Ismail H, Raja Azidin RMF (2018b) Knee and hip extension responses to prolonged simulated soccer match-play: a 2D study. Int J Eng Technol (UAE) 7(4):6–14

Hamdan M, Ismail SI, Hassan H, Ismail H, Raja Azidin RMF (2018) The effects of A 90-minute simulated soccer match-play on knee and hip kinematics. Mov, Health Exerc (MoHE) 7(1):59–72

Hamdan M, Ang GY, Sharir R, Yeo WK, Azidin RMFR (2020) Changes in hamstring eccentric peak torques and angles of peak torque following 90 minutes of soccer specific exertions. Malays J Mov, Health & Exerc 9(2)

Hashemi J, Breighner R, Chandrashekar N, Hardy DM, Chaudhari AM, Shultz SJ, Slauterbeck JR, Beynnon BD (2011) Hip extension, knee flexion paradox: a new mechanism for non-contact ACL injury. J Biomech 44(4):577–585

Kristianslund E, Krosshaug T, Van Den Bogert AJ (2012) Effect of low pass filtering on joint moments from inverse dynamics: implications for injury prevention. J Biomech 45(4):666

Lim YY, Sterzing T, Teo CJY, Alonzo R, Pan JW, Teng PSP, Kong PW (2020) Between-limb asymmetry in kinetic and temporal characteristics during bilateral plyometric drop jumps from different heights. J Sport Sci 38(14):1605–1614. https://doi.org/10.1080/02640414.2020.1752535

O'Connor KM, Monteiro SK, Hoelker IA (2009) Comparison of selected lateral cutting activities used to assess ACL injury risk [Article]. J Appl Biomech 25(1):9–21. https://doi.org/10.1123/jab.25.1.9

Raja Azidin RMF, Sankey SP, Drust B, Robinson MA, Vanrenterghem J (2015) Effects of treadmill versus overground soccer match simulations on biomechanical markers of anterior cruciate ligament injury risk in side cutting. J Sport Sci 33(13):1332–1341. https://doi.org/10.1080/02640414.2014.990491

Robinson MA, Vanrenterghem J (2012) An evaluation of anatomical and functional knee axis definition in the context of side-cutting. J Biomech 45(11):1941–1946

Sanna G, O'Connor KM (2008) Fatigue-related changes in stance leg mechanics during sidestep cutting maneuvers. J Clin Biomech 23(7):946–954. https://www.clinbiomech.com/article/S0268-0033(08)00128-9/fulltext

Vanrenterghem J, Gormley D, Robinson M, Lees A (2010) Solutions for representing the whole-body centre of mass in side cutting manoeuvres based on data that is typically available for lower limb kinematics. Gait Posture 31(4):517–521

Ball Speed Assessment of a New Sepaktakraw Ball Launcher

Burhan Basyiruddin, Zulfakri Mohamad, Mohad Anizu Mohd Nor, and Shariman Ismadi Ismail

Abstract The purpose of this study is to assess a new prototype of a ball launcher that is capable of generating serves characterized by uniform speed. A bespoke ball launcher (Launcher mechanism; 2 units of 24 V 10A 7000RPM sponge roller-diameter 15 cm, width 11.5 cm; Total height; 2.5 m; Total weight; 40 kg) was developed to replicate real-world sepaktakraw ball speed during sepaktakraw serve in elite matches. The ball speed at four different motor speeds (70, 90, 95, and 100% capacity) produced by the ball launcher was analysed using a high-speed camera (CASIO- EXILIM EX-ZR70). A speed radar gun (BUSHNELL-model 101,911) was also used to measure the launcher's ball speed. Data obtained from the high-speed camera were then processed using motion analysis software [Kinovea (V.0.9.5)] to calculate the ball speed. In addition, maximum ball launch speed from the ball launcher was compared to the maximum ball speed during serve execution produced by ten male elite sepaktakraw players. It was found that maximum ball speed analysed using the high-speed camera and the speed radar gun was highly similar (high-speed camera: 56.2 km/h; speed gun radar: 55 km/h). However, the average maximum ball speed produced by the ball launcher (56.1 ± 1.7 km/h) did not effectively replicate the average maximum ball speed produced by elite players (64.1 ± 5.3 km/h) and was significantly different [$t(9) = -4.57$, $p = 0.0002$]. Despite this setback, the developed ball launcher prototype demonstrates the capability to generate serves with consistent speed in line with competitive benchmarks. Upgrading the launcher specifications to improve its peak ball speed will be the next progression of this study.

Keywords Ball launcher · Sepaktakraw · High-speed camera · Speed radar

---

B. Basyiruddin (✉) · M. A. M. Nor · S. I. Ismail
Faculty of Sports Science and Recreation, Universiti Teknologi MARA, Shah Alam, Selangor, Malaysia
e-mail: burhanbasyiruddin@unj.ac.id

S. I. Ismail
e-mail: shariman_ismadi@uitm.edu.my

Z. Mohamad
College of Engineering, Universiti Teknologi MARA, Shah Alam, Selangor, Malaysia

M. H. A. Hassan et al. (eds.), *Proceedings of the 3rd International Conference on Innovation and Technology in Sports*, Lecture Notes in Bioengineering,
https://doi.org/10.1007/978-981-95-0804-4_2

1 Introduction

Sepaktakraw is a Malay cultural heritage sport that combines high acrobatic skills with speed and complex strategies (Lhee 2021; Musa 2020a). With its remarkable abilities, sepaktakraw has gained international recognition and is regularly contested in prestigious events such as the SEA Games and Asian Games. The uniqueness and beauty of this game lie not only in the techniques used but also in the skills demanded from the players (Hassan 2023; Tohidin et al. 2021; Al-Amien et al. 2021).

The main challenge in sepaktakraw is the ability to effectively receive the first ball or defend (Hassan 2024; Musa 2020b). In international competitions, especially against top countries like Thailand, the extremely high serve speed poses one of the biggest challenges for players (Hassan 2022; Ibrahim and Nasution 2021; Musa 2020c). According to previous research (Zarei 2021), the serve speed in sepaktakraw can reach up to 80 km/h, which requires quick reactions and excellent defensive techniques from players.

In training contexts, traditional methods such as wall drills, pairing with a partner, or using manual aids often face limitations (Amalia et al. 2023). Manual training can lead to instability in the ball delivery process, affecting the quality of the training. Tan (2022) noted that the limitations in manual methods are often due to the instability of the coach's condition and suboptimal training aids, impacting the effectiveness of the training.

Furthermore, current training methods may not fully reflect actual game conditions (Abdillah 2021). For example, training with manual aids such as wooden rackets or tennis rackets often cannot replicate the speed and direction variations faced in actual matches. Padli (2023) demonstrated that technical abilities like good service and ball reception greatly influence match outcomes, making it essential to use aids that more accurately reflect game conditions.

With technological advancements, many other sports have adopted advanced training aids, such as ball launching machines, to improve training consistency and quality (Musa 2020d). This technology allows for greater variation in speed, direction, and ball throw patterns, helping athletes better prepare for real game conditions. Zayar (2024) asserts that a consistent and reliable ball launcher is crucial for effective training.

However, in sepaktakraw, current ball launchers still face several challenges. Many of these devices have impractical designs, such as the height of the device making it difficult for coaches to operate and adjust during training. Xin (2021) observed that impractical device designs can hinder training effectiveness and cause discomfort for coaches.

Another challenge is the lack of validation regarding the effectiveness of ball launchers in improving players' defensive abilities. Although these devices are designed to assist in training, there has been insufficient research to show how effective they are in enhancing players' performance in matches. Previous research (Jufrianis. 2021) indicates that existing ball launchers have not been designed with optimal specifications for sepaktakraw training needs.

Given these challenges, there is an urgent need to develop a more effective and efficient sepaktakraw ball launcher (Irawan 2021; Musa 2020e; Ghany 2023). An optimal ball launcher should replicate actual game conditions, offer suitable variations and challenges for training needs, and enhance the effectiveness of the training process. This study aims to address these issues by developing a ball launcher optimized for defensive training. Prior to using the ball launcher, device validation in terms of ball launch speed is considered one of the most critical parts in its development.

The objective of this study is to assess a new prototype of a sepaktakraw ball launcher that is capable of generating serves characterized by uniform speed. The ball speed launched from the ball launcher was analysed using a speed radar gun and high-speed camera. These data were then compared to the ball speed from executed serve measured from elite players during a competitive elite sepaktakraw tournament. It is expected that the sepaktakraw ball speed measured using speed radar gun and high-speed camera will be comparable. It is also expected that the ball speed produced by the new prototype will be able to replicate the maximum ball speed produced by the elite player's serve.

2 Methodology

2.1 Research Procedure

A bespoke ball launcher (Launcher mechanism; 2 units of 24 V 10A 7000RPM sponge roller-diameter 15 cm, width 11.5 cm; Total height; 2.5 m; Total weight; 40 kg) as shown in Fig. 1 was developed to replicate real-world sepaktakraw ball speed during sepaktakraw serve in elite matches. The specifications of the ball launcher are shown in Table 1.

The initial ball speed at 4 different motor speeds (70, 90, 95 and 100% capacity) produced by the ball launcher was analysed using a high-speed camera (CASIO-

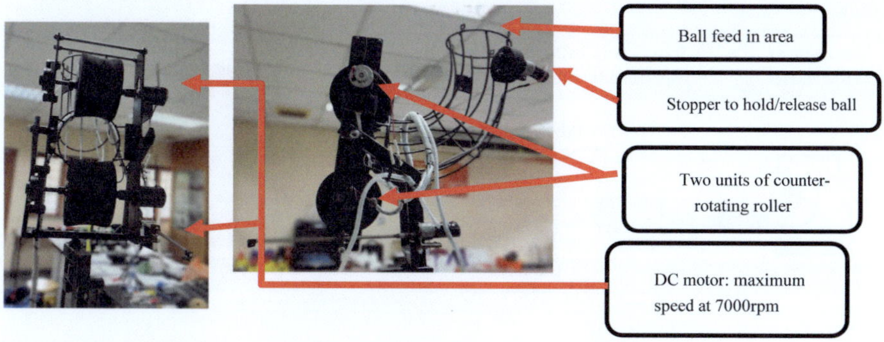

Fig. 1 The prototype of sepaktakraw ball launcher

Table 1 Specifications of the sepaktakraw ball launcher

Parameter	Details
Launcher roller + motor	2 units of Eva-Roller, diameter 15 cm, width 11.5 cm with 24V DC motor: maximum speed at 7000rpm
Roller arrangement	Two counter-rotating rollers positioned on top and bottom in-between the ball (top-spin)
Expected maximum ball launch speed	100km km/h
Feeding Ball	Feeding ball from top to bottom
Energy Source	Electrical or battery-operated 24V 20Ah 7s3p 18,650 battery lithium battery 20000mAh electric
Controller	Android application remote control to change the roller rotational speed (range 0 to 100%)

EXILIM EX-ZR70). Simultaneously, the ball launch speed for each different motor speed was measured using a speed radar gun (BUSHNELL-model 101,911). Data obtained from the high-speed camera were then processed using a motion analysis software [Kinovea (V.0.9.5)] to calculate the ball speed. As shown in Fig. 2, the high-speed camera center of lens was positioned 15 cm from the center of the ball initial position before launch. Meanwhile, a 30 cm ruler, was positioned 40 cm from the camera to calibrate the travel distance of ball in motion. Upon launch, the ball motion was captured by the high-speed camera. The travel time and distance of the ball from initial position until 30 cm (length of calibration ruler) were analysed to calculate the ball speed by dividing the travel distance by the time taken to cover that distance. The experimental setup and motion analysis interface using Kinovea are also shown in Fig. 2.

In addition, a speed radar gun was used to measure the maximum ball speed of the ball launcher. The maximum ball speeds from ten different trials conducted by the ball launcher were selected for analysis (Fig. 3). Furthermore, ball speeds produced by ten elite male sepaktakraw servers were also measured using a speed radar gun at an elite sepaktakraw tournament that was recently organized in 2024.

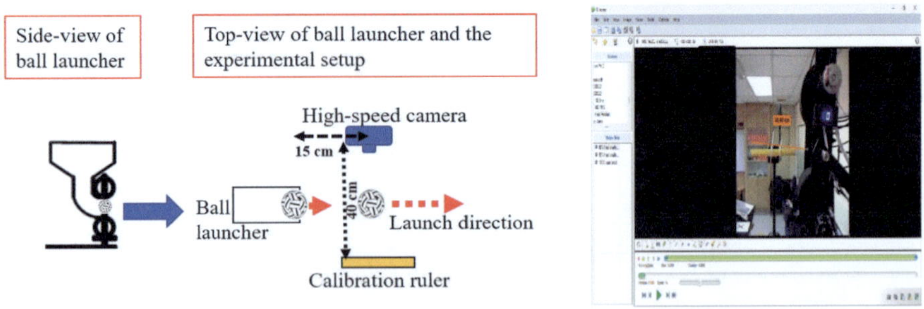

Fig. 2 **a** Experiment set-up of ball-speed verification during launch, and **b** Kinovea interface to measure ball-speed based on high-speed video (480 Hz)

The speed radar gun was positioned 5 m from the server position during the serve execution for the ball speed measurement (Corporation and USA 2024) and the maximum ball speed produced by each server was selected and averaged for further analysis (Fig. 4). Normality of the data was analysed using Shapiro–Wilk Test ($p < 0.05$). Differences of the maximum ball speed produced by the ball launcher and the maximum ball speed produced by elite sepaktakraw's serve were statistically analysed using independent samples t-test ($p < 0.05$). All statistical analyses were performed using IBM SPSS Statistics 26 (SPSS Inc., Chicago, IL, USA).

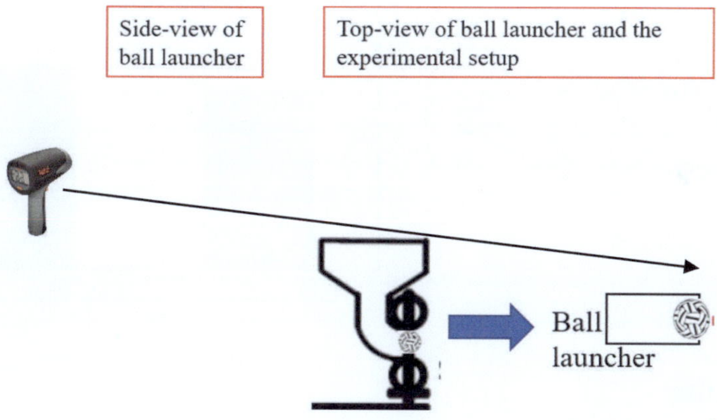

Fig. 3 Measurement of the ball launcher's maximum ball speed using speed radar gun

Fig. 4 Measurement of elite players' maximum serve ball speed using speed radar gun

Table 2 Ball speed: high-speed camera versus speed radar gun measurement

Launched ball speed at various motor capacity				
	70%	90%	95%	100%
High-speed camera (km/h)	33.64	48.05	53.39	56.2
Speed radar gun (km/h)	32.92	47.03	52.25	55
Measurement difference (km/h)	0.72	1.02	1.14	1.2

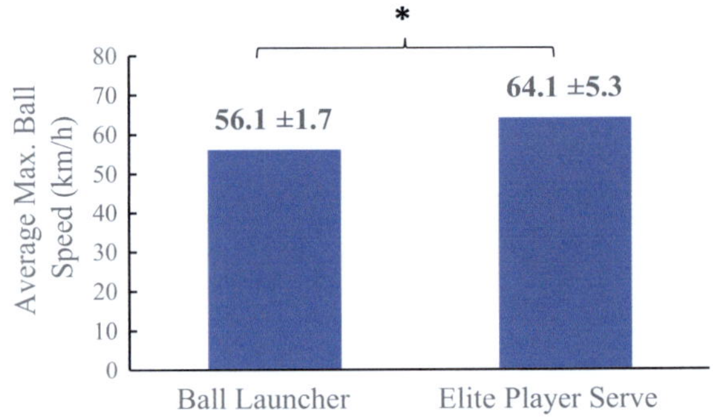

Fig. 5 Comparison between the maximum ball speed produced by the ball launcher and elite sepaktakraw players' serves (*$p < 0.05$)

3 Results

The comparison between the maximum ball speed launched by the ball launcher measured using high-speed camera and speed radar gun is shown in Table 2. The ball speed measurements from both methods was relatively similar, recorded between 33.64 and 33.92 km/h during 70% launcher's motor capacity and 55–56.2 km/h at 100% motor capacity. It was also found that differences in measurement between high-speed camera and speed radar gun ranged between 0.72 and 1.2% across various launcher's motor capacity.

Meanwhile, Fig. 5 illustrates the comparison between maximum ball speed produced by the ball launcher and elite sepaktakraw players' serves. Results from independent sample t-test indicated that there was a significant difference in the average maximum ball speed produced by the ball launcher (56.1 ± 1.7 km/h) when compared to the average maximum ball speed (64.1 ± 5.3 km/h) produced by the elite sepaktakraw players's serve [$t(9) = -4.57$, $p = 0.0002$].

4 Discussions

This study assessed a new prototype of a sepaktakraw ball launcher capability to replicate the maximum ball speed of serve executed by elite male sepaktakraw players during competitive sepaktakraw tournament. It was expected that the ball speed measured using speed radar gun and high-speed camera would be comparable. It was also expected that the ball speed produced by the new prototype would be able to replicate the maximum ball speed produced by the elite player's serve.

The study results highlighted the comparable ball speed measurement capabilities of both the high-speed camera at 480 Hz and the speed radar gun. It was found that the average ball speed analysed using the high-speed camera and the speed gun radar was similar (high-speed camera: 56.2 km/h; speed radar gun: 55 km/h) at maximum launcher motor capacity. In addition, the overall measurement difference between the high-speed camera and the speed radar gun was also relatively low across all launcher motor capacities, ranging only between 0.72 and 1.2 km/h. This small range of difference (~2.18%) suggests that the speed radar gun is a valid and reliable tool to measure sepaktakraw ball speed. A similar outcome was previously reported, where speed radar gun produced by Bushnell (model 101,911) would generate percentage difference ranging between 1.9 and 2.17 km/h when compared to other reliable speed measurement devices (Makar et al. 2024). This outcome likely indicates that the speed radar gun is suitable for sports related and ball's speed measurement, including sepaktakraw ball. Furthermore, sepaktakraw ball speed during competitive match typically range between 90 and100 km/h. Therefore, this maximum speed range is also suitable for Bushnell speed radar gun application as it has been shown that speed range between 29 and 102 km/h fall within its effective range. Utilization of speed radar gun in both laboratory and outdoor setting would be beneficial due to its simple setup, compact size and affordability.

Meanwhile, the results of this study also indicated a significant difference between the average maximum ball speed produced by the ball launcher (56.1 $\pm$ 1.7 km/h) and the average ball speed (64.1 $\pm$ 5.3 km/h) produced by the elite sepaktakraw players [$t(9) = -4.57$, $p = 0.0002$] during an elite competitive sepaktakraw tournament. While this difference may limit the ball launcher's ability to replicate the maximum ball speed of elite serves at higher speed range, it still managed to produce ball speed from a server at the lower speed range, which is typically between 56 and 58 km/h. In addition, the ball launcher would still be appropriate for lower-level athletes, such as those at the school or college level. However, the ball launcher prototype requires further improvement before it can be considered suitable for training high-level and elite athletes. One of these improvements includes the requirement to operate the desired ball speed launcher at a moderate launcher's motor speed, for example, at 50–70%. This is to ensure that the ball speed produced by the ball launcher will be sustained throughout the training session. Operating an electrical motor at its maximum capacity could potentially damage the motor and cause other issues.

5 Conclusions

The new prototype ball launcher did not successfully replicate the average maximum ball speed produced by elite sepaktakraw server, ranging from 57–71 km/h with an average speed of 64.1 km/h, as the average maximum ball speed produced by the ball launcher only reached 56.1 km/h at roller capacity of 100%. Despite this setback, the developed ball launcher prototype demonstrates the capability to generate serves with consistent speed that meets competitive benchmarks. Upgrading the launcher specification to improve its maximum ball speed will be the next progression of this study. In addition, it was also confirmed that Bushnell speed radar gun is suitable for measuring sepaktakraw ball speed, particularly related to the serve executed by elite male players.

Acknowledgements The authors would like to thank Persatuan Sepaktakraw Malaysia (PSM) for providing access and permission to conduct data collection during the Tun Ali Sepaktakraw Championship 2024.

References

Abdillah AA (2021) Pengembangan Model Latihan Sepak Sila Dalam Permainan Sepaktakraw. Jurnal Pendidikan Olahraga 10(1):94–102. https://doi.org/10.31571/jpo.v10i1.2789

Al-Amien MR, Iyakrus I, Sumarni S (2021) Improvement learning outcomes of Sepaktakraw basic techniques using audiovisual media during Covid-19 pandemic in PJOK students of Sriwijaya university. J Educ Res Eval 10(2):74–85. https://doi.org/10.15294/jere.v10i2.52352

Amalia R, Nur S, Zainuri MI (2023) Development of learning applications for physical education, sports, and health Sepak Takraw material for high school students meutia. J Phys Educ, Sport, Health Recreat 12(3):247–253. https://doi.org/10.15294/active.v13i1.76207

Bushnell Corporation USA (2024) Product manual- 101911. https://www.bushnell.com/on/demandware.static/-/Library-Sites-HuntShootAccessoriesSharedLibrary/default/dwb3731ad0/productPdfFiles/bushnellPdf/Product%20Manuals/Outdoor-Technology/PDF/101911VelocitySpeedGun_1LIM_web.pdf?srsltid=AfmBOory1EsauCr9PBU7MOEGA9iPtseONtKDoFKJFq67NB3LHmRsZQnU

Ghany RMA (2023) Nutritional status and performance of sepak takraw players following intervention by a sports dietitian. J Health Transl Med 26(2):133–138. https://doi.org/10.22452/jummec.vol26no2.18

Hassan NMHM (2022) Head injury evaluation strategy for Sepaktakraw ball heading: a systematic review. IET Conf Proc 2022(22):484–492. https://doi.org/10.1049/icp.2022.2667

Hassan NMHM (2023) Experimental investigation of mechanical properties of Sepak Takraw ball based on different ball orientation. Lect Notes Bioeng 203–211. https://doi.org/10.1007/978-981-99-0297-2_16

Hassan, N. M. H. M. (2024). Head Injury During Heading of Two Types of Sepak Takraw Balls: Analytical Approach. *Lecture Notes in Mechanical Engineering*, 199–208. https://doi.org/10.1007/978-981-99-6890-9_16

Ibrahim IZ, Nasution U (2021) Development of application-based Sepak Takraw teaching model and E-learning in the faculty of sports sciences Medan state university. https://doi.org/10.2991/assehr.k.211110.195

Irawan R (2021) Developing of top serve accuracy test on Sepak takraw sport game. J Hum Sport Exerc 16. https://doi.org/10.14198/jhse.2021.16.Proc3.48

Jufrianis (2021) The effect of knowledge level (Iq) and physical conditions (power, flexibility and coordination) on smash technique learning skill in sepaktakraw. Phys Educ Theory Methodol 21(3):264–272. https://doi.org/10.17309/TMFV.2021.3.10

Lhee SH (2021) Sports injury and illness incidence among South Korean elite athletes in the 2018 Asian Games: a single-physician prospective study of 782 athletes. BMJ Open Sport Exerc Med 7(1). https://doi.org/10.1136/bmjsem-2019-000689

Makar P, Silva AF, Silva RM, Janusiak M, Smoter M, Clemente FM (2024) The agreement between Bushnell and stalker radar guns for measuring ball speed in throwing and kicking. Appl Sci 14(22):10476. https://doi.org/10.3390/app142210476

Musa RM (2020d) Physical fitness parameters in the identification of high-potential sepak takraw players. SpringerBriefs Appl Sci Technol 41–48. https://doi.org/10.1007/978-981-15-3219-1_5

Musa RM (2020e) Relationship between psycho-maturity and performance of sepak takraw. SpringerBriefs Appl Sci Technol 49–58. https://doi.org/10.1007/978-981-15-3219-1_6

Musa RM (2020a) An overview of beach soccer, sepak takraw and the application of machine learning in team sports. SpringerBriefs Appl Sci Technol 1–12. https://doi.org/10.1007/978-981-15-3219-1_1

Musa RM (2020c) Identifying talent in sepak takraw via anthropometry indexes. SpringerBriefs Appl Sci Technol 29–39. https://doi.org/10.1007/978-981-15-3219-1_4

Musa RM (2020b) Concluding remarks. SpringerBriefs Appl Sci Technol 59–61. https://doi.org/10.1007/978-981-15-3219-1_7

Padli (2023) Evaluation of the Sepak Takraw training program in the student sports education and training center (SSETC). J Phys Educ Sport 23(12):3332–3340. https://doi.org/10.7752/jpes.2023.12381

Tan FY (2022) Classification of Sepak Takraw kicks using machine learning. Lect Notes Mech Eng 321–331. https://doi.org/10.1007/978-981-16-4115-2_26

Tohidin D, Sepdanius E, Putra AA (2021) Study of service ability in Sepaktakraw (effect leg muscle power, flexibelity and self confidence of service abilities). Int J Hum Mov Sport Sci 9(5):905–911. https://doi.org/10.13189/saj.2021.090511

Xin JPZ (2021) Ankle range of motion and dynamic balance in recreational Sepak Takraw players with and without ankle injury-A comparative study. Pedagog Phys Cult Sport 25(6):355–360. https://doi.org/10.15561/26649837.2021.0603

Zarei A (2021) Sport Tourists' preferred event attributes and motives: a case of Sepak Takraw, Malaysia. J Hosp Tour Res 45(7):1188–1213. https://doi.org/10.1177/1096348020913091

Zayar H (2024) Inter-limb asymmetry in female sepak takraw players: an observational study. Symmetry 16(7). https://doi.org/10.3390/sym16070902

Kinematics Tracking Performance of In-Match Football Activities Using Smartphone-Based Radar Gun

Siti Nur Aqilah Rusli and Khairul Imran Sainan

Abstract This study evaluates the capability of a smartphone-based radar gun to track player movement during a football match. Measurement accuracy was assessed by comparing data from the smartphone-based radar gun with that obtained from a global positioning system (GPS) device. Additionally, data were collected from team-sport-specific conditioning (TSSC) activities for comparative analysis. Seven football players participated in the study: three during match play and four during TSSC activities. All players wore GPS sports vests, and their movements were simultaneously recorded using a smartphone-based radar gun and a 10 Hz GPS device. Match data were segmented into four 20 min halves with 5 min rest intervals, while the TSSC protocol included four movement patterns: two straight-line runs (25 m and 50 m), a 30 m zig-zag course, and two L-shaped runs (25 m and 10 m). Each circuit was repeated twice with a 1 min rest between repetitions. Results indicated that in-match speeds (mean $\pm$ SD $= 1.39 \pm 1.61$ m/s) were better than those recorded during TSSC activities (mean $\pm$ SD $= 1.79 \pm 0.48$ m/s), contradicting the initial hypothesis that radar guns would yield more accurate readings in controlled TSSC environments. This discrepancy may be attributed to the structured nature of TSSC activities, which fail to replicate the dynamic and unpredictable movements of real match play. In conclusion, smartphone-based radar guns show promise for tracking player movement in both match and TSSC settings. However, the limited in-match data collected in this study may not fully represent real-world performance. Future research should incorporate a larger dataset to enhance the reliability of these findings. This study provides an initial step toward validating smartphone-based radar guns as a viable tool for football performance tracking.

S. N. A. Rusli · K. I. Sainan (✉)
Faculty of Mechanical Engineering, Universiti Teknologi MARA, 40450 Shah Alam, Selangor, Malaysia
e-mail: imransainan@uitm.edu.my

K. I. Sainan
Sports Engineering Artificial Intelligence Centre, Universiti Teknologi MARA, 40450 Shah Alam, Selangor, Malaysia

M. H. A. Hassan et al. (eds.), *Proceedings of the 3rd International Conference on Innovation and Technology in Sports*, Lecture Notes in Bioengineering,
https://doi.org/10.1007/978-981-95-0804-4_3

Keywords Smartphone-based radar gun · Global positioning system (GPS) · In-match · Team-sport specific conditioning (TSSC)

1 Introduction

Tracking systems play a crucial role in monitoring athlete performance by providing data on total distance covered during training sessions and matches. This information is essential for optimizing fitness levels and understanding the transient patterns of acceleration and deceleration, which serve as indicators of in-match fatigue (Akenhead et al. 2013). Additionally, tracking player movement during team-sport-specific conditioning (TSSC) activities enables coaches to analyze positional characteristics and tailor training programs for improved performance specificity (Russell 2016; Tierney et al. 2016).

Global Positioning Systems (GPS) are widely used to assess kinematic performance and player movement patterns. Metrics such as speed and total distance covered during training are frequently compared with in-match data (Tierney and Clarke 2019). GPS technology is generally favored due to its accuracy, ease of use, and reliability in both training and competitive settings (De Silva 2018; Bayliff et al. 2019). Beyond GPS, alternative tracking systems have been developed (Tierney and Clarke 2019; Ponsano et al. n.d.), including local positioning systems (LPS) (Tierney et al. 2016), local position measurement (LPM) (Rawstorn et al. 2014), ultra-wideband (UWB), and orbital tracking systems (OTS). These technologies have been extensively studied in both in-match (Pons 2019; Schulze et al. 2021; Rago et al. 2019) and TSSC contexts (Johnston et al. 2014; Huggins 2020; Brosnan et al. 2021) within football. Other tracking methods, such as wrist trackers (Makar et al. 2023), photoelectric barriers (Uhlmann et al. 2023), and radar guns, offer additional options for movement analysis.

Radar guns, traditionally used to measure ball speed in baseball, have potential applications in tracking player movement. However, their effectiveness in football, particularly in dynamic match scenarios and complex training drills involving multidirectional and curvilinear circuits, remains largely unexplored (Rawstorn et al. 2014).

To date, only two studies (Bataller-Cervero et al. 2019; Beato et al. 2018) have examined the use of radar guns for tracking instantaneous player speed, focusing on straight-line 40 m runs and 20 m sprints in TSSC settings. Notably, no prior research has investigated the feasibility of smartphone-based radar guns for tracking player performance in either football matches or TSSC activities.

This study aims to evaluate the effectiveness of smartphone-based radar guns for tracking player movement in football matches, assess their measurement accuracy, and compare their performance against GPS systems in both in-match and TSSC conditions.

2 Overview of Previous Studies

Previous studies primarily employed GPS (Rawstorn et al. 2014) and NdGPS (Akenhead et al. 2013) as tracking systems, often without a dedicated reference system for comparison. Additionally, semi-automatic multiple-camera video technology (VID) systems like Prozone (Makar et al. 2023) and radio-based local positioning systems (LPS) such as Inmotio (Makar et al. 2023) have been used alongside GPS for data collection during games. Notably, the VICON system recognized for its high precision has been utilized with the Center of Mass (COM) as a reference criterion. However, due to its reliance on stationary infrared cameras, VICON is only suitable for small-area tracking, making it impractical for full-scale field applications. In addition, factors such as venue and satellite reception have also been considered potential sources of error. Various tracking systems, including video-based time-motion analysis (Chen et al. 2021), semi-automatic multiple-camera systems (Chen et al. 2021), and GPS, have been employed to capture player movement data. However, these studies frequently lack a dedicated reference system, relying instead on comparisons between different tracking technologies.

Participant recruitment for data collection varies across studies, with sample sizes and player categories differing based on study methodology. Examples include 36 professional male football players (Akenhead et al. 2013), 11 professional under-21 soccer players (Russell 2016), 14 male soccer players, 20 highly trained outfield football players (Chen et al. 2021), full 11-versus-11 professional football matches (Tierney et al. 2016), and 82 highly trained soccer players (Makar et al. Oct. 2023). In all studies, participants are informed about safety procedures, potential risks or discomfort, and the research protocol.

The matches were played on outdoor natural grass pitches (Akenhead et al. 2013; Russell 2016). Detailed player information was collected, including age, height, body mass, and body fat (Akenhead et al. 2013; Tierney et al. 2016). The football games followed a 4–4–2 formation (Akenhead et al. 2013), consisting of four defenders, four midfielders, and two forwards. Before each game, players completed a 15 min warm-up monitored by a global navigation satellite system (GNSS) device. Data from each 45 min half were subsequently analyzed, with each half further divided into nine 5 min segments (Akenhead et al. 2013). Prior to the competitive season, players underwent a period of physical and technical preparation, including ten friendly matches. Each game consisted of two 45 min halves, totaling 90 min of playtime (Russell 2016). The study was conducted over two consecutive days, with ten of the fourteen male soccer players participating each day. Four players from the first day were replaced by four different players on the second day. Before each 5-a-side friendly match, players completed a sport-specific course (SSC), consisting of a predefined circuit of prescribed movements under controlled conditions. Following the SSC, they performed the 20-m shuttle run test (SHU), a two-minute test requiring repeated 20-m sprints. Finally, they participated in small-sided games (SSGs), specifically 5-a-side matches.

The data were organized into three 15 min periods per half to analyze overall temporal patterns (Akenhead et al. 2013). Player data were included in the analysis only if the following criteria were met; completion of the full game, absence of injury during the game (Akenhead et al. 2013; Russell 2016), consistent play in a single, comfortable position throughout the game, and a first half extended by no more than two minutes (Akenhead et al. 2013). This study primarily collected data to analyze positional information, instantaneous speed, acceleration and deceleration, total distance (TD) covered, high-speed running (HSR), and high metabolic load distance (HMLD). Data are presented as mean ± standard deviation (SD) (Russell 2016; Tierney et al. 2016; Chen et al. 2021). Statistical analysis was conducted using analysis of variance (ANOVA) (Akenhead et al. 2013; De Silva 2018; Chen et al. 2021) and multivariate analysis of variance (MANOVA) (Tierney et al. 2016). Root mean square error (RMSE) was also calculated. All statistical analyses were performed using IBM SPSS Statistics (formerly known as the Statistical Package for the Social Sciences, SPSS) (Russell 2016; Tierney et al. 2016).

3 Methodology

3.1 Experimental Approach to the Problem

This study involved seven male participants to assess the agreement between a smartphone-based radar gun and 10 Hz GPS data for measuring speed during both football matches and team-sport-specific conditioning (TSSC) activities. The aim was to compare the data collected by these two tracking devices (Giménez and Gomez 2019).

Data collection occurred over two days. On Day 1, three participants were monitored during an actual football match. On Day 2, four different participants completed the TSSC protocol (Johnston et al. 2014), which included four circuits, each repeated twice (Huggins 2020). All activities were conducted at the UiTM Shah Alam Sports Centre in Selangor, Malaysia.

3.2 Participants

Seven male football players (age: 20 ± 3 years; body mass: 65 ± 7.0 kg; height: 165 ± 6.5 cm) voluntarily participated in this study after providing written informed consent (Brosnan et al. 2021). Participants were thoroughly briefed on all study procedures prior to providing consent. All players were injury-free and regularly participated in competitive football matches. Additionally, each participant completed an information sheet approved by the Research Ethics Committee (REC) at Universiti Teknologi MARA (UiTM), Selangor, Malaysia (Paprancová et al. 2024; Calderón-Pellegrino

2022). This study is part of the Sports Engineering and Artificial Intelligence Centre (SEA-IC) program, which focuses on the evaluation of various tracking devices and sensors for athletes.

3.3 Procedures

All participants were familiar with the testing protocols, which were overseen and supervised by the researcher and a qualified football trainer during both the in-match and TSSC activities. Participants wore a 10 Hz GPS device, positioned in a sports vest on their upper back, between the shoulder blades. Prior to participation, they performed a 5 min warm-up (Willmott et al. 2019; Reilly et al. 2015). The match was divided into four 20 min quarters, with 5 min rest periods between each quarter. GPS devices were activated 10 min before the match began. A smartphone-based radar gun, mounted on a tripod, was positioned outside the field to monitor player movement. During each match, the radar gun was directed towards a selected player and continuously tracked to capture speed data. Players in left or right back or wing positions were prioritized, as these players typically positioned themselves along the longer sides of the field, facilitating more effective data collection due to their proximity to the radar gun. In the TSSC activities, participants completed each circuit twice, with a one-minute rest period between repetitions.

3.4 Circuit

The TSSC circuits (shown in Fig. 1) were marked using measuring tape and their accuracy was verified with a theodolite. The TSSC consisted of two straight lines (25 and 50 m), an L-shaped line (25 m × 10 m), and a zig-zag track (30 m). A smartphone-based radar gun, mounted on a tripod for stability, was positioned 5 m away from the circuit, at the end of each segment.

3.5 Data Analysis

Agreement between smartphone-based radar guns and GPS were statistically calculated and analyzed using Jeffrey's Amazing Statistics Program (JASP) (version 0.18.3.0, for Windows, University of Amsterdam, Netherland), MedCalc software (version 22.001, for Windows), and Microsoft Excel (version 2111, for Windows).

For the in-match activities, linear regression analysis was performed, with GPS data as the independent variable and radar gun data as the dependent variable (Chen et al. 2021; Healy et al. 2016). Linear regression equations were generated to allow for conversion between the speed measurements obtained from the two systems.

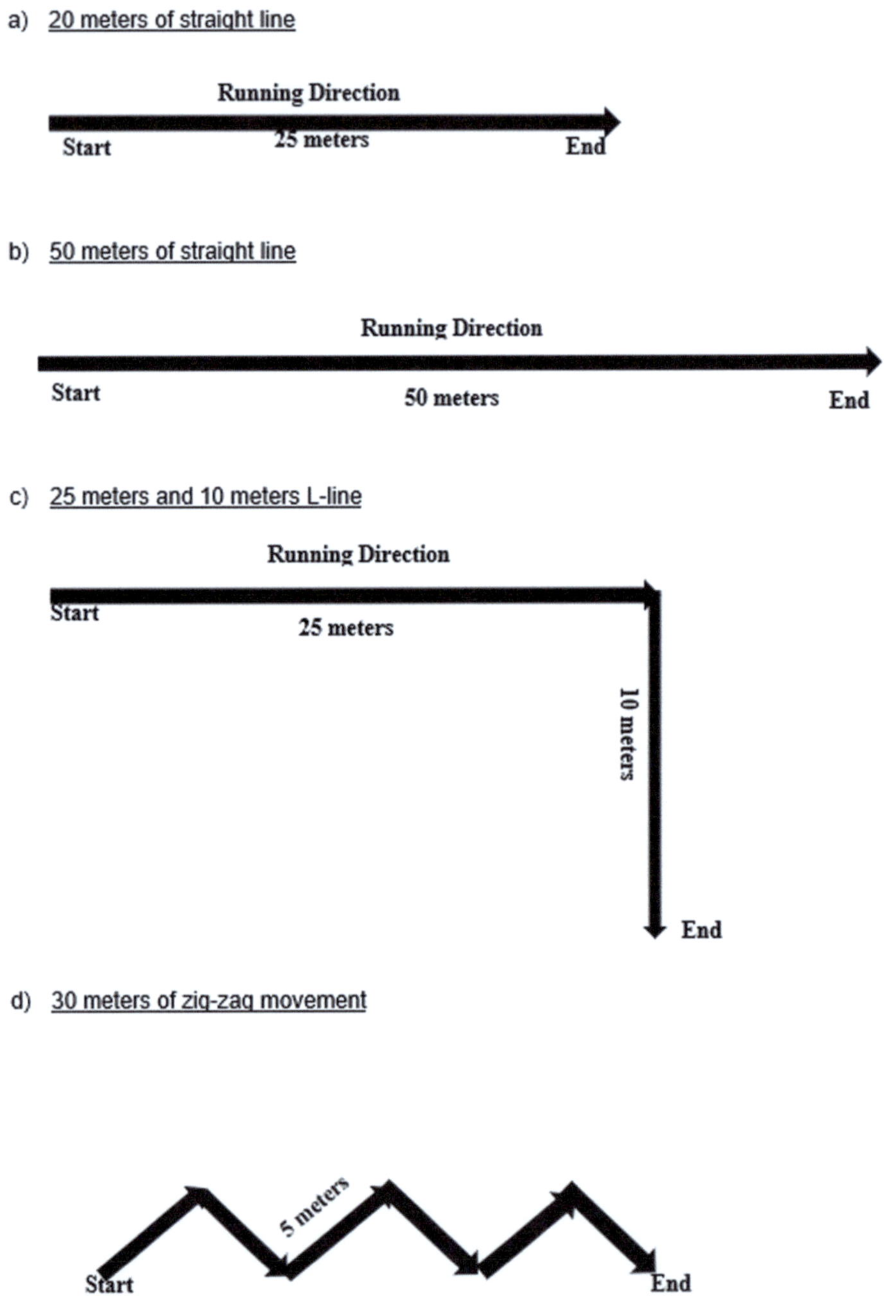

Fig. 1 Team Sport Simulation Circuit (TSSC) design to assess the ability of smartphone-based radar gun and Global Positioning System (GPS) units

The multiple R, R-squared, and adjusted R-squared values were calculated. For the TSSC, the mean difference and standard deviation of the speed data were calculated and reported to quantify the error between the devices, using GPS as the criterion measure. In addition, the upper and lower limits of agreement (LOAs), as well as the upper and lower 95% standard error (SE) and LOA intervals, were calculated.

To assess the agreement between GPS and radar gun measurements, a Bland–Altman plot was used to visually examine the magnitude of any bias (Schulze et al. 2021; Castellano et al. 2019). A paired t-test was conducted to determine if the mean difference between the two devices was statistically significant. The significance level for both the t-test and the regression analysis was set a priori at $p < 0.05$ (Makar et al. 2023).

4 Results

During data collection, there was no cloud cover, and the number of satellites was sufficient to provide a good gathering of data for valid and reliable GPS. The weather was dry with light breeze and an ambient temperature of 25 °C (Buchheit et al. 2014).

Table 1 presents the regression statistics for the relationship between speed measurements from GPS and a smartphone-based radar gun during a football match. Figure 2 visualize the trends of the relationship. The correlation coefficient ($r = 0.659$) indicates a moderate to strong positive correlation between the two devices, with a standard error of the estimate (SEE) of 0.175. The mean difference between the two devices was 1.39 ± 1.61 ($p < 0.05$), indicating a statistically significant, though small, difference in measured speeds. The **slope of the regression line (0.8543)** suggests that the smartphone-based radar gun slightly underestimates speed compared to GPS.

Table 2 presents the results of paired t-tests comparing speed measurements obtained from GPS and a smartphone-based radar gun across four different circuits. Mean speed ± standard deviation for each circuit was as follows: Circuit 1 = 1.986 ± 0.493; Circuit 2 = 1.97 ± 0.187; Circuit 3 = 1.954 ± 0.166; and Circuit 4 = 1.279 ± 0.559. Limits of agreement (LOAs) were calculated for each circuit to construct Bland–Altman plots and assess systematic bias between the GPS and smartphone-based radar gun measurements. The Bland–Altman plots in Fig. 3 revealed similar bias for Circuits 1, 2, and 3, with mean differences close to 2.00 (Circuit 1 = 1.986, Circuit 2 = 1.97, and Circuit 3 = 1.954). Circuit 4 (mean difference = 1.279) demonstrated the least bias, with data points closer to zero, but exhibited lower precision compared to the other circuits. Circuits 2 (LOA = 2.335 to −1.604) and 3 (LOA =

Table 1 Regression statistics in football match

R	R^2	Adjusted R^2	Standard error	Mean ± SD
0.659	0.435	0.383	0.175	1.39 ± 1.61

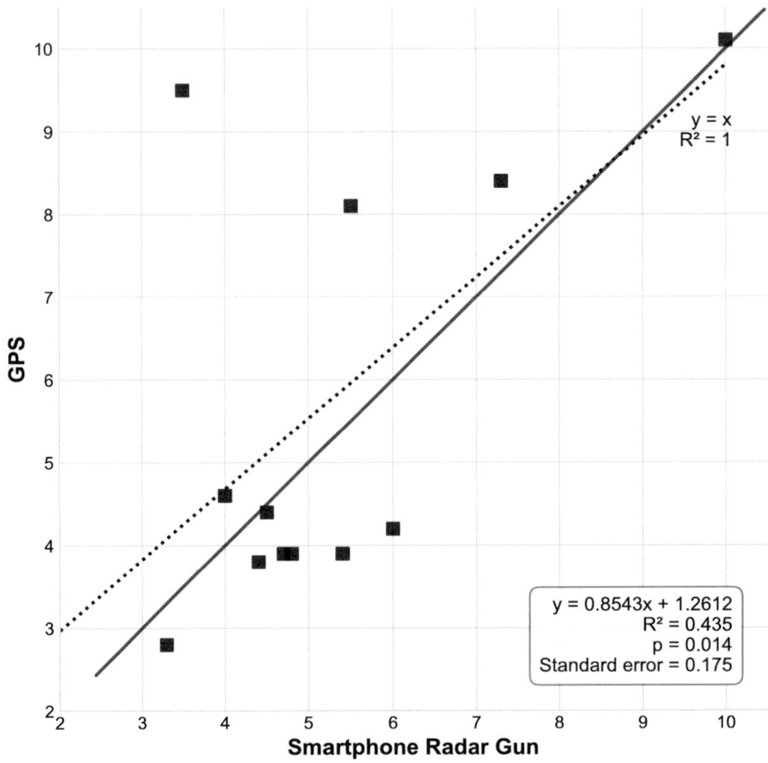

Fig. 2 Relationship of the speed between GPS and smartphone-based radar guns during football matches. The black line is the line of identity, and the dotted line is the trend line

2.279 to −1.629) showed similar patterns: relatively high precision, but lower accuracy due to the consistent bias. Circuit 1 (LOA = 2.953 to −1.019) demonstrated both lower precision and lower accuracy.

The paired t-tests indicated statistically significant differences between the two measurement methods for all TSSC circuits (Circuit 1: $p = 0.001$; Circuit 2: $p = 0.001$; Circuit 3: $p = 0.001$; and Circuit 4: $p = 0.003$). A statistically significant difference was also found for the in-match comparison ($p = 0.014$) (Table 3).

5 Discussion

This study evaluated the feasibility of using **smartphone-based radar guns** to measure speed during **football matches** and **team-sport-specific conditioning (TSSC) activities,** comparing its accuracy with **GPS measurements.** The findings indicate a **positive correlation (r = 0.659, R^2 = 0.435)** between **GPS and smartphone-based radar gun speed data,** supporting its potential as a viable

Table 2 Analysis of GPS and smartphone radar gun in TSSC

Measure 1–Measure 2	Mean difference	p	Standard deviation	SE difference	95% CI for mean difference		LOA upper	95% CI for LOA upper		LOA Lower	95% CI for LOA lower	
					Lower	Upper		Lower	Upper		Lower	Upper
GPS C1 - Radar Gun C1	1.986	0.001	0.493	0.201	1.468	2.504	2.953	2.056	3.85	1.019	0.122	1.916
GPS C2 - Radar Gun C2	1.97	0.001	0.187	0.076	1.774	2.166	2.335	1.996	2.675	1.604	1.265	1.943
GPS C3 - Radar Gun C3	1.954	0.001	0.166	0.068	1.78	2.128	2.279	1.978	2.581	1.629	1.328	1.931
GPS C4 - Radar Gun C4	1.279	0.003	0.559	0.228	1.866	0.692	2.376	1.2	0.834	0.183	3.393	1.359

95% CI: confidence interval at 95%, LOA: limit of agreement.

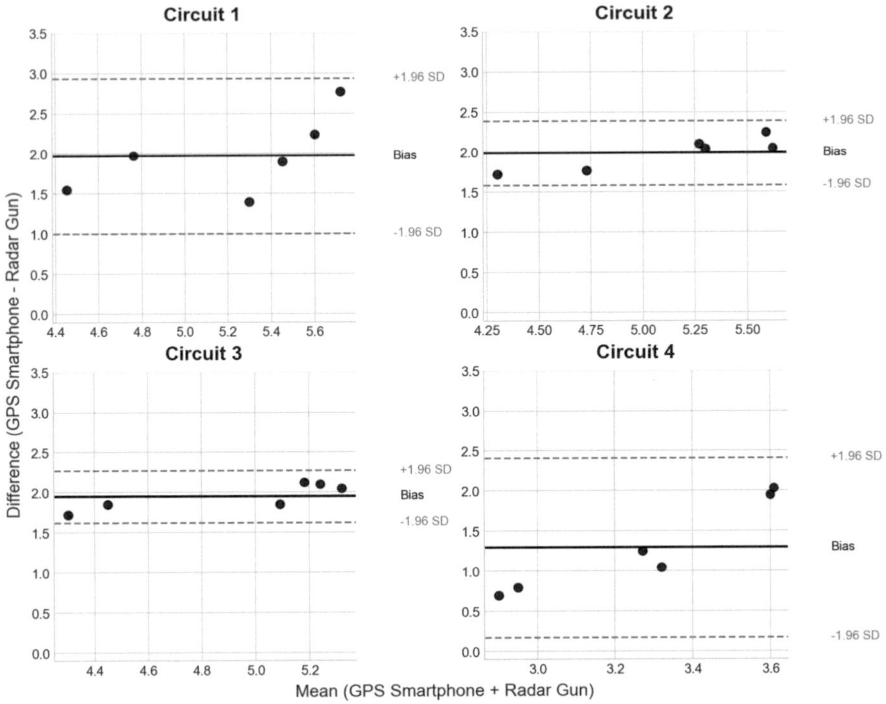

Fig. 3 Bland–Altman plots GPS vs Smartphone-based radar gun in T

Table 3 Agreement analysis between football match and TSSC

Variable	Standard error	Mean ± SD	P-value	LOA upper	95% CI for LOA upper		LOA lower	95% CI for LOA lower	
					Lower	Upper		Lower	Upper
Match	0.175	1.39 ± 1.61	0.014	3.532	1.348	5.716	−4.649	−6.833	−2.464
TC	0.477	1.79 ± 0.48	0.0015	2.486	1.808	2.485	1.657	1.527	1.787

95% CI: confidence interval at 95%, LOA: limit of agreement.

tracking tool. However, notable differences were observed between the two measurement methods, highlighting key **methodological and technical considerations.**

The results revealed that **GPS overestimated running speed by approximately 32%** compared to the **smartphone-based radar gun**, aligning with previous research (Roland et al. 2023). However, both methods deviated from the **true speed**, with GPS **underestimating actual values by 10–15%**. This discrepancy is consistent with prior findings, which suggest that **GPS accuracy diminishes a higher running speed (VHSR) compared to lower running speeds (LSR)** (Calderón-Pellegrino 2022). Interestingly, previous studies reported an **almost perfect correlation (R = 0.97)** between **GPS and radar gun measurements** (Bataller-Cervero et al. 2019), which differs from the **moderate-to-strong correlation (R = 0.659)** found in this study. These differences may be attributed to **variations in experimental conditions, player movement complexity, and tracking system limitations.** The reduced accuracy at higher speeds may also contribute to the observed discrepancies.

Despite demonstrating **reasonable agreement** with GPS data, the **smartphone-based radar gun** presents several **limitations.** One major challenge is its **stationary nature,** which restricts tracking to **a single player at a time.** Unlike GPS, which provides continuous data on all players throughout a match (Little and Williams 2005; Randers et al. 2010), **radar guns require manual adjustments and precise positioning** to capture speed measurements accurately. Additionally, **distance constraints** affect the effectiveness of radar guns in **real-time tracking scenarios.** The detection range is limited, preventing it from capturing movements occurring beyond a specific **field of view** (Harley et al. 2011). This limitation is particularly relevant in football, where **players frequently change direction, speed, and position,** making it difficult to maintain continuous tracking without **interruption or data loss.**

The findings suggest that **smartphone-based radar guns** can serve as **alternative speed-tracking tools** for **TSSC activities,** where movement patterns are more controlled and predictable. However, their application in **football match analysis** remains **challenging** due to limitations discussed previously. Future studies should explore **hybrid tracking approaches,** combining **radar guns with GPS or optical tracking systems** to improve **accuracy and data completeness.**

Further research is also needed to **assess performance across different playing positions, environmental conditions, and player skill levels.** Increasing **sample sizes** and incorporating **higher-resolution radar technology** may help refine measurement accuracy and reduce systematic bias in **football-specific movement tracking.**

6 Conclusion and Suggestions

In summary, while **smartphone-based radar guns** demonstrate potential for measuring **player speed**, their **stationary nature, tracking limitations, and range constraints** hinder their effectiveness in **dynamic match scenarios.** Nevertheless, their affordability and ease of use make them a **valuable tool for controlled training environments.** Future research should focus on **enhancing radar gun technology, integrating complementary tracking systems, and optimizing data processing techniques** to improve their **applicability in professional football performance analysis**.

References

Akenhead R, Hayes PR, Thompson KG, French D (2013) Diminutions of acceleration and deceleration output during professional football match play. J Sci Med Sport 16(6):556–561. https://doi.org/10.1016/j.jsams.2012.12.005

Bataller-Cervero AV et al (2019) Validity and reliability of a 10 Hz GPS for assessing variable and mean running speed. J Hum Kinet 67(1):17–24. https://doi.org/10.2478/hukin-2018-0084

Bayliff GE, Jacobson BH, Moghaddam M, Estrada C (2019) Global positioning system monitoring of selected physical demands of NCAA division i football players during games. J Strength Cond Res 33(5):1185–1191. https://doi.org/10.1519/JSC.0000000000003137

Beato M, Devereux G, Stiff A (2018) Validity and reliability of global positioning system units (STATSports Viper) for measuring distance and peak speed in sports. J Strength Cond Res 32(10):2831–2837. https://doi.org/10.1519/jsc.0000000000002778

Brosnan RJ, Watson G, Stuart W, Twentyman C, Kitic CM, Schmidt M (2021) The validity, reliability, and agreement of global positioning system units—can we compare research and applied data? J Strength Cond Res. Publish Ahead of Print. https://doi.org/10.1519/jsc.0000000000004139

Buchheit M et al (2014) Monitoring accelerations with GPS in football: time to slow down? Int J Sports Physiol Perform 9(3):442–445. https://doi.org/10.1123/ijspp.2013-0187

Calderón-Pellegrino G (2022) Physical demands during the game and compensatory training session (MD + 1) in elite football players using global positioning system device. Sensors 22(10):3872. https://doi.org/10.3390/s22103872

Castellano J, Casamichana D, Campos Vázquez MA, LangarikaRocafort A (2019) Interchangeability of two tracking systems to register physical demands in football: multiple camera video versus GPS technology. Archivos De Medicina Del Deporte 36(3):268–269

Chen Z, Bian C, Liao K, Bishop C, Li Y (2021) Validity and reliability of a phone app and stopwatch for the measurement of 505 change of direction performance: a test-retest study design. Front Physiol 12. https://doi.org/10.3389/fphys.2021.743800

De Silva V (2018) Player tracking data analytics as a tool for physical performance management in football: a case study from Chelsea football club academy. Sports 6(4). https://doi.org/10.3390/sports6040130

Giménez JV, Gomez MA (2019) Relationships among circuit training, small-sided and mini goal games, and competition in professional soccer players. J Strength Cond Res 33(7):1887–1896. https://doi.org/10.1519/jsc.0000000000002804

Harley JA, Lovell RJ, Barnes CA, Portas MD, Weston M (2011) The interchangeability of global positioning system and semiautomated video-based performance data during elite soccer match play. J Strength Cond Res 25(8):2334–2336. https://doi.org/10.1519/jsc.0b013e3181f0a88f

Healy R, Norris M, Kenny IC, Harrison AJ (2016) A novel protocol to measure short sprint performance. In: Procedia engineering, Elsevier Ltd, pp 706–711. https://doi.org/10.1016/j.proeng.2016.06.252

Huggins RA (2020) The validity and reliability of global positioning system units for measuring distance and velocity during linear and team sport simulated movements. J Strength Cond Res 34(11):3070–3077. https://doi.org/10.1519/jsc.0000000000003787

Johnston RJ, Watsford ML, Kelly SJ, Pine MJ, Spurrs RW (2014) Validity and interunit reliability of 10 Hz and 15 Hz GPS units for assessing athlete movement demands. J Strength Cond Res 28(6):1649–1655. https://doi.org/10.1519/jsc.0000000000000323

Little T, Williams AG (2005) Specificity of acceleration, maximum speed, and agility in professional soccer players. J Strength Cond Res 19(1):76. https://doi.org/10.1519/14253.1

Makar P, Kawczyński A, Silva RM, Yildiz M, Silva AF, Akyildiz Z (2023) Validity and reliability of Polar M400 GPS watches for measuring distances covered by team sports players. Heliyon 9(10):e20920. https://doi.org/10.1016/j.heliyon.2023.e20920

Paprancová A, Šimonek J, Paška Ľ, Krčmár M (2024) Impact of smallsided games and runningbased highintensity interval training on physical performance in female football players. J Phys Educ Sport 24(1):113–122. https://doi.org/10.7752/jpes.2024.01015

Pons E (2019) A comparison of a GPS device and a multi-camera video technology during official soccer matches: agreement between systems. PLoS One 14(8). https://doi.org/10.1371/journal.pone.0220729

Ponsano N, Hosseini Varde'i C, Izzo RE, Cejudo A (n.d.) The role of fatigue in football matches, performance model analysis and evaluation during quarters using live global positioning system technology at 50Hz [Online]. https://www.researchgate.net/publication/338829411

Rago V et al (2019) Methods to collect and interpret external training load using microtechnology incorporating GPS in professional football: a systematic review. Res Sports Med 28(3):1–22. https://doi.org/10.1080/15438627.2019.1686703

Randers MB et al (2010) Application of four different football match analysis systems: a comparative study. J Sports Sci 28(2):171–182. https://doi.org/10.1080/02640410903428525

Rawstorn JC, Maddison R, Ali A, Foskett A, Gant N (2014) Rapid directional change degrades GPS distance measurement validity during intermittent intensity running. PLoS One 9(4). https://doi.org/10.1371/journal.pone.0093693

Reilly B, Akubat I, Lyons M, Collins DK (2015) Match-play demands of elite youth gaelic football using global positioning system tracking. J Strength Cond Res 29(4):989–996. https://doi.org/10.1519/jsc.0000000000000714

Roland FGP, Lagestad P (2023) Reliability and validity of polar team pro measurements in running at different velocities in an indoor setting. Front Sport Act Living 5. https://doi.org/10.3389/fspor.2023.1165801

Russell M (2016) Changes in acceleration and deceleration capacity throughout professional soccer match-play. J Strength Cond Res 30(10):2839–2844. https://doi.org/10.1519/JSC.0000000000000805

Schulze E, Julian R, Skorski S (2021) The accuracy of a low-cost gps system during football-specific movements. J Sport Sci Med 126–132. https://doi.org/10.52082/jssm.2021.126

Tierney P, Clarke N (2019) A comparison of a smartphone app with other GPS tracking type devices employed in football. Exerc Med 3:4. https://doi.org/10.26644/em.2019.004

Tierney PJ, Young A, Clarke ND, Duncan MJ (2016) Match play demands of 11 versus 11 professional football using global positioning system tracking: variations across common playing formations. Hum Mov Sci 49:1–8. https://doi.org/10.1016/j.humov.2016.05.007

Uhlmann T, Bräuer S, Zaumseil F, Brunnett G (2023) A novel inexpensive camera-based photoelectric barrier system for accurate flying sprint time measurement. Sensors 23(17):7339. https://doi.org/10.3390/s23177339

Willmott AGB, James CA, Bliss A, Leftwich RA, Maxwell NS (2019) A comparison of two global positioning system devices for team-sport running protocols. J Biomech 83:324–328. https://doi.org/10.1016/J.JBIOMECH.2018.11.044

Optimization of Retinoic Acid Loaded Nanosponge Formulation for Active Individuals: Impact of Drying Method and Homogenization Speeds

Sharifah Shakirah binti Syed Omar and Hazrina Ab Hadi

Abstract Active individuals, particularly those frequently engaged in outdoor activities, are at a heightened risk of skin damage due to increased exposure to ultraviolet (UV) radiation. This study focuses on optimizing a retinoic acid-loaded nanosponge formulation, aimed at protecting the skin from UV damage. Retinoic acid, a potent derivative of vitamin A, is known to combat photoaging, hyperpigmentation, and other harmful effects of UV exposure. The study investigates the influence of homogenization speeds and drying techniques on the particle size of this formulation, which directly affects its absorption and efficacy. Nanosponge formulations were prepared with retinoic acid and subjected to oven drying methods and freeze-drying methods. The nanosponge was then added into a base emulgel for optimization with homogenization speeds of 2000, 3000, and 4000 rpm and subjected to a stability study at real-time and accelerated specifications. The results revealed significant variations in particle size, polydispersity index, and zeta potential with different drying methods and different homogenization speeds. The optimized formulation with optimal characteristics can deliver higher efficiency of the actives to enhance the skin's defense against UV-related damage and maintain a healthier skin. In conclusion, the optimization of this formulation represents a significant consideration in formulating advance formulations. These small adjustments can result in formulations that could provide active individuals with a more effective solution to combat UV effects on skin.

Keywords Retinoic acid · Nanosponge · UV skin damage

S. S. S. Omar (✉)
Department of Pharmaceutical Technology, Faculty of Pharmacy, Universiti Malaya, Kuala Lumpur, Malaysia
e-mail: shakirah@um.edu.my

H. A. Hadi
Pharmaceutical Technology Department, Kulliyyah of Pharmacy, International Islamic University Malaysia, Kuantan, Pahang, Malaysia

M. H. A. Hassan et al. (eds.), *Proceedings of the 3rd International Conference on Innovation and Technology in Sports*, Lecture Notes in Bioengineering,
https://doi.org/10.1007/978-981-95-0804-4_4

1 Introduction

Physical activity, although known to be beneficial for health, also comes with downsides. And one particular downside happens to athletes and active individuals that are exposed to the sun for a long time. Outdoor recreation exposes humans to ultraviolet radiation which can lead to sunburn, solar damage and higher risk of skin cancer (Kliniec et al. 2023). Specifically, ultraviolet radiation (UVR) causes photoaging (premature ageing) characterized by wrinkles, fine lines, rough and leathery textured skin. It also causes hyperpigmentation whereby the skin overstimulates production of melanin as a defense mechanism that leads to uneven skin tone, dark patches, sunspots and freckles. It can also lead to chronic inflammation and DNA mutation which can eventually lead to the worst-case scenario which is skin cancer. Beside UV related skin disorder, athletes are also at risk of acne mechanica a.k.a sport-induced acne which can be caused by heat, pressure, occlusion and repetitive friction (Pujalte et al. 2023). Athletes usually train for long hours under the sun, particularly endurance athletes involved in cycling, marathon running, rowing, swimming and triathlons. Athletes in water sports garners higher sensitivity to UVR due to the hydration of the outer skin layer that shift the absorption spectrum of UV to a shorter wavelength & decreases the dispersion and reflection (Luca et al. 2012).

Retinoic acid, a derivative of vitamin A, is known for its role in skin health and regeneration, making it a potential candidate for active individuals seeking to optimize their dermal well-being (Szymański et al. 2020). Retinoic acid can help repair UVR damage by accelerating skin cell turnover which help replace damage cells with healthier one, fades hyperpigmentation, improve skin texture and repair DNA damage through its interaction with retinoic acid receptors (RARs) in the skin cells. Retinoic acid also plays a role in boosting collagen production and reducing its degradation through fibroblast activation and inhibition of enzymes involved in degradation. The melanin producing pathways is also inhibited by retinoic acid further contributing the lightening of hyperpigmentation. As retinoic acid is an anti-oxidant, it can help neutralize free radicals which is helpful to protect the skin against further oxidative stress and damage (Syed Omar et al. 2022). Studies have also found topical retinoic acid can be helpful in the prevention and treatment of skin cancer progression (Micali et al. 2010; Ramchatesingh et al. 2022). However, the effective delivery of retinoic acid topically remains a challenge due to its hydrophobic nature and sensitivity to air, light and heat which not only makes it difficult to formulate, but also difficult to ensure the delivery of an effective dose. Besides that, it is also known for its side effects whereby it can potentially cause skin irritation (Kandhari 2024; Siddique 2024).

Nanotechnology has emerged as a promising approach in the field of drug delivery, offering various benefits such as improved bioavailability, targeted delivery, and enhanced stability of therapeutic agents (Dhapte and Pokharkar 2019). One such nanoparticulate system that has garnered significant attention is the nanosponge, a porous and polymeric carrier capable of encapsulating a wide range of active ingredients, including hydrophobic drugs like retinoic acid (Jain et al. 2020). Nanosponges have been explored as a versatile platform to overcome these limitations, offering improved solubility, controlled release, and enhanced skin tolerability (Jain et al. 2020). The sponge-like structure is able to entrap the active ingredient within it and around it making it more soluble, reducing its degradation whilst also controlling the rate of release making it much slower and controlled. This slow-release mechanism coupled with the reduction of direct contact of the active ingredient to the skin is able to reduce the occurrence of side effects too. The formulation and optimization of retinoic acid-loaded nanosponges are crucial to ensure optimal performance and suitability for the target population of active individuals. In this study, we investigated the impact of homogenization speeds and drying methods on the physicochemical characteristics of retinoic acid-loaded nanosponge formulations. Active individuals, especially those who are frequently exposed to the sun, are prone to various forms of UV-related skin damage. Some common examples among this population include premature aging, sunspots, accelerated wrinkling, and increased risk of skin cancer (Syed Omar et al. 2022; Damiani and Puglia 2019). The use of retinoic acid in a more efficient manner through an advanced delivery system such as nanosponges can help mitigate these concerns and provide better protection and rejuvenation for the skin of active individuals (Ferreira et al. 2020).

2 Methodology

2.1 Preparation of Nanosponge

Nanosponge was prepared according to previous methods (Omar et al. 2024). The nanosponge was prepared at a ratio of 1:4 polymer to crosslinker. The polymer used was β-cyclodextrin with carbonydiimidazole as the crosslinker. The final product was then subjected to different drying methods, namely through oven drying and freeze dryer and assessed for optimized characteristics. The selected nanosponge was then added into a base emulgel prepared at different speeds which were 2000, 3000 and 4000 rpm respectively.

2.2 Characterizations

pH, Particle Size, Polydispersity Index and Zeta Potential Analysis. For pH analysis, formulation was diluted with distilled water to 10% w/w before being analyzed with a calibrated pH meter (Metler Toledo, Switzerland). While particle size, zeta potential and polydispersity index were analysed using Zetasizer (ZEN 1600, Malvern, UK). Samples placed in capillary cells were diluted 1:1000 using deionized water to prevent multiple scattering effects. Measurements were taken in triplicates at a temperature of 25 °C.

Stability Assessment. Stability study was done over the span of 3 months with the nanosponge in emulgel formulation. The study was carried out at temperatures of 30 °C ± 2 and 40 °C ± 2 at 75% ± 5% RH in real-time and accelerated stability chamber, respectively (Binder KBF 240, Fisher Scientific, Germany). Properties of the formulation were assessed at time points 0 weeks, 2 weeks, 4 weeks and 12 weeks. The assessment includes pH, particle size, polydispersity index, and zeta potential.

Data Analysis. All the data obtained were recorded in triplicates and presented in the form of mean ± standard deviation. The data obtained from the development and optimization of gel-cream formulation underwent analysis using GraphPad Prism. To ascertain significance in the optimization process, t-test was applied. The stability study which spanned over a three-month period, was subjected to analysis using repeated measure ANOVA.

3 Results and Discussion

3.1 Effect of Drying Methods on Nanosponge

From Fig. 1 it is evident that the application of different drying methods for the developed nanosponge results in different characteristics. Particle size especially shows a stark significance ($p > 0.005$) of 2115.0 ± 69.91 nm for oven dry to 254.8 ± 4.90 nm for the freeze-dried sample. Smaller particle sizes are preferred as they enhance skin retention, allowing for more efficient and prolonged drug delivery. Studies have shown that particle sizes ranging from 700 nm to below 100 nm achieve high skin retention which leads to improved drug delivery (Micali et al. 2010; Ramchatesingh et al. 2022).Freeze drying causes the particle to become smaller compared to oven drying due to the removal of water under vacuum conditions, which creates a more porous structure and smaller particles (Bhardwaj et al. 2020). The polydispersity index also shows the oven dried sample having a significantly ($p > 0.005$) higher value of 0.764 ± 0.035 in comparison to the freeze-dried sample at 0.317 ± 0.007. A previous study done on a different powder found that oven-drying caused an increase in particle size as well as a bimodal size distribution in comparison to freeze-dried sample which produced a smaller size particle with a unimodal size distribution most likely an effect of using heat (Buzera et al. 2022). These findings are consistent with the results of the current study whereby freeze drying resulted in more desirable particle characteristic in comparison heat drying method.

The zeta potential although as found to be significantly different ($p > 0.005$), did not show as a big gap as the other parameters with oven dried sample at −21.70 ± 0.10 and freeze-dried sample at −28.10 ± 0.10. Freeze dried sample being more negative, indicates it is further from the agglomeration threshold indicating a better stability in comparison to the oven dried sample. This may be due to the increased surface area effects of the smaller particles making it more affected by Brownian motion hence the larger zeta potential values (Ramchatesingh et al. 2022; Kandhari

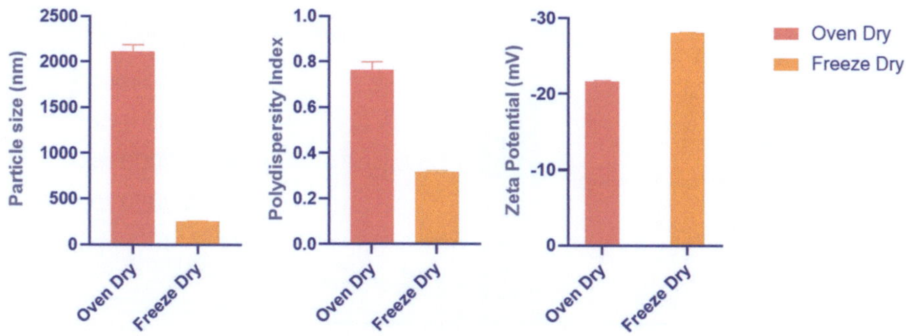

Fig. 1 Particle size, polydispersity index and zeta potential for nanosponge prepared with different drying methods (mean ± SD, n = 3)

2024). The higher surface area of the freeze dried nanosponge can also lead to greater drug loading and ultimately higher encapsulation efficiency (Patel et al. 2020). In general freeze drying was able to improve the properties of the nanosponge for better controlled and targeted drug delivery.

3.2 *Effect of Speed on Loaded Nanosponge in Gel-Cream*

The changes over time and homogenization speed for both accelerated and stability studies did show significant differences ($p < 0.005$). There was a slight increase in pH with increasing speed, which is more evident in real-time stability study as seen in Fig. 2. The general trend over all groups is a decrease in pH to become more acidic whereby the values started in week 0 was above pH 6 an ended at week 12 with a value below pH 6. An acidic pH is preferred to help restore the skin's natural acidic mantle on the outer layer of the skin, which can be disrupted by frequent washing. Applying an acidic product aid will help return the skin to its normal state of being acidic (Siddique 2024; Dhapte and Pokharkar 2019). It is widely recommended for topical products to have a pH between 4 and 6, aligning with the typical pH of adult skin, which ranges from 4.1 to 5.8 (mean 4.9). The acidic pH of the stratum corneum is essential for skin barrier function, lipid synthesis, epidermal differentiation, desquamation, microbiome balance, inflammation reduction, and antimicrobial protection (Pujalte et al. 2023; Jain et al. 2020). Despite the changes, the formulation remained stable with an acceptable pH. A previous study examining the impact of homogenization pressure on the pH of sesame vegetable cream found no significant changes. Similarly, the findings from this study and prior research on cream formulations indicate that pH is not readily affected by technical parameters, in this case, the homogenization speed (Samani and Naji 2019).

From Fig. 3, it is evident that all speeds show variation in size at each time point from week 0, week 2, week 4 and week 12. Analysis of the data showed that both

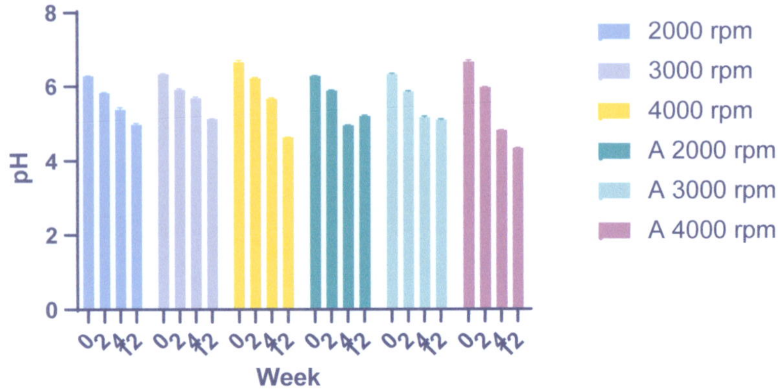

Fig. 2 pH Values of retinoic acid loaded nanosponge gel-cream prepared at different speeds stored in 30 °C/75%RH and 40 °C/75%RH at week 0, week 2, week 4 and week 12 (Mean ± SD, n = 3). (Abbreviations: 0 is Week 0; 2 is week 2; 4 is Week 4; 12 is Week 12; A refers to accelerated)

speed and time caused significant changes on the particle size value ($p < 0.005$). However, formulation at 3000 rpm does show less fluctuations in value over time in both conditions indicating a stable formulation. Figure 4 depicts the particle size distribution in the formulation. These results relate to the particle size in which the speeds with larger fluctuation in particle size over time show a larger polydispersity index that indicated the various sizes present in the formulation. Speed 2000 and 4000 rpm generally have larger PDI values in comparison to 3000 rpm which reflects on the fluctuations of their particle size over time as well. PDI was also found to have significant differences across different speeds and time points ($p < 0.005$).

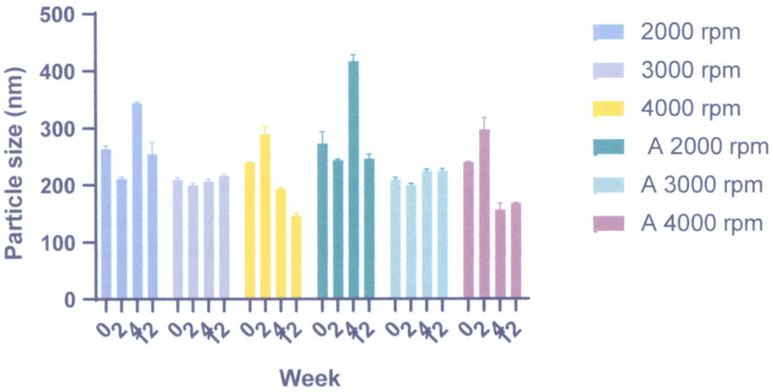

Fig. 3 Particle size of retinoic acid loaded nanosponge gel-cream prepared at different speeds stored in 30 °C/75%RH and 40 °C/75%RH at week 0, week 2, week 4 and week 12 (Mean ± SD, n = 3). (Abbreviations: 0 is Week 0; 2 is week 2; 4 is Week 4; 12 is Week 12; A refers to accelerated)

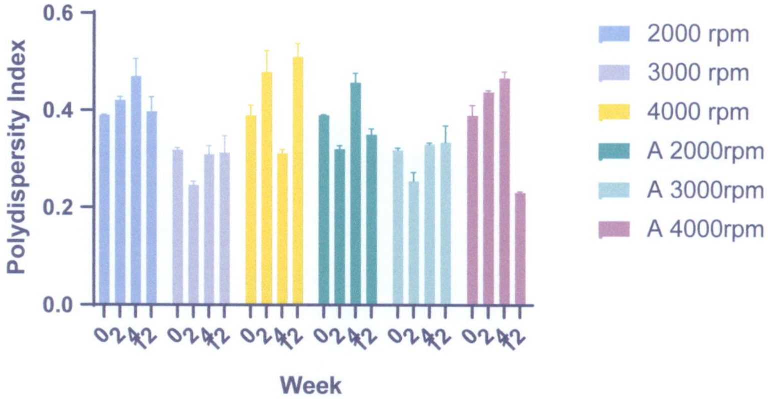

Fig. 4 Polydispersity index of retinoic acid loaded nanosponge gel-cream prepared at different speeds stored in 30 °C/75%RH and 40 °C/75%RH at week 0, week 2, week 4 and week 12 (Mean ± SD, n = 3). (Abbreviations: 0 is Week 0; 2 is week 2; 4 is Week 4; 12 is Week 12; A refers to accelerated)

The results in Fig. 3 illustrates the optimal speed for achieving the smallest particle size is 3000 rpm, producing particles of 209.53 ± 1.56 nm at week 0, compared to 263.5 ± 5.91 nm at 2000 rpm. Smaller particle sizes enhance formulation stability by minimizing Ostwald ripening and improving skin retention for controlled drug delivery (Omar et al. 2024; Argenziano et al. 2019). The reduction in particle size with increased homogenization speed is attributed to higher energy density and shear stress, which disrupt cohesive forces within droplets (Liu et al. 2018; Bhardwaj et al. 2020). At 3000 rpm, sufficient mechanical energy breaks droplets into smaller sizes, but speeds above this, such as 4000 rpm (239.47 ± 1.56 nm), increase coalescence due to higher collision rates (Bamba et al. 2018). Over time in both real-time and accelerated chambers, there is increase in particle size which is likely due to Ostwald ripening, a process where smaller particles dissolve and redeposit onto larger ones to minimize the surface area ratio and achieve a thermodynamically stable state (Westen and Groot 2018). This phenomenon often leads to creaming, flocculation, or phase separation. However, nano-sized formulations can delay or prevent Ostwald ripening due to their small size, which enhances stability over a longer duration (Simonazzi et al. 2018). Additionally, smaller particles improve stability and contribute to increased viscosity (Khor et al. 2014).

Similarly, the polydispersity index (PDI), indicating size distribution, decreases from 0.3903 ± 0.0012 at 2000 rpm to 0.3187 ± 0.0046 at 3000 rpm, reflecting more uniform particles. The PDI increases again at 4000 rpm (0.3900 ± 0.0209), likely due to droplet coalescence and size variability. With higher initial PDI values, fluctuations of particle size and PDI over time are more evident due to the presence of varying sizes and higher occurrence of coalescence. These findings suggest 3000 rpm provides the optimal balance for small, uniform particles that remain consistent throughout the study in both real-time and accelerated chamber proving its ability to remain stable.

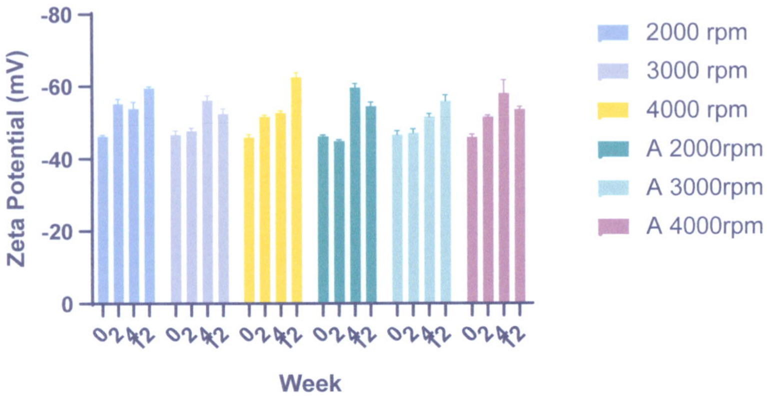

Fig. 5 Zeta potential of retinoic acid loaded nanosponge gel-cream prepared at different speeds stored in 30 °C/75%RH and 40 °C/75%RH at week 0, week 2, week 4 and week 12 (Mean ± SD, n = 3). (Abbreviations: 0 is Week 0; 2 is week 2; 4 is Week 4; 12 is Week 12; A refers to accelerated)

For zeta potential, the values in week 0 did not show any significant difference between speeds for both real-time and accelerated implying that the speeds did not have a significant impact on zeta potential in comparison to the other parameters. Maintaining a zeta potential outside the range of −30 mV to +30 mV is crucial for formulation stability. As values approach the agglomeration threshold (0 mV), the risk of particles agglomerating or coalescing increases due to attractive van der Waals forces, potentially causing physical instabilities like creaming or phase separation (Samimi et al. 2019). The data in Fig. 5 indicates that over time, zeta potential values move further from the agglomeration threshold, enhancing stability. This trend may be influenced by a slight decrease in the formulation's pH from week 0 to week 12, as pH is known to affect zeta potential (Romdhane et al. 2015).

4 Conclusion

The study successfully demonstrated the optimization of a retinoic acid-loaded nanosponge formulation. By evaluating the effects of homogenization speeds and drying methods, it was established that freeze-drying and homogenization at 3000 rpm produced the most favorable characteristics, including smaller particle size, lower polydispersity index, and enhanced zeta potential. These attributes contribute to better skin penetration, increased stability, and controlled drug delivery. The findings underscore the importance of balancing mechanical energy and drying conditions to achieve stable, effective nanosponge formulations. This optimized formulation holds significant potential for commercial and clinical applications in sun aftercare and therapeutic skincare products, offering a targeted solution for mitigating UV-induced skin damage for active individuals and athletes frequently exposed to UV radiation. Prevention and management of the risk associated with UV exposure, especially among athletes, needs to be of greater concern to prevent any discomfort related to the skin which may impact performance. Future work should address potential challenges in scalability and environmental stability to facilitate broader adoption. Additionally, exploring the formulation's compatibility with other active ingredients could expand its applicability and use in dermatological care.

References

Argenziano M, Haimhoffer A, Bastiancich C et al (2019) In vitro enhanced skin permeation and retention of imiquimod loaded in β-cyclodextrin nanosponge hydrogel. Pharmaceutics 11:138

Bamba BSB, Shi J, Tranchant CC, Xue SJ, Forney CF, Lim LT, Xu W, Xu G (2018) Coencapsulation of polyphenols and anthocyanins from blueberry pomace by double emulsion stabilized by whey proteins: effect of homogenization parameters. Mol 23:2525

Bhardwaj B, Singh P, Kumar A, Kumar S, Budhwar V (2020) Eco-friendly greener synthesis of nanoparticles. Adv Pharm Bull 10:566–576

Buzera A, Gikundi E, Orina I, Sila D (2022) Effect of pretreatments and drying methods on physical and microstructural properties of potato flour. Foods 11:507

Damiani E, Puglia C (2019) Nanocarriers and microcarriers for enhancing the UV protection of sunscreens: an overview. J Pharm Sci 108:3769–3780

De Luca JF, Adams BB, Yosipovitch G (2012) Skin manifestations of athletes competing in the summer Olympics: what a sports medicine physician should know. Sport Med 42:399–413

Dhapte V, Pokharkar V (2019) Nanosystems for drug delivery: design, engineering, and applications. Green Synth Charact Appl Nanoparticles 321–345

Ferreira R, Napoli J, Enver T, Bernardino L (2020) Advances and challenges in retinoid delivery systems in regenerative and therapeutic medicine. Nat Commun 111(11):1–14

George JM, Aswani MT, Pavan Kumar MV, Varghese B (2020) Green synthesis of ZnO nanoparticles. AIP Conf Proc. https://doi.org/10.1063/5.0017250/774499

Jain A, Prajapati SK, Kumari A, Mody N, Bajpai M (2020) Engineered nanosponges as versatile biodegradable carriers: an insight. J Drug Deliv Sci Technol 57:101643

Kandhari R (2024) Side effects of anti-acne medications: a narrative review. Egypt J Dermatology Venereol 44:1–6

Khan BA, Akhtar N, Khan H, Braga V de A (2013) Development, characterization and antioxidant activity of polysorbate based O/W emulsion containing polyphenols derived from hippophae rhamnoides and cassia fistula. Brazilian J Pharm Sci 49:763–773

Khor YP, Koh SP, Long K, Long S, Ahmad SZS, Tan CP (2014) A comparative study of the physicochemical properties of a virgin coconut oil emulsion and commercial food supplement emulsions. Molecules 19:9187–9202

Kliniec K, Tota M, Zalesińska A, Łyko M, Jankowska-Konsur A (2023) Skin cancer risk, sun-protection knowledge and behavior in athletes—a narrative review. Cancers 15:3281

Kumar S, Prasad M, Rao R (2021) Topical delivery of clobetasol propionate loaded nanosponge hydrogel for effective treatment of psoriasis: Formulation, physicochemical characterization, antipsoriatic potential and biochemical estimation. Mater Sci Eng C 119:111605

Liu X, Shen B, Shen C, Zhong R, Wang X, Yuan H (2018) Nanoparticle-loaded gels for topical delivery of nitrofurazone: effect of particle size on skin permeation and retention. J Drug Deliv Sci Technol 45:367–372

Lukić M, Pantelić I, Savić SD (2021) Towards optimal pH of the skin and topical formulations: from the current state of the art to tailored products. Cosmet 8:69

Micali G, Lacarrubba F, Dinotta F, Massimino D, Nasca MR (2010) Treating skin cancer with topical cream. Expert Opin Pharmacother 11:1515–1527

Mulia K, Safiera A, Pane I, Krisanti E (2019) Effect of high speed homogenizer speed on particle size of polylactic acid. J Phys Conf Ser. https://doi.org/10.1088/1742-6596/1198/6/062006

Omar SSS, Hadi HA, Doolanea AA, Zulkifli N (2024) Nanosponge as a novel delivery modulator of retinoic acid. Pharm Sci Asia 51:106–117

Omar SSS, Hadi H, Hanif NM, Ahmad HMA, Ng S-F (2021) Lightening effect of skin lightening cream containing piper betle L. extract in human volunteers. Cosmet 8:32

Patel D, Zode SS, Bansal AK (2020) Formulation aspects of intravenous nanosuspensions. Int J Pharm 586:119555

Peng J, Dong WJ, Li L, Xu JM, Jin DJ, Xia XJ, Liu YL (2015) Effect of High-pressure homogenization preparation on mean globule size and large-diameter tail of oil-in-water injectable emulsions. J Food Drug Anal 23:828–835

Proksch E (2018) PH in nature, humans and skin. J Dermatol 45:1044–1052

Pujalte GGA, Costa LMC, Clapp AD, Presutti RJ, Sluzevich JC (2023) More than skin deep: dermatologic conditions in athletes. Sports Health 15:74–85

Ramchatesingh B, Martínez Villarreal A, Arcuri D, Lagacé F, Setah SA, Touma F, Al-Badarin F, Litvinov IV (2022) The use of retinoids for the prevention and treatment of skin cancers: an updated review. Int J Mol Sci 23:12622

Romdhane A, Aurousseau M, Guillet A, Mauret E (2015) Effect of pH and ionic strength on the electrical charge and particle size distribution of starch nanocrystal suspensions. Starch—Stärke 67:319–327

Samani SA, Naji MH (2019) Effect of homogenizer pressure and temperature on physicochemical, oxidative stability, viscosity, droplet size, and sensory properties of sesame vegetable cream. Food Sci Nutr 7:899

Samimi S, Maghsoudnia N, Eftekhari RB, Dorkoosh F (2019) Lipid-based nanoparticles for drug delivery systems. In: Charact Biol Nanomater Drug Deliv Nanosci Nanotechnol Drug Deliv Elsevier, Tehran, pp 47–76

Siddique I (2024) Comparing vitamin c and vitamin e in improving the photostability of a retinol serum exposed to heat degradation. J South Carolina Acad Sci 22

Simonazzi A, Cid AG, Villegas M, Romero AI, Palma SD, Bermúdez JM (2018) Nanotechnology applications in drug controlled release. In: Drug target. Stimuli Sensitive Drug Deliv Syst Elsevier, pp 81–116

Syed Omar SS, Hadi H (2022) Advancement of all-trans retinoic acid delivery systems in dermatological application. Cosmet 9:140

Szymański Ł, Skopek R, Palusińska M, Schenk T, Stengel S, Lewicki S, Kraj L, Kamiński P (2020) Retinoic acid and its derivatives in skin. Cells 9:2660

Van Westen T, Groot RD (2018) Effect of temperature cycling on ostwald ripening. Cryst Growth des 18:4952

Optimizing Athletic Abilities: Exploring the Effects of Complex Training on Speed, Change of Direction and Explosive Power in Youth Soccer Players

Muhammad Iftiqar Ahmad Termizi, Raja Mohammed Firhad Raja Azidin, and Raihana Sharir

Abstract The aim for this study is to investigate the effects of 8 weeks of complex training (CT) on speed, change of direction and power performance between complex training group and control group. Thirty soccer players ($n = 30$) were recruited from Sultan Idris Shah II Secondary School, Perak, Malaysia, comprising two age groups: under 15 ($n = 15$; aged 13–15 years) and under 18 ($n = 15$; aged 16–18 years). The participants underwent three physical tests assessing speed, change of direction (COD) and explosive power for the pre-test. Then, they were divided into two groups which were complex training (CT) group and control group. They underwent 8 weeks of intervention which comprised of CT program. Paired Sample T-Test revealed that all 4 assessments had significant results which were 20 m Sprint Test ($p < 0.005$, 0.001), Arrowhead Test (Right) ($p < 0.05$, $P = 0.001$), Arrowhead Test (Left) ($p < 0.005$, $p = 0.001$) and Counter-Movement Jump Test (CMJ) ($p < 0.005$, $p = 0.001$). In aspect of differences between the two groups in speed, COD and explosive power after 8 weeks of intervention, Independent T-Test revealed a significant difference ($p < 0.05$, $p = 0.03$) only in power performance (CMJ Test) between CT and control group. Current findings indicated that CT training method is a good strategy to be implemented to improve sport specific power performance and at the same time, the current findings may help in understanding and provides valuable insights for coaches, highlighting the effectiveness of CT in improving sport performance.

Keywords Complex training · Soccer · Youth

M. I. A. Termizi · R. M. F. R. Azidin · R. Sharir (✉)
Faculty of Sports Science and Recreation, Universiti Teknologi MARA, Shah Alam, Malaysia
e-mail: raihanasharir@uitm.edu.my

R. Sharir
School of Sport and Exercise Sciences, Liverpool John Moores University, Liverpool, UK

R. M. F. R. Azidin
Faculty of Sport Science, Malang State University, Malang, Indonesia

M. H. A. Hassan et al. (eds.), *Proceedings of the 3rd International Conference on Innovation and Technology in Sports*, Lecture Notes in Bioengineering,
https://doi.org/10.1007/978-981-95-0804-4_5

1 Introduction

In recent years, modern soccer has undergone a significant transformation, evolving into a fast-paced, dynamic sport characterized by intermittent bursts of activity (Michailidis et al. 2021; Yu et al. 2021; Andrašić et al. 2021). The game now involves a combination of short, high-intensity movements such as sprinting, running, jumping, duels, tackling, change of direction (COD), as well as walking and standing at various points throughout (Pavillon et al. 2021; Gualtieri et al. 2023). On average, players engage in high-intensity actions every 60 s, with maximal efforts occurring every 4 min across the course of the match (Pavillon et al. 2021). Studies show that players make over 1,200 unpredictable changes in activity during a game, which include up to 11–12% sprinting of total distance covered, 1,200–1,400 changes in direction, 30–40 tackles and jumps, and between 90 and 140 ball interactions, such as passing, kicking, and dribbling (Yu et al. 2021; Negra et al. 2022; Hulton et al. 2022). Most of these actions are highly explosive in nature (Križaj 2020).

By possessing great physiological abilities could play a key role in optimizing performance and improving tactical decisions made by coaches (Zghal et al. 2019; Yu et al. 2021; Keiner et al. 2021). The importance of physiological abilities like speed, change of direction, and explosive power are crucial not only for individual players but also for the overall team performance, in both adult and youth soccer (Zghal et al. 2019; Andrašić et al. 2021). Attackers rely on these abilities to apply pressure and execute tackles to gain goal scoring opportunities, while defenders use them to close down opponents' space and limit attacking opportunities, thus minimizing threats to the team and conceiding goals (Andrašić et al. 2021). This trend emphasizes how modern soccer has become more and more athletic, with key attributes such as strength, power, sprinting, jumping, and changing direction play crucial roles in determining success in invasive sport like soccer (Sonesson et al. 2020; Sariati et al. 2020; Keiner et al. 2021). Križaj (2020) further reinforces this, stating that there is a strong correlation between sprinting, COD and sufficient explosive power in the lower limbs.

Speed is a key performance indicator of success in soccer which defined as the ability to run at high velocity and cover a distance from one point to another as fast as possible (Yildiz et al. 2018). In practical terms, it refers to the body's capacity to move swiftly and cover ground as quickly as possible (Krosta et al. 2020). Speed is essential throughout both halves of a soccer game (Field et al. 2022). In game situations, players typically cover between 110–330 m of sprinting or high-speed running, depending on their position (Yildiz et al. 2018; Krosta et al. 2020). Saeidi & Khodamoradi (2017) note that professional players cover 5–10% less distance in the second half, especially at medium (11.1–19.0 km/h) and high (18.0–30.0 km/h) intensities. However, Reinhardt et al. (2019) found that elite soccer players cover 24–36% of their total distance at high-intensity running and 36–63% during linear sprints throughout the game. For youth soccer players, it has been reported that they cover approximately 6–8 km per match and perform an average of 80 accelerations ($>1\ m{\cdot}s^{-2}$), with each lasting up to 2–4 s (Fernández-Galván et al. 2022). Sprint performance in youth players is closely linked to lower body muscle strength and power, with players who possess strong sprinting abilities often having a greater advantage in scoring (Sonesson et al. 2020).

Over the past few decades, change of direction (COD) and reactive agility were often considered synonymous (Andrašić et al. 2021). However, ongoing research has distinguished COD as a movement where no immediate reaction to a stimulus is required, meaning the direction change is pre-planned or deliberate (Andrašić et al. 2021). COD, a key component of speed, involving rapid, whole-body movements that include phases of acceleration, deceleration, and re-acceleration in a different direction, which relevant in typical soccer game (Falch et al. 2019). COD is vital in invasive sports like soccer, where players frequently need to change direction (Sariati et al. 2020; Michailidis et al. 2021). CODs in soccer are determined by the rate of stride adjustment, sprint pace, and characteristics of leg muscles such as strength, power, and reactive strength according to Horníková and Zemková 2021. A soccer player usually changes direction every 2–4 s during a match, resulting up to 1,200–1,400 changes in total during a match Keiner et al. 2021. This frequency defines the importance of COD for efficient movement during soccer play. Players with higher COD capabilities are better equipped to gain an advantage on the field (Keiner et al. 2021). Trecroci et al. (2020) further underline that COD skills such as side-steps, swerves, turns, and crossover steps are essential for evading opponents, creating space, and positioning may lead to higher scoring opportunities.

Explosive power is the ability to produce strength and speed in muscle contractions, resulting in fast and forceful movements (Darni and Elkadiowanda 2019). Power plays a crucial role in many invasive sports, including soccer (Joksimović et al. 2019). This component relies on a combination of muscular strength and the speed of movement, during movements execution (Darni and Elkadiowanda 2019). In soccer, lower limb explosive power is important which it influence speed and the ability to change direction. Besides that, player who had appropriate lower limbs explosive power could enhance their jumping, turning, and sprinting capabilities which then may influence and improving overall performance (Franca et al. 2021; Cossio-Bolaños et al. 2021). This is supported by Joksimović et al. (2019) highlight that explosive power is vital for executing high-speed actions such as sprinting and jumping. In soccer, power is crucial for explosive muscle contractions also play crucial role during movements like shooting, making passes, jumping for headers, and quickly changing direction in one-on-one duels (Behm 2018) which then may influence the team's advantages during the game. Jusoh et al. (2019) emphasize that power is one of the most important physical factors for success in modern soccer, with power (strength-speed), speed, and related components being essential for a team's performance.

When working with youth, it's important to prioritize their long-term development, which makes managing training loads and fatigue are crucial in training interventions (Magallanes et al. 2022). While the precise physiological and specific mechanisms behind training adaptations are still being researched, specific and appropriate training program must be implemented to gradually increase athletes' physical capabilities by the time they reach full maturity, (Trecroci et al. 2020; Magallanes et al.

2022). Therefore, when designing training programs for youth, it's essential to ensure the program is not only effective in enhancing performance but also efficient in improving their overall physical development.

One of the method to improve all the three components is complex training (CT) (Greenwood and Carter 2014; Wang et al. 2023). CT could be used to fulfill the requirements for soccer players and it is stated that it suitable to be used to improve muscular strength, explosive power and its' demands (Wang et al. 2023). CT is a training which combining both resistance training (RT) and plyometric training (PT) in the same training session (Greenwood and Carter 2014; Wang et al. 2023). This training regime is supported by the assumption of post-activation potential of the neuromuscular system which it helps in increasing the muscle contraction related to its' previous contraction (Wang et al. 2023). Research has shown that CT can significantly improve individual and team performance by enhancing motor unit recruitment and stored energy in the muscle–tendon unit during the stretch–shortening cycle (SSC). This leads to better muscle efficiency in producing explosive power following maximum or near-maximum contractions during CT training (Freitas et al. 2017) which hese characteristics may help and give advantages towards the athletes to success in their sport as in this current study context, in soccer.

Previous studies have shown that CT method can be an effective strategy for enhancing sport-specific power performance, though some studies have failed to observe these effects (Ali et al. 2017). Additionally, there are limited studies that have tried to promote the importance and effectiveness of complex training (CT) towards athlete physical development and performance (Talpey et al. 2016b). As such, this study aims to contribute to a deeper understanding of CT's impact on the physical development of youth athletes, particularly in youth soccer players.

2 Method

2.1 Experimental Approach to the Problem

This study emphasized pre- and post-test research design, incorporating three physical tests: the 20 m sprint test, the arrowhead test, and the counter-movement jump (CMJ) test. The aim was to assess differences in speed, change of direction, and explosive power between U-15 and U-18 soccer players. Participants were asked to avoid any physical activity for 48 h prior to testing. On the test day, all subjects received identical instructions as a group to ensure consistency. Testing began following a standardized warm-up, and the sequence was as follows: 20 m sprint test (3 trials), arrowhead test (3 trials in each direction—right and left), and CMJ test (3 trials). A 20 min rest interval was provided between each test. One week after the pre-test, the subjects underwent 8 weeks of CT intervention. The post-test was conducted after the intervention finished.

2.2 Subjects

A total of 30 soccer players (n = 30) from Sultan Idris Shah II Secondary School, Malaysia, participated in the study. The participants were divided into two groups: the CT group (n = 15; mean age = 15.33 ± 1.08 years; height = 1.65 ± 0.05 m; body mass = 53.45 ± 6.38 kg) and the control group (n = 15; mean age = 15.40 ± 1.14 years; height = 1.67 ± 0.07 m; body mass = 58.30 ± 12.90 kg). Participants had to be injury-free in the lower limbs for at least six months prior to testing. All participants were thoroughly briefed on the overall study's procedures and provided written informed consent for their parents/guardian to approved. The experimental protocol was approved by the Ethics Committee of Universiti Teknologi Mara (UiTM) prior to the commencement of the study and was conducted in accordance with the university's ethical guidelines.

2.3 Test Protocols

20 m Sprint Test

The test was designed to assess maximal sprint effort (Madarsa et al. 2020) and is commonly used to measure acceleration and maximum speed qualities (Brini et al. 2021). Timing gates (Brower Timing System, USA) were used to record sprint times to the nearest 0.001 s (Murtagh et al. 2018), ensuring accuracy and eliminating potential bias. The timing gates were placed at point A, which was 0 m, the starting line, and point B, which was 20 m, the finish line. The time recorded was to an accuracy of 1 ms. The subjects were asked to stand with their bodies leaned forward and started when they were ready. They did three maximal sprints over a 20 m distance (Brini et al. 2021) (Fig. 1). 3 min recovery was allowed after each sprint and the best time taken to two decimal places from the three efforts.

Arrowhead Test

The arrowhead test was used to assess change of direction (COD) and is a reliable assessment to evaluate soccer player's ability to perform longer sprints involving changing direction (Michailidis et al. 2021; Rago et al. 2020). Subjects needed to complete six trials (Three to the right and three to the left). They started at cone A, 0.20 m from the gate. On command, they sprinted to the middle cone (B) and then, depending on the trial, turned either left (E) or right (C), continuing with a sprint around the side marker and top cone (D). The test was finished by sprinting back to cone A, with times recorded through the Brower Timing System (Rago et al. 2020). The test was done on the right side first which then followed by the left. Subjects did the test rotated alphabetically by surname upon called for each trial (Fig. 2). 3 min rest interval were given to the subjects to allow sufficient recovert before next trial. The fastest time from the trials was recorded to be analyzed.

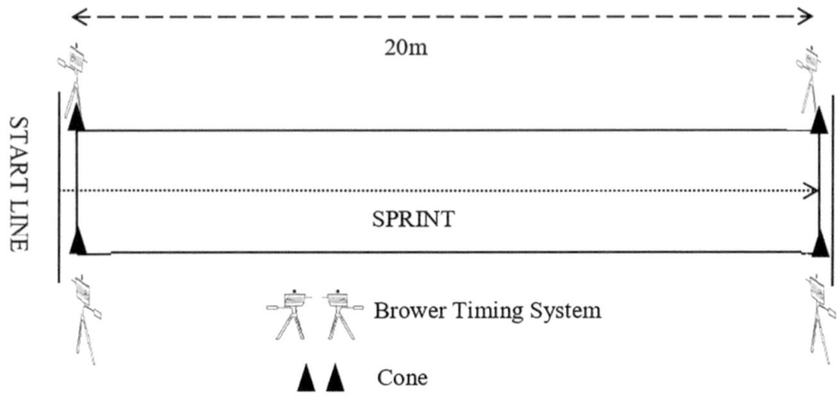

Fig. 1 Diagram of a 20-meter sprint setup. The start and finish lines are marked with cones. A Brower Timing System is positioned at both ends to measure sprint time. The word "SPRINT" is labeled in the center of the track.

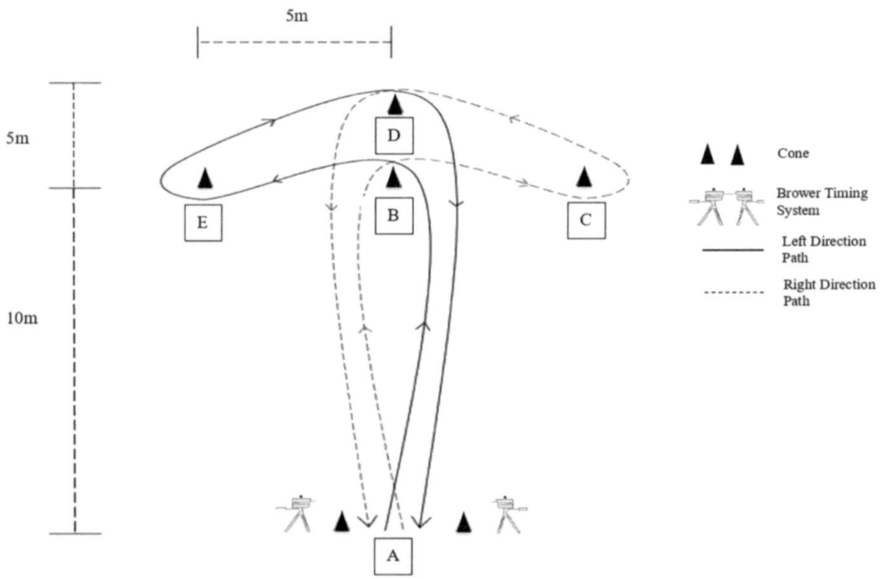

Fig. 2 Diagram of a running drill setup with cones and timing systems. The layout includes five cones labeled A to E, arranged in a T-shape. Cone A is at the base, with cones B and C forming the top line, and cones D and E at the ends. Solid and dashed lines indicate left and right running paths, respectively. Distances between cones are marked as 5 meters horizontally and 10 meters vertically. Timing systems are positioned near cone A.

Counter-Movement Jump Test (CMJ)

Counter-movement jump (CMJ) was used to analyze explosive power performance which according to Barker et al. (2018), is a great assessment to assess lower body power performance. Instead of using the traditional vertex, current study used CMJ mobile application (My Jump 2) on an iPhone XR to analyze CMJ (Haynes et al. 2019) which according to Bogataj et al. (2020b, a) this application is a valid, reliable, and affordable tool for measuring jump performance in youth athletes. The application calculated CMJ flight time by detecting the take-off and landing sequences from the video. All participants' personal data was pre-entered into the app. During the test, subjects started in a standing position with their torso straight, knees fully extended, feet shoulder-width apart, and hands on their hips. Upon instruction, they performed a quick downward movement (approximately 90° knee flexion) followed by an explosive upward jump (Bogataj et al. 2020a, b) (Fig. 3). Participants needed to complet three maximal jumps while keeping their hands on their hips. The My Jump 2 app automatically calculated the flight time by identifying take-off and landing moments in the video (Fig. 4), then converted this into jump height using the equation $h = t^2 \times 1.22625$. The researcher recorded all jumps from the same position, 1.5 m away from the subject, following the manufacturer's guidelines. The highest jump from the three attempts was used for analysis.

Fig. 3 A series of three photos showing a person performing a squat jump exercise. In the first image, the person stands upright with hands on hips. The second image captures the person in a squatting position, knees bent, and leaning slightly forward. The third image shows the person mid-air, having jumped, with feet off the ground. The setting is an outdoor area with a concrete floor, a red pillar, and green foliage in the background. Various objects, including a green bucket and a small tripod, are visible on the ground.

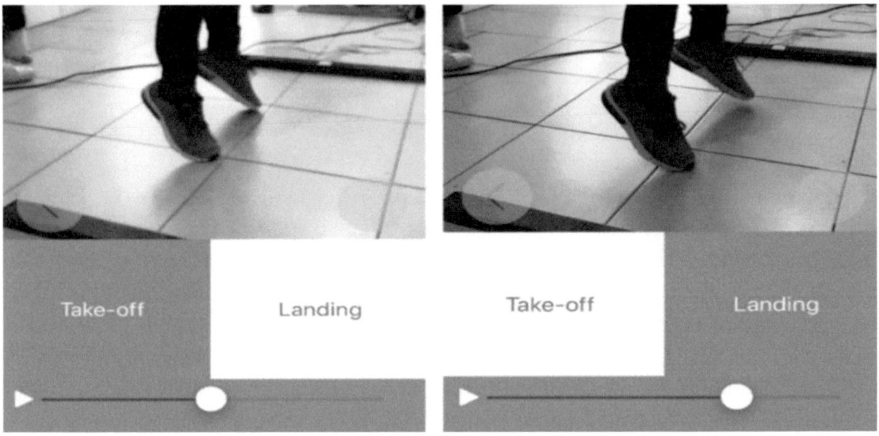

Fig. 4 side-by-side screenshot shows two stages of a jump. The left image is labeled "Take-off" with a slider at the start position, depicting feet leaving the ground. The right image is labeled "Landing" with a slider at the end position, showing feet about to touch the ground. Both images have a tiled floor and cables in the background.

2.4 Intervention

The CT group underwent complex training (CT) three times per week, based on a program adapted from Abdelkader (2016), while the control group continued with standard soccer training routines. The CT program integrated both resistance training (RT) and plyometric training (PT). RT targeted lower-body strength using equipment such as dumbbells, barbells, and Olympic bars, with participants performing 2 sets of 3–8 repetitions at 70–95% of their one-repetition maximum (1RM). Following the RT component, the PT exercises included box and hurdle jumps with heights ranging from 40–60 cm. These PT exercises were performed in 2 sets of 6–10 repetitions within the same session. Rest intervals between sets were set at 2–3 min to ensure adequate recovery (Table 1).

3 Result

See Table 2.

This study aimed to evaluate the differences in speed, change of direction (COD), and explosive power performance between the complex training (CT) group and the control group after an 8-week intervention. Paired Sample T-Tests revealed statistically significant changes in all four performance measures. For the 20 m Sprint Test, the mean difference post-intervention was [Mean difference = 0.08, SD = 0.03, 95% CI (0.06, 0.09)], with a significant result at the 0.05 level ($t = 10.72$, $df = 14$, $p < 0.05$). In the Arrowhead Test (Right), the mean difference following the 8 weeks of

Table 1 CT training program

Day/Week	Monday	Tuesday	Wednesday	Thursday	Friday	Saturday	Sunday
0	No training	No training	No training	No training	No training	No training	Pre-test
1	2 × 8 Olympic bar back squat (70% 1RM) 2 × 6 alternate lunge jump	Soccer team training	2 × 8 dumbbell standing calf raise at 70% 1RM 2 × 8 depth jump 40 cm	Soccer team training	2 × 8 leg extension at 70% 1RM 2 × 8 drop jump at 40 cm	Rest day/No training	Rest day/No training
2	2 × 8 single leg dumbbell Lunge at 70% 1RM 2 × 8 Alternate Lunge Jump	Soccer team training	2 × 8 Dumbbell single leg standing calf raise at 70% 1RM 2 × 8 single leg depth jumps 20 cm box	Soccer team training	2 × 8 leg curl at 70% 1RM 2 × 8 single leg hurdle jumps 30 cm (4 Hurdles)	Rest day/No training	Rest day/No training
3	2 × 8 leg press at 70% 1RM 2 × 10 box jump (50 cm)	Soccer team training	2 × 6 barbell split squat at 80% 1RM 2 × 10 split squat jump	Soccer team training	2 × 6 barbell box step up at 80% 1RM (40 cm Box) 2 × 8 box step up jump	Rest day/No training	Rest day/No training

(continued)

Table 1 (continued)

Day/Week	Monday	Tuesday	Wednesday	Thursday	Friday	Saturday	Sunday
4	1 × 6 Olympic bar back squat at 80% 1RM 1 × 8 vertical hurdle jump (4 Hurdles) 1 × 6 dumbbell single leg lunge at 80% 1RM 1 × 8 alternate lunge jump	1 × 6 dumbbell standing calf raise at 80% 1RM 1 × 10 depth jumps to second box (40 cm) 1 × 5 Olympic Bar Split Squat at 85% 1RM 1 × 5 Split Squat Jumps followed by 5 Vertical Hurdles Jumps (40 cm)	Soccer team training	1 × 6 leg extension at 80% 1RM 1 × 10 box jump at 40 cm 1 × 6 leg curl at 80% 1RM 1 × 10 vertical side hurdle jumps 40 cm (4 Hurdles)	Soccer team training	Light training/ Recovery	Post-Test 1
5	Soccer team training/light training/recovery	1 × 8 Leg Press at 70% 1RM 1 × 8 Box Jump (50 cm) then Depth Jumps to Second Box 40 cm 1 × 6 Barbell Single Leg Lunge at 80% 1RM 1 × 10 Alternate Lunge Jump	Soccer team training	1 × 5 dumbbell calf raise at 85% 1RM 1 × 8 side depth jump 50 cm followed by side jumping over 4 hurdles 40 cm 1 × 6 leg curl at 80% 1RM 1 × 6 vertical counter-movement jumps followed by forward jump	Soccer team training	1 × 5 Olympic bar box step up at 85% 1RM (40 cm Box) 1 × 8 step up jumps 1 × 5 leg extension at 85% 1RM 1 × 8 single leg side hurdle jumps 40 cm (4 Hurdles) followed by Forward Jump	Rest day/No training

(continued)

Table 1 (continued)

Day/ Week	Monday	Tuesday	Wednesday	Thursday	Friday	Saturday	Sunday
6	1 × 5 Olympic bar split squat at 85% 1RM 1 × 8 box jump then 2 vertical hurdle jumps 1 × 4 Olympic bar standing calf raise at 90% 1RM 1 × 10 depth jumps to second box 40 cm then 3 vertical hurdle jumps	Soccer team training	1 × 4 Olympic bar squat at 90% 1RM 1 × 8 backward depth jump 45 cm 1 × 6 leg press at 90% 1RM 1 × 10 box jump (50 cm) to Depth jumps then 2 hurdle vertical jumps	Soccer team training	1 × 4 Olympic bar box step up at 90% 1RM (40 cm Box) 1 × 8 step up jumps 1 × 4 Olympic bar standing calf raise at 90% 1RM 2 × 10 side depth jump followed by 3 hurdles jumps	Rest day/No training	Rest day/No training

(continued)

Table 1 (continued)

Day/ Week	Monday	Tuesday	Wednesday	Thursday	Friday	Saturday	Sunday
7	1 × 4 leg extension at 90% 1RM 1 × 8 depth jump to second box to drop jump 40 cm 1 × 4 Leg curl at 90% 1RM 1 × 6 single leg hurdle jump 30 cm (4Hurdles)	Soccer team training	1 × 4 leg press a 90% 1RM 1 × 6 box jump (60 cm) then depth jump to second box 40 cm 1 × 4 leg extension at 90% 1RM 1 × 10 box jump 40 cm to drop jump	Soccer team training	1 × 4 Olympic bar back squat at 90% 1RM 1 × 6 squat jump followed by 2 vertical hurdle jumps (40 cm) then 2 side vertical hurdle jumps 1 × 4 Olympic bar split squat at 90% 1RM 1 × 4 vertical hurdle jump (4 Hurdles) then box jump (40 cm) to depth jump to vertical jump	Rest day/No training	Rest day/No training
8	1 × 3 Olympic bar standing calf raise at 95% 1RM 1 × 6 depth jumps to second box 40 cm 1 × 3 Olympic bar single leg lunge at 95% 1RM 1 × 6 alternating lunge jump then box jump to depth jump	1 × 3 Olympic bar box step up at 95% 1RM (40 cm Box) 1 × 8 step up jump 1 × 3 back squat at 95% 1RM 1 × 4 vertical hurdle jump (4 Hurdles) then box jump (40 cm) to depth jump to vertical jump	Soccer team training	1 × 3 leg curl at 95% 1RM 1 × 6 single leg hurdle jump 30 cm (4 Hurdles) 1 × 3 split squat at 95% 1RM 1 × 8 split squat jump then 2 vertical hurdle jumps then vertical jump	Soccer team training	Light training/ Recovery	Post-Test 2

Table 2 Effects of complex training (CT) on speed, change of direction (COD) and explosive power

	Mean	Std	S.E Mean	95% confidence interval of the difference		Paired t test		
				Upper	Lower	t	df	p
Speed performance (20 m sprint test)	0.08	0.03	0.01	0.06	0.09	10.72	14	0.001*
Change of direction (COD) Arrowhead test (Right)	0.14	0.11	0.03	0.07	0.20	4.58	14	0.001*
Change of direction (COD) Arrowhead test (Left)	0.09	0.05	0.01	0.06	0.11	7.30	14	0.001*
Explosive power (CMJ test)	−6.72	2.76	0.71	−8.25	−5.19	−9.42	14	0.001*

* Indicate significant difference, $p < 0.005$

CT training was [Mean difference = 0.14, SD = 0.11, 95% CI (0.07, 0.20)], which was also statistically significant at the 0.05 level ($t = 4.58$, $df = 14$, $p < 0.05$). For the Arrowhead Test (Left), the mean difference after the CT intervention was [Mean difference = 0.09, SD = 0.05, 95% CI (0.06, 0.11)], with statistical significance at the 0.05 level ($t = 7.30$, $df = 14$, $p < 0.05$). Finally, the CMJ Test showed a mean difference of [Mean difference = −6.72, SD = 2.76, 95% CI (−8.25, −5.19)] post-intervention, with a statistically significant result at the 0.05 level ($t = -9.42$, $df = 14$, $p < 0.05$).

Regarding the differences between the complex training (CT) group and the control group in speed, change of direction (COD), and power performance after 8 weeks of intervention, Independent T-Tests revealed a significant difference in power performance between the two groups ($p < 0.05$, $p = 0.003$) (Table 3). However, no significant differences were found for the other three performance measures. Specifically, the 20 m Sprint Test ($p = 0.16$), Arrowhead Test (Right) ($p = 0.07$), and Arrowhead Test (Left) ($p = 0.12$) all showed p-values greater than 0.05, indicating no significant differences between the groups.

Table 3 Differences of performance between CT and control group after 8 weeks

	CT group (n = 15)	Controlled group (n = 15)	t	df	p
	Mean ± Std	Mean ± Std			
Speed performance (20 m sprint test)	3.12 ± 0.11	3.18 ± 0.11	−1.434	28	0.16
Change of direction (COD) Arrowhead test (Right)	8.18 ± 0.20	8.37 ± 3.33	−1.920	28	0.07
Change of direction (COD) Arrowhead test (left)	8.28 ± 0.24	8.45 ± 0.31	−1.600	28	0.12
Explosive power (CMJ test)	44.39 ± 5.37	40.00 ± 5.21	2.271	28	0.03*

* Indicate significant difference, $p < 0.05$

4 Discussion

4.1 *Effects of Complex Training (CT) on Speed, Change of Direction (COD) and Explosive Power in CT Group*

The findings of our study are similar with those in Thapa et al. (2020), who demonstrated that complex training (CT) resulted in moderate to large improvements in linear sprinting, change of direction (COD) performance, and countermovement jump (CMJ) performance. These improvements are likely due to three key factors: Post-Activation Potentiation (PAP), morphological changes, and hormonal and cellular adaptations that influence muscle function. The first factor, PAP, is a neuromuscular phenomenon that occurs following an initial bout of heavy resistance exercise, as highlighted by Thapa et al. (2020). PAP has been shown to enhance the rate of force development in specific muscle groups, leading to increased acceleration and velocity during movement (Lim and Barley 2016). This effect contributes to both immediate and long-term improvements in muscular strength and power of the lower limb. PAP is particularly enhanced by a higher percentage of type II muscle fibers, which generate greater maximal muscle contractions, a faster anaerobic force rate, and increased metabolite production linked to fatigue. These factors together enhance the PAP response, thereby supporting the outcomes of CT intervention (Gołaś et al. 2016).

The use of extremely heavy loads is associated with greater recruitment of type IIx muscle fibers, significantly enhancing the PAP effects during maximal-intensity, short-duration movements (Gołaś et al. 2016). This is align with the findings in Thapa et al. (2020), who reported that an increase in type IIx muscle fibers leads to higher explosive power production which characterized by increased muscle contraction velocity and rate of force development compared to type IIA and type I fibers. These adaptations are thought to be mediated by changes in myosin light chain phosphorylation, which enhances the sensitivity of actin-myosin to calcium ions (Ca2+) which then modifies the structure of myosin heads, and increases cross-bridge force output. Moreover, muscle pre-contraction has been shown to increase the excitation potential across the spinal cord, resulting in greater motor unit recruitment (Chen 2017). Jusoh et al. (2019) further emphasized that CT may improves motor unit coordination and recruitment, which caused higher peak jump heights and shorter contact times during the stretch–shortening cycle (SSC), ultimately improving performance in jump-related tasks such as CMJ. The combined effects of resistance training (RT) and plyometric training (PT) further enhance the storage and release of elastic energy during the deceleration and concentric phases of movements, increasing muscle force and power output. Over time, this leads to improved maximal strength, contributing to greater force production during movement execution (Abdelkader 2016; Zghal et al. 2019).

Besides that, CT has been shown to improve morphological changes in muscle structure. CT protocols which combined both resistance training (RT) and plyometric training (PT) in the same training regime has been demonstrated to significantly enhance muscle size, including increases in muscle thickness and cross-sectional area (Grgic et al. 2021). This effect is attributed to the nature of RT and PT, which if combined together may stimulate hypertrophic adaptations in muscle fibers. These findings are align with Schneiker et al. (2023), who reported that strength-power training protocols lead to improvements in muscle architecture, hypertrophy, and tendon stiffness which contribute to enhanced performance in explosive movements. The morphological adaptations resulting from CT also involve increases in the recruitment of type IIx muscle fibers, which are crucial for the production of explosive power. Grgic et al. (2021) noted that these adaptations, along with the PAP effect, promote greater strength and power output during dynamic movements. Despite some debate in the literature regarding the correlation between muscle hypertrophy and maximal strength production which some studies indicated that strength improvements could occur without the increases in muscle size, it is widely acknowledged that muscle hypertrophy could play a vital role in improving maximal strength (Vigotsky et al. 2018). Reggiani and Schiaffino (2020) further support this idea by suggesting that changes in muscle architecture, such as the enlargement of muscle fibers and muscle size may lead to increased contractile performance and a higher capacity for force generation during movements. Moreover, Thapa et al. (2020) emphasized that the increase in type IIx muscle fibers associated with CT enhances not only muscle power but also the rate of force development and contraction velocity, which are critical for explosive movement execution. The ability to produce higher rates of force development, especially during short-duration, high-intensity efforts, is a key factor in improving athletic performance across various sports. Thus, the combination of both RT and PT in the same training program could give impact on muscle hypertrophy, fiber recruitment, and the PAP effect results in significant improvements in explosive power, strength, and overall athletic performance.

Lastly, the increase of hormonal response specifically in testosterone levels following 8 weeks of training. This aligns with the study by Greenwood and Carter (2014), which demonstrated that a combined RT and PT regimen results in a significant increase in testosterone production which may lead in enhancing power output in youth athletes. The elevated testosterone levels associated with the 8-week CT intervention are likely to contribute to changes in the physical development of the athletes, including increases in muscle mass and strength, as indicated by Vanny and Moon (2015). These hormonal changes may, in turn, facilitate higher strength and power production during physical activities or during movements execution (Ali et al. 2019; Thapa et al. 2020). Testosterone plays a crucial role in regulating several physiological processes, including its ergogenic, anabolic, and anti-catabolic effects on skeletal muscle that caused these phenomenon. These mechanisms are essential for enhancing muscle strength, power, and endurance, thus supporting overall athletic performance (Gharahdaghi et al. 2020). Furthermore, Zouhal et al. (2022) suggest that high-intensity exercise regimens, such as CT may lead to an increase in testosterone production, further contributing to muscle mass and strength development. Consequently, the hormonal adaptations observed in response to the CT intervention likely played a significant role in the observed improvements in performance and physical capabilities in the athletes.

4.2 The Differences Between the Complex Training (CT) Group and Control Group in Speed, Change of Direction (COD) and Explosive Power After 8 Weeks of Intervention

The significant difference in counter-movement jump (CMJ) performance between the two groups could be primarily attributed to the same three key factors as before which are Post-Activation Potentiation (PAP), morphological changes, and hormonal and cellular adaptations within the muscle tissue. These findings align with those of Abdelkader (2016), where significant improvements in CMJ performance were observed in young soccer players after underwent CT intervention, with an increase in CMJ height of at least 5.08 cm. The current study's results suggest that the enhancement in CMJ power following 8 weeks of combined training may be a consequence of the PAP response triggered by CT (Thapa et al. 2020). PAP is a neuromuscular phenomenon that occurs after an initial bout of high-intensity resistance exercise, leading to improved performance in subsequent movements (Lim and Barley 2016). Specifically, PAP enhances the rate of force development in the targeted muscle groups, which facilitates an increase in movement velocity and acceleration, contributing to both short-term and long-term gains in muscular strength and power (Lim and Barley 2016). This improvement can be attributed to a higher proportion of type II muscle fibers, which facilitate maximal muscle contractions, exhibit an elevated anaerobic force production rate, and increase the generation of metabolites associated with fatigue. These physiological responses enhance the PAP effect and align with the changes induced by the CT intervention (Gołaś et al. 2016).

Golaś et al. (2016) proposed that extremely heavy resistance loads employed in trained athletes could facilitate a greater recruitment of type IIx muscle fibers, thereby enhancing the Post-Activation Potentiation (PAP) response. This notion is supported by Thapa et al. (2020), who suggested that increased type IIx muscle fiber activation contributes to enhanced power production, characterized by faster contraction velocities, greater explosive power, and superior rates of force development compared to type IIA and type I fibers. These physiological changes are mediated by myosin light chain phosphorylation, which increases the sensitivity of the actin-myosin complex to calcium ions (Ca2+), alters the myosin head structure, and promotes higher cross-bridge force generation. Additionally, muscle pre-contraction enhances the excitation potential across the spinal cord, thereby improving motor unit recruitment (Chen 2017). Jusoh et al. (2019) further emphasized that combined training (CT) can optimize motor unit activation coordination, which enhances peak jump height and shortens contact time during the stretch–shortening cycle (SSC), leading to superior jumping performance. Similarly, Abdelkader (2016) highlighted that CT enhances the storage of elastic energy during the deceleration phase, which is subsequently released during the shortening phase, thereby boosting muscle force and power output. This sequence of adaptations is believed to contribute to maximal strength improvements, facilitating greater force production and enhanced jump performance in the CT group (Zghal et al. 2019).

The improvement in jumping performance observed in the CT group in the current study may be partly associated with morphological changes in the muscle tissue. So far, a combined RT and PT intervention within one training program has exhibited an additive effect on muscle hypertrophy in several studies (Grgic et al. 2021). Indeed, in some studied papers, such as Grgic et al. 2021, muscle thickness and, accordingly, the size of muscles increased notably after weeks of combined training including RT and PT. The same study also stated that the young athletes could develop a significant muscle cross-sectional area and muscle size through combination of RT and PT. Further, Schneiker et al. (2023) went ahead to propose that improvements in the performance of CMJ after strength-power training may be related to increases in cross-sectional area, architectural changes, hypertrophy, and tendinous adaptations of muscle, which in turn would improve neuromuscular function and jumping performance. Grgic et al. (2021) indicated that combined training increases the development of type IIx muscle fibers, which have higher responsiveness to the movements entailed in CT which this caused superior levels of strength and power output. Reggiani and Schiaffino (2020) indicated that the change in muscle architectures and size can elicit an improvement in the contractile performance and force generation, hence helping to enhance the ability of performance in dynamic movements such as jumping.

Hormonal and cellular adaptations also played a significant role in the observed changes in performance, particularly through the increase in testosterone levels among participants, which may have facilitated which induced the enhancement of strength and power production (Ali et al. 2019; Thapa et al. 2020). Greenwood and Carter (2014) suggested that a combined resistance training (RT) and plyometric training (PT) regimen could stimulate higher testosterone production, potentially leading to improved power outcomes in young athletes. This idea is supported by Vanny and Moon (2015), who emphasized testosterone's crucial role in various physiological processes during the transition from adolescence to adulthood, such as the maintenance of muscle mass, strength, and bone density, all of which can contribute to maximizing athletic performance. Testosterone levels are considered a key determinant of athletic performance, with studies indicating that testosterone-related differences account for up to 10–20% of performance variability among athletes (Bezuglov et al. 2023). The implementation of explosive exercises as part of the combined training (CT) protocol over the 8-week intervention period likely elicited a significant hormonal response, specifically increasing testosterone levels. Zouhal et al. (2022) further noted that physical training can enhance testosterone production, which may lead to an increase in muscle mass. The regulation of testosterone thus plays a central role in athletes' physiological responses, contributing to a range of ergogenic, anabolic, and anti-catabolic functions in skeletal muscle, ultimately enhancing muscle strength, power, and endurance (Gharahdaghi et al. 2020).

While the current study demonstrated significant effects of combined training on speed and change of direction (COD) performance, no significant differences were found between the CT and control groups following 8 weeks of intervention. This contrasts with previous studies showing a positive correlation between leg extensor strength gained from CT and improved sprint performance (Hammami et al. 2018). However, the current study found no significant differences between the two groups analyzed. This outcome is similar to the findings of Cavaco et al. (2014), who reported no significant improvements in speed between the CT and control groups after a similar duration of training. One possible explanation for this lack of difference could be the absence of a speed-specific training component within the CT program, as suggested by Hammami et al. (2018). This notion is supported by Falch et al. (2019), who emphasized that the inclusion of specific speed and COD movements in training programs is essential for yielding optimal improvements in linear speed and COD performance. This was demonstrated in Poulos et al. (2010), where the study found that combining training with speed-specific drills led to improved performance times. This is supported by Carvajal and Salazar (2021), who stated that by adding speed-specific exercises to a training regimen has been consistently shown to enhance performance. Therefore, if speed and COD-specific exercises and drills been incorporated into the CT protocol, the differences between the CT and control groups might have been more pronounced, as suggested by prior research (Hammami et al. 2018; Falch et al. 2019; Carvajal and Salazar 2021).

Another explanation for the non-significant results observed in the current study may be related to technical factors among the subjects and the athletes' training status throughout the intervention and testing periods (Hammami et al. 2018). The quality of running technique which is one of the key determinant of speed and COD performance, could have influenced the outcomes in this study (Pojskic et al. 2018). It is possible that some subjects may have improved their technique during the intervention, leading to a reduction in body mass or experiencing psychological or hormonal changes that could have positively impacted their performance during the testing (Cavaco et al. 2014). Despite the differences in training protocols between the experimental and control groups, both groups participated in soccer-specific training, which included elements of speed and COD training. This may explain the lack of significant differences between the two groups. Kirby et al. (2010) supported this idea, suggesting that sprint training outside of weight training sessions could also improve muscle function and enhance pre-activation in the eccentric phase of the stretch–shortening cycle (SSC), which could contribute to improved performance on final final findings. To better differentiate the effects between the experimental and control groups, future research could include more specific sprint and COD skill development within the CT program. This approach could lead to more pronounced differences between the two groups and improve performance techniques among the subjects which lead to better findings (Pojskic et al. 2018; Hammami et al. 2018; Falch et al. 2019).

This study has several limitations that must be acknowledged. Firstly, CT protocol did not incorporate speed and agility-specific training component, which these components may have influenced the final outcomes. As stated before, previous research, such as in Poulos et al. (2010), has demonstrated that if speed-specific drills were included in CT training could lead to significant improvements in maximal sprint performance. Therefore, the lack of such a component in the current study may have contributed to the absence of significant differences between the two groups. Another limitation is the relatively small sample size in the study which may have affected the study's final findings (Brysbaert 2019). Future research should consider increasing the sample size to enhance the robustness of the results and provide more definitive and concrete conclusions. Despite these limitations that been found, the current study offers valuable insight of CT for the development of youth soccer players, serving as an important contribution to the existing body of evidence in this field.

5 Conclusion

Soccer is a high-intensity, intermittent contact sport game that demands a broad spectrum of physical and physiological capabilities for successful performance either in adult or youth. Besides technical and tactical skills, players must have high levels of aerobic and anaerobic fitness, speed, agility, change of direction (COD) ability, and power to excel in the game. The findings of this study suggest that CT significantly enhances linear sprinting, COD performance, and countermovement jump (CMJ) height, with moderate to large improvements observed among subjects. Other than traditional isolate training regime, these results indicate that CT could be an effective training method for improving sport-specific power performance and its' derivatives factors. Moreover, the insights provided by this study offer valuable guidance for coaches, reinforcing the utility of CT in enhancing key physical capabilities necessary for optimal soccer performance.

References

Abdelkader MA (2017) Effect of complex training on jumping variables during heading the ball for soccer players

Ali K, Verma S, Ahmad I, Singla D, Saleem M, Hussain ME (2019) Comparison of complex versus contrast training on steroid hormones and sports performance in male soccer players. J Chiropr Med 18(2):131–138

Ali K, MEH, Verma S, Ahmad I, Singla D, Jha P (2017b) Complex training: an update. J Athl Enhanc 06(03)

Andrašić S, Gušić M, Stanković M, et al (2021) Speed, change of direction speed and reactive agility in adolescent soccer players: age related differences. Int J Environ Res Public Health 18(11)

Barker LA, Harry JR, Mercer JA (2018) Relationships between countermovement jump ground reaction forces and jump height, reactive strength index, and jump time. J Strength Cond Res 32(1):248–254

Behm DG (2018) Neuromuscular Physiology, Exercise, and Training during Youth—The year that was 2017. Pediatr Exerc Sci 30(1):35–37

Bekris E, Komsis G, Zacharoula P, Ioannis G (2017) The effect of a strength and speed training program in speed improvement of female soccer players. ResearchGate

Bezuglov E, Ahmetov II, Lazarev A, Mskhalaya G, Talibov O, Устинов B, Shoshorina M, Bogachko E, Azimi V, Morgans R, Hackney AC (2023) The relationship of testosterone levels with sprint performance in young professional track and field athletes. Physiol Behav 271:114344

Bogataj Š, Pajek M, Andrašić S, Trajković N (2020a) Concurrent validity and reliability of My Jump 2 app for measuring vertical jump height in recreationally active adults. Appl Sci 10(11):3805

Bogataj Š, Pajek M, Hadžić V, Andrašić S, Padulo J, Trajković N (2020b) Validity, reliability, and usefulness of My Jump 2 app for measuring vertical jump in primary school children. Int J Environ Res Public Health 17(10):3708

Brini S, Boullosa D, Calleja-González J, Delextrat A (2021) Construct validity and reliability of a new basketball multidirectional reactive repeated sprint test. Int J Environ Res Public Health/ Int J Environ Res Public Health 18(20):10695

Brysbaert M (2019) How many participants do we have to include in properly powered experiments? a tutorial of power analysis with reference tables. J Cogn 2(1)

Carvajal R, Salazar W (2021) The combined effect of strength and speed training on change of direction performance in soccer players. Avrasya Spor Bilimleri Ve Eğitim Dergisi 3(1):1–11

Cavaco B, Sousa N, Reis VM et al (2014) Short-term effects of complex training on agility with the ball, speed, efficiency of crossing and shooting in youth soccer players. J Hum Kinet 43(1):105–112

Chan HC-K-M (2011) The difference of physical ability between youth soccer player and professional soccer player: a training implication. In: Proceedings of the 3rd HKASMSS student conference on sport medicine, rehabilitation and exercise science, Hong Kong 53

Chen Z, Lo S, Wang M, Yu C, Peng H (2017) Can different complex training improve the individual phenomenon of Post-Activation potentiation? J Hum Kinet 56(1):167–175

Cossio-Bolaños M, Vidal-Espinoza R, Albornoz CU, Portella DL, Vega-Novoa S, Mendez-Cornejo J, Lopez JF, Gomez-Campos R (2021) A systematic review of intervention programs that produced changes in speed and explosive strength in youth footballers. Eur J Transl Myol 31(3)

Darni N, Elkadiowanda I (2020) Overview of the physical condition of football players of high school students. Adv Soc Sci, Educ HumIties Res 460
Falch HN, Rædergård HG, Van Den Tillaar R (2019) Effect of different physical training forms on change of direction (COD) ability: a systematic review and meta-analysis. Sport Med-Open 5(1)
Fernández-Galván LM, Jiménez-Reyes P, Cuadrado-Peñafiel V, Casado A (2022) Sprint performance and mechanical force-velocity profile among different maturational stages in young soccer players. Int J Environ Res Public Health/Int J Environ Res Public Health 19(3):1412
Field A, Naughton RJ, Haines M, Lui S, Corr LD, Russell M, Page R, Harper LD (2022) The demands of the extra-time period of soccer: a systematic review. J Sport Health Sci 11(3):403–414
França C, Élvio Gouveia R, Caldeira R, Marques A, Martins J, Lopes H, Costa M, Ihle A (2021) Predictors of explosive leg power in elite football players. J Phys Educ Sport 21(06)
Freitas TT, AM-R-G (2017) Short-term adaptations following complex training in team-sports: a meta-analysis. PLoS One 12(6):1–19
Gharahdaghi N, Piasecki M, Szewczyk NJ, Smith K, Wilkinson DJ, Atherton PJ (2021) Links between testosterone, oestrogen, and the growth Hormone/Insulin-Like growth factor axis and resistance exercise muscle adaptations. Front Physiol 11
Giudicelli BB, Luz LGO, Sogut M, Sarmento H, Massart AG, Júnior AC, Field A, Figueiredo AJ (2021) Chronological age, somatic maturation and anthropometric measures: association with physical performance of Young Male Judo athletes. Int J Environ Res Public Health/Int J Environ Res Public Health 18(12):6410
Gołaś A, Maszczyk A, Zając A, Mikołajec K, Šťastný P (2016) Optimizing post activation potentiation for explosive activities in competitive sports. J Hum Kinet 52(1):95–106
Greenwood JC (2014) Complex training reexamined: review and recommendations to improve strength and power. Strength Cond J 36(2):11–19
Grgić J, Schöenfeld BJ, Mikulić P (2021) Effects of plyometric vs. resistance training on skeletal muscle hypertrophy: a review. J Sport Health Sci 10(5):530–536
Gualtieri A, Rampinini E, Iacono AD, Beato M (2023) High-speed running and sprinting in professional adult soccer: current thresholds definition, match demands and training strategies: a systematic review. Front Sport Act Living 5
Hammami MM, Gaamouri N, Shephard RJ, Chelly MS (2019) Effects of contrast strength vs. plyometric training on lower-limb explosive performance, ability to change direction and neuromuscular adaptation in soccer players. J Strength Cond Res 33(8):2094–2103
Haynes T, Bishop C, Antrobus M, Brazier J (2019) The validity and reliability of the My Jump 2 app for measuring the reactive strength index and drop jump performance. J Sport Med Phys Fit 59(2)
Horička P, Hianik J, Šimonek J (2014) The relationship between speed factors and agility in sport games. J Hum Sport Exerc 9(1):49–58
Horníková H, Zemková E (2021) Relationship between physical factors and change of direction speed in team sports. Appl Sci 11(2):655
Hulton AT, Malone JJ, Clarke ND, MacLaren DPM (2022) Energy requirements and nutritional strategies for male soccer players: a review and suggestions for practice. Nutrients 14(3):657
Joksimović M, Pavlović R, Pantović M, Eler N, Nikšić E, Bijelić S (2019) Manifestations of explosive power: differences in the leg springs between footballers of the different competitive level. Eur J Phys Educ Sport Sci 5(10). https://doi.org/10.5281/zenodo.3342021
Jusoh A, Ishak AA, James J (2019) Integrated power training mode in soccer: merging complex training with cluster set. Int J Acad Res Bus & Soc Sci 9(1)
Keiner M, Kapsecker A, Stefer T, Kadlubowski B, Wirth K (2021) Differences in squat jump, linear sprint, and change-of-direction performance among youth soccer players according to competitive level. Sports 9(11):149
Kirby TJ, Erickson TM, McBride JM (2010) Model for progression of strength, power, and speed training. Strength Cond J 32(5):86–90

Krosta R, Drozd M, Goralczyk R, Kostrzewa M, Fischerova P, Ignatieva A (2020) Models of speed development in soccer. Balt J Health Phys Act 12(2):27–34

Lim J, Barley C (2016) Complex training for power development: practical applications for program design [review of complex training for power development: practical applications for program design]. ResearchGate; Strength Cond J

Madarsa NI, Mohamad NI, Malek NFA, Chinnasee C, Nadzalan AM (2021) Profiling and relationship between sprint time and cardiovascular fitness during in-season's training among professional soccer players. J Phys: Conf Ser 1793(1):012058

Magallanes A, Feye ASP, Magallanes C, Gonzalez-Ramírez A (2022) Strength training in youth soccer players: towards an optimal relationship between load magnitude and performance enhancement. Sport TK 2

Michailidis Y, Karydopoulos C, Kapralos D, Mitrotasios M, Ispirlidis I, Margonis K, Metaxas T (2021) Arrowhead agility test in Elite U-19 soccer players: positional differences and relationships with other performance tests. Trends Sport Sci 28(1):37–44

Murtagh CF, Brownlee TE, O'Boyle A, Morgans R, Drust B, Erskine RM (2018) Importance of speed and power in elite youth soccer depends on maturation status. J Strength Cond Res 32(2):297–303

Negra Y, Sammoud S, Ramirez-Campillo R, Bouguezzi R, Moran J, Chaabene H (2022) The effects of repeated sprint training with versus without change of direction on measures of physical fitness in youth male soccer players. J Sport Med Phys Fit 63(1)

Nikolaidis PT, Knechtle B, Clemente FM, Torres-Luque G (2016) Reference values for sprint performance in male soccer players aged 9–35 years old. ResearchGate

Pavillon T, Tourny C, Aabderrahman AB, Salhi I, Zouita S, Rouissi M, Hackney AC, Granacher U, Zouhal H (2021) Sprint and jump performances in highly trained young soccer players of different chronological age: effects of linear vs. change–of–direction sprint training. J Exerc Sci Fit 19(2):81–90

Pojskic H, Åslin E, Krolo A, Jukic I, Uljević O, Spasic M, Sekulic D (2018) Importance of reactive agility and change of direction speed in differentiating performance levels in junior soccer players: reliability and validity of newly developed soccer-specific tests. Front Physiol 9

Poulos N, Kuitunen S, Buchheit M (2010) Effect of preload squatting on sprint performance in adolescent athletes. New Stud Athl 4(No.3):95–103

Rago V, Brito J, Figueiredo P, Ermidis G, Barreira D, Rebelo A (2020) The arrowhead agility test: reliability, minimum detectable change, and practical applications in soccer players. J Strength Cond Res 34(2):483–494

Reggiani C, Schiaffino S (2020) Muscle hypertrophy and muscle strength: dependent or independent variables? a provocative review. Eur J Transl Myol 30(3)

Reinhardt L, Schwesig R, Lauenroth A, Schulze S, Kurz E (2019) Enhanced sprint performance analysis in soccer: new insights from a GPS-based tracking system. PLoS ONE 14(5):e021778

Saeidi A, Khodamoradi A (2017) Physical and physiological demand of soccer player based on scientific research. ResearchGate

Sariati D, Hammami R, Chtara M, Zagatto A, Boullosa D, Clark C, Hackney AC, Granacher U, Souissi N, Zouhal H (2020) Change-of-direction performance in elite soccer players: preliminary analysis according to their playing positions. Int J Environ Res Public Health/Int J Environ Res Public Health 17(22):8360

Schneiker KT, Fyfe JJ, Teo SY, Bishop DJ (2023) Comparative effects of contrast training and progressive resistance training on strength and power-related measures in subelite Australian rules football players. J Strength Cond Res 37(7):1440–1448

Sonesson S, Lindblom H, Hägglund M (2020) Performance on sprint, agility and jump tests have moderate to strong correlations in youth football players but performance tests are weakly correlated to neuromuscular control tests. Knee Surg Sports Traumatol Arthrosc 29(5):1659–1669

Talpey SW, Young WB, Saunders N (2016) Is nine weeks of complex training effective for improving lower body strength, explosive muscle function, sprint and jumping performance? Int J Sports Sci Coach 11(5):736–745

Thapa RK, Lum D, Moran J, Ramírez-Campillo R (2021) Effects of complex training on sprint, jump, and change of direction (COD) ability of soccer players: a systematic review and meta-analysis. Front Psychol 11

Trecroci A, Rossi A, Dos'Santos T, Formenti D, Cavaggioni L, Longo S, Iaia FM, Alberti G (2020) Change of direction asymmetry across different age categories in youth soccer. PeerJ 8:e9486

Turner AN, Stewart PF (2014) Strength and conditioning for soccer players. Strength Cond J 36(4):1–13

Vanny P, Moon JR (2015) Physiological and psychological effects of testosterone on sport performance: a critical review of literature. Sport J

Vigotsky AD, Schöenfeld BJ, Than C, Brown JR (2018) Methods matter: the relationship between strength and hypertrophy depends on methods of measurement and analysis. PeerJ 6:e5071

Wang X, Lv C, Xin Q, Ji S, Dong D (2023) Effectiveness of plyometric training vs. complex training on the explosive power of lower limbs: a systematic review. Front Physiol 13

Yıldız S, Ateş O, Gelen E, Çırak E, Bakıcı D, Sert V, Kayıhan G (2018) The relationship between start speed, acceleration and speed performances in soccer. Univers J Educ Res 6(8):1697–1700

Warren B, Young BD (2015) Agility and change-of-direction speed are independent skills: implications for training for agility in invasion sports. Int J Sport Sci Coach 10(1):159–169

Yu L, Altieri C, Bird SP, Corcoran G, Jiuxiang G (2021) The importance of in-season strength and power training in football athletes: a brief review and recommendations. Int J Strength Cond 1(1)

Zghal F, Colson SS, Blain GM, Behm DG, Granacher U, Chaouachi A (2019) Combined resistance and plyometric training is more effective than plyometric training alone for improving physical fitness of pubertal soccer players. Front Physiol 10

Zouhal H, Jayavel A, Parasuraman K, Hayes LD, Tourny C, Rhibi F, Laher I, Abderrahman AB, Hackney AC (2021) Effects of exercise training on anabolic and catabolic hormones with advanced age: a systematic review. Sports Med 52(6):1353–1368

The Effect of Physical Activity Brain Breaks Among University Students

Ngien Siong Chin, Boon Sian Teo, and Matheus Yan Yu Chin

Abstract The prevalence of overweight and obesity rates are rising throughout Asia, with Malaysia being at the forefront for the past years. The lack of recent data on young people, particularly university students in Malaysia has resulted in this study which sought to examine on the prevalence of overweight and obesity among university students and effect of physical activity brain breaks at one public university in Sarawak, Malaysia. The health-related fitness test (HRFT) was administered while physical activity brain breaks (PABB) videos were used as an intervention tool. Descriptive analysis and paired sample t-test were used. For the Body Mass Index (BMI), results revealed that the frequency of participants increased significantly from pre (52.1%) to post (75.0%) in the normal weight category. Besides, the frequency of the participants decreased from pre-to -post in the underweight (2.1–0.0%), overweight (20.8–14.6%) and obese (25.0–10.4%) category. In terms of HRFT test, the frequency of participants increased from pre to post in superior fitness level (18.8–52.1%) and excellent fitness level (12.5–20.8%). Conversely, the participants' frequency declined from pre to post in the fit fitness level (31.2–16.7%), poor fitness level (37.5–10.4%). Moreover, paired sample t-test on HRFT test revealed that there was significant increase from pre to post test in step test, $p = .002$, curl-ups, $p = .004$, and sit and reach, $p = .005$. Additionally, the paired sample t-test for HRFT test for pre and post-test based on groups showed statistically significant increase on the intervention group, whereby there was significant increase from pre to post on step test, $p = .001$, push-ups, $p = .004$, curl-ups, $p = .011$, and sit and reach, $p = .001$. The study revealed important findings that can be applied to devise strategic interventions at improving the health and fitness of the university students.

N. S. Chin (✉)
Institute of Teacher Education Malaysia Batu Lintang Campus, Kuching, Malaysia
e-mail: ipgm-2785@moe-dl.edu.my

B. S. Teo
SJKC Chung Hua Undop, Sri Aman, Malaysia

M. Y. Y. Chin
University of Malaya, Kuala Lumpur, Malaysia

M. H. A. Hassan et al. (eds.), *Proceedings of the 3rd International Conference on Innovation and Technology in Sports*, Lecture Notes in Bioengineering,
https://doi.org/10.1007/978-981-95-0804-4_6

Keywords Physical activity · Brain breaks · University

1 Introduction

Regular physical activity is essential for global health through improved cardiovascular health, stronger muscles, enhanced mental well-being, and reduced risk of chronic diseases such as diabetes and obesity (Piercy et al. 2018). Despite these well-established benefits and World Health Organization recommendations for weekly moderate- or vigorous-intensity exercise, a significant portion of the global population, including a large majority of adolescents and nearly one in four adults, fails to meet these guidelines (World Health Organization 2021; Institute for Public Health 2019). This lack of physical activity is a major risk factor for non-communicable diseases such as diabetes, cancer, hypertension, and cardiovascular problems. In Malaysia, only a quarter of adults, adolescents and children meet recommended activity levels (Santos et al. 2023). This global inactivity crisis is predicted to cost hundreds of billions of dollars in healthcare expenses and has been exacerbated by the COVID-19 pandemic, which led to a substantial decrease in physical activity levels among young people (Neville et al. 2022).

The World Health Organization's 2016 report highlighted a concerning trend of widespread physical inactivity among young people globally (World Health Organization 2016). This issue is particularly pronounced among adolescents, with studies indicating that over 80% of students aged 11–17 do not engage in enough physical activity (Guthold et al. 2020). This pattern of insufficient exercise often persists into young adulthood, significantly impacting university students (Sharma et al. 2019). Studies have shown a strong correlation between increased sedentary behaviors, such as excessive screen time, and decreased participation in sports and other physical activities. The COVID-19 pandemic has further intensified this problem by shifting to on-line learning environments that have drastically reduced opportunities for students to engage in physical activity. A longitudinal study demonstrated the pandemic's negative impact, revealing a significant decrease in exercise levels among university students (Yang and Koenigstorfer 2020). This decline in physical fitness has far-reaching consequences as it increases the risk of developing non-communicable chronic diseases, negatively affects mental health, potentially leading to anxiety and depression, and have contribute to lower academic performance due to decreased focus and cognitive function. Therefore, it is crucial to develop and implement effective intervention strategies on a global scale to promote and improve physical fitness across all age groups, especially among youth.

Incorporating technology-driven brain breaks, particularly those involving physical activity has paved a promising way to improve teaching and learning across various subjects and disciplines. Studies indicated that these short bursts of exercise can positively impact mental well-being, fitness levels, and motivation (Hajar et al. 2020). Systematic reviews have demonstrated that such interventions especially those that integrate physical activity with curriculum content and cognitive engagement can lead to improvements in students' physical activity levels and classroom behavior (Masini et al. 2020; Maillot et al. 2012). The benefits appear to be greater when the active breaks are directly related to the contents of the subjects.

Recognizing the widespread use of technology by children, studies have examined technology-based approaches to promote physical fitness and activity (Biljon and Longhurst 2012; Gao 2013; Tumynaite et al. 2014; Ding et al. 2021). Mobile health apps, wearable fitness trackers, and online fitness communities which have been evolving are current applications for encouraging physical activity, particularly among young adults. These applications offer features like real-time feedback, goal setting, and social support, all of which can motivate individuals to achieve their fitness goals (McMahon et al. 2017). In addition, studies have demonstrated the effectiveness of gamified fitness apps in keeping students engaged through interactive and enjoyable workouts, potentially leading to the development of long-term exercise habits (Chen and Pu 2018). In a study using a Dance Dance Revolution program have showed significant increases in daily physical activity levels compared to a control group (Gao 2013). Another study (Tumynaite et al. 2014) using an online combined exercise intervention for obese children found significant improvements in body fat percentage, cognitive function, quality of life, and mental health after just three months.

While studies have demonstrated the positive effects of physical activity interventions on primary and secondary school students, there is a relative lack of study focusing on university, institute, and college students. This study aims to address this gap by investigating the potential impact of Physical Activity Brain Breaks (PABBs) on university students. Therefore, the purpose of this study was to assess the effectiveness of PABBs on the health-related fitness of university students.

2 Materials and Methods

2.1 Participants

A total of 48 university students comprising of 24 males and 24 females aged between 18 and 20 years old from one university in Sarawak volunteered to participate. In terms of ethnicity, Bumiputera Sarawak formed the largest percentage (64.6%), followed by Malay with 14 (29.2%) and Chinese with 3 (6.3%) of the population. The participants were divided into an intervention group (IG) and a control group (CG), with each group consisting of twelve males and twelve females. The health-related fitness test was conducted by the lecturers. The Institution Research Ethics Committee approved the study. Informed consent was obtained from all participants to ensure their confidentiality before the study commenced.

2.2 *Measures*

2.2.1 Health-Related Fitness Test (HRFT)

The health-related fitness test (HRFT) was administered to evaluate health-related fitness components, including body composition, muscle strength, muscle endurance, flexibility and cardiovascular endurance. The body composition was assessed using the Body Mass Index (BMI) test, which is calculated by dividing weight (kg) by height (m2), following the guidelines established by the World Health Organization (WHO 2004). According to WHO standards, a BMI $\leq$ 18.5 indicates underweight, 18.5 $\leq$ BMI $\leq$ 24.9 signifies normal weight, 25.0 $\leq$ BMI $\leq$ 29.9 represents overweight, 30.0 $\leq$ BMI $\leq$ 34.9 denotes obesity, and BMI $\geq$ 35 is classified as extreme obesity. The HRFT encompasses four specific assessments: step-up, push-ups, curl-ups and sit-and-reach test. Step-up test was used to measure cardiovascular endurance, push-up test assessed muscle strength, curl-up test gauged muscle endurance and sit and reach test was performed to evaluate flexibility.

2.2.2 Physical Activity Brain-Break

The physical activity brain breaks (PABB) videos which were developed by the researchers were used as an intervention tool to train the participants. The PABB videos consisted of twenty different types of simple and low-intensity structured physical activity and movements in which each activity only last around 3 to 5 min. It was designed to improve the participants' physical fitness. The PABB was conducted at the beginning of their lectures twice per week. The physical activities were led by the appointed leader of the classes who were guided on how to conduct the PABB to ensure that the orderly implementation of the physical activities. The pre-test on health-related fitness test was conducted before the intervention, while post-test was conducted after the 12-week intervention. The participants in the intervention group were involved in the programwhile participants in the control group were not involved in the program.

2.3 *Data Analysis*

Data analysis was conducted using IBM's Statistical Package for the Social Science (SPSS) version 28.0. Descriptive analysis was used to show the demographic characteristics of the participants. Paired sample t-test was performed to compare the mean score differences between pre-test and post-test on SEGAK test in the intervention group and control group. A p-value threshold of less than 0.05 has been previously adopted to determine statistical significance.

3 Results

Table 1 shows the demographic characteristics of the participants. There were 48 participants involved in the study, with a total of 24 (50.0%) males and 24 (50.0%) females from Year 4, 5 and 6. The age groups categories showed that 10 (20.8%) were 18 years old, 20 (41.7%) were 19 years old and 18 (37.5%) were 20 years old. In terms of ethnicity, the majority of the participants were Bumiputera Sarawak with a total of 31 (64.6%), followed by Malay with 14 (29.2%) and Chinese with 3 (6.3%) of the population, respectively. Based on the groups, there were 24 participants in the intervention group and the control group, respectively.

Table 2 shows the pre-test and post-test of BMI and HRFT test. For the Body Mass Index (BMI), results revealed that the frequency of participants increased significantly from pre (52.1%) to post (75.0%) in the normal weight category. Conversely, the frequency of the participants declined slightly from pre (2.1%) to post (0.0%) in the underweight category. Besides, the frequency of the participants decreased slightly from pre (20.8%) to post (14.6%) in the overweight category. The frequency of participants in the obese category decreased significantly from pre (25.0%) to post (10.4%). Whereas, for the HRFT test, the frequency of participants increased significantly from pre (18.8%) to post (52.1%) in superior fitness level. Next, the frequency of participants increased slightly from pre (12.5%) to post (20.8%) in the excellent fitness level. Furthermore, the frequency of the participants declined sharply in the fit fitness level from pre (31.2%) to post (16.7%). Lastly, the frequency of participants in the poor fitness level decreased gradually from pre (37.5%) to post (10.4%). In addition, there was no participants in the below average fitness level from pre to post.

Table 1 Demographic characteristics of the 48 participants (N = 48)

Characteristics	Frequency (F)	Percentage (%)
Gender		
Male	24	50.0
Female	24	50.0
Age groups		
18	10	20.8
19	20	41.7
20	18	37.5
Ethnicity		
Bumiputera Sarawak	31	64.6
Malay	14	29.2
Chinese	3	6.3
Groups		
Intervention	24	50.0
Control	24	50.0

Table 2 Pre-test and post-test of BMI and HRFT test

Variables	Pre-test		Post-test	
	frequency (F)	Percentage (%)	frequency (F)	Percentage (%)
Body mass index (BMI)				
Normal weight (18.5 to 25)	25	52.1	36	75.0
Underweight (16.00 to 18.5)	1	2.1	0	0.0
Overweight (25 to 20)	10	20.8	7	14.6
Obese (30 above)	12	25.0	5	10.4
HRFT test				
Superior	9	18.8	25	52.1
Excellent	6	12.5	10	20.8
Fit	15	31.2	8	16.7
Poor	18	37.5	5	10.4
Below average	0	0.0	0	0.0

Table 3 shows the mean difference of HRFT test variables for pre and post-test. The mean score of step test was significant increase from pre (3.50 ± 1.07) to post (4.00 ± 0.99), $\mu = -0.50$, $t(47) = -3.293$, $p = 0.002$. Conversely, the mean score of push-ups test was no statistically significant from pre (3.50 ± 1.61) to post (3.63 ± 1.66), $\mu = -0.13$, $t(47) = -0.590$, $p = 0.558$. Moreover, the mean score of curl-ups test was significant increase from pre (3.69 ± 1.39) to post (4.27 ± 1.22), $\mu = -0.58$, $t(47) = -3.029$, $p = 0.004$. Last but not least, the mean score of sit and reach test was also statistically significant from pre (2.75 ± 1.41) to post (3.38 ± 1.52), $\mu = -0.63$, $t(47) = -2.950$, $p = 0.005$.

Table 4 shows the paired sample t-test for HRFT test for pre and post-test based on the intervention and control group. For the intervention group, results revealed that there was significant increase on step test from pre (3.92 ± 1.02) to post (4.70 ± 0.46), $\mu = -0.79$, $t(23) = -3.651$, $p = 0.001$. Besides, there was significant increase on push-ups test from pre (4.29 ± 1.08) to post (5.00 ± 0.01), $\mu = -0.71$, $t(23) = -3.205$, $p = 0.004$. Likewise, there was significant increase on curl-ups test from pre (4.25 ± 1.33) to post (5.00 ± 0.01), $\mu = -0.75$, $t(23) = -2.769$, $p =$

Table 3 Paired sample T-Test on HRFT test

Variables	Pre-test		Post-test		μ (95% CI)	*t*-test	
	M	SD	M	SD		*t*(df = 47)	*p*
Step test	3.50	1.07	4.00	0.99	−0.50	−3.293	0.002*
Push-ups	3.50	1.61	3.63	1.66	−0.13	−0.590	0.558
Curl-ups	3.69	1.39	4.27	1.22	−0.58	−3.029	0.004*
Sit and reach	2.75	1.41	3.38	1.52	−0.63	−2.950	0.005*

* $p < 0.05$

Table 4 Paired sample T-Test of HRFT test based on the intervention and control group

Variables	Test	Intervention group				Control group			
		M	SD	t (df = 23)	p	M	SD	t (df = 23)	p
Step test	Pre	3.92	1.02	−3.651	0.001*	3.08	0.97	−1.405	0.307
	Post	4.70	0.46			3.29	0.86		
Push-ups	Pre	4.29	1.08	−3.205	0.004*	2.71	1.68	1.414	0.171
	Post	5.00	0.01			2.25	1.29		
Curl-ups	Pre	4.25	1.33	−2.769	0.011*	3.13	1.23	−1.514	0.144
	Post	5.00	0.01			3.54	1.38		
Sit and reach	Pre	3.33	1.40	−3.938	0.001*	2.17	1.17	−0.171	0.866
	Post	4.54	0.51			2.21	1.28		

* $p < 0.05$

0.011. Similarly, there was significant increase on sit and reach test from pre (3.33 ± 1.40) to post (4.54 ± 0.51), $\mu = -1.21$, $t(23) = -3.938$, $p = 0.001$. Whereas, for the control group, results revealed that there was no significant increase on step test from pre (3.08 ± 0.97) to post (3.29 ± 0.86), $\mu = -0.21$, $t(23) = -1.405$, $p = 0.307$. Moreover, there was no significant increase but slightly decrease on push-ups test from pre (2.71 ± 1.68) to post (2.25 ± 1.29), $\mu = 0.46$, $t(23) = 1.414$, $p = 0.171$. Next, there was no significant increase on curl-ups test from pre (3.13 ± 1.23) to post (3.54 ± 1.38), $\mu = -0.42$, $t(23) = -1.514$, $p = 0.144$. Lastly, there was also no significant increase on sit and reach test from pre (2.17 ± 1.17) to post (2.21 ± 1.28), $\mu = -0.42$, $t(23) = -0.171$, $p = 0.866$.

4 Discussion

The study showed that there were increase in the percentage of normal weight and decrease in the overweight and obese BMI categories as well as increase in superior, excellent and decrease in the poor levels in the health-related fitness test. In addition, the intervention group showed higher improvement in the step test, curl-ups and sit and reach as compared to the push-up test. These implied that the impact of PABB on students' physical activity behavior have resulted in an increase in the fitness categories and tests. This also indicated that the PABB encouraged students to engage in physical activity during their daily lecture hours. The findings confirmed the systematic review of Maillot et al. (2012) which found that introducing physical activity breaks raised students' physical activity levels during the school day and enhanced their attentiveness in the lessons immediately after these breaks. The findings were also consistent with (Biljon and Longhurst 2012) study who showed that engaging in physically simulated sports games improved cognitive and physical skills crucial for one's everyday functional abilities. The PABB offered a comprehensive pause that stimulated both the mind and body, encouraging more physical activity and overall movement during academic lessons.

The findings also reflected upon past studies which showed students who practice, engage or train on their own after their academic hours and free time due to the enjoyment and fun derived from the brain breaks. They can also follow the PA and exercises by playing the PABB videos during their leisure time easily according to the time and needs of the students.

The PABB which were conducted during the lecture classes had shown cooperativeness, fostering teamwork and social interaction among students. These factors contribute to the reasoning of students often love PABB and physical activity programs, as they provide both immediate enjoyment and long-term benefits for physical and mental well-being. These also promotes the students to engage in physical activities beyond regular academic hours.

The study had limitations in which the 12-weeks intervention period only showed short-term effects of the PABB on students' physical activity engagement. A longer intervention period, such as 24 weeks, would be necessary to determine if these brain breaks could positively impact students' fitness levels. In addition, the results may not be representative of all university students in Malaysia, as the samples were only drawn from one university in Sarawak. Future studies could broaden the scope by including participants from a more diverse range of universities across Malaysia that could strengthen the generalizability of the findings.

5 Conclusions

The study revealed that short breaks of physical activity can help to students to engage in physical activities can promote social interaction and emotional well-being, contributing to a holistic approach to physical well-being during, after lectures and leisure times. This study suggests a more tailored, accessible, and effective approach to addressing obesity and lack of fitness among university students. The PABB can be personalized and vary that allow students to choose what they enjoy most or what suits their preferences that can increase their intrinsic motvation to appreciate the values and benefits of the health and sense of achievement for physical health in the long run.

References

Chen Y, Pu P (2018) Health together: exploring social incentives for mobile fitness applications. In: Proceedings of the 12th international conference on persuasive technology, pp 85–97

Ding M, Yi X, Yan P, McDonough DJ, Gao Z, Dong X (2021) Application of an online combination exercise intervention to improve physical and mental health in obese children: a single arm longitudinal study. Front Psychol 12:1–9. https://doi.org/10.3389/fpsyg.2021.638618

Gao Z (2013) The impact of an exergaming intervention on urban school children's physical activity levels and academic outcomes. Asian J Exerc Sport Sci 10(2):1–10

Guthold R, Stevens GA, Riley LM, Bull FC (2020) Global trends in insufficient physical activity among adolescents: A pooled analysis of 298 population-based surveys with 1.6 million participants. Lancet Child Adolesc Health 4(1):23–35

Hajar MS, Rizal H, Muhamad AS, Kuan G (2020) The effects of brain-breaks on short-term memory among primary school children in Malaysia. In: Lecture notes in bioengineering, pp 493–502. https://doi.org/10.1007/978-981-15-3270-2_50

Institute for Public Health (2019) National health and morbidity survey 2019: non-communicable diseases. Healthcare demand, and health literacy. Ministry of Health Malaysia

Maillot P, Perrot A, Hartley AA (2012) Effects of interactive physical-activity video-game training on physical and cognitive function in older adults. Psychol Aging 27(3):589–600. https://doi.org/10.1037/a0026268

Masini A, Marini S, Gori D, Leoni E, Rochira A, Dallolio L (2020) Evaluation of school-based interventions of activity breaks in primary schools: a systematic review and meta-analysis. J Sci Med Sport 23:377–384

McMahon E, Lewis BA, Oakes JM, Wierenga CE, Baer JT, Boersma F (2017) A conceptual framework for wearable physical activity tracking devices: validation against strategies for behavioral change targets. Transl Behav Med 7(4):647–660

Neville RD, Lakes KD, Hopkins WG, Tarantino G, Draper CE, Beck R, Madigan S (2022) Global changes in child and adolescent physical activity during the COVID-19 pandemic: a systematic review and meta-analysis. JAMA Pediatrics 176:886–894. https://doi.org/10.1001/jamapediatrics.2022.2313

Piercy KL, Troiano RP, Ballard RM, Carlson SA, Fulton JE, Galuska DA, George SM, Olson RD (2018) The physical activity guidelines for Americans. JAMA 320(19):2020–2028

Santos AC, Willumsen J, Meheus F, Ilbawi A, Bull FC (2023) The cost of inaction on physical inactivity to public health-care systems: a population-attributable fraction analysis. Lancet Glob Health 11(1):e32–e39. https://doi.org/10.1016/S2214-109X(22)00464-8

Sharma M, Sharma M, John J (2019) Influence of parent-child relationship on sedentary behavior and physical activity in the 21st century: a systematic review. J Phys Act Health 16(8):682–690

Tumynaite L, Miežien B, Ching MMM, Chin M, Putriute V, Rupainiene V, Stankeviciene G, Emeljanovas A (2014) Effects of intervention "HOPSports Brain Breaks" program on physical fitness and sedentary behaviour in primary school. Balt J Sport Health Sci 3(94):57–66

Van Biljon A, Longhurst GK (2012) The influence of exergaming on the functional fitness in overweight and obese children. Afican J Phys Act Health Sci 18(4):984–991

WHO (2004) Appropriate body-mass index for Asian populations and its implications for policy and intervention strategies. The Lancet 363(9403);157–163

World Health Organization (2016) Global health observatory (GHO) data: prevalence of insufficient physical activity. https://www.who.int/gho/ncd/risk_factors/physical_activity_text/en/

World Health Organization (2021) Global action plan on physical activity 2018–2030: More active people for a healthier world. https://www.who.int/publications/i/item/9789241514187

Yang Y, Koenigstorfer J (2020) Determinants of physical activity maintanice during the COVID-19 pandemic: a focus on fitness apps. Transl Behav Med 10(3):835–842

Correlation of Strength Characteristics with Running Economy in Well-Trained Male Distance Runners

Tham Thaiyanont, Thareerat Sairaya, Paweena Tanapin, Purinat Haowhan, Nattawut Meepol, Benthiwa Surasartpisal, Panupong Patarachao, Phitchapa Konthasing, Taspol Keerasomboon, and Parunchaya Jamkrajang

Abstract The objective of the study was to investigate the relationship between strength characteristics and running economy (RE) at speeds of 10, 12, 14, and 16 km·h^{-1} in runners. Twenty male well-trained male distance runners, aged between

T. Thaiyanont · T. Sairaya · P. Tanapin · P. Haowhan · N. Meepol · P. Patarachao · P. Konthasing
Sport and Health Science Program, School of Health Science, Mae Fah Luang University, Chiang Rai, Thailand
e-mail: tham.thaiyanont@gmail.com

T. Sairaya
e-mail: thareerat.sry@gmail.com

P. Tanapin
e-mail: paweenacoachandcraft@gmail.com

P. Haowhan
e-mail: radar.purinat@gmail.com

N. Meepol
e-mail: cnior2002@gmail.com

P. Patarachao
e-mail: panupong.pat@mfu.ac.th

P. Konthasing
e-mail: phitchapa.kon@mfu.ac.th

T. Keerasomboon (✉) · P. Jamkrajang
College of Sports Science and Technology, Mahidol University, Nakhon Pathom, Thailand
e-mail: taspol.kee@mahidol.edu

P. Jamkrajang
e-mail: parunchaya.jam@mahidol.edu

B. Surasartpisal
Faculty of Sport and Health Science, Thailand Sports National University Lampang Campus, Lampang, Thailand
e-mail: benthiwa.s@gmail.com

M. H. A. Hassan et al. (eds.), *Proceedings of the 3rd International Conference on Innovation and Technology in Sports*, Lecture Notes in Bioengineering,
https://doi.org/10.1007/978-981-95-0804-4_7

18 and 25 years. Subsequently, all participants completed the testing involving Countermovement Jump (CMJ), Drop Jump (DJ), Isometric Mid-Thigh Pull (IMTP), and 1 Repetition Maximum (RM) strength at the beginning of the testing session. After a 15 min rest, 10, 12, 14, and 16 km·h^{-1} RE were measured during running. Pearson's correlation was used for statistical analysis with a significance level set at 0.05. The results revealed that CMJ peak power and CMJ height, along with some DJ variables, DJ height, and reactive strength index (RSI), were not significantly correlated with RE at any speed. Additionally, there was a non-significant relationship between the peak force, along with 1RM strength and RE. However, RFD 0–100 and RFD 0-150 ms showed a large significant correlation with RE. Explosive strength (e.g., RFD) may be a significant determinant of higher running speeds, which are defined as speeds above 12 km·h^{-1}. The findings indicate that the capacity for rapid high-force production in the lower limbs is a significant determinant of an endurance runner's speed.

Keywords Rate of force development · Reactive strength · Running performance · Strength quality · Neuromuscular characteristics

1 Introduction

Running economy (RE), the energy expended at submaximal running speeds, is a key determinant of endurance performance (Llanos-Lagos and Ramirez-Campillo 2023). Alongside maximal oxygen uptake ($\dot{V}O_2$max), anaerobic threshold, and anaerobic capacity, RE accounts for 89–95% of the variation in long-distance running performance (Daniels 1985). Efficient RE enables runners to sustain higher speeds with lower oxygen consumption, with variations of up to 30% observed even among well-trained runners with similar aerobic capacities (Prampero et al. 1993). Notably, RE often correlates more strongly with running performance than VO2max in homogeneous populations, highlighting its critical role (Li et al. 2021). Understanding factors influencing RE to enhance the design of training programs and improve endurance performance (Hooren et al. 2024; Saunders et al. 2004).

RE is influenced by physiological, biomechanical, environmental, and training factors (Blagrove et al. 2018). Strength training, including heavy resistance, plyometric, and complex training, has been shown to improve RE through neuromuscular adaptations (Llanos-Lagos and Ramirez-Campillo 2023; Beattie et al. 2014a, 2017a; Llanos-Lagos et al. 2024; James et al. 2022). Strength characteristics, as categorized by James et al. (James et al. 2022), including (1) Maximal isometric strength can be assessed using the isometric mid-thigh pull (IMTP), which is commonly measured by peak force (2). Explosive strength, referred to as rapid force production or rate of force development, measures the ability to generate force quickly. (3) Heavy dynamic strength is the primary strength attribute that is measured through 1RM testing. While (4) fast dynamic strength and (5) reactive strength are linked to the stretch–shortening cycle, the low SSC and fast SSC are derived from the counter-movement jump and drop jump, respectively. These interrelated attributes can inform training programs and athlete profiling (Lum et al. 2020).

While prior researches have linked strength characteristics to endurance performance (Hooren et al. 2024; Zhang et al. 2022; Barnes and Kilding 2015), specific relationships with RE remain underexplored. For example, Lum et al. (Lum et al. 2020) identified correlations between RE and Rate of Force Development (RFD) but not peak force in recreational runners, while Zhang et al. (Zhang et al. 2022) found moderate to large relationships between RE and explosive strength metrics. However, investigations on strength characteristics at higher running speeds (>12 km·h^{-1}) and their relevance to well-trained runners are limited.

A fast dynamic strength and a heavy dynamic strength, respectively, are parameters that represent muscle force production and power output, as represented by the Countermovement Jump (CMJ) and One Repetition Maximum Strength (1RM Strength) However, limited research has investigated the relationships between these types of markers and RE (Llanos-Lagos and Ramirez-Campillo 2023; Hooren et al. 2024; Spurrs et al. 2003). Barnes et al. (2015) reported moderate to large correlations between CMJ height and RE, while others, (Hooren et al. 2024; Spurrs et al. 2003) observed no significant relationship. Reactive strength, measured via the Reactive Strength Index, has shown potential correlations with RE at various speeds (Hooren et al. 2024; Beattie et al. 2017b) yet evidence remains sparse.

Despite these findings, a comprehensive examination of the correlation between strength characteristics and RE in well-trained runners is lacking. Identifying these relationships could provide valuable insights for strength and conditioning professionals to design effective training regimens. This study aims to evaluate the relationship between strength characteristics (maximal isometric strength, explosive strength, heavy dynamic strength, fast dynamic strength, and reactive strength) and RE in well-trained distance runners. We hypothesized that there is an inverse relationship between strength characteristics at different speeds (10, 12, 14, and 16 km·h^{-1}) in well-trained distance runners.

2 Methodology

Twenty male healthy distance runners (age of 24.45 $\pm$ 5.10 years, height of 172.09 $\pm$ 5.42 cm, body mass of 64.33 $\pm$ 10.50 kg, running experience of 4.09 $\pm$ 2.22 years, and weekly running distance of 61.8 $\pm$ 18.7 km.) were recruited for participation in this study. They are presently abstaining from engaging in any strength training program. The sample size is determined to be 17 participants, calculated using the G*Power software (version 3.1.9.4) based on previous studies investigating the correlation between RSI and RE at km·h^{-1}(Hooren et al. 2024) with an effect size of 1.19, α (significance level) of 0.05, and power (1-β) of 0.70. In addition, we recruited more from 20% of participants calculated to avoid drop-offs. Prior to their participation, each participant completed an informed consent form after being fully told about the study's methods, possible risks, and advantages. There were no injuries to any of the subjects that would have impacted on their physical performance. The ethics committees of Mae Fah Luang University, Thailand, granted ethical permission for this study (ID number: 2017047). In Tables 1 and 2, the basic physical, physiological, and strength characteristics of the participants are presented in detail.

Table 1 Presents the participants' physical and physiological characteristics

Variable	Mean ± SD
Age (y)	24.45 ± 5.10
Height (cm)	172.09 ± 5.42
Body mass (kg)	64.33 ± 10.50
FFM (kg)	30.92 ± 4.25
% Fat (%)	13.50 ± 4.44
$\dot{V}O_2max$ (mL·kg^{-1}·min^{-1})	65.28 ± 8.36
RE at 10 km·h^{-1} (mL·kg^{-1}·min^{-1})	40.25 ± 3.27
RE at 12 km·h^{-1} (mL·kg^{-1}·min^{-1})	47.54 ± 3.58
RE at 14 km·h^{-1} (mL·kg^{-1}·min^{-1})	55.05 ± 4.24
RE at 16 km·h^{-1} (mL·kg^{-1}·min^{-1})	61.81 ± 4.71
Training experience (y)	4.10 ± 2.22

* FFM = fat-free mass; FM = fat mass; RE = running economy

Table 2 Presents the participants' strength characteristics

Variable	Mean ± SD
CMJ peak power (W/kg)	48.57 ± 6.21
CMJ height (cm)	33.15 ± 5.07
DJ height (cm)	29.07 ± 4.96
RSI (cm/s)	62.50 ± 17.64
Relative peak force (N/kg)	28.16 ± 2.89
RFD0-100 (N/s)	1701.25 ± 870.14
RFD0-150 (N/s)	2405.05 ± 1201.87
RFD0-200 (N/s)	2126.70 ± 883.70
RFD0-250 (N/s)	2501.65 ± 736.95
1RM squat strength (kg)	84.04 ± 30.23

* CMJ = countermovement jump; DJ = drop jump; IMTP = isometric mid-thigh pulls; RFD = rate of force development; RSI = reactive strength index

2.1 Procedure

Before the test, participants were instructed to fast for at least 3 h and arrive at the lab rested (no strenuous exercise 48 h prior, more than 8 h of sufficient sleep, and no soreness in their muscles). When the participants arrived at the laboratory for testing, the investigator in charge used a questionnaire to verify that they were following the rules. The NSCA's testing battery and the previous study (Hooren et al. 2024) were used for all testing procedures, with a 15 min rest in between.

Body Composition Test. A bioimpedance analyzer (ACCUNIQ BC380; Selva Health Care, Seoul, Korea) was used to measure the subject's body mass, fat mass, and fat-free mass.

Countermovement Jump (CMJ) and Drop Jump (DJ) Test. All participants executed CMJ and DJ from a 30 cm box. CMJ Peak Power, CMJ Jump height, DJ Height, and RSI were recorded using a dual force platform sampling at 1,000 Hz (VALD Forcedecks, VALD Performance, Brisbane, QLD, Australia). The participants maintained their hands on their hips while positioned on the force platform to execute the CMJ. They were subsequently instructed verbally to promptly perform a downward squat and leap vertically to achieve maximum height. Following CMJ testing, all participants executed a drop jump from a 30 centimeter box. To achieve maximum height and minimal ground contact time, the participants were instructed to leap vertically and explosively. They were also instructed to step down to the force plate from the box, rather than jumping, while maintaining their hands on their hips. Regarding all jumps test, 3 trials were conducted, separated by a one-minute passive recovery period. The data analysis was conducted using the highest jump height. The jumping height (cm) divided by the contact time (s) yielded the reactive strength index.

Isometric Mid-Thigh Pull (IMTP). The IMTP was performed on using a customized rack and dual force platform sampling at 1,000 Hz (VALD Forcedecks, VALD Performance, Brisbane, QLD, Australia). Following the guidelines outlined in Comfort et al. (2018), the procedures of IMTP were implemented. To mimic the second pull of the clean, participants were asked to stand with a knee flexion angle of 125°–145° and a hip flexion angle of 140°–150°. Participants must fully extend their elbows and hold the bar with lifting straps to avoid grip issues. The relative peak force was defined as the maximum force produced during IMTP. The RFD at 0–100, 0–150, 0–200, and 0–250 ms (RFD 0–100, RFD 0–150, RFD 0–200, and RFD 0–250) from the onset of pull was also determined for each trial.

1RM Squat Test. The classic protocol was employed to measure the 1RM squat strength in a power rack. Four warm-up trials were conducted following dynamic stretching, with the participants undertaking 50% (10 repetitions), 70% (5 repetitions), 80% (3 repetitions), and 90% (1 repetition) of their estimated 1RM strength. The warm-up 1RM load for each individual was calculated by the researchers using body mass, training experience, and age. The participants' actual 1RM was determined by performing 3–4 maximal-effort trials following the warm-up trials, followed by a 3–5 min recovery period. The goal was to have each participant squat through their whole range of motion while keeping their thighs parallel to the floor. The subjects were kept safe during the test by two certified spotters.

Running Economy. Running economy was conducted on a motorized treadmill (Venus; HP-Cosmos, Nussdoff-Traunstein, Germany). Concentrations of O_2 and CO_2 in expired air were analyzed continuously during the RE using the gas analyzer (Vacumed VISTAMAX, United States of America.), which was calibrated before each trial in accordance with the manufacturer's specifications. Using a Polar H10 HR monitor strap (Polar, Finland), heart rate (HR) was quantified. Participants were initially instructed to warm up at a speed of 7 km·h^{-1} for 10 min. Following a 5 min period of rest, they were instructed to run at speeds of 10, 12, 14, and 16 km·h^{-1} for 4 min each. The average of the $\dot{V}o_2$ (mL·kg^{-1}·min^{-1}) during the final minute of each speed was used to determine RE. This RE testing protocol was similar to the previous study (Hooren et al. 2024). Following the final phase of the RE test, the treadmill speed was established at 17 km·h^{-1}, which was systematically increased by 1 km·h^{-1} every 2 min until the participant attained exhaustion. To determine exhaustion, the following criteria were applied: peakHR at the end of the test >90% of the age-predicted maximal HR (calculated as 220—runner's age); respiratory exchange ratio >1.10; and rating of perceived exertion >18 (Hooren et al. 2024). $\dot{V}O_2$max was determined as the highest $\dot{V}O_2$ value using a 30 s moving window after reaching the criteria (Hooren et al. 2024).

2.2 *Statistical Analysis*

All data are reported as mean ± standard deviation. Normality was verified through the Shapiro–Wilk normality test. Pearson's correlation coefficients were selected to determine the association between the strength characteristics and RE (8, 10, 12, 14, 16 km/h). Correlations were classified as follows: small (0.1–0.29), moderate (0.3–0.49), large (0.5–0.69), very large (0.7–0.89), extremely large (0.9–1.0) (Hopkins et al. 2009). A level of $p < 0.05$ was used to determine statistical significance.

3 Results

For the CMJ test, non-significant, small correlations were found between the CMJ peak power ($r = 0.140, 0.238, 0.278, 0.241, p > 0.05$, respectively) and RE at 10, 12, 14, and 16 km·h^{-1}. Whereas CMJ height ($r = 0.095, 0.159, 0.173, 0.064, p > 0.05$, respectively) has a non-significant, not too small correlation with RE at 10, 12, 14, 16 km·h^{-1}.

Regarding the DJ test, we found a non-significant small correlation between DJ height ($r = -0.21, -0.19, -0.18, -0.29, p > 0.05$, respectively) and RE at 10, 12, 14, 16 km·h^{-1}. Reactive strength, as measured by RSI ($r = -0.37, p > 0.05$), showed a non-significant moderate correlation with RE at 10 km·h^{-1}. Furthermore, there were no significant and small correlations observed between RSI ($r = -0.28, -0.14, -0.10, p > 0.05$, respectively) and RE at 12, 14, and 16 km·h^{-1}.

In the IMTP Test, we discovered a non-significant positive moderate correlation between the peak force (r = 0.33, 0.31, $p > 0.05$, respectively) and RE at 10 and 12 km·h^{-1}. But there was a non-significant, small correlation ($r = 0.13, p < 0.05$) between peak force and RE at 14 km·h^{-1}, and a substantial no correlation ($r = 0.06, p > 0.05$) between peak force and RE at 16 km·h^{-1}. Regarding the RFD, there were significantly large correlations between RFD 0–100 ($r = -0.59, -0.53, -0.50, -0.50, p < 0.008$, 0.017, 0.029, 0.033, respectively) and RE at 10–16 km·h^{-1} (Fig. 1). The relationships between RFD0-150 ($r = -0.57, -0.55, -0.45, p < 0.011$, 0.013, 0.045, respectively) and RE at 10, 12, 14 km·h^{-1} is statistically significant, moderate to large relationship (Fig. 2) as well as there were insignificant and moderate correlations between RFD 0–150 ($r = -0.43$, $p = 0.061$) and RE at 16 km·h^{-1}. Furthermore, there were insignificant and small correlations between RFD 0–200 ($r = -0.27, -0.33, -0.27, p > 0.05$) and RE at 10, 12, 14 km·h^{-1}. However, the relationship between RE at 16 km·h^{-1} and RFD0-200 is statistically significant, with moderate correlation ($r = -0.46, p = 0.048$). Meanwhile, RFD 0–250 ($r = -0.49, p = 0.038$) has only a significantly moderate correlation with RE at 16 km·h^{-1} (Fig. 3). On the other hand, there were insignificant, small to moderate correlations between RFD 0–250 ($r = -0.28, -0.38, -0.34, p > 0.05$) and RE at 10, 12, 14 km·h^{-1}.

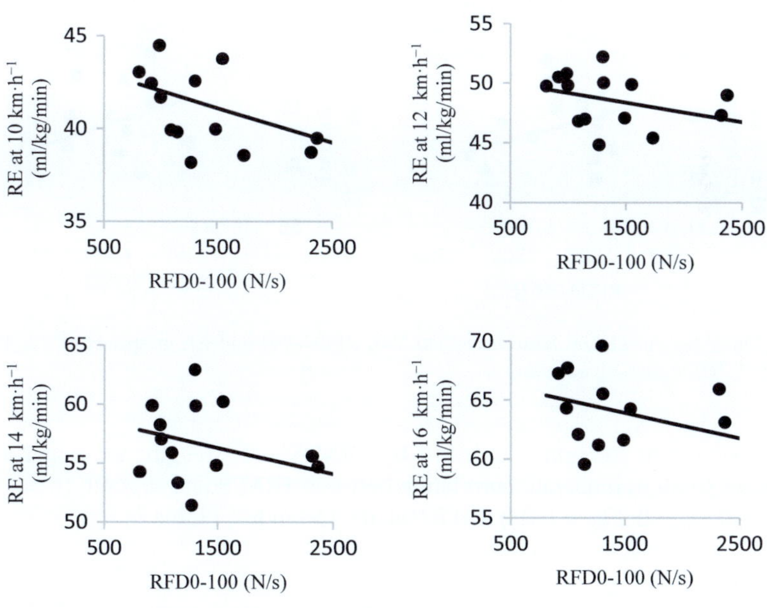

Fig. 1 Illustrates correlation between RFD0-100 and RE at speeds of 10, 12, 14 and 16 km·h^{-1}. RFD = rate of force development; RE = running economy

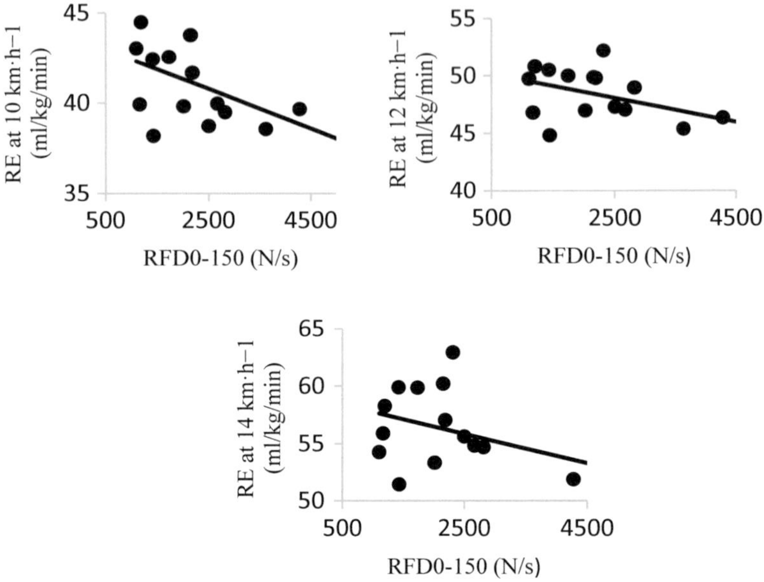

Fig. 2 Illustrates correlation between RFD0-150, RE at speeds of 10, 12, and 14 km·h^{-1}, RE = running economy

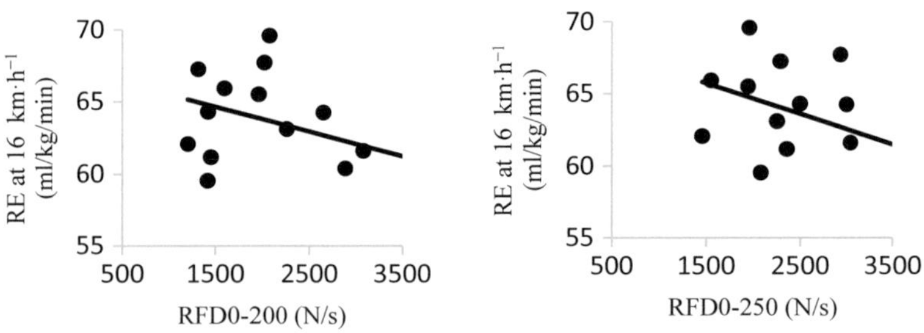

Fig. 3 Illustrates correlation between RFD0-200, RFD0-250 and RE at speeds of 10, 12, and 14 km·h^{-1}, RE = running economy

The maximum strength was tested by 1RM Squat strength; we found a non-significant small to moderate correlation between 1RM Squat strength ($r = -0.27$, -0.37, -0.14, $-0.39, p > 0.05$) and RE at 10–16 km·h^{-1} (Table 3).

Table 3 Correlation between strength characteristics and RE at 10, 12, 14, 16 km·h^{-1}

Variables	RE at 10 km·h^{-1}	RE at 12 km·h^{-1}	RE at 14 km·h^{-1}	RE at 16 km·h^{-1}
CMJ peak power	$r = 0.140$ $p = 0.732$	$r = 0.238$ $p = 0.856$	$r = 0.278$ $p = 0.889$	$r = 0.241$ $p = 0.832$
CMJ jump height	$r = 0.095$ $p = 0.664$	$r = 0.159$ $p = 0.760$	$r = 0.173$ $p = 0.774$	$r = 0.064$ $p = 0.599$
DJ jump height	$r = -0.21$ $p = 0.214$	$r = -0.19$ $p = 0.244$	$r = -0.18$ $p = 0.259$	$r = -0.29$ $p = 0.154$
RSI	$r = -0.37$ $p = 0.077$	$r = -0.28$ $p = 0.143$	$r = -0.14$ $p = 0.307$	$r = -0.10$ $p = 0.365$
Relative peak force	$r = 0.33$ $p = 0.896$	$r = 0.31$ $p = 0.880$	$r = 0.13$ $p = 0.674$	$r = 0.06$ $p = 0.576$
RFD0-100	$r = -0.59$** $p = 0.008$**	$r = -0.53$* $p = 0.017$*	$r = -0.50$* $p = 0.029$*	$r = -0.50$* $p = 0.033$*
RFD0-150	$r = -0.57$* $p = 0.011$*	$r = -0.55$* $p = 0.013$*	$r = -0.45$* $p = 0.045$*	$r = -0.43$ $p = 0.061$
RFD0-200	$r = -0.27$ $p = 0.154$	$r = -0.33$ $p = 0.104$	$r = -0.27$ $p = 0.676$	$r = -0.46$* $p = 0.048$*
RFD0-250	$r = -0.28$ $p = 0.144$	$r = -0.38$ $p = 0.071$	$r = -0.34$ $p = 0.104$	$r = -0.49$* $p = 0.038$*
1RM squat strength	$r = -0.27$ $p = 0.154$	$r = -0.37$ $p = 0.077$	$r = -0.14$ $p = 0.311$	$r = -0.39$ $p = 0.086$

Note * $p < 0.05$, ** $p < 0.01$, *** $p < 0.001$

4 Discussion

Our primary finding partly supports the aim of this study, there were only IMTP variables; RFD 0–100, RFD 0–150, RFD 0–200 and RFD 0–250 has showed a moderate to large significant correlation with RE at 10–16 km·h^{-1}. CMJ variables; CMJ peak power and CMJ height along with some DJ variables; DJ height and RSI were not significantly correlated with RE at any speed. Additionally, there is no significant relationship between 1RM squat strength and RE.

The maximal isometric strength and explosive strength are measured by the IMTP, which is used to evaluate the force–time characteristics of various sports. This assessment enables the hip, knee, and ankle joints to be positioned at an optimal biomechanical angle during running (Zhang et al. 2022; McGuigan 2019). According to Zhang et al. (2022) there was a strong correlation between RE and the RFD (0–100, 0–150, 0–200, and 0–250 ms) at 10 km·h^{-1}, suggesting that runners who produce more force had better RE. This finding corresponds with Lum et al. (2020) who indicated a correlation between RFD and RE at 10 km·h^{-1}, so partially supporting our hypothesis. To the best of our knowledge, this investigation is the first to study the relationship between RFD and RE beyond a speed of 12 km·h^{-1}. We found that RFD 0–100 and RFD 0–150 have moderately inversed correlation with RE at every speed (10, 12, 14, 16 km·h^{-1}). RFD is mainly determined by the maximal force production (i.e., peak force) (Maffiuletti et al. 2016; Andersen and Aagaard 2006), the weaker individuals tend to have lower force output over a shorter duration of time compared to the stronger individuals, resulting in a lower RFD (Stone et al. 2024; Suchomel et al. 2016; Aagaard et al. 2002; Cormie et al. 2010). The subjects' lack of background in resistance training, as well as their average values of peak force, which are lower than in previous studies (Zhang et al. 2022; Barnes and Kilding 2015), may affect their ability to quickly generate larger forces from a resting position because the amount of force produced during 0–200 and 0–250 ms are consequently lower. A greater RFD would allow athletes to generate higher levels of force in short periods of time, allowing a rapid transition from the braking phase to the propulsion phase of the gait cycle (James et al. 2022; Suchomel et al. 2016; Blazevich et al. 2020). Having no significant correlations between RFD 0–200 and 0–250, with RE at 10, 12, and 14 km·h^{-1} suggest the need to increase the subjects' ability for force generation at these speeds through resistance training. We found that RFD 0–200 and RFD 0–250 have a significantly moderate correlation with RE at 16 km·h^{-1}. Our findings indicate that increased running speed correlates more strongly with the late RFD (0–200, 0–250). Despite the participants' lower strength, force generation was required to achieve and maintain a high running speed (i.e., 16 km·h^{-1} = 3.45 min·km pace); therefore, it is possible that their strength level for generating greater force is characterized by late RFD. However, as running relies on the fastest durations to complete their distances, it is advantageous for these athletes to develop greater RFD earlier or upon gun start for winning performances. This can be best improved with improving strength characteristics through strength training.

Theoretically, runners can maintain a particular speed or execute running movements with less effort because of their muscles' capacity to contract and produce high force (Fletcher and MacIntosh 2017). The results of our study revealed no significant correlation between peak force and RE at any speed, supporting the findings of Lum et al. (2020) and Zhang et al. (2022) who also reported no significant correlation between peak force and RE or running performance in male runners at RE speeds of 10 and 12 km·h^{-1}, respectively.

This can be explained by the fact that the muscular behavioral process that stimulates energy savings must be precisely timed during the support phase of running (Barnes and Kilding 2015), whereas the IMTP peak force necessitates a relatively lengthy period to achieve its maximum force (Chavda et al. 2020). Consequently, in contrast to the RFD, the peak force is not indicative of the energy expenditure associated with running (Barnes and Kilding 2015). As a heavy dynamic strength that allows runners to keep a consistent speed or perform each running movement with a relatively lower force ratio. Even so, there exists a lack of research investigating the correlation between 1RM Squat and RE. Our study demonstrated no correlation between 1RM Squat strength and RE at any velocity. This finding aligns with the studies by Li et al. (2021) which indicated no correlation between 1RM squat strength and RE at speeds of 12, 14, 16 km·h^{-1}. Recent research has demonstrated a substantial correlation between maximal isometric strength (IMTP peak force) and heavy dynamic strength (1RM Strength). These two attributes may be employed interchangeably, but they should be implemented with cautiously (Lum et al. 2020; Suchomel et al. 2016). This suggests that the squat movement speed was relatively slow in comparison to the lower limb velocity during running, which may have contributed to the limited correlation with RE and lower performance in races, similar to the peak force derived from IMTP (Hooren et al. 2024).

One of the key factors influencing metabolic energy-saving strategies during running is elastic energy utilization. A runner's ability to effectively utilize the elastic energy and SSC generated by the muscle–tendon complex is characterized by their reactive strength (Llanos-Lagos et al. 2024). The relationship between distance running performance and jump ability has been investigated in numerous studies using a variety of jumping tests (Hooren et al. 2024; Cole et al. 2006; Hudgins et al. n.d.). The RSI is not substantially correlated with RE at any speed, as demonstrated by our findings. Contrary to the findings of Li et al. (2021) stated that the RSI has moderate ($r = -0.419$) and large ($r = -0.559$ to -0.572) associations with running economy at 12, 14, and 16 km·h^{-1}, suggesting that persons with higher RSI values have greater efficiency in prolonged running activities. The possible reason for this could be because the experience in distance running varies between 1 and 8 years as well as the box height that we used is different from the previous study, which may affect the correlation. Furthermore, individuals with greater strength are likely to exhibit greater fast SSC (stretch–shortening cycle) or reactive strength compared to those with lesser strength. This leads to more efficient movement and a greater contribution to the task at that moment (Suchomel et al. 2016). Plyometric or complex training in combination with endurance training has been shown in numerous studies to dramatically increase RE (Denadai et al. 2017; Ramirez-Campillo et al. 2021; Eihara et al. 2022). Nevertheless, we did not observe any significant correlation between CMJ Peak Power, CMJ height, and RE. Spurrs et al. (2003) and Li et al. (2021) have also reported that a change in RE is not significantly correlated with a change in CMJ but is correlated with a change in five consecutive horizontal jumps (Hooren et al. 2024; Spurrs et al. 2003). This finding is consistent with these findings. The slow SSC of fast dynamic strength derived from CMJ contrasts with DJ, which is defined by a fast SSC. Additionally, the eccentric phase is delayed in CMJ, and the reflex contribution to SSC potentiation is less in CMJ than in other activities, such as hopping (Hooren et al. 2024). Consequently, elastic energy plays a minor role in CMJ. Plyometric activities (reactive strength and fast dynamic strength) may have a more significant impact on athletes with higher performance levels, according to numerous studies (James et al. 2022; Suchomel et al. 2016; Denadai et al. 2017; Ramirez-Campillo et al. 2021; Eihara et al. 2022). This may lead to a restricted connection with RE, based on our individuals' backgrounds.

The study has to adhere to certain limitations that must be recognized. Firstly, the current data suggests that the strength characteristics serve as a reliable predictor of running economy in well-trained distance runners without resistance training experience. Further investigation is required to explore the potential relationships among highly trained runners and female runners. Secondly, the IMTP assessment may not immediately correspond to running biomechanics. It is valuable to investigate different isometric tests (e.g., isometric squat test). A significant correlation may not necessarily indicate a cause-effect relationship as the underlying mechanisms and the cause-effect relationship are still unknown. Future research should investigate the impact of resistance training on distance runners to acquire a better understanding of how strength characteristics affect RE. Specifically, the use of electromyography during running can help identify direct neuromuscular markers and determine the relationship between a distance runner's strength and RE. It is necessary to conduct longitudinal research to confirm the impact of neuromuscular characteristics and how they might be improved through training to enhance running performance.

5 Conclusion

This study found that most of the RFD time-specific were inversely correlated with RE across speeds of 10, 12, 14, and 16 km·h^{-1} in well-trained male distance runners without prior resistance training experience. Notably, RFD at 0–200 and 0–250 ms demonstrated significant inverse correlations with RE only at higher speeds (14 and 16 km·h^{-1}), Explosive strength (i.e., RFD) may be a significant determinant of higher running speeds, which are defined as speeds above 12 km·h^{-1} (i.e., 4.59 min·km pace). These findings underscore the importance of lower-limb force production in sustaining faster running speeds. The result suggests that distance runners aiming to enhance their speed should concentrate on the capacity for rapidly generating force, particularly during foot contact times of less than 250 ms, which is associated with both heavy-load resistance training and weightlifting derivative methods.

From a practical perspective, the results suggest that incorporating resistance training, particularly heavy-load resistance training and weightlifting derivatives, may improve explosive strength and, in turn, enhance RE at higher running speeds. Coaches and strength and conditioning practitioners should consider designing training regimens focusing on rapid force production to optimize running performance in distance runners. This study highlights the potential of strength training to improve RE and running performance, providing a foundation for future investigations into how neuromuscular adaptations can be harnessed to benefit distance runners. By bridging the gap between strength training and endurance performance, these efforts can contribute to more effective training programs tailored to the needs of competitive runners.

Acknowledgements The authors would like to thank all the runners who volunteered in this research and the research team from Mae Fah Luang University, Mahidol University, and Thailand National Sports University Lampang Campus, Thailand.

Funding This study was sponsored by Mae Fah Luang University (Fund Number: 671A05004).

Conflict of Interest All authors declare that they have no conflicts of interest relevant to the content of this review.

References

Aagaard P, Andersen JL (2010) Effects of strength training on endurance capacity in top-level endurance athletes. Scand J Med Sci Sports 20(s2):39–47. https://doi.org/10.1111/j.1600-0838.2010.01197.x

Aagaard P, Simonsen EB, Andersen JL, Magnusson P, Dyhre-poulsen P, Dyhre-Poulsen P (2002) Increased rate of force development and neural drive of human skeletal muscle following resistance training. J Appl Physiol 93:1318–1326. https://doi.org/10.1152/japplphysiol.00283.2002.-The

Aagaard P et al (2011) Effects of resistance training on endurance capacity and muscle fiber composition in young top-level cyclists. Scand J Med Sci Sports 21(6):e298–e307. https://doi.org/10.1111/j.1600-0838.2010.01283.x

Andersen LL, Aagaard P (2006) Influence of maximal muscle strength and intrinsic muscle contractile properties on contractile rate of force development. Eur J Appl Physiol 96(1):46–52. https://doi.org/10.1007/s00421-005-0070-z

Barnes KR, Kilding AE (2015) Running economy: measurement, norms, and determining factors. Springer. https://doi.org/10.1186/s40798-015-0007-y

Bartolomei S, Nigro F, Lanzoni IM, Masina F, Di Michele R, Hoffman JR (2020) A comparison between total body and split routine resistance training programs in trained men [Online]. www.nsca.com

Beattie K, Kenny IC, Lyons M, Carson BP (2014a) The effect of strength training on performance in endurance athletes. Sports Med 44(6):845–865. https://doi.org/10.1007/s40279-014-0157-y

Beattie K, Carson BP, Lyons M, Kenny IC (2017b) The relationship between maximal strength and reactive strength. Int J Sports Physiol Perform 12(4):548–553. https://doi.org/10.1123/ijspp.2016-0216

Beattie K, Kenny IC, Lyons M, Carson BP (2014) The effect of strength training on performance in endurance athletes. Adis International Ltd. https://doi.org/10.1007/s40279-014-0157-y

Beattie K, Carson BP, Lyons M, Rossiter A, Kenny IC (2017) The effect of strength training on performance indicators in distance runners. J Strength Cond Res [Online]. https://journals.lww.com/nsca-jscr

Blagrove RC, Howatson G, Hayes PR (2018) Effects of strength training on the physiological determinants of middle- and long-distance running performance: a systematic review. Sports Med 48(5):1117–1149. https://doi.org/10.1007/s40279-017-0835-7

Blazevich AJ, Wilson CJ, Alcaraz PE, Rubio-Arias JA (2020) Effects of resistance training movement pattern and velocity on isometric muscular rate of force development: a systematic review with meta-analysis and meta-regression. Springer. https://doi.org/10.1007/s40279-019-01239-x

Chavda S, et al (2020) A practical guide to analyzing the force-time curve of isometric tasks in excel. Strength Cond J 42(2) [Online]. https://journals.lww.com/nsca-scj/fulltext/2020/04000/a_practical_guide_to_analyzing_the_force_time.4.aspx

Cole A, Woodruff M, Lee M, Mahon A (2006) Strength, power, and aerobic exercise correlates of 5-km cross-country running performance in adolescent runners. Pediatr Exerc Sci 18:374–384. https://doi.org/10.1123/pes.18.3.374

Comfort P, Dos T, Beckham GK, Stone MH, Guppy SN, Gregory Haff G (2018) Standardization and methodological considerations for the isometric midthigh pull. Strength Cond J [Online]. http://journals.lww.com/nsca-scj

Contreras B, et al (2017) Effects of a six-week hip thrust vs. front squat resistance training program on performance in adolescent males: a randomized controlled trial [Online]. www.nsca.com

Cormie P, McGuigan MR, Newton RU (2010) Adaptations in athletic performance after ballistic power versus strength training. Med Sci Sports Exerc 42(8):1582–1598. https://doi.org/10.1249/MSS.0b013e3181d2013a

Cuadrado-Peñafiel V, Castaño-Zambudio A, Martínez-Aranda LM, González-Hernández JM, Martín-Acero R, Jiménez-Reyes P (2023) Microdosing sprint distribution as an alternative to achieve better sprint performance in field hockey players. Sensors 23(2). https://doi.org/10.3390/s23020650

Daniels JT (1985) A physiologist's view of running economy. Med Sci Sports Exerc 17(3):332–338 [Online]. http://europepmc.org/abstract/MED/3894870

Denadai BS, de Aguiar RA, de Lima LCR, Greco CC, Caputo F (2017) Explosive training and heavy weight training are effective for improving running economy in endurance athletes: a systematic review and meta-analysis. Springer International Publishing. https://doi.org/10.1007/s40279-016-0604-z

Di Prampero PE et al (1993) Energetics of best performances in middle-distance running. J Appl Physiol 74(5):2318–2324. https://doi.org/10.1152/jappl.1993.74.5.2318

Dumke CL, Pfaffenroth CM, McBride JM, McCauley GO (2010) Relationship between muscle strength, power and stiffness and running economy in trained male runners. Int J Sports Physiol Perform 5(2):249–261. https://doi.org/10.1123/ijspp.5.2.249

Eihara Y et al (2022) Heavy resistance training versus plyometric training for improving running economy and running time trial performance: a systematic review and meta-analysis. Sports Med Open 8(1):138. https://doi.org/10.1186/s40798-022-00511-1

Fletcher JR, MacIntosh BR (2017) Running economy from a muscle energetics perspective. Frontiers Media S.A. https://doi.org/10.3389/fphys.2017.00433

Hopkins WG, Marshall SW, Batterham AM, Hanin J (2009) Progressive statistics for studies in sports medicine and exercise science. https://doi.org/10.1249/MSS.0b013e31818cb278

Hudgins B, Scharfenberg J, Triplett NT, Mcbride JM (n.d.) Relationship between jumping ability and running performance in events of varying distance [Online]. www.nsca.com

James LP, Gregory Haff G, Kelly VG, Connick MJ, Hoffman BW, Beckman EM (2018) The impact of strength level on adaptations to combined weightlifting, plyometric, and ballistic training. Scand J Med Sci Sports 28(5):1494–1505. https://doi.org/10.1111/sms.13045

James LP, Talpey SW, Young WB, Geneau MC, Newton RU, Gastin PB (2022) Strength classification and diagnosis: not all strength is created equal [Online]. www.nsca-scj.com

Li F, Newton RU, Shi Y, Sutton D, Ding H (2021) Correlation of eccentric strength, reactive strength, and leg stiffness with running economy in well-trained distance runners. J Strength Cond Res 35(6):1491–1499. https://doi.org/10.1519/JSC.0000000000003446

Llanos-Lagos C, Ramirez-Campillo R, Moran J, Sáez De Villarreal E (2023) Effect of strength training programs in middle- and long-distance runners' economy at different running speeds: a systematic review with meta-analysis. Sports Medicine 1–38, https://doi.org/10.1007/S40279-023-01978-Y

Llanos-Lagos C, Ramirez-Campillo R, Moran J, Sáez de Villarreal E (2024) The effect of strength training methods on middle-distance and long-distance runners' athletic performance: a systematic review with meta-analysis. Springer Science and Business Media Deutschland GmbH. https://doi.org/10.1007/s40279-024-02018-z

Lum D, Chua K, Rashid Aziz A (2020) Isometric mid-thigh pull force-time characteristics: a good indicator of running performance [Online]. http://trainology.org/

Maffiuletti NA, Aagaard P, Blazevich AJ, Folland J, Tillin N, Duchateau J (2016) Rate of force development: physiological and methodological considerations. Springer Verlag. https://doi.org/10.1007/s00421-016-3346-6

McGuigan M (2019). Testing and evaluation of strength and power. https://doi.org/10.4324/9780429028182

Ramirez-Campillo R, Andrade DC, García-Pinillos F, Negra Y, Boullosa D, Moran J (2021) Effects of jump training on physical fitness and athletic performance in endurance runners: a meta-analysis. J Sports Sci 39(18):2030–2050. https://doi.org/10.1080/02640414.2021.1916261

Saunders PU, Pyne DB, Telford RD, Hawley JA (2004) Factors affecting running economy in trained distance runners. Sports Med 34(7):465–485. https://doi.org/10.2165/00007256-200434070-00005/FIGURES/3

Schoenfeld B, N-NS, Association C, Snarr RL (2021) NSCA's essentials of personal training. Human Kinetics [Online]. https://books.google.co.th/books?id=A3dQEAAAQBAJ

Spurrs RW, Murphy AJ, Watsford ML (2003) The effect of plyometric training on distance running performance. Eur J Appl Physiol 89(1):1–7. https://doi.org/10.1007/s00421-002-0741-y
Stone MH, et al (2024) The use of free weight squats in sports: a narrative review-squatting movements, adaptation, and sports performance: physiological [Online]. www.nsca.com
Suchomel TJ, Nimphius S, Stone MH (2016) The importance of muscular strength in athletic performance. Springer International Publishing. https://doi.org/10.1007/s40279-016-0486-0
Turner AN et al (2020) Developing powerful athletes, Part 1: mechanical underpinnings. Strength Cond J 42(3):30–39. https://doi.org/10.1519/SSC.0000000000000543
Van Hooren B, Jukic I, Cox M, Frenken KG, Bautista I, Moore IS (2024) The relationship between running biomechanics and running economy: a systematic review and meta-analysis of observational studies. Sports Med. https://doi.org/10.1007/s40279-024-01997-3
Zhang Q, Nassis GP, Chen S, Shi Y, Li F (2022) Not lower-limb joint strength and stiffness but vertical stiffness and isometric force-time characteristics correlate with running economy in recreational male runners. Front Physiol 13. https://doi.org/10.3389/fphys.2022.940761

EMG Monitoring of Motorcyclists' Muscle Activation During Two Different Riding Postures

Wan Zaim Hazim Wan Hilmi, Nur Muhamad Farhan Mohd Johan, Mohammad Azzeim Mat Jusoh, and Helmi Rashid

Abstract Motorcycle riding is a physically challenging task in which certain actions can lead to muscle tension while seated, potentially contributing to road accidents. Maintaining a consistent riding posture over a lengthy period can induce lower back discomfort. This research aims to monitor the muscle activation of motorcyclists' back muscles, specifically the latissimus dorsi and erector spinae (longissimus) sections. Additionally, it compares the total average root mean square (RMS) amplitude of muscle activities between two types of motorcycles: the standard versus the sport motorcycle. The data was collected using surface electromyography (sEMG) with the electrode sensor placed on specific skin locations. The experiment was performed on five respondents in static riding posture. As a result, the erector spinae (longissimus) muscles showed a significantly higher average RMS value when the respondent adopted an upright riding posture compared to a forward-leaning posture. Furthermore, the average RMS of the latissimus dorsi muscle was high (above 50% rate) when the respondent adopted a forward-leaning riding posture. These results suggest that riding postures can influence muscle activation, and prolonged use may lead to fatigue and further injuries.

Keywords Surface electromyography (sEMG) · Erector spine (longissimus) muscles · Latissimus dorsi muscle · Static riding posture

1 Introduction

The motorcycle is a useful method of transportation, especially in developing countries. It has become a standard mode of transportation in Asia, with the number of registrations gradually increasing, particularly for rural roads with low traffic volumes (Starkey 2016). In 2020, Malaysian transportation statistics stated that all modes of

W. Z. H. W. Hilmi · N. M. F. M. Johan · M. A. M. Jusoh (✉) · H. Rashid
School of Mechanical Engineering, College of Engineering, Universiti Teknologi MARA (UiTM), Shah Alam, Selangor, Malaysia
e-mail: m_azzeim@uitm.edu.my

M. H. A. Hassan et al. (eds.), *Proceedings of the 3rd International Conference on Innovation and Technology in Sports*, Lecture Notes in Bioengineering,
https://doi.org/10.1007/978-981-95-0804-4_8

transportation experienced a 3.73% increase compared to the previous year. It also estimates that, over the next two years, the need for less expensive modes of transportation will increase, mostly due to the oil crisis (Young 2013). Most road users prefer to use motorcycles rather than driving. Others are more interested in alternative transportation options that are fuel-efficient and easy to maintain. Nonetheless, we cannot ignore motorcycles' adaptability in reducing traffic in highly populated Asian cities and rural areas. According to this report, motorcyclists have surpassed cars as the most popular mode of transportation on the roads. Unfortunately, with the growth in motorcycle usage, there has also been a rise in motorcycle traffic accidents (Lanka 2015).

In sports activity, comfortable riding posture is one of the essential factors for a rider's experience. Indirectly, negative conditions such as discomfort, fatigue, and even lower back pain can occur during prolonged periods (Mathurkar 2016). Discomfort results from physiological elements such as joint angles, postures, stressors, muscle contractions, and pressure distribution during seating (Ospina-Mateus and Quintana Jiménez 2019). Determining the best posture for various motorcycles is important, as it affects the rider's back muscles and can lead to muscular disorders. Moreover, the rider's posture is crucial in determining muscle activation. Currently, there is no single riding position that fits all riding conditions (Velagapudi et al. 2010). Based on the motorcycle's design and operating conditions, we can identify the optimal position for the rider, one with the least stress on the muscles. For example, the rider of a sports bike assumes a forward-leaning position, whereas the rider of a standard motorcycle finds an upright seating posture to be the best option.

In short, different types of motorcycles require different seating positions and riding postures, which result in varying levels of discomfort and back pain. Individuals with a history of physical illness, particularly those related to musculoskeletal problems (MSDs), are most affected. While seated, the intradiscal forces on the lumbar area of the spine are more strained compared to standing (Karmegam et al. 2008). When the body is tilted slightly forward, it is supported by the latissimus dorsi muscle, one of the broadest muscles in the human body. This muscle stems from the lower thoracic spines and thoracolumbar fascia and attaches to the bicipital groove of the arm (Laitung and Peck 1985). The latissimus dorsi is the muscle that connects the upper and lower body muscles. However, the preferable riding posture can only be determined by analyzing fatigue in selective muscles over a prolonged period of time (Velagapudi et al. 2010).

Therefore, this study will directly assess the muscle activation of the latissimus dorsi and erector spinae (longissimus) using wireless surface electromyography (sEMG) measurements by reading the RMS amplitude. The average RMS value is the parameter that represents the intensity of the physiological activity in the motor unit during contraction. This value is generally utilized in the analysis and processing of electric muscle signals (Fukuda et al. 2008). Moreover, sEMG has been extensively addressed for various clinical and human factors-related activities in the literature. Currently, published research related to the effect of posture on muscle activation during motorcycle riding among teenagers is very minimal.

2 Methodology

In this experiment, firstly, a series of questionnaires were set up to gather information on the public's perceptions of motorcycle riding postures before conducting the experimental testing. The questionnaire consists of two sections, in which the personal and general information of the riders was taken in Sect. 1. Whereas in Sect. 2, the survey was focusing on riders' experiences, especially the discomfort at the back part of muscle when riding at different posture and with different motorcycle type.

Next, a quantitative method was used, employing wireless surface electromyography (sEMG) as a tool. Next, the sEMG measurement was applied to analyze the muscle activation data of two selected muscles: the latissimus dorsi and the erector spinae (longissimus). These muscles were selected due to their association with back pain and discomfort issues in motorcyclists. The guidelines from SENIAM (Surface Electromyography for the Non-Invasive Assessment of Muscles) were used for muscle palpation and electrode placement. SENIAM was proposed by the Biomedical Health and Research Program (Hermens et al. 2000).

The experiment was carried out in the Automotive Lab of UiTM Shah Alam. This location was selected for its privacy and closed environment, as the respondents needed to expose their upper bodies for the transmitter to be attached to the skin. The surrounding conditions of the lab was in ambient state, with a temperature of 31 °C and 75% humidity, which suited the ambient conditions during riding. The noise level of the room was also adequate.

The number of respondents who volunteered to participate in this research was five. It was made clear to the respondents that they could withdraw from the experiment at any time and that all collected data would be kept strictly confidential. Before proceeding with the selection of respondents, criteria were set to standardize the research. Only male respondents with a minimum of 1 year of motorcycle riding experience were selected. The age of the respondents ranged from 23 to 25 years old. Respondents needed to be within the normal range of body mass index (BMI), physically healthy, and able to participate in the experiment session. They must not have any history of serious physical or physiological pain and must not engage in heavy physical training for 12–24 h prior to the experiment, ensuring they also had enough sleep before the experiment (Snijders et al. 1995).

Figure 1 shows the flow of the experiment with a step-by-step procedure to collect the average RMS amplitude. This was done to compare two different riding postures with two different types of muscles. Figure 2 shows the sequence of connectivity from the electrode transmitter to the receiver, then to the converter (analog to digital), and finally to the laptop using proEMG software. Finally, Fig. 3 shows the preparation of the respondents before performing any of the sEMG experiments, indicating the electrode positioning on the latissimus dorsi and the erector spinae (longissimus) muscles.

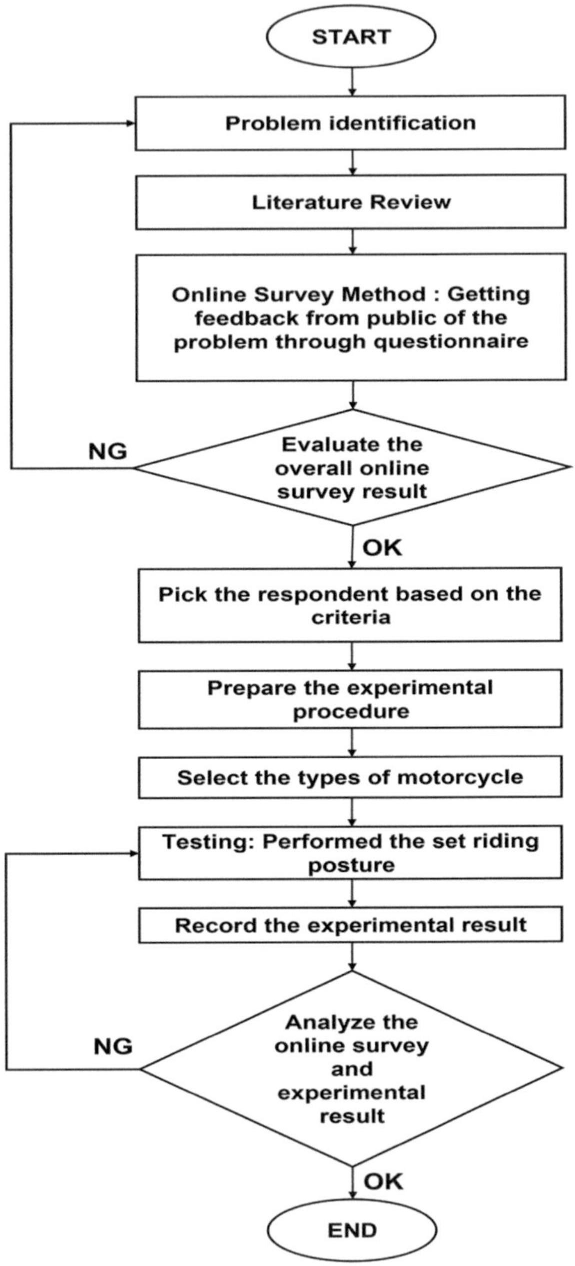

Fig. 1 Overall procedure for the research

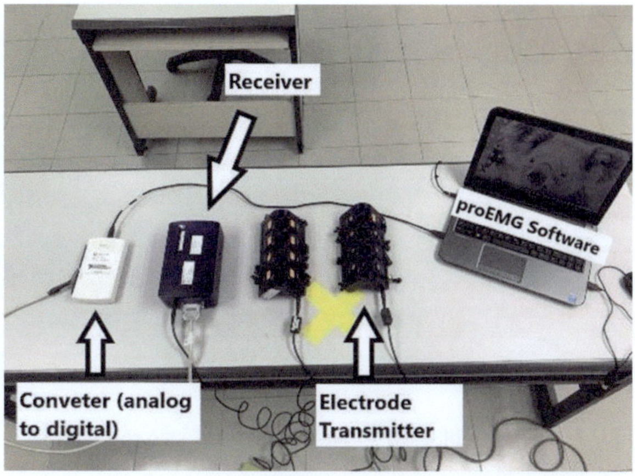

Fig. 2 The setup of the Wireless Myon sEMG experiment

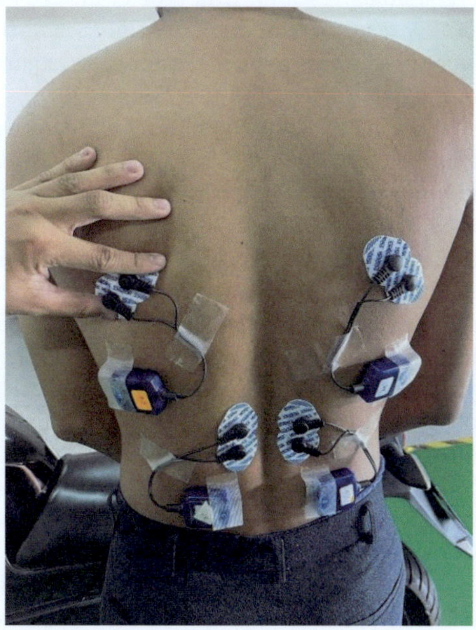

Fig. 3 Preparation of respondent with placement of electrode sensor on the latissimus dorsi (upper-side) and erector spinae longissimus (lower-center) muscle area

The experimental procedure has been established for this research to propose a guideline for handling experimentation data using wireless Myon Surface Electromyography (sEMG) for motorcyclists. The details of the procedures were initially presented, with each respondent performing two different riding postures throughout the session. The respondents needed to adopt the standard motorcycling riding posture according to standard guidelines (Ma'arof and Ahmad 2012). Moreover, motorcycles are frequently classified and selected for study purposes under several classification systems, such as RIPOC (Riding Posture Classification) used in

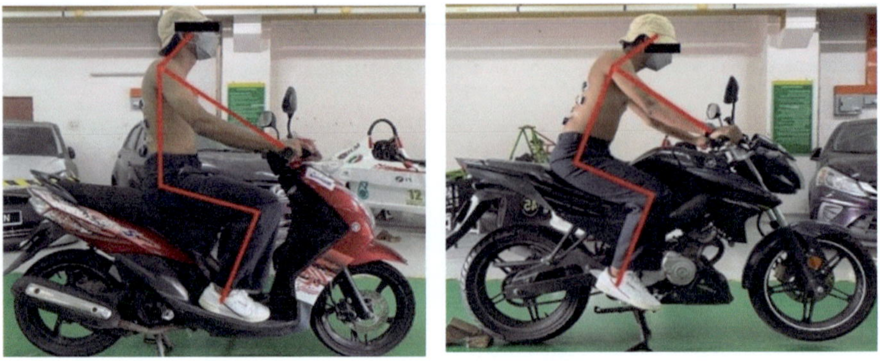

Fig. 4 The upright (left) versus the forward lean riding posture (right)

ergonomics research (Ma'arof et al. 2017). Two types of motorcycles were used in this research: the Yamaha Ego S and the Yamaha FZ150i. These motorcycles were placed on a center stand as the experiment was based on static posture positioning, as shown in Fig. 4. Each of the respondents was instructed to remove their shirt during the experiment, following the same procedure used in previous research (Rashid et al. 2015). The sEMG device was used to read the muscle activities from the transmitter, which was attached to the muscle points bilaterally (left and right) for 12 min. The first 2 min were a trial period to position the respondent correctly in a riding posture without collecting any data. Data collection started after the initial 2 min, with the remaining 10 min timed accurately using a stopwatch. The proEMG software collected signals for 10 s at 2-min intervals, resulting in a total of five measurements over the 10-min period.

Data was recorded every 1 s at 2000 Hz. The short reading duration aimed to observe and compare muscle activation from different specific postures at the selected muscle points. The 10-min duration was chosen as a baseline to consider riders involved in e-hailing food delivery (based on survey result). Throughout the experiment, two back muscles were monitored: the latissimus dorsi and the erector spinae (longissimus).

In this research, the electrode sensor placement for surface electromyography (sEMG) followed the steps recommended by SENIAM (Hermens et al. 2000). There are six steps in the standard procedure. Steps one and two generally involve selecting the sensor placement points on the muscles, while steps three through six provide detailed guidelines for the experimenter to follow for each muscle. The following sequential techniques were used for placing the sEMG sensors:

Procedure 1: Determine the characteristics of sEMG sensors.

In accordance with SENIAM's recommendations, the sensor selection was prepared. The size of each electrode used was 10 mm. In this project, the electrode tape was supplied by Universiti Teknologi Mara (UiTM). Two electrode tapes were needed to attach the electrodes, ensuring the distance between the centers of the electrode gels was approximately 20 mm. The electrode gel material was designed to ease the connection with the respondent's skin and provide effective electrode–skin interaction. The gel was made of silver (Ag) or silver chloride (AgCl), which does not chemically react with the skin and has low impedance between the electrodes and human skin.

Procedure 2: Skin preparation of the respondents.

The skin preparation procedure was necessary to ensure a strong connection between the electrode gel and the skin. Good contact with this material will influence the accuracy of surface electromyography (sEMG) readings. Proper skin preparation reduces electrical interference, lowers noise, and minimizes electrode fluctuation. Removing unnecessary hair at the muscle point is also necessary before applying the electrode gel. The skin area was cleaned with alcohol swabs to remove dust and dead skin cells, ensuring clean contact between the skin and gel. The alcohol was allowed to dry through vaporization before placing the electrodes and sEMG transmitter.

Procedure 3: Setting up the initial riding posture for the respondents.

The precise placement of the electrodes on the muscles was determined by the initial positions adopted by the respondents before the electrodes were attached. Therefore, when respondents were in their starting positions, the correct positioning of the electrodes was ascertained using both anatomical attachment and physical palpation of the muscles. The procedures for the experiment included detailed explanations of the starting positions taken by respondents.

Procedure 4: Placing the electrode sensor and transmitter of the sEMG.

The electrodes were positioned on the muscles to align with the sensors' longitudinal placement. Although EMG is considered a unique and practical approach to minimizing crosstalk in an experiment, exposure to this variable significantly reduces the reliability of sEMG recordings (Talib et al. 2019). The space between the centers of the conductive sections of the two bipolar electrodes is called the inter-electrode distance. For the surface electromyography (sEMG) procedure, the inter-electrode distance was set at 20 mm between the two electrodes.

Procedure 5: Validating the placement of the electrode sensor on muscles.

In this experiment, the procedure to position the electrode sensors on the specific muscles followed the guidelines of past research. The muscles selected for posture positioning were the latissimus dorsi (LD) and the erector spinae (ES) longissimus. On the latissimus dorsi, the lateral border of the scapula can be palpated by placing three fingers distal and parallel to the posterior axillary fold (Lehman et al. 2006). For the erector spinae longissimus, the sensor electrodes should be placed two finger-widths to the side of L1's spine (Guo et al. 2012). These electrodes were placed on the upper layer of skin or as bony prominence (Blanc and Dimanico 2014). When a specific muscle is to be analyzed, bipolar sEMG electrodes are positioned at the sensor site perpendicular to the muscle fibers. The electrode sensor placement was verified to ensure the correct procedure was followed before the experiment was conducted.

Procedure 6: Evaluating the connection between the skin-muscle contact and the electrode.

To evaluate the contact between the electrode and the muscle, it is important to refer to the reading data on the ProEMG software, which needs to be integrated from sEMG into the provided laptop. The software will show the connectivity from electrode sensors to the EMG through OFF and ON indicators. The recorded data was tested with a trial run to see the muscle activation signals captured on ProEMG software. To determine the reliability of the recorded signal, examine the raw graph data over time in the software. If there is noise or loss of signal, the electrode sensor would be attached with a different code number from the transmitter. Since the experiment involved static posture, each muscle tested was also subjected to an isometric contraction test to determine its level of muscle activation.

3 Results

At first, a questionnaire session was conducted via a google form, where a total of 117 respondents has taken part. There were ten categories of respondents, from 19 to 30 years of age. 66.6% of the respondents were teenagers, between the age of 23 to 25 years old. From Fig. 5 (left) below, 50.4% of them have been working in e-hailing delivery for less than one year. Whereas 36.8% had between one until three years of experience and the rest were more than three years. From Fig. 5 (right), 64.9% of e-hailing riders estimated that the average delivery time per order was between 10 to 30 min, compared to 34.1%, assuming the delivery was done in less than 10 min, and finally 0.85% for more than 30 min.

Figure 6 depicts the feedback from respondents when highlighting the body part with discomfort feeling. 73.5% highlighted the lower back muscle at the erector spinae longissimus with the highest discomfort feel compared to the latissimus dorsi muscle (around 21.4%). By pinpointing the specific area, the respondents able to locate the muscles. The maximum rating of discomfort for selected muscle was 4 out of 5 rates.

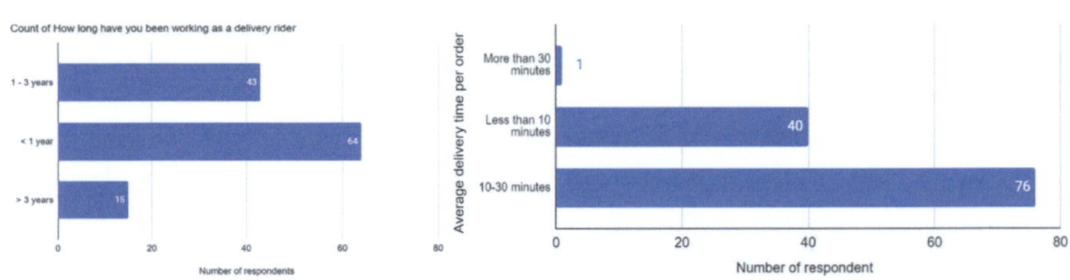

Fig. 5 Rider's working experience (Left) and Average time delivery per order (Right)

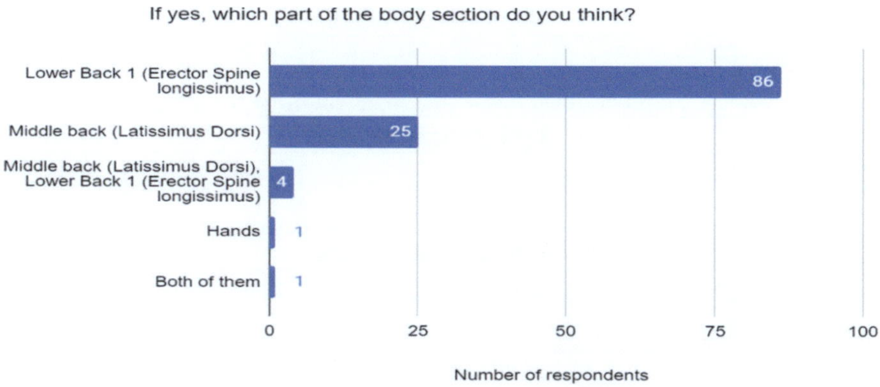

Fig. 6 Region of muscle discomfort area (while riding a motorcycle)

Next is to proceed with the EMG experimentation. Five respondents participated in the experiment. All respondents were confirmed to be in a healthy physical and mental state. The average RMS values are reported for each selected muscle: latissimus dorsi (left and right) and erector spinae longissimus (left and right). The variation between left and right sides depends on factors such as the body's weight balance and adjustments made due to discomfort or unstable positions during the experiment. Figures 7 and 8 display the RMS voltage signal readings over 10 s for forward lean and upright postures, respectively. The tabulated data is from one of the respondents.

The total average RMS for each respondent is summarized in Tables 1 and 2. Each table presents the average RMS value, calculated by dividing the summation

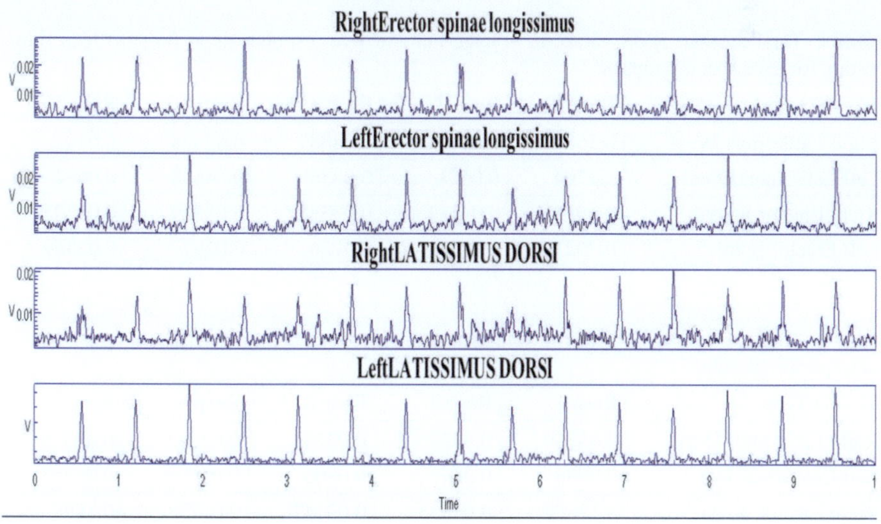

Fig. 7 RMS signal results (in 10 s) for forward lean posture

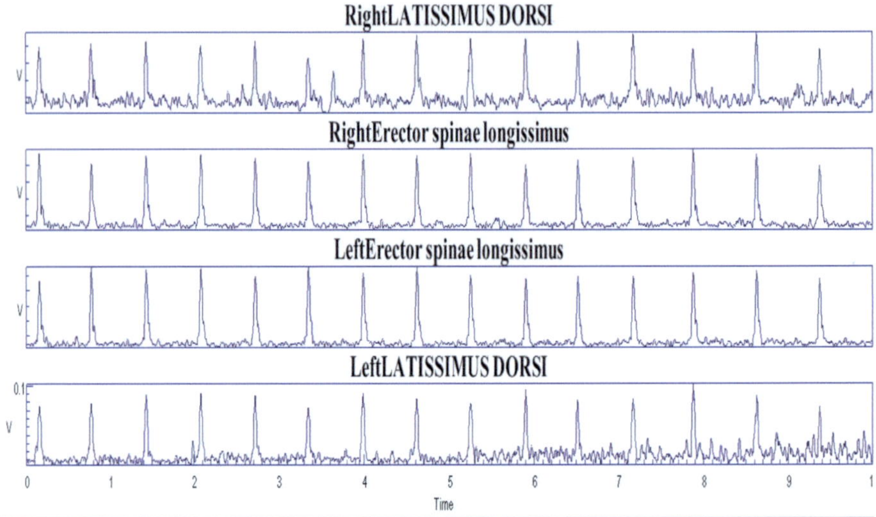

Fig. 8 RMS signal results (in 10 s) for upright posture

of RMS values (in voltage V) captured over time (10 s) by 2000 Hz and multiplying by the time taken. As detailed in the methodology section, data was collected in 5 sets per respondent at 2-min intervals.

Figures 9 and 10 display the total average RMS from five different respondents, comparing percentages to determine differences between two riding postures: forward lean and upright. Figure 9 shows that during the forward lean posture, the

Table 1 Total average RMS value (in voltage unit) of five respondents at forward lean riding posture for each specific muscle

Muscle Type	Resp.1	Resp.2	Resp.3	Resp.4	Resp.5
Right Latissimus Dorsi	0.558	0.02741	0.04889	0.02828	0.04
Left Latissimus Dorsi	1.1704	0.0621	0.0836	0.04638	0.0632
Right Erector Spinae	0.1958	0.04111	0.03585	0.03219	0.03705
Left Erector Spinae	0.3921	0.02819	0.0736	0.0297	0.03369

Table 2 Total average RMS value (in voltage unit) of five respondents at upright riding posture for each specific muscle

Muscle Type	Resp.1	Resp.2	Resp.3	Resp.4	Resp.5
Right Latissimus Dorsi	0.05769	0.03265	0.03645	0.02833	0.03679
Left Latissimus Dorsi	0.0998	0.0642	0.0806	0.0569	0.05283
Right Erector Spinae	0.17898	0.04833	0.09147	0.07278	0.1886
Left Erector Spinae	0.2411	0.145	0.0535	0.05125	0.1157

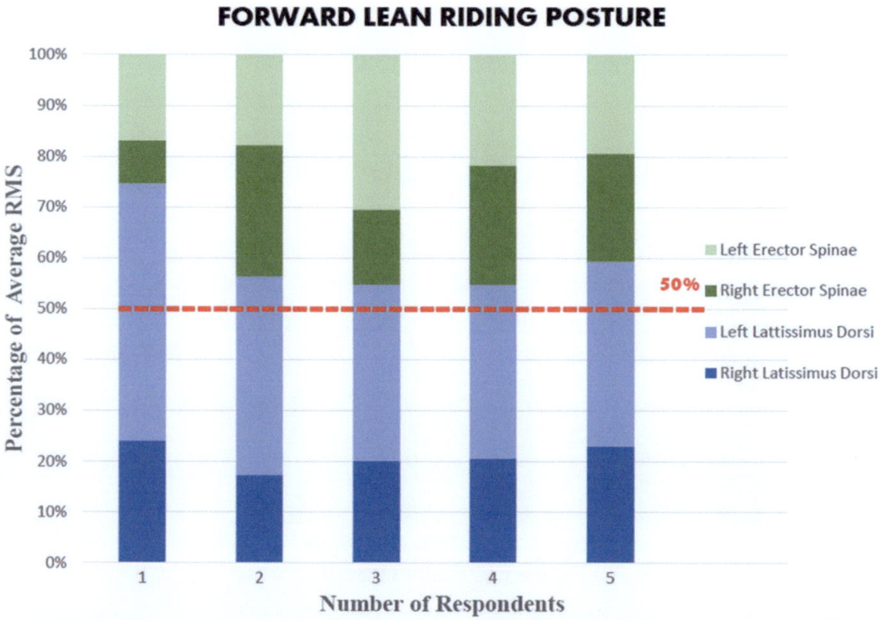

Fig. 9 Percentage of total average RMS for forward lean riding posture

latissimus dorsi muscles bilaterally are dominant, with rates exceeding 50% for all respondents. Respondent 1 recorded the highest total average RMS value of 1.7284 V for the latissimus dorsi muscles bilaterally, representing a 74.6% rate compared to the erector spinae longissimus.

Figure 10 presents the results for the upright riding posture, where the erector spinae longissimus muscles bilaterally dominate over the latissimus dorsi. This is attributed to the body's weight distribution, which emphasizes the lower back muscles more than the middle section during this posture. Respondent 5 shows the highest percentage of average RMS compared to other respondents, at 77.2%, with a total average RMS value of 0.3043 V for the erector spinae longissimus muscles bilaterally. Overall, the results also confirm that during the upright riding posture, the average RMS percentage for the latissimus dorsi muscle is low, less than 50% across all data. Thus, these quantitative findings demonstrate that different riding postures significantly influence the total average RMS values of specific muscles.

Regarding the post-experiment feedback, after completing the experimental sessions, all five respondents were asked to fill out a feedback form to share their honest responses and experiences. According to the report, all respondents agreed that the forward lean riding position was their preferred choice after experiencing it during the experiment. They unanimously found it more comfortable and convenient compared to the upright riding posture. One respondent particularly favored the forward lean position for its stability, feeling that leaning forward aligned well with the motorcycle.

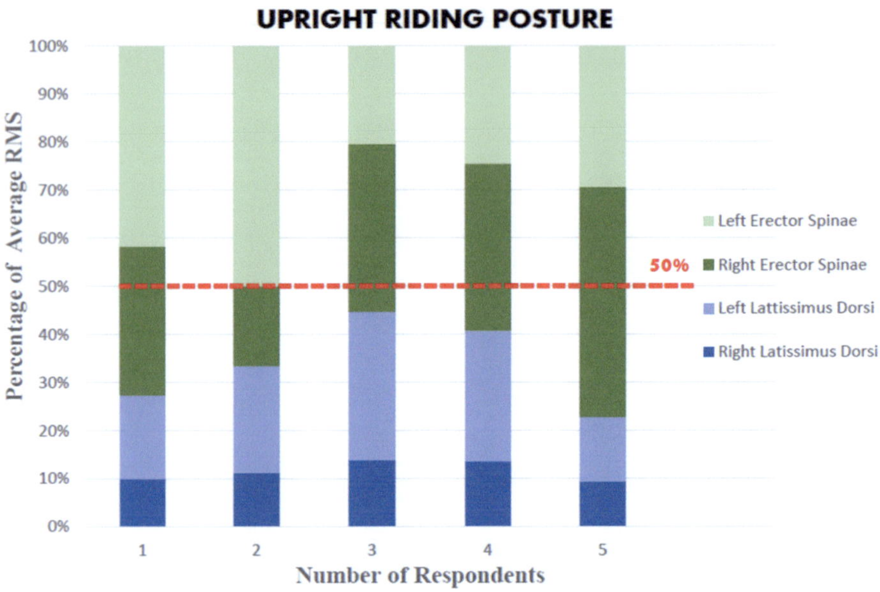

Fig. 10 Percentage of total average RMS for upright riding posture

Additionally, all respondents expressed discomfort with their back muscles. After two experimental sessions, 60% of them reported experiencing more discomfort when in the upright posture compared to the forward lean posture. Furthermore, respondents provided suggestions for alleviating back pain while riding motorcycles. The majority acknowledged the importance of adopting the correct riding posture after the experiment, and they recognized that riding comfort depends significantly on the motorcycle itself.

4 Discussion and Conclusion

The objective of this research was to monitor muscle activation specifically in the latissimus dorsi and erector spinae longissimus muscles and compare the RMS amplitude between different riding postures of motorcyclists. High readings of muscle activity data indicate that motorcycling is physically and physiologically challenging for riders. The evidence demonstrates that discomfort associated with specific motorcycle postures is not merely speculative.

The results revealed that the upright riding posture recorded the lowest average RMS percentage for the latissimus dorsi muscle compared to the forward lean posture. Conversely, for the erector spinae longissimus muscle, the forward lean posture showed the lowest percentage compared to the upright riding posture. This data confirms that adopting a forward lean posture reduces muscular activity, particularly in the lower back muscles. Motorcyclists often experience seated positions that place significant pressure on the discs in their lower back compared to other postures (Amol and Amrutkar 2011). The findings suggest that discomfort experienced in the upright riding posture is related to muscular activity.

As a result, it was discovered that the forward-leaning motorcycle posture offers potential benefits for motorcyclists. The forward-leaning posture allows riders to shift their body weight distribution more effectively during motorcycling maneuvers by adjusting their body position. A typical motorcyclist's riding position involves flexion of the spine exceeding 20 degrees, with both feet positioned beneath the knees (Ma'arof et al. 2016), enabling effective motorcycle control and maneuverability. Additionally, occasional changes in sitting position are beneficial, as no single posture should be maintained for an extended period (O'Sullivan et al. 2012). Furthermore, the forward-leaning posture enhances the aerodynamics of motorcycling by streamlining the motorcycle. Motorcyclists achieve this through extreme spinal flexion, as seen in motorcycle racers.

Although this research clarified that the forward lean riding posture does not directly affect the erector spinae longissimus muscle, which is associated with back pain, it still contributes to physical and physiological discomfort. Some speculations may attribute this discomfort to factors such as age or body weight. Further studies should focus on these elements to better understand discomfort levels and raise public awareness about motorcyclists. Consequently, ergonomic interventions could be developed to improve working conditions and safety for motorcycle riders.

In summary, the sEMG experiment aimed to determine muscle activation in the latissimus dorsi and erector spinae longissimus muscles with two different static riding postures and types of motorcycles. The forward lean posture showed the lowest average RMS reading for the erector spinae longissimus, whereas the upright riding posture showed the highest. Since the research evaluated muscle activity related to back pain in motorcyclists, particularly pressure on the lower back muscles, the forward lean posture is recommended as the ideal riding posture.

Additionally, two questionnaires were distributed: one to respondents after the experiment session and another to e-hailing riders. It was found that most respondents practiced upright riding posture and reported high discomfort levels in the lower back muscles, which correlated with muscle activation in those areas.

However, the discomfort level could be alleviated through ergonomic modifications. This research able to support those various types of physical and physiological discomfort in the lower back, associated with the forward-leaning posture, are not caused by muscular actions involved in adopting and maintaining that posture. It is crucial to promote the most ergonomically safe motorcycle riding practices to reduce rider discomfort and consequently minimize human error-related road accidents. Ultimately, the objectives of this research were achieved.

For future work, several recommendations can enhance ergonomic studies related to motorcyclists' posture. Firstly, it is advisable for first-time riders to choose a sport-type motorcycle over a regular motorcycle to achieve a more comfortable riding posture and reduce the likelihood of back pain while riding. Additionally, riders should take breaks when fatigued to allow their muscles to rest and recover. Furthermore, to enhance overall riding comfort, further research could investigate forearm strain while using sport-type motorcycles. As for the experimental procedure, by expanding the study to include a larger sample size with diverse ages and different muscle selection points would validate its reliability and explore how muscle activity changes over time. Previous research has suggested that muscle activity may decrease over time due to fatigue. Finally, enhancing the sEMG test from static to dynamic state able to improve the result, with strict conditions that the noise level needs to be closely controlled.

References

Amol NRR, Amrutkar S (2011) Ergonomic posture for motorcycle riding, 5 [Online]. Available: https://www.researchgate.net/publication/260657740_Ergonomic_Posture_for_Motorcycle_Riding

Blanc Y, Dimanico U (2014) Electrode Placement in Surface Electromyography (sEMG)" Minimal Crosstalk Area "(MCA). Open Rehabil J 3(1):110–126. https://doi.org/10.2174/1874943701003010110

Fukuda TY, Alvarez AS, Nassri LFG, Gurjão de Godoy CM (2008) Quantitative electromyographic assessment of facial muscles in cross-bite female children. Rev Bras Eng Biomed, 121–129

Guo LY et al (2012) Comparison of the electromyographic activation level and unilateral selectivity of erector spinae during different selected movements. Int J Rehabil Res 35(4):345–351. https://doi.org/10.1097/MRR.0b013e32835641c0

Hermens HJ, Freriks B, Disselhorst-Klug C, Rau G (2000) Development of recommendations for SEMG sensors and sensor placement procedures. J Electromyogr Kinesiol 10(5):361–374. https://doi.org/10.1016/S1050-6411(00)00027-4

Karmegam K, Ismail MY, Sapuan SM, Ismail N (2008) Conceptual design and prototype of an ergonomic back-leaning posture support for motorbike riders. J Sci Ind Res (India) 67(8):599–604

Laitung JKG, Peck F (1985) Shoulder function following the loss of the latissimus dorsi muscle. Br J Plast Surg 38(3):375–379. https://doi.org/10.1016/0007-1226(85)90245-0

Lanka S (2015) IDF South-East Asia region, pp 2011–2012

Lehman GJ, MacMillan B, MacIntyre I, Chivers M, Fluter M (2006) Shoulder muscle EMG activity during push up variations on and off a Swiss ball. Dyn Med 5:1–7. https://doi.org/10.1186/1476-5918-5-7

Ma'arof MIN, Ahmad IN (2012) Proposed standard method for motorcycle nomenclature system. 2012 Southeast Asian Netw Ergon Soc Conf Ergon Innov Leveraging User Exp Sustain SEANES, pp 2–7. https://doi.org/10.1109/SEANES.2012.6299579

Ma'arof MIN, Omar AR, Haron R,Husain H, Wahab MA (2017) Muscular activities measurements of forward lean and upright sitting motorcycling postures via surface electromyography (sEMG). MATEC web conf, vol 135. https://doi.org/10.1051/matecconf/201713500046

Ma'arof MIN, Omar AR, Mahmud Z, Senin N (2016) Original article preliminary study on the best working posture for," Malaysian J Hum Factor Ergon 1(2):39–47

Mathurkar SR (2016) Design of test rig for motorcycle seat for human comfort. Int J Adv Eng Res Dev 3(10):891–898. https://doi.org/10.21090/ijaerd.03101

Ospina-Mateus H, Quintana Jiménez LA (2019) Understanding the impact of physical fatigue and postural comfort experienced during motorcycling: a systematic review. J Transp Heal 12:290–318. https://doi.org/10.1016/j.jth.2019.02.003

O'Sullivan K, O'Sullivan P, O'Sullivan L, Dankaerts W (2012) What do physiotherapists consider to be the best sitting spinal posture? Man Ther 17(5):432–437

Rashid H, Nor Maâ MI, Haron R, Syahmi Wan Fauzi WM, Omar AR, Jaafar R (2015) Selection of muscle groups for surface electromyography (sEMG) measurement in analyzing motorcycling activity. Jurnal Teknologi 76(7)

Snijders CJ, Bakker MP, Vleeming A, Stoeckart R, Stam HJ (1995) Oblique abdominal muscle activity in standing and in sitting on hard and soft seats. Clin Biomech 10(2):73–78. https://doi.org/10.1016/0268-0033(95)92042-K

Starkey P (2016) The benefits and challenges of increasing motorcycle use for rural access. In: International conference on transport and road research, pp 16–18

Talib I, Sundaraj K, Lam CK, Hussain J, Ali MA (2019) A review on crosstalk in myographic signals. Eur J Appl Physiol 119(1):9–28. https://doi.org/10.1007/s00421-018-3994-9

Velagapudi SP, Balasubramanian V, Adalarasu K,Babu R, Mangaraju V (2010) Muscle fatigue due to motorcycle riding. SAE Tech. Pap.https://doi.org/10.4271/2010-32-0100

Young E (2013) Petrol prices: why are they rising?. https://www.bbc.com/news/business-21547684. Accessed Jun. 28, 2022

The Association of *ACE I/D* and *ACTN3 R577X* Polymorphisms with Athlete Status, Playing Status and Vertical Jump Performance in Malaysian Footballers

Nur Fatihah Saidi, Chee Hian Tan, Raja Mohammed Firhad Raja Azidin, Choo Yee Yu, and Geik Yong Ang

Abstract The primary aim of the present study is to analyze the association of *ACE I/D* and *ACTN3 R577X* with athlete status, playing status and lower limb explosive power performance in Malaysian footballers. A total of 77 Malay male footballers (professional, n = 22; non-professional, n = 55) were recruited and their anthropometry, body composition and lower limb explosive power performance (vertical jump and 30-m sprint) were assessed. Genotyping of *ACE I/D* and *ACTN3 R577X* polymorphisms were carried out by allele-specific polymerase chain reaction and the results showed that *ACE I/D* polymorphism, but not *ACTN3 R577X* polymorphism, was significantly associated with athlete status, playing status as well as vertical jump performance. More specifically, the *ACE I/D* polymorphism was associated with athlete status under the *I*-allele dominant model (II + ID vs. DD: OR = 4.21; 95% CI: 1.16–15.23). However, professional players were found to be more likely to possess a *D* allele as compared to non-professional players (*D* vs. *I*: OR = 2.74; 95% CI: 1.24–6.02). The *ACE I/D* polymorphism accounted for ≈6% to 14% of the genetic variance for jump performance with the additive allelic model (13.9%) explaining a larger portion of the genetic variance as compared to the *I*-allele dominant (9.3%) and *D*-allele dominant (6.4%) models. The association of *ACE I/D* polymorphism with athlete status, playing status and vertical jump performance found in the present study extend the knowledge in the area of genetic association research in football with specific reference to Malaysian footballers.

Keywords Football · Performance · Genetic polymorphisms

N. F. Saidi · R. M. F. Raja Azidin · G. Y. Ang (✉)
Faculty of Sports Science and Recreation, Universiti Teknologi MARA, Selangor, Malaysia
e-mail: geikyong@uitm.edu.my

C. H. Tan
Department of Sport Science, Faculty of Applied Sciences, Tunku Abdul Rahman University of Management and Technology, Kuala Lumpur, Malaysia

C. Y. Yu
Institute of Bioscience, Universiti Putra Malaysia, Selangor, Malaysia

M. H. A. Hassan et al. (eds.), *Proceedings of the 3rd International Conference on Innovation and Technology in Sports*, Lecture Notes in Bioengineering,
https://doi.org/10.1007/978-981-95-0804-4_9

1 Introduction

Football is a complex intermittent team sport that involves numerous technical and tactical actions that vary in form and intensity throughout a 90-min game (Turner and Stewart 2014). The demanding performance requirement of a football match is known to tax both the anaerobic and aerobic energy systems. Explosive actions such as jumps, short sprints and shooting, which are integral in goal-scoring situations, rely heavily on the anaerobic system whereas a strong aerobic system is also essential to maintain the endurance needed for full matches and to facilitate a faster recovery between bouts of high-intensity actions (Dolci et al. 2020). Furthermore, studies have shown that the physical attributes, match-activity profiles and energetic demands of football are also position and competitive-level dependent (Slimani and Nikolaidis 2019; Bush et al. 2015). Specific training programs are usually designed to ensure footballers have a high level of tactical, technical and physical ability to succeed but interindividual differences in training responses meant that adherence to a well-designed training program does not necessarily guarantee the attainment of elite status (Jones et al. 2016).

As evidence of significant association between genetics and athletic performance continues to grow in recent years, the number of football genetic research has also been increasing in tandem (McAuley et al. 2021a). Some genetic association studies seek to elucidate whether genetic polymorphisms were associated with footballer status (Ulucan et al. 2015a; Galeandro et al. 2017) while other investigated if there were competitive-level differences (Coelho et al. 2018) and whether power- or endurance-orientated genotype was more prevalent in footballers (Egorova et al. 2014a; Gineviciene et al. 2014a). A recent systematic review of genetic association research in football revealed that the *ACTN3* is the most widely studied gene followed by the *ACE* gene (McAuley et al. 2021a). A premature stop codon (*X*) instead of arginine (*R*) at position 577 due to a *C*-to-*T* allele substitution in *ACTN3* gene attracts significant attention as it leads to the absence of alpha-actinin-3 (ACTN3) protein that aids in stabilizing the contractile apparatus in type-II muscle fibers that are responsible for explosive and powerful movements such as sprinting and jumping (Garton et al. 2014). On the other hand, the *ACE I/D* polymorphism is investigated due to its association with angiotensin I converting enzyme (ACE) activity level as the enzyme plays a central role in the renin-angiotensin system that is involved in the blood pressure and fluid-electrolyte homeostasis (Rigat et al. 1990).

Generally, the *ACE D* and *ACTN3 R* alleles tend to be more prevalent in strength/power-oriented athletes whereas *ACE I* and *ACTN3 X* alleles are more pronounced among endurance athletes (Ahmetov et al. 2022). However, the implications of the collective findings on the associations of the *ACE I/D* and *ACTN3 R577X* polymorphisms with athletic performance are complicated in sports that require both power- and endurance-related traits such as football (McAuley et al. 2021b). Furthermore, previous football genetic studies had yielded positive, negative and even contrasting allelic associations and these discrepancies could be attributed, in part, to numerous inter-individual variations such as ethnicity and competitive playing level (McAuley et al. 2021b). Therefore, this study aimed to investigate the association of *ACE I/D* (rs4343) and *ACTN3 R577X* (rs1815739) with athlete status, playing status, sprint and jump performance in Malaysian footballers.

2 Materials and Methods

2.1 Participants

Ethical approval was obtained from the Universiti Teknologi MARA Research Ethics Committee (REC/07/2022 (ST/MR/151) and participants were informed about the procedures, possible risks and benefits of the study before they provided written informed consent. The present study consisted of 77 male footballers of Malay ethnicity (age 19.6 ± 1.4 years; height 170.6 ± 6.3 cm, body mass 65.6 ± 7.4 kg) that competed in the Malaysia football leagues (professional, n = 22; non-professional, n = 55). Classification of footballers into professional and non-professional was performed as described previously (McAuley et al. 2021b). All the participants had no family history of admixture or inter-marriage for at least three generations.

2.2 Genotyping

Saliva sample (2 ml) of each participant was collected using the SalivaGene Collection Module II kit (Invitek Molecular, Germany). The quality and quantity of the DNA that was extracted with the PSP SalivaGene DNA Kit (Invitek Molecular, Germany) from the saliva samples were determined with the NanoDrop 2000 spectrophotometer (Thermo Fisher Scientific, Massachusetts, USA). Genotyping by allele-specific polymerase chain reaction for *ACE I/D* (rs4343) and *ACTN3 R577X* (rs1815739) were carried out as described previously (Ang et al. 2023; Saidi et al. 2024). The amplified products were separated via electrophoresis and a gel documentation system (Uvitec, Cambridge, UK) was used to visualize the presence of the targeted band that corresponds to the expected size of each allele of the tested genetic polymorphism.

2.3 Performance Measurement

The lower limb explosive power of each participant was measured via the countermovement vertical jump (cm) using the yardstick vertical jump device (Swift Performance Equipment, Australia; intraclass correlation coefficient (ICC) = 0.98) as described previously (Ang et al. 2023). Speed was assessed with a 30-m sprint test and the elapsed time (s) was measured using 2 pairs of photoelectric cells (Speed Trap

II, Brower Timing Systems, USA; ICC = 0.99; coefficient of variation (CV) = 0.9%) (Saidi et al. 2022) that were set at a height of 1.2 m and positioned at the starting and finishing lines. Subjects started the sprint when they felt ready from a standing start that was set at 0.3 m behind the first timing gate. They were instructed to run as fast as possible once they have initiated their sprint. Each participant completed two trials of the vertical jump and 30-m sprint with the best performance being used for further analysis. All the participants wore a standard training football kit.

2.4 Data Analysis

All statistical analyses were conducted using the Statistics Package for Social Sciences for Windows, version 29.0 (IBM Corporation, New York, USA). The genotype and allele frequencies of *ACE I/D* and *ACTN3 R577X* of the footballers were tested for Hardy–Weinberg equilibrium before differences between the footballers and a control cohort comprising 96 Southeast Asian Malays (Wong et al. 2013) and then, between professional and non-professional players, were assessed using four genetic models: allele contrast; recessive; dominant; and over-dominant. Pairwise comparison between genotypes were also conducted. The strength of association was measured using odds ratios (ORs) and significance of the ORs was calculated using chi-square test and 95% confidence intervals (CI). The comparison of vertical jump height and 30-m sprint time between genotypes were assessed using ANOVA and significant results were further assessed to investigate the relationship between genotype and phenotype distribution using correlation analysis. The coefficient of determination (r^2) from the correlation analysis was used to estimate the validity of the genetic model tested and was expressed as a percentage of the variance explained by the genotype effects. The p value for significance testing was set at $p < 0.05$.

3 Results

3.1 Genotype and Allele Frequency Analysis

Representative amplicons of *ACE I/D* (rs4343) and *ACTN3 R577X* (rs1815739) polymorphisms are shown in Fig. 1. The bands obtained were consistent with the expected sizes and presence of the specific allele was confirmed by sequencing. Genotype and allele frequencies were manually calculated, and the results are shown in Table 1. The distribution of genotypes and alleles of *ACE I/D* and *ACTN3 R577X* in the footballers conformed with Hardy–Weinberg equilibrium ($p > 0.05$).

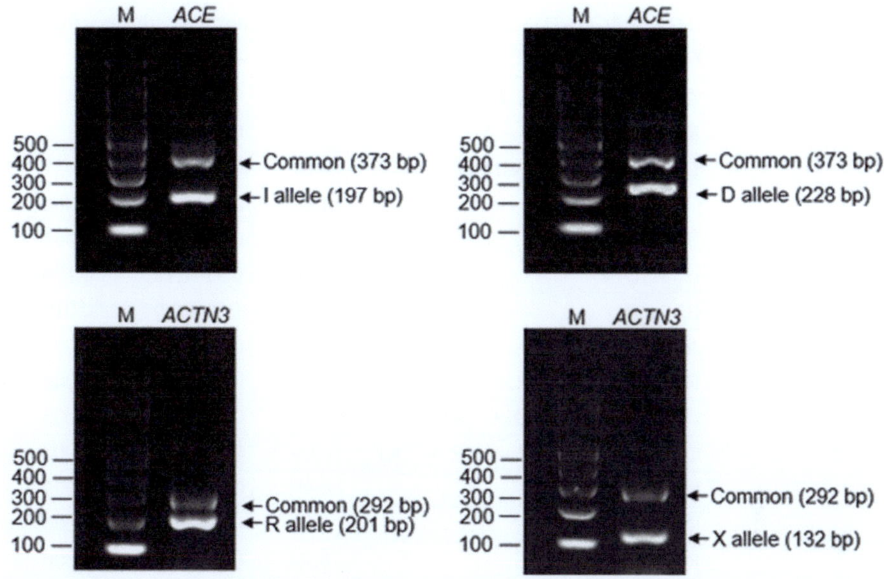

Fig. 1 Representative amplicons of *ACE I/D* (top) and *ACTN3 R577X* (bottom) polymorphisms electrophoresed in 2% agarose gel. *Note* Lane M contains the GeneRuler 100 bp DNA ladder

Table 1 Genotype and allele frequencies of *ACE I/D* and *ACTN3 R577X* polymorphisms in Malaysian footballers

Gene (rsID)	Genotype	Frequency (n = 77)	Allele	Frequency (n = 77)	Hardy–Weinberg p value
ACE	DD	0.039	*D*	0.227	0.526
(rs4343)	ID	0.377	*I*	0.773	
	II	0.584			
ACTN3	RR	0.208	*R*	0.494	0.209
(rs1815739)	RX	0.571	*X*	0.506	
	XX	0.221			

3.2 Association Analysis

The results of the genetic polymorphism-athlete status association analysis are shown in Table 2. Only the *I*-allele dominant model (II + ID vs. DD: OR = 4.21; 95% CI: 1.16–15.23) of *ACE I/D* polymorphism was significantly associated with athlete status and pairwise comparison between *ACE I/D* genotypes revealed a significant association of ID vs. DD (OR = 5.88; 95% CI: 1.51–22.97). When professional players were compared to non-professional players, several significant associations were observed between the *ACE I/D* polymorphism and genetic comparison models with playing status (Table 3). Specifically, the allele contrast (D vs. I: OR = 2.74;

Table 2 Association tests of *ACE I/D* and *ACTN3 R577X* polymorphisms with athlete status

Gene	Genetic model	OR	95% CI	p
ACE I/D	Allele contrast (*I* vs. *D*)	1.23	0.75–2.02	0.454
	Recessive (II vs. ID + DD)	0.88	0.48–1.63	0.755
	Dominant (II + ID vs. DD)	4.21	1.16–15.24	0.021
	Over-dominant (ID vs. II + DD)	1.92	0.99–3.70	0.066
	II vs. DD	3.56	0.96–13.14	0.061
	II vs. ID	0.60	0.31–1.18	0.174
	ID vs. DD	5.88	1.51–22.97	0.010
ACTN3 R577X	Allele contrast (*R* vs. *X*)	0.99	0.65–2.52	1.000
	Recessive (RR vs. RX + XX)	1.00	0.48–2.09	1.000
	Dominant (RR + RX vs. XX)	0.99	0.48–2.04	1.000
	Over-dominant (RX vs. RR + XX)	0.99	0.54–1.82	1.000
	RR vs. XX	0.99	0.40–2.47	1.000
	RR vs. RX	1.00	0.46–2.15	1.000
	RX vs. XX	0.99	0.47–2.10	1.000

95% CI: 1.24–6.02) and *D*-allele dominant model (DD + ID vs. II: OR = 3.60; 95% CI: 1.28–10.13) were significantly associated with playing status and pairwise comparison between *ACE I/D* genotypes revealed a significant association of ID vs. II (OR = 3.26; 95% CI: 1.13–9.45). Conversely, the *ACTN3 R577X* was neither associated with athlete status nor playing status under the genetic comparison models that were tested. The vertical jump and 30-m sprint performance by genotype groups are shown in Table 4. Following the observation of a significant difference in vertical jump performance between *ACE* genotypes (p = 0.013), the underlying genetic mechanism was investigated further using genetic models. The additive allelic model was found to have the best fit and accounted for 13.9% of the genetic variance for vertical jump performance (p = 0.008) as compared to *I*-allele dominant (9.3%) and *D*-allele dominant (6.4%) models.

4 Discussion

The present study found that *ACE I/D* polymorphism was associated with athlete status in football and the overrepresentation of the *ACE* ID genotype in footballers was similar that observed by Gineviciene et al. (Gineviciene et al. 2014b) among Lithuanian professional male footballers. However, when professional players were separated from the non-professional players, a significant association between *ACE I/D* polymorphism and playing status was observed whereby professional players was more akin to power athletes in their tendency to possess the *D* allele. This observation was contradictory to that reported by McAuley et al. (2021b) in which

Table 3 Association tests of *ACE I/D* and *ACTN3 R577X* polymorphisms with playing status

Gene	Genetic model	OR	95% CI	p
ACE I/D	Allele contrast (*D* vs. *I*)	2.74	1.24–6.02	0.018
	Recessive (DD vs. ID + II)	5.40	0.46–62.87	0.195
	Dominant (DD + ID vs. II)	3.60	1.28–10.13	0.020
	Over-dominant (ID vs. II + DD)	2.68	0.97–7.41	0.070
	DD vs. II	9.25	0.74–114.87	0.106
	DD vs. ID	2.83	0.23–34.92	0.568
	ID vs. II	3.26	1.13–9.45	0.034
ACTN3 R577X	Allele contrast (*R* vs. *X*)	0.71	0.35–1.43	0.375
	Recessive (RR vs. RX + XX)	1.18	0.36–3.89	1.000
	Dominant (RR + RX vs. XX)	0.34	0.11–1.05	0.072
	Over-dominant (RX vs. RR + XX)	0.40	0.14–1.09	0.080
	RR vs. XX	0.51	0.12–2.12	0.481
	RR vs. RX	1.77	0.49–6.40	0.492
	RX vs. XX	0.29	0.09–0.96	0.056

Table 4 Mean vertical jump and 30-m sprint performance within genotype groups

Gene (rsID)	Genotype	Vertical Jump		30-m sprint	
		Mean (cm)	p	Mean (s)	p
ACE	DD	71.0 ± 6.0^{a}	0.013	4.00 ± 0.18^{a}	0.108
(rs4343)	ID	63.0 ± 5.1^{ab}		4.21 ± 0.15^{a}	
	II	60.6 ± 6.7^{b}		4.19 ± 0.16^{a}	
ACTN3	RR	65.0 ± 6.5^{a}	0.063	4.14 ± 0.14^{a}	0.219
(rs1815739)	RX	60.7 ± 6.1^{a}		4.19 ± 0.17^{a}	
	XX	62.3 ± 6.7^{a}		4.24 ± 0.13^{a}	

Note The values with different superscript letters in a column are significantly different ($p < 0.05$)

non-professional players, who played at amateur or youth level, were instead more likely to possess a *D* allele or DD genotype. In contrast to non-professional players, the specific overrepresentation of the *D* allele in the current cohort of professional players may be attributed to the greater frequency of high-intensity actions performed by footballers playing at a higher level as compared to their lower level counterparts (Mohr et al. 2003). Given that football involves frequent bursts of high intensity activity that includes jumps, tackles, shots and sprints, particularly in goal-scoring situations, possession of the *D* allele may be favorable for footballers as prior studies have reported of its association with greater strength and muscle volumes (Ahmetov et al. 2013) as well as power-oriented phenotypes (vertical jump and 10-30m sprint performance) (Pimenta et al. 2013; Dionísio et al. 2017).

Likewise, it was hypothesized that the *R* allele of *ACTN3 R577X* polymorphism would be associated with athlete or playing status due to similar associations that were found between the *R* allele and power-oriented phenotype in football (Pimenta et al. 2013; Dionísio et al. 2017). Furthermore, multiple studies have also shown footballers at professional and elite levels tend to have power/strength-oriented genetic variations (Galeandro et al. 2017; Coelho et al. 2018; Ulucan et al. 2015b; Egorova et al. 2014b). The *ACE D* and *ACTN3 R* alleles were associated with a higher proportion of fast-twitch muscle fibres and the greater amount of phosphocreatine stored in these muscle fibres may benefit the carriers of these alleles by enabling a quicker ATP resynthesis via the phosphagen system which predominates during short intense bursts of activities (Turner and Stewart 2013). A recent meta-analysis of 15 *ACTN3* genetic association studies in football has demonstrated significant association between *ACTN3 R* allele and professional footballer status (McAuley et al. 2021b) but this association, along with the association between *ACE D* allele and non-professional footballers, were not replicated in the present study. Nevertheless, homozygotes for the *ACE D* and *ACTN3 R* alleles had better 30-m sprint and vertical jump performance than other genotypes but only the differences in vertical jump performance between *ACE I/D* genotypes reached statistical significance. The *ACE I/D* polymorphism accounted for 13.9% of the variance in vertical jump performance with the *D* allele contributing to a higher vertical jump height in an additive manner. Therefore, the present study confirmed an advantageous effect on vertical jump performance for the *ACE D* allele but not for the *ACTN3 R* allele. The *ACE I/D* and *ACTN3 R577X* polymorphisms produced no observable effect on 30-m sprint performance and these results were inconsistent with earlier findings (Pimenta et al. 2013; Dionísio et al. 2017).

One of the possible causes of discrepancy in the findings between studies is the ethnicities and nationality of the footballers investigated. Footballing status in Brazilians and Caucasians footballers were found to be associated with contrasting *ACTN3* genotypes with RX genotype being associated with footballing status for the former and RR genotype for the latter (McAuley et al. 2021b). Hence, the results of association studies are expected to differ for footballers of different geographical ancestry. Other plausible explanation for the non-replication of the prior findings may lie in the competitive playing level of the footballers. Although the present study adopted the categorization of players into professional and non-professional as described by McAuley et al. (2021b), the competitive playing level of these players would differ greatly if international rankings were taken into consideration. Therefore, the association of *ACTN3 R577X* polymorphism with athlete status may only be apparent in top-level professional footballers. The present study is not without its limitations. Genetic association studies generally require a large homogenous sample size to attain sufficient statistical power and to demonstrate significant associations and replications given that the contribution from each gene or genotype to sporting performance is likely very small (McAuley et al. 2021a). However, football genomics studies tend to suffer from small sample sizes as professional footballers are a unique population and available access to this population can be limited (McAuley et al. 2021a). Furthermore, the professional footballers recruited in this study may also be a limiting factor that influenced the study outcomes as they are not the highest performing footballers in Malaysia. In addition to expanding the number of genetic polymorphisms beyond those that have been explored in the present study, future research should also aim to explore other footballing performance characteristics including technical, tactical and psychological attributes of the footballers.

5 Conclusion

In conclusion, findings from the present study on the association of *ACE I/D* and *ACTN3 R577X* with athlete status, playing status and vertical jump performance extend the knowledge and contribute towards the current evidence base of genetic association research in football. The present study observed that the *I*-allele of *ACE* gene was associated, in a dominant manner, with athlete status in football but this homogenous team-sport cohort could be further differentiated in accordance to their playing level by the *ACE I/D* polymorphism as the present study showed that *D*-allele is overrepresented in professional players. The *D* allele was also shown to contribute towards a better vertical jump performance and in view of the relative contribution of power-oriented phenotype during decisive moments in a match, the *ACE I/D* polymorphism is likely a contributing factor that influences football performance.

References

Ahmetov II et al (2013) The association of ACE, ACTN3 and PPARA gene variants with strength phenotypes in middle school-age children. J Physiol Sci 63(1):79–85

Ahmetov II et al (2022) Advances in sports genomics. Adv Clin Chem 107:215–263

Ang GY et al (2023) Combined influence of ACE and ACTN3 polymorphisms on vertical jump performance following resistance training in professional footballers. Sport Sci Health 20(2):593–600

Bush M et al (2015) Evolution of match performance parameters for various playing positions in the English Premier League. Hum Mov Sci 39:1–11

Coelho DB et al (2018) Evidence for a role of ACTN3 R577X polymorphism in football player's career progression. Int J Sports Med 39(14):1088–1093

Dionísio TJ et al (2017) The influence of genetic polymorphisms on performance and cardiac and hemodynamic parameters among Brazilian soccer players. Appl Physiol Nutr Metab 42(6):596–604

Dolci F et al (2020) Physical and energetic demand of soccer: a brief review. Strength Conditioning J 42(3):70–77

Egorova ES et al (2014a) The polygenic profile of Russian football players. J Sports Sci 32(13):1286–1293

Galeandro V et al (2017) ACTN3/ACE genotypes and mitochondrial genome in professional soccer players performance. J Biol Regul Homeost Agents 31(1):207–213

Garton FC et al (2014) α-Actinin-3 deficiency alters muscle adaptation in response to denervation and immobilization. Hum Mol Genet 23(7):1879–1893

Gineviciene V et al (2014a) Variation in the ACE, PPARGC1A and PPARA genes in Lithuanian football players. Eur J Sport Sci 14(sup1):S289–S295

Gineviciene V et al (2014b) Variation in the ACE, PPARGC1A and PPARA genes in Lithuanian football players. Eur J Sport Sci 14(Suppl 1):S289–S295

Jones N et al (2016) A genetic-based algorithm for personalized resistance training. Biol Sport 33(2):117–126

McAuley ABT et al (2021a) Genetic association research in football: a systematic review. Eur J Sport Sci 21(5):714–752

McAuley ABT et al (2021b) The association of the ACTN3 R577X and ACE I/D polymorphisms with athlete status in football: a systematic review and meta-analysis. J Sports Sci 39(2):200–211

Mohr M, Krustrup P, Bangsbo J (2003) Match performance of high-standard soccer players with special reference to development of fatigue. J Sports Sci 21(7):519–528

Pimenta EM et al (2013) Effect of ACTN3 gene on strength and endurance in soccer players. J Strength Cond Res 27(12):3286–3292

Rigat B et al (1990) An insertion/deletion polymorphism in the angiotensin I-converting enzyme gene accounting for half the variance of serum enzyme levels. J Clin Invest 86(4):1343–1346

Saidi NF et al (2022) Associations between speed, power and endurance performance in super league soccer players. Malaysian J Sport Sci Recreation 18(2):150–160

Saidi NF et al (2024) Association and predictive ability of ACTN3 C/T, ACE I/D and PPARGC1A A/G polymorphisms on jump height of elite Malaysian footballers. Malaysian J Sport Sci Recreation 20(2):1–8

Slimani M, Nikolaidis PT (2019) Anthropometric and physiological characteristics of male soccer players according to their competitive level, playing position and age group: a systematic review. J Sports Med Phys Fitness 59(1):141–163

Turner AN, Stewart PF (2013) Repeat sprint ability. Strength Conditioning J 35(1):37–41

Turner AN, Stewart PF (2014) Strength and conditioning for soccer players. Strength Conditioning J 36(4):1–13

Ulucan K, Sercan C, Biyikli T (2015) Distribution of angiotensin-1 converting enzyme insertion/deletion and α-actinin-3 codon 577 polymorphisms in Turkish male soccer players. Genetics & epigenetics 7:GEG. S31479

Ulucan K, Sercan C, Biyikli T (2015) Distribution of angiotensin-1 converting enzyme insertion/deletion and α-actinin-3 codon 577 polymorphisms in Turkish male soccer players. Genetics & Epigenetics 7:GEG.S31479

Wong LP et al (2013) Deep whole-genome sequencing of 100 southeast Asian Malays. Am J Hum Genet 92(1):52–66

Match Performance Variables of Thai Football Players in the First Division and the Second Division Teams

Sirawut Pukaror, Arom Treeraj, Sarit Suwanmana, and Weerawat Limroongreungrat

Abstract Understanding running performance during a soccer match is essential so that coaches can use this information to prepare the athlete's physical readiness. Several studies investigated match running performance in many leagues, such as the English Premier League and La Liga League. However, match-running performance studies in the Thai soccer leagues are minimal. Therefore, the current study compared the match-running performances between different Thai soccer league levels. The study recruited thirty-one professional male soccer players from Thai League 1 ($n = 17$) and Thai League 2 ($n = 14$) during the Thai League competition from 2022–2023. Total distance (TD) and high metabolic load distance (HMLD) data were collected using the STATSports Apex Units (GPS) from 8 home matches (4 matches for each level). Statistical analysis revealed no differences between Thai Leagues 1 and 2 in both TD (7.78 ± 0.2 km vs. 7.64 ± 0.43 km, $p = 0.64$) and HMLD (1.33 ± 0.09 km vs. 1.22 ± 0.08 km, $p = 0.157$), respectively. Despite no significant differences observed, these data still provide valuable information for coaches regarding the match-running performances of Thai soccer leagues. Further studies with more match comparisons in different playing positions and other metrics are warranted for a better understanding of the demands of the Thai soccer league.

Keywords Thai league · GPS tracking · Thai football · Football

1 Introduction

Performance analysis has gained popularity in football since it provides many benefits, including performance monitoring, injury prevention and management, and training optimization. Global positioning system (GPS) technology has gained popularity for performance analysis during training and match plays (Tierney et al. 2016). Several metrics from GPS analysis were measured, such as total distance (TD), speed

S. Pukaror · A. Treeraj · S. Suwanmana · W. Limroongreungrat (✉)
College of Sports Science and Technology, Mahidol University, Nakhon Pathom, Thailand
e-mail: weerawat.lim@mahidol.edu

M. H. A. Hassan et al. (eds.), *Proceedings of the 3rd International Conference on Innovation and Technology in Sports*, Lecture Notes in Bioengineering,
https://doi.org/10.1007/978-981-95-0804-4_10

and acceleration, high-intensity run, and high metabolic load distance (HMLD). TD is one of the most commonly reported metrics during analysis. Malo et al. (2015) investigated running TD in the Spanish La Liga friendly match. They found the average of TD is approximately 10.8 km (Mallo et al. 2015). HMLD was another standard parameter, representing the distance a player covers when the metabolic load was greater than 25.5 W·kg-1 (Tierney et al. 2016), corresponding to running at a constant speed of 5.5 m·s-1 or during acceleration or deceleration.

Comparison of players' performance between different leagues or divisions has been previously investigated in European professional football leagues. Salvo et al. (2013) compared physical demands during match play between the English Premier League (EPL) and the Championship League (CL) (Salvo et al. 2013). They reported that the EPL players cover less TD and high-intensity running than CL. Similar results were reported by Bradley and colleagues (Bradley et al. 2013). On the other hand, Thomas et al. did not observe significant differences in HMLD between the first and the second Spanish La Liga players (García-Calvo et al. 2022).

Although GPS tracking systems have been widely used worldwide, the use of GPS systems in Thai professional football leagues is minimal, particularly the football teams in Thai League 2. As a result, there is a lack of information about player performance during match play between different Thai football professional leagues. Understanding match load demands in the Thai professional football league provides more information for coaches to track and monitor their player performance. Thus, this study compared player performance during match play between Thai League 1 (T1) and Thai League 2 (T2). We are specifically interested in 2 performance variables: TD and HMLD. The study hypothesized that the T1 would have greater TD and HMLD than T2.

2 Material and Methods

2.1 Participants

One professional football team in each division, T1 and T2, were selected for this study. Data were collected during the Thai League 2022–2023 season. During the data collection period, the teams from T1 and T2 were ranked in the top five of the table. Seventeen players from each T1 and T2 were recruited for the study. Players who play less than 45 min were excluded from this study. The goalkeepers were also excluded from this study. All players were fully informed of all experimental procedures and provided written informed consent to participate. This study has been approved by the Mahidol University Central Institutional Review Board (MU-CIRB 2023/116.03034).

2.2 *Methods*

The data were collected during four home matches of each team from T1 and T2. During the matches, players wore the GPS pod (STATSports, Apex Units, STATSPORTS Inc., UK). These units consisted of GPS data, accelerometer, gyroscope sensors, and magnetometer, which collected data at the sampling rate of 18 Hz, 10 Hz, 952 Hz, and 10 Hz, respectively. Each participant was assigned a dedicated GPS unit and consistently used the same microsensor device throughout the data collection period to minimize inter-variation between units (Castellano et al. 2011; Buchheit et al. 2013). The GPS units were activated outdoors, following the manufacturer's guidelines, allowing them to acquire satellite signals before the pre-match warm-up, which occurred at least 30 min before kickoff. After the end of the match, data was downloaded using a customized software package and exported in the text file format. Microsoft Excel was later used to process and analyze data.

2.3 *Statistical Analysis*

Descriptive data were reported as mean ± SD. Normal distribution of data controlled by Q-Q plots and tested for normality assumption using the Shapiro–Wilk test. An independent t-test or Mann–Whitney U test was performed to determine differences in player performance variables between two Thai league teams. Statistical significance was set at $p < 0.05$. All statistical analyses were performed using JAMOVI statistical software (Jamovi version 2.2, Sydney, Australia).

3 Results

3.1 *Participant Characteristics*

This study included 31 players, 17 from T1 and 17 from T2. Two players from T2 were excluded during the analysis since their play time was less than 45 min, and one player refused to wear a GPS pod. Only 14 players were left for analysis. Table 1 presents the characteristics of each team's participants.

Table 1 Participant characteristics (Mean ± Standard Deviation)

	T1 (n = 17)	T2 (n = 14)
Age (year)	28.47 ± 4.96	24.64 ± 4.30
Weight (Kg)	78.16 ± 8.13	74.77 ± 7.77
Height (cm)	178.29 ± 8.24	178.36 ± 8.77
BMI (Kg/m^2)	24.61 ± 2.25	23.47 ± 1.41

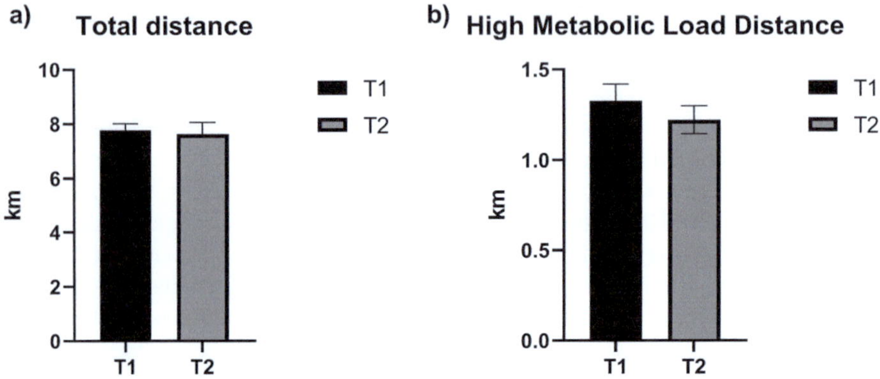

Fig. 1 Match performance (total distance and high metabolic load distance) comparison between Thai league1 team (T1) and Thai league2 team (T2)

3.2 *Performance Variables*

Figure 1 shows match performance variables between two different Thai league teams. The average total distance (TD) covered by T1 and T2 team players was 7.78 ± 0.2 km and 7.64 ± 0.43 km, respectively. There was no significant difference in TD between the two groups ($p = 0.64$). Similarly, the independent t-test revealed no statistically significant difference in high metabolic load distance (HMLD) between the two teams ($p = 0.157$). Nevertheless, the average HMLD of players in T1 (1.33 ± 0.09 km) was slightly higher compared to T2 (1.22 ± 0.08 km). Figure 2 illustrates individual performance in TD and HMLD in both Thai league teams.

4 Discussion

This study examined Thai professional football players' performance variables (TD and HMLD) between two Thai professional leagues. This is the first study that compared performance metrics during four home match plays. We selected the team from the top five of the table during the data collection period since they represented the top of their league. We hypothesized that the players from T1 would have greater TD and HMLD than players from T2. However, the result of this study did not support the hypothesis since no statistically significant differences were found in performance variables between the two groups.

The results of this study contrasted with previous research (Salvo et al. 2013). In a comparison of English football leagues and English championship leagues, different performance data between the two English football leagues were identified. Players in the English championship league cover greater total distances and engage in more high-intensity running than players in higher-tier English leagues. This discrepancy

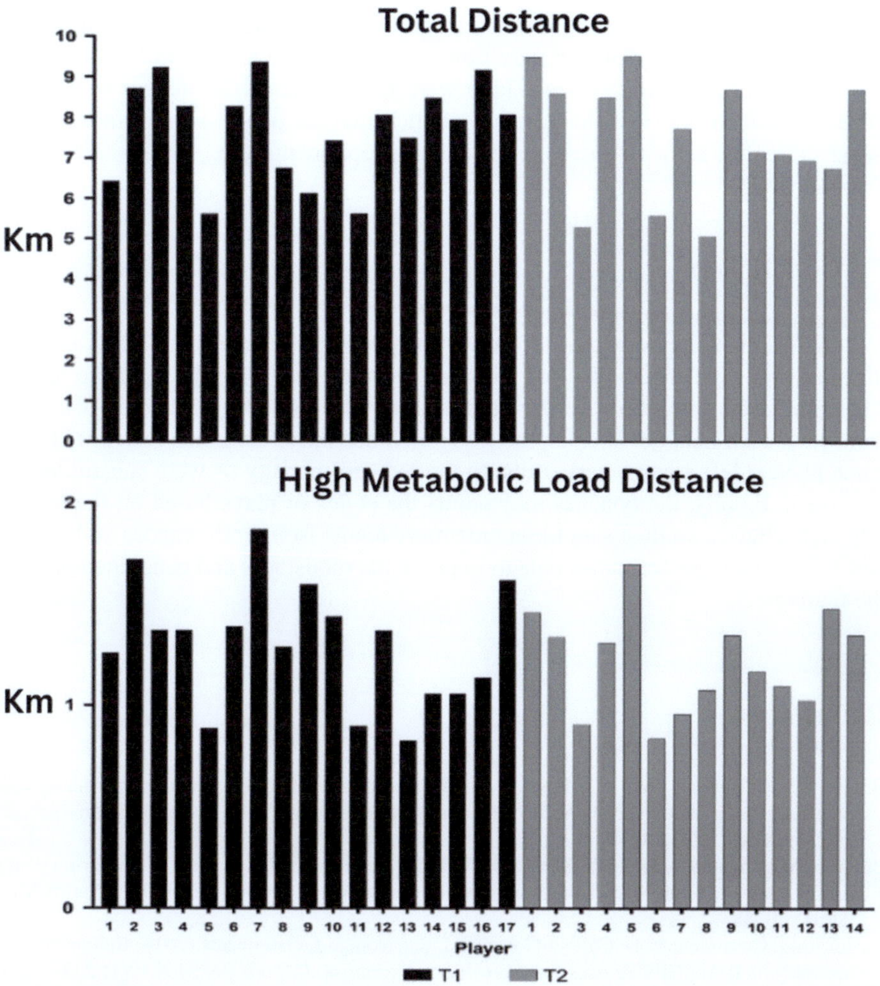

Fig. 2 Comparison of match running performance during player T1 and T2 in Total distance and High metabolic load distance

may be attributed to game intensity, market values, and weather conditions. The English league may be more competitive and intense due to its market values, which lead to variations in formation and style of play that affect movement in each game. Previous studies revealed a difference in the total match distance in CL and English Premier League (EPL) at 11.1 ± 0.9 km and 10.8 ± 1.0 km, respectively (Salvo et al. 2013).

In contrast, Thai league teams T1 and T2 had distances of 7.78 ± 0.2 km and 7.64 ± 0.43 km, respectively. The total distance covered is very similar between the two teams. One of the results may be due to both teams being among the top 5 in their respective leagues, with similar play styles and formations. However, direct comparison between Thai and English leagues is challenging due to differences in hot and humid weather, which may limit players' ability to sustain high-intensity activity and affect performance. There were no significant differences in TD and HMLD between the two Thai Leagues, which may be due to several factors, i.e., game strategies, game intensity, styles of plays, opponents, etc.

5 Conclusion

In conclusion, there were no differences in match performance in terms of total distance and high metabolic load distance between the T1 and T2 league teams. There are some limitations in this study. First, this study only collected one team from each Thai League. Second, the study included only four home matches. Third, some players left a game earlier due to the red card penalty or were substituted by the coach. Finally, the coaches may adjust the styles of plays based on the team's opponent. Future studies should include more teams from both leagues and collect additional match performance data to improve the robustness and generalizability of the findings.

References

Bradley PS, Carling C, Diaz AG, Hood P, Barnes C, Ade J et al (2013) Match performance and physical capacity of players in the top three competitive standards of English professional soccer. Hum Mov Sci 32(4):808–821

Buchheit M, Simpson B, Mendez-Villanueva A (2013) Repeated high-speed activities during youth soccer games in relation to changes in maximal sprinting and aerobic speeds. Int J Sports Med 34(01):40–48

Castellano J, Casamichana D, Calleja-González J, San Román J, Ostojic SM (2011) Reliability and accuracy of 10 Hz GPS devices for short-distance exercise. J Sports Sci Med 10(1):233

Di Salvo V, Pigozzi F, Gonzalez-Haro C, Laughlin M, De Witt J (2013) Match performance comparison in top English soccer leagues. Int J Sports Med 34(06):526–532

García-Calvo T, Ponce-Bordón JC, Pons E, Del Campo RL, Resta R, Raya-González J (2022) High metabolic load distance in professional soccer according to competitive level and playing positions. PeerJ 10:e13318

Mallo J, Mena E, Nevado F, Paredes V (2015) Physical demands of top-class soccer friendly matches in relation to a playing position using global positioning system technology. J Hum Kinet 47(1):179–188

Tierney PJ, Young A, Clarke ND, Duncan MJ (2016) Match play demands of 11 versus 11 professional football using global positioning system tracking: variations across common playing formations. Hum Mov Sci 49:1–8

Biomechanical Effects of Golf Club Length on Address Position Among Collegiate Golfers

Muhamad Safiq Saiful Annur, Hosni Hasan, Shariman Ismadi Ismail, and Muhamad Noor Mohamed

Abstract The address position is a key factor in golf, influencing overall performance and accuracy, and is often considered a main indicator of a successful swing. This study aimed to compare differences when using One-Length (OL) and Normal Length (NL) irons among skilled and less-skilled golfers during address position. Ten golfers (five skilled, five less skilled) each completed 20 swings using four different club setups (5NL, 5OL, 6NL, 6OL). A three-dimensional (3D) motion analysis system, with 8 infrared cameras operated at 240 Hz framerate were used to record the kinematic of address position in different conditions of iron clubs. A mixed-model ANOVA revealed significant differences in stance width, elbow flexion, and wrist positioning between skilled and less-skilled golfers ($p < 0.05$). Skilled golfers consistently had a wider stance, particularly with the 5NL club. Less-skilled golfers showed less lead elbow flexion, while skilled golfers had less trail wrist flexion across all conditions ($p < 0.05$). Both groups maintained consistent lead wrist flexion, regardless of the club used. The values are consistent across all club conditions for both groups. No significant differences were found in the stance-to-ball position across any of the clubs between the groups. These findings highlight the importance of addressing specific elements in the address position to improve golf performance, especially in differentiating between skill levels.

Keywords Golf swing · Address position · One-length iron · Normal length iron

---

M. S. S. Annur (✉) · M. N. Mohamed
Faculty of Sports Science and Recreation, Universiti Teknologi MARA (UiTM) Kampus Seremban, Seremban, Negeri Sembilan, Malaysia
e-mail: msafiq@uitm.edu.my

H. Hasan · S. I. Ismail
Faculty of Sports Science and Recreation, Universiti Teknologi MARA (UiTM) Shah Alam, Shah Alam, Selangor, Malaysia

M. H. A. Hassan et al. (eds.), *Proceedings of the 3rd International Conference on Innovation and Technology in Sports*, Lecture Notes in Bioengineering,
https://doi.org/10.1007/978-981-95-0804-4_11

1 Introduction

Golf is a sport that requires accuracy and detailed swing technique (Huzaifah et al. 2024). Golf swings are highly complex because a golfer must coordinate body movement and the need to control the golf clubs (Yamamoto et al. 2023) for them to be executed skillfully (Lin et al. 2023). However, it is impossible for every golfer to have the 'perfect' swing since it was difficult for a golfer to remain invariant in their golf swing regardless of what club they are using. The main problem golfers face each time during the address position is how to get the clubface back during contact with the ball with the correct sequence of the impact factors in achieving the optimal outcome and ball flight (Langdown 2012).

The address or set-up phase involves gripping the golf club and positioning the body parallel to the ball's intended direction. Most studies in golf swing have centered towards the downswing action which sub divided into four critical events (starting, acceleration, impact and follow through) which the action took only in 300 ms (Fuyu et al. 2019; Zheng et al. 2008). However, the address position in golf is also a critical component of the swing and has been studied extensively to identify its impact on overall performance and accuracy and is considered as the main indicator of success in golf swing (Bradley 2013; Faldo and Simmons 2012) Adjustments must occur during address position, as changes during the downswing are difficult and can negatively impact shot outcomes (Bradley 2013; Faldo and Simmons 2012). The kinematics and kinetics of the address position among highly skilled golfers has been said to improve their swing skills (Kim et al. 2018). If this is the case, it is possible that players who can create consistent shot outcomes would also be constant in their address position (Jones 2019). Smith et al. (2015) emphasized the importance of maintaining a consistent spine angle and foot stance throughout the swing which highlights the critical role of the address position in ensuring proper body alignment and swing mechanics. Additionally, experts have also highlighted the influence of the address position on the entire golf swing, emphasizing its impact on ball contact, clubhead speed, and overall shot accuracy (Zhang and Shan 2014). Another study by Hwang et al. (2021) found that maintaining a consistent shape and position of the feet before swinging is essential for a correct impact shoulders, knees, and waist, is crucial in achieving a correct swing.

However, the research findings on the address variability have been mixed where Bradshaw et al. (2009) found that skilled golfers are more consistent in their address position compared to less skilled golfers, while Langdown et al. (2013) results showed the otherwise. The succession shot rate for skilled golfers was higher compared to non-skilled golfers and that the skilled golfers did have more consistency in foot stance and body position at address which may contribute to success in shot execution (Bradshaw et al. 2009). Meanwhile, Langdown et al. (2013) found that the only significant difference was found in shoulder alignment in relation to stance during the address position. The misalignment of the shoulder would have a negative impact on the swing plane, club head path and angle of approach to the golf ball (Langdown et al. 2013). Jones (2019) studied the address position variability and swing time over five different golfers. The findings from Jones (2019) indicated that skilled golfers have a tendency to produce lower variability in their address position but the differences in variability tend to be small. The author also found that the address position variability has low relationship with the shot outcome and there are indeed no perfect swing for everyone and coaches should focus on the individual's swing outcome, rather than focus on the movement itself.

Therefore, accurately identifying and assessing the body information of the golfer is crucial for analyzing posture and determining the appropriate address position (Hwang et al. 2021. Here we sought to understand whether the use of One-Length iron can induce the consistency of address position. The One-Length iron may not be developed to produce further ball displacement as compared to normal iron clubs, but it promotes more consistent swing and eliminates the need for stance adjustment, ball position adjustments and swing plane adjustments (Schupak 2017). To date, there is only one proceeding paper that has been compared between the one length iron and normal iron regarding the biomechanical analysis of a golf swing. Kashiwagi et al. (2023) found there were no significant differences in carry distances between the normal irons and the custom-made one length irons which the differences are only accounted for about 4.2% only, for which the normal irons are reported to have longer distances. Therefore, the study is needed to provide scientific kinematic analysis evidence based on the effect of using the One-Length iron versus Normal Length iron on the address position.

2 Methodology

2.1 Participants

Five skilled golfers (n = 5; age = 21.6 ± 1.51 years; height = 172 ± 0.06 cm; mass = 69.4 kg ± 11.37 kg; handicap = 4.2 ± 0.84) and five less skilled golfers (n = 5; age = 20 ± 1.41 years; height = 169.4 ± 0.04 cm; mass = 69.8 kg ± 8.34 kg; handicap = 12.8 ± 2.04) were recruited from the National Golf Academy, Universiti Utara Malaysia (UUM), Sintok, Kedah. All participants were recruited after ethical approval has been obtained. The sample was drawn using purposive sampling technique and they have declared that they are free from any diseases or injuries at the point of the testing. The current study employed two different groups, five highly skilled golfers which have handicap of plus four and below while another five less skilled golfers have handicap of eleven and above (McHugh et al. 2024). All participants are right-handed golfers, and they were examined in a controlled environment within a laboratory setting. Their participation in the study was based on a voluntary basis and informed consent was also obtained before the experiment being conducted.

Table 1 Properties of one-length iron and normal length iron

Type	Club label	Club	Loft	Lie angle	Offset	Length	Steel shaft weight	Shaft flex
One-Length	5OL	5	21.5°	64.5°	5.1	37.25″	D0	Stiff
	6OL	6	24.0°	64.0°	4.2	37.25″	D0	Stiff
Normal Length	5NL	5	20.5°	62.5°	4.6	38.5″	D0	Stiff
	6NL	6	23.5°	63.0°	3.9	37.75″	D0	Stiff

2.2 Instrumentation

The golf club used in the study was from Cobra Golf™ for both the normal length and One-Length irons. The models are LTDx Irons and LTDx One-Length iron which both utilize the same club head design with the only difference between these two types of iron is the shaft length. The reason behind this is to avoid internal validity issues, which will be challenging to determine which club property influenced the performance if they used different club head. The utilization of a consistent clubhead throughout testing would yield potentially meaningful data for subsequent analysis (MacKenzie 2017). The properties of both One-Length iron and Normal Length iron are presented in Table 1. The standard length for the Normal Length irons; 5 and 6 irons which are 38.5 inches and 37.75 inches, respectively. While the length for the One-Length iron would be 37.5 inches for both the 5 and 6 irons, respectively. Both the Normal Length iron and One-Length iron were fitted with steel shaft with stiff flexibility as suggested by Melton (2020). The swing weight for the steel shaft was a D0 which indicated a standard swing weight for men's iron golf club. The 5 and 6 irons were selected in the study because the majority of golf clubs used in previous study were drivers (57%), 5 iron (27%), 6 iron (5%), 7 iron (7%) and pitching wedges (3%) (Bourgain et al. 2022). All testing golf irons in the study were for the right-handed golfer only.

2.3 Procedure

The participants underwent 3D motion analysis system utilizing 39 spherical reflective markers with a diameter of 10 mm that was in line with the recommendation of International Society of Biomechanics recommendations (Derrick et al. 2020; Wu and Cavanagh 1995; Wu et al. 2002; Wu et al. 2005) and OptiTrack conventional full body landmarks (OptiTrack 2020). All markers were applied to the participant using hypoallergenic double-sided tape and a headband. An additional 4 markers were placed on clubhead, mid-shaft and superior aspect of the grip. Two reflective marker tape was placed on the golf ball. Markers on the body were placed on the left and right anterior of the head and left and right posterior of the head. Other markers were placed on the sternum jugular notch, xiphoid process, cervical spine process 7 (C7), thoracic spine process 7 (T7), bilateral sides on acromion (shoulders), epicondyle (elbow), upper arms (between the groove of triceps muscle), second metacarpal (knuckle of index finger), ulnar styloid and medial point on radius styloid (wrist), anterior superior iliac spine, lateral side of both legs aligned with the joint center of the knee, both left and right thigh near the midline (asymmetrical) and both left and right shin bone near the midline of lower leg. On the foot area, markers were placed on the ankle region (lateral malleolus) both sides, 2nd metatarsal both left and right foot and both left and right foot at the middle of the calcaneus. Markers were placed directly on the skin of the participants using double sided tape for the upper extremity, hand, and lower leg. Other markers were placed on skin-tight clothing, headband, or shoes. Kinematic data (recorded at 240 Hz) were captured by using eight high speed infrared cameras (Primex 13W, OptiTrack, United States) connected to the Motive software to obtain the 3D marker data. Each participant's static trial posture was taken to record the relative positions of each marker (Carson et al. 2019; Hasan et al. 2017). Visual3D (C-Motion 23, V3D, USA) was employed to construct ten segment model consisting of head, upper arms (left and right), thorax, pelvis, hip, knee (left and right), and foot (left and right). All recorded data was filtered using a fourth order low-pass Butterworth with a cut-off frequency of 6 Hz digital filter (Fuyu et al. 2019; Fedorcik et al. 2012; Kwon et al. 2013).

An adapted dynamic warm up was performed by all participants before the data collection (Park et al. 2021) followed by ten practice swings using 5NL and 6NL, respectively, and continued by 5 shots with the 5OL and 6OL irons. The participants were required to use two normal length irons (NL5 and NL6) and two One-Length irons (OL5 and OL6) that were provided by the researcher. Thereafter, each participant hit 5 shots with each club condition in the testing session, resulting in 20 shots per session, where the order of the club conditions was randomized. They are required to hit a straight ball towards the target as they were practicing or in a competition with maximum effort. Participants were given a short break of 1 min between each shot to minimize the risk of fatigue (Park et al. 2021). The study protocol required participants to maintain consistency in their shot selection throughout the testing period. To minimize confounding variables and prevent the adoption of varied shooting strategies, researchers provided ongoing reminders to participants regarding the importance of executing the same shot type across all attempts (Langdown et al. 2012; Morrison et al. 2016). The grip styles used during the golf swing are not controlled as it is up to the golfers to choose which grip styles suit them comfortably. The data that was measured is at the address position for every shot.

2.4 Data Analysis

The collection of kinematic measurements was initiated at the address position, which was operationally defined as the final frame preceding the club's backward movement from the ball (Kim et al. 2021) and when the velocities of the clubhead is >−0.47 mm·s^{-1} which indicate as the start of the backswing (Langdown et al. 2012). The kinematic variables that were measured during the address position are a) stance width (cm)—defined as the distance between the right and left head of metatarsal markers, b) ball position (cm)—defined as the distance between the left second metatarsal marker a projection of the ball onto the line connecting right and left second metatarsal markers, c) lead and trail elbow flexion—defined as the angle between the forearm and the upper arm for both left and right, and d) lead and trail wrist flexion—defined as the angle between the wrist markers and the hand marker, also both left and right hand. Movement in extension were defined as positive and flexion defined as negative (Carson et al. 2019).

A mixed model ANOVA with one between-participant (less skilled; skilled) and one within-participant factor (5NL; 5OL; 6NL; 6OL) was used to compare the address kinematic variables (using mean value in each club condition). Bonferroni corrections were applied to control the Type I errors and violations of the sphericity assumption for repeated measure variables were checked using Mauchley's test of sphericity. The Bonferroni method post hoc test was used to further analyze the significant main effects and interactions to determine the location of difference between (skilled groups) and within (club conditions) factors. The significant values were set at $p < 0.05$. All data were analyzed using the Statistical Package for Social Science (SPSS V23.0, Chicago, IL, USA).

3 Results

Stance width at address. There was no statistically significant interaction between Group*Club conditions, F (2.46, 117.83) = 0.299, with $p = 0.786$, partial $\eta^2 = 0.006$. However, there was a significant main effect of club conditions on stance width, F (2.46,117.83) = 37.18, with $p < 0.005$, partial $\eta^2 = 0.716$. The main effect of group also showed there was a statistically significant difference in mean stance width between skilled versus less skilled groups, F (1, 48) = 25.70, with $p < 0.005$, partial $\eta^2 = 0.349$. The skilled group showed a significant larger stance width compared to less skilled group ($p < 0.005$). 5NL club conditions have significantly bigger stance width values compared to 5OL, 6NL and 6OL club conditions, for both groups ($p < 0.05$) (Fig. 1 and Table 2).

Stance to ball position. There was no statistically significant interaction between Group*Club conditions, F (2.82, 135.36) = 0.448, with $p = 0.707$, partial $\eta^2 = 0.009$. There were also no significant main effects of club conditions on stance to ball position, F (2.82, 135.36) = 0.437, with $p = 0.715$, partial $\eta^2 = 0.009$ nor the main effect of groups on stance to ball positions, F (1,48) = 0.617, with $p = 0.436$, partial $\eta^2 = 0.013$.

Lead and trail elbow flexion. For lead elbow flexion, there was no significant interaction between Group*Club conditions, F (2.258, 108.379) = 0.219, with $p = 0.829$,

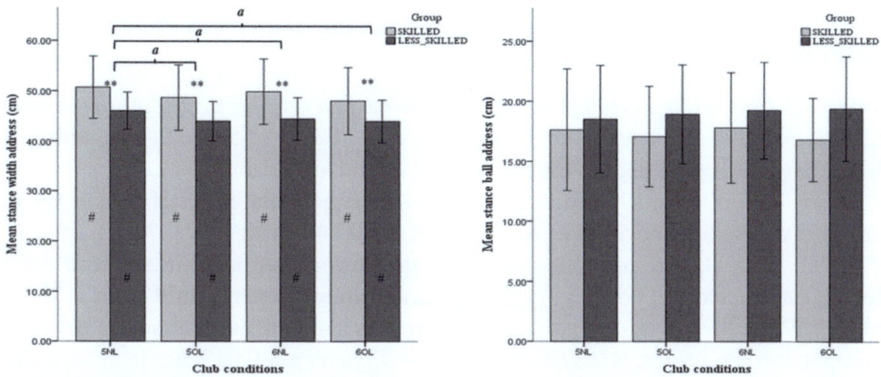

Fig. 1 Mean (SD) stance width and stance to ball during address position for skilled and less-skilled golfers under four club conditions. Note: *a* denotes a statistically significant difference between club conditions, # denotes a significant difference between club conditions (by group), ** denotes a significant difference between skilled and less-skilled golfers

Table 2 Means and standard deviations of kinematic variables

Conditions	Group	Lead Elbow (°)	Trail Elbow (°)	Lead Wrist (°)	Trail Wrist (°)
5NL	Skilled	18.6 ± 5.6	21.4 ± 9.2*	−9.6 ± 6.2	−2.2 ± 1.6 [b]
	Less skilled	14.6 ± 8.6	24.3 ± 3.9	−11.8 ± 15.4	−12.3 ± 5.9
5OL	Skilled	18.2 ± 4.7	19.9 ± 9.2 *	−12.6 ± 5.8	−2.2 ± 1.7 [b]
	Less skilled	14.2 ± 8.6	24.2 ± 3.9	−11.1 ± 16.4	−9.2 ± 6.8
6NL	Skilled	19.1 ± 5.2 [a]	20.2 ± 9.4 *	−12.5 ± 6.4	−3.1 ± 1.8 [b]
	Less skilled	14.5 ± 8.7	23.6 ± 4.6	−10 ± 16.5	−9.6 ± 7.3
6OL	Skilled	18.7 ± 5.1 [a]	19.7 ± 9.2 *	−11.1 ± 8.1	−3.1 ± 2.1 [b]
	Less skilled	14.6 ± 8.3	24.2 ± 3.3	−8.4 ± 12.4	−8.7 ± 7.3

[a] indicates a significant difference in club conditions (6NL and 6OL, between groups). [b] indicates a significant difference in club conditions (5NL, 5OL, 6NL and 6OL, between groups). * indicates a significant difference between 5NL versus 5OL, 5NL versus 6NL and 5NL versus 6OL (in skilled golfers)

partial $\eta^2 = 0.005$. For the main effect of club conditions, there was also no significant effect on the lead elbow flexion values, F (2.258, 108.379) = 0.555, with $p = 0.597$ partial $\eta^2 = 0.011$. However, there are significant main effects of group on the lead elbow flexion, F (1, 48) = 4.675, with $p = 0.036$, partial $\eta^2 = 0.089$. The skilled group have significantly higher lead elbow flexion when they are using the 6NL ($p = 0.028$) and 6OL ($p = 0.044$) club conditions compared to the less skilled group. For trail elbow flexion, there was no significant interaction between Group*Club conditions, F (1, 48) = 3.959, with $p = 0.052$, partial $\eta^2 = 0.076$, nor groups on the trail elbow flexion, F (1, 48) = 3.668, with $p = 0.061$ partial $\eta^2 = 0.071$. For the main effect of club conditions on the trail elbow flexion, there was a significant main effect, F (1, 48) = 5.464, with $p = 0.024$ partial $\eta^2 = 0.102$. 5NL club conditions have significant differences on the trail elbow flexion values compared to 6NL (p = 0.018) and 6OL (p = 0.005) club conditions.

Lead and trail wrist flexion. For the lead wrist flexion, there was no significant interaction between Group*Club conditions, F (2.493, 119.659) = 2.417, with p = 0.109, partial $\eta^2 = 0.043$, nor the main effect of club conditions, F (2.493, 119.659) = 1.175, with p = 0.319, partial $\eta^2 = 0.024$. For the main effect of groups, there was also no significant main effect, F (1,48) = 0.140, with p = 0.710, partial η^2 = 0.003. Both skilled and less skilled group showed consistent in the lead wrist flexion during address position, with the mean difference range only from 1.51° to 2.809°. While for the trail wrist flexion, there was a significant interaction between Group*Club conditions, F (1.698, 81.492) = 4.93, with p = 0.013, partial $\eta^2 = 0.093$ and the main effect of groups, F (1,48) = 0.6.529, with $p < 0.005$, partial $\eta^2 = 0.432$. There was no significant main effect of club conditions, F (1.698, 81.492) = 2.41, with p = 0.104, partial $\eta^2 = 0.048$. In group comparisons, less skilled group have higher flexion for their trail wrist compared to skilled group ($p < 0.005$) in all club conditions. The skilled group showed very consistent trail lead wrist flexion with the mean difference range only from 0.016° to 0.926° for all club conditions.

4 Discussion

The present study investigated the differences in golf address between skilled and less skilled golfers across various club conditions. Our findings reveal several important insights into the biomechanics of golf address position and how expertise level influences these parameters.

Stance Width and Ball Position: One of the most notable findings was the significant main effect of club conditions on stance width. This suggests that players adjust their stance based on the club they are using, which is consistent with the biomechanical demands of different club lengths (You et al. 2023). Notably, the 5NL club condition resulted in significantly wider stances compared to other club conditions for both skill groups. This wider stance may be attributed to the need for increased stability when using longer clubs (Olizarowicz 2024). The significant difference in stance width between skilled and less skilled is particularly intriguing. Skilled golfers consistently demonstrated wider stances, which could be indicative of a more stable foundation for the swing. This finding supports the work of Bradshaw et al. (2009) who proposed that expert golfers prioritize stability in their setup to enhance swing consistency. However, the effect of club conditions on stance width was consistent across different skill levels, as evidenced by the lack of a significant interaction effect. Interestingly, stance width was also not affected by the 5OL and 6OL club conditions as both these clubs presented the same length. This supports the notion that the use of One-Length iron eliminates the need for stance adjustment (Schupak 2017). Contrary to stance width, we found no significant differences in ball position relative to stance between groups or across club conditions. This uniformity suggests that ball positioning might be a more standardized aspect of golf setup, possibly due to widespread coaching practices and the variation of ball positioning as small as 1% of stance length can lead to unsuccessful golf swing (Zhang and Shan 2014). In this study, both skilled and less skilled golfers positioned the ball further forward towards the lead foot.

Lead and Trail Elbow Flexion: The analysis of elbow flexion revealed some noteworthy patterns. While there was no significant interaction or main effect of club conditions on lead elbow flexion, we observed a significant main effect of group. Skilled golfers exhibited greater lead elbow flexion, particularly with the 6NL and 6OL club conditions. This increased flexion among skilled golfers may contribute to better control and consistency in the backswing. Interestingly, this finding contrasts with the study by Zheng et al. (2008), which found no significant difference in lead elbow flexion at address across different playing abilities. This discrepancy may be attributed to the use of different clubs in the studies; Zheng et al. (2008) used a driver, whereas the current study employed a mid-iron, suggesting that club type influences motor strategies. Both groups demonstrated consistent lead elbow flexion across all club conditions, indicating that even less skilled golfers may have developed a stable motor pattern for this aspect of their swing. For trail elbow flexion, a significant main effect of club conditions was found, with increased trail elbow flexion observed in both groups when using the 5NL club. This suggests that players adjust their arm positioning based on club length, an indication of adaptability in motor control. However, the significant difference in trail elbow flexion was only evident in the skilled group across all club conditions. This may be explained by a learning effect in the less skilled participants may take place, as proposed by Kenny et al. (2008), where they maintained consistent trail elbow flexion regardless of the club used, potentially reflecting an early stage of motor learning where the focus is on maintaining a consistent movement pattern. Although the elbows are often studied, only Zheng et al. (2008) and Egret et al. (2006) highlighted the kinematic differences between skilled and less skilled players (Bourgain et al. 2022).

Lead and Trail Wrist Flexion: The findings regarding wrist flexion were particularly interesting. While lead wrist flexion showed remarkable consistency across groups and conditions, trail wrist flexion demonstrated a significant interaction between group and club conditions $F\ (1.698, 81.492) = 4.93$, $p = 0.013$, partial $\eta2 = 0.093$). Both skilled and less skilled groups showed consistency in lead wrist flexion during the address position, with mean differences ranging only from 1.51° to 2.809°. This supports Zheng et al. (2008) findings where they did not find any significant difference in both lead and trail wrist flexion among Professional, low, mid and high handicap players. Current work, O'Brien et al. (2021) and Zheng et al. (2008) findings were contrary to the findings by Carson et al. (2019) where the lead wrist was extended at address position using three different grip style (ranging from 33° to 22.2°). A possible explanation is that Carson et al. (2019) used a driver in his study, which the ball position is near to the lead foot thus affecting the wrist position at address. The wrist data consistency across skill levels and club conditions suggests that lead wrist positioning might be a more standardized aspect of the golf setup, possibly due to common instructional practices or biomechanical requirements of the golf swing. The higher trail wrist flexion observed in less skilled golfers (9.9° vs 2.6°, less skilled vs skilled, respectively) could indicate a less optimal wrist position at address, potentially leading to inconsistencies in club face control. The data also shows that only less skilled golfers have significant difference in trail wrist flexion between 5NL versus 5OL ($p = 0.009$) and 5NL versus 6OL ($p = 0.029$). The consistency in trail wrist flexion among skilled golfers across all club conditions (mean difference range only from 0.016° to 0.926°) further supports the notion that maintaining a consistent wrist position may be a characteristic of expertise in golf.

5 Conclusion

This study provides valuable insights into the biomechanical adjustments made by golfers of varying skill levels when using different golf club types. The findings reveal significant differences in stance width, elbow flexion, and wrist positioning between skilled and less skilled golfers, as well as adaptations made in response to different club conditions. Overall, neither Normal Length nor One Length irons can be definitively declared "best" for all golfers. One Length irons seem to offer advantages in terms of setup consistency, which could benefit some players, especially those with higher skill levels. One Length irons may provide more consistency in the stance setup and consistent wrist positioning across other club conditions during address position. However, Normal Length irons allow for more customization in setup, which some players might prefer. The choice between One Length and Normal Length irons would likely depend on individual player preferences, skill level, and specific performance goals. Further research directly comparing performance outcomes between these two types of clubs would be necessary to make a more definitive judgment on which is "best" for overall golf performance. Future recommendations also to include a dynamic swing analysis consisting of the entire golf swing, not only the address position.

Acknowledgements The authors would like to thank Universiti Teknologi MARA for funding under the Geran Penyelidikan MyRA (RMC/RD/70/015/2022) and National Golf Academy (NGA), Universiti Utara Malaysia (UUM) for providing the participants in the study.

References

Bourgain M, Rouch P, Rouillon O, Thoreux P, Sauret C (2022) Golf swing biomechanics: a systematic review and methodological recommendations for kinematics. Sports (Basel, Switzerland) 10(6):91. https://doi.org/10.3390/sports10060091

Bradley N (2013) The 7 laws of the golf swing. Harry N. Abrams; Revised ed. edition (April 2, 2013)

Bradshaw EJ, Keogh JW, Hume PA, Maulder PS, Nortje J, Marnewick M (2009) The effect of biological movement variability on the performance of the golf swing in high- and low-handicapped players. Res Q Exerc Sport 80(2):185–196. https://doi.org/10.1080/02701367.2009.10599552

Carson HJ, Richards J, Mazuquin B (2019) Examining the influence of grip type on wrist and club head kinematics during the golf swing: benefits of a local co-ordinate system. Eur J Sport Sci 19(3):327–335. https://doi.org/10.1080/17461391.2018.1508504

Derrick TR, van den Bogert AJ, Cereatti A, Dumas R, Fantozzi S, Leardini A (2020) ISB recommendations on the reporting of intersegmental forces and moments during human motion analysis. J Biomech 99:109533. https://doi.org/10.1016/j.jbiomech.2019.109533

Egret CI, Nicolle B, Dujardin FH, Weber J, Chollet D (2006) Kinematic analysis of the golf swing in men and women experienced golfers. Int J Sports Med 27(6):463–467. https://doi.org/10.1055/s-2005-865818

Faldo N, Simmons R (2012) A swing for life: revised and updated (First Atria Books hardcover edition.). Atria Books

Fedorcik GG, Queen RM, Abbey AN, Moorman CT 3rd, Ruch DS (2012) Differences in wrist mechanics during the golf swing based on golf handicap. J Sci Med Sport 15(3):250–254. https://doi.org/10.1016/j.jsams.2011.10.006

Fuyu M, Lee HM, Chen PY, Chou LW, Wei SH (2019) A biomechanical approach to investigate swing characteristics in elite golfers. J Chin Med Assoc JCMA 82(7):589–594. https://doi.org/10.1097/JCMA.0000000000000121

Hasan H, Davids K, Chow JY, Kerr G (2017) Changes in organisation of instep kicking as a function of wearing compression and textured materials. Eur J Sport Sci 17(3):294–302. https://doi.org/10.1080/17461391.2016.1241829

Huzaifah A, Omar MN, Johari NH, Hassan MHA, Ishak M, Aziz Z (2024) Development of motion tracking device to determine swing profile of a golf player. In: Mohamed Z, Ngali MZ, Sudin S, Ibrahim MF, Casson A (eds) Proceedings of the 2nd international conference on innovation and technology in sports, ICITS 2023, 27–28 November, Malaysia. ICITS 2023. Lecture Notes in Bioengineering. Springer, Singapore. https://doi.org/10.1007/978-981-97-3741-3_17

Hwang S, Ko K-R, Pan SB (2021) Motion data acquisition method for motion analysis in golf. Concurrency Computat Pract Exper 33:e5215. https://doi.org/10.1002/cpe.5215

Jones K (2019) Movement and outcome variability of the golf swing of amateur golfers with driver and iron clubs Ulster Univeristy

Kashiwagi Y, Hirano T, Wada T, Shimatani S, Funato K (2023) Biomechanical analysis of the golf swing in one-length and conventional irons. XXIX Conference of the international society of biomechanics (ISB) and the XXIX Conference of the Japanese Society of Biomechanics (JSB). Fukuoka University

Kenny IC, Wallace ES, Otto SR (2008) Influence of shaft length on golf driving performance. Sports Biomech 7(3):322–332. https://doi.org/10.1080/14763140802233249

Kim SE, Koh YC, Cho JH, Lee SY, Lee HD, Lee SC (2018) Biomechanical effects of ball position on address position variables of elite golfers. J Sports Sci Med 17(4):589–598

Kim SE, Lee J, Lee SY, Lee HD, Shim JK, Lee SC (2021) Small changes in ball position at address cause a chain effect in golf swing. Sci Rep 11(1):2694. https://doi.org/10.1038/s41598-020-79091-7

Kwon YH, Han KH, Como C, Lee S, Singhal K (2013) Validity of the X-factor computation methods and relationship between the X-factor parameters and clubhead velocity in skilled golfers. Sports Biomechanics 12(3):231–246. https://doi.org/10.1080/14763141.2013.771896

Langdown BL, Bridge M, Li FX (2012) Movement variability in the golf swing. Sports Biomech 11(2):273–287. https://doi.org/10.1080/14763141.2011.650187

Langdown BL, Bridge MW, Li FX (2013) Address position variability in golfers of differing skill level. Int J Golf Sci 2:1–9

Lin ZJ, Peng YC, Yang CJ, Hsu CY, Hamill J, Tang WT (2023) Lower limb biomechanics during the golf downswing in individuals with and without a history of knee joint injury. Bioengineering (Basel, Switzerland) 10(5):626. https://doi.org/10.3390/bioengineering10050626

MacKenzie S (2017) How the shaft of a golf club influences performance. In: Toms M, MacKenzie S, Robertson S, Lochbaum M, Kingston KFA (Eds), Routledge international handbook of golf science, pp 64–74. Routledge

McHugh MP, O'Mahoney CA, Orishimo KF, Kremenic IJ, Nicholas SJ (2024) Kinematic, kinetic, and temporal metrics associated with golf proficiency. J Strength Cond Res 38(3):599–606. https://doi.org/10.1519/JSC.0000000000004663

Melton Z (2020) Here's the shaft flex you should play based on your swing speed. Retrieved September 8, 2022 from https://golf.com/instruction/shaft-flex-you-should-play-based-on-swing-speed/

Morrison A, McGrath D, Wallace ES (2016) Motor abundance and control structure in the golf swing. Hum Mov Sci 46:129–147. https://doi.org/10.1016/j.humov.2016.01.009

O'Brien B, Juhas B, Bieńkiewicz M, Buloup F, Bringoux L, Bourdin C (2021) Sonification of golf putting gesture reduces swing movement variability in novices. Res Q Exerc Sport 92(3):301–310. https://doi.org/10.1080/02701367.2020.1726859

Olizarowicz B (2024) 6 Strategies to increase distance with your irons. Retrieved 28 August 2024 from https://mygolfspy.com/news-opinion/6-strategies-to-increase-distance-with-your-irons/

OptiTrack (2020) PrimeX series: quick start guide. Retrieved 5th October 2022 from https://d111sr qycjesc9.cloudfront.net/PrimeX%20Series%20Quick%20Start%20Guide.pdf

Park D, Kwon I, Yoon JH (2021) The effects of different types of warm-up exercises on golf performance. J Mens Health 17(3):132–138. https://doi.org/10.31083/jomh.2021.036

Schupak A (2017) Golf: golf pro aims to revolutionize an industry with one-size-fits-all irons. Retrieved February 1st 2022, from https://www.nytimes.com/2017/01/28/sports/golf/single-len gth-irons-cobra.html

Smith A, Roberts J, Wallace E, Kong PW, Forrester S (2015) Golf coaches' perceptions of key technical swing parameters compared to biomechanical literature. Int J Sports Sci Coach 10(4):739–755. https://doi.org/10.1260/1747-9541.10.4.739

Wu G, Siegler S, Allard P, Kirtley C, Leardini A, Rosenbaum D, Whittle M, D'Lima DD, Cristofolini L, Witte H, Schmid O, Stokes I, Standardization and Terminology Committee of the International Society of Biomechanics (2002). ISB recommendation on definitions of joint coordinate system of various joints for the reporting of human joint motion--part I: ankle, hip, and spine. International society of biomechanics. J Biomechanics 35(4):543–548. https://doi.org/10.1016/s0021-9290(01)00222-6

Wu G, van der Helm FC, Veeger HE, Makhsous M, Van Roy P, Anglin C, Nagels J, Karduna AR, McQuade K, Wang X, Werner FW, Buchholz B, International Society of Biomechanics (2005) ISB recommendation on definitions of joint coordinate systems of various joints for the reporting of human joint motion—Part II: shoulder, elbow, wrist and hand. J Biomechanics 38(5):981–992.https://doi.org/10.1016/j.jbiomech.2004.05.042

Yamamoto K, Hasegawa Y, Suzuki T, Suzuki H, Tanabe H, Fujii K (2023) Extracting proficiency differences and individual characteristics in golfers' swing using single-video markerless motion analysis. Frontiers Sports Active Living 5:1272038. https://doi.org/10.3389/fspor.2023.1272038

You X, Xu Y, Liang M, Baker JS, Gu Y (2023) The relationship between ground reaction forces, foot positions and type of clubs used in golf: a systematic review and meta-analysis. Appl Sci 13(12):7209. https://www.mdpi.com/2076-3417/13/12/7209

Zhang X, Shan G (2014) Where do golf driver swings go wrong? Factors influencing driver swing consistency. Scand J Med Sci Sports 24(5):749–757. https://doi.org/10.1111/sms.12061

Zheng N, Barrentine SW, Fleisig GS, Andrews JR (2008) Kinematic analysis of swing in pro and amateur golfers. Int J Sports Med 29(6):487–493. https://doi.org/10.1055/s-2007-989229

Optical Tracking System Review: Results of Validity and Reliability Studies on Prescribed Circuits in Football

Khairul Imran Sainan, Ahmad Khushairy Makhtar, and Zulkifli Mohamed

Abstract Optical tracking systems (OTS) offer significant advantages over wearable-based tracking devices. By recording player movements as videos, OTS provide a non-contact method for collecting data, including the movements of opposing teams. This paper presents a comprehensive review of the results of validity and reliability studies conducted using OTS on prescribed circuits in football. A total of eight relevant articles were identified and analyzed, covering a wide range of prescribed circuits and various activities. The results consistently demonstrate high levels of validity and reliability in measuring football-specific movements, making them an invaluable tool for coaches and researchers to enhance performance analysis, training loads, and injury prevention in football.

Keywords OTS · Validity and reliability · Review on results

1 Introduction

Professional football has increasingly relied on technology for elite-level preparation, performance, and recovery. Optical tracking systems (OTS) have emerged as one of invaluable technology for analyzing player movements and training loads (Robertson et al. 2023). Therefore, the accuracy and reliability of OTS data are crucial for making informed decisions regarding player development and team performance. The system typically consists of cameras and computers. The cameras capture video footage of player movements, while the computers process the data to extract information such as position, velocity, and acceleration.

K. I. Sainan (✉) · A. K. Makhtar · Z. Mohamed
Faculty of Mechanical Engineering, Universiti Teknologi MARA, 40450 Shah Alam, Selangor, Malaysia
e-mail: imransainan@uitm.edu.my

K. I. Sainan · Z. Mohamed
Sports Engineering Artificial Intelligence Centre, Universiti Teknologi MARA, 40450 Shah Alam, Selangor, Malaysia

M. H. A. Hassan et al. (eds.), *Proceedings of the 3rd International Conference on Innovation and Technology in Sports*, Lecture Notes in Bioengineering,
https://doi.org/10.1007/978-981-95-0804-4_12

Using the video data, the OTS's tracker detects, tracks, and calculates the local position of the footballers. In general, these processes involve identifying the players within the video footage and then following their movements over time. The process commonly begins with image preprocessing to enhance image quality and extract relevant features (Wu et al. 2022). Feature-based methods detect players based on their color, shape, and motion characteristics, while deep learning models, such as convolutional neural networks, are trained to directly recognize players as region of interest (ROI) from image data. Tracking maintains the identity of the players over time as it moves within a video sequence. Correlation-based methods compare the player's appearance in subsequent frames to maintain their identity, while Kalman filtering (Schlipsing et al. 2017) uses a predictive model and measurement updates to estimate the player's position and motion over time. Data obtained is filtered for camera distortion or occlusions and calibrated (Manafifard 2024) to account for the camera's perspective to ensure accurate localizations (Karungaru et al. 2022; Imran Sainan et al. 2018). The localized x and y data on each frame over time provide position, speed, acceleration, and trajectories of the players.

OTS offer several practical advantages over other tracking systems. First, OTS use match recordings as the source of data, allowing coaches to retrace results and analyze specific plays. Second, OTS are often accompanied by powerful video analysis features that enable coaches to identify tactical errors, analyze player positioning, and evaluate team strategies. Third, OTS are non-contact systems, eliminating the need for players to wear uncomfortable vests or belts, which can be a significant advantage for players who prefer a less restrictive feel. Unlike GPS (Guitart et al. 2022; Brosnan et al. 2022) and LPS (Pino-Ortega et al. 2021; Hoppe et al. 2018), which require wearable devices, OTS can collect data on both the player's team and the opposing team, providing a more comprehensive understanding of the match.

While OTS offer several advantages over wearable-based tracking devices, they are not without limitations. OTS face limitations due to low source data quality, detection and tracking errors, and inaccurate localization. Low source data quality can result from abrupt motion and zoom, calibration inaccuracy, image noise, inconsistent background and environmental conditions, lack of pixel resolution for small distant players, clutter, and motion blur. This, in turn, increases detection and tracking failures, especially when images have similar player appearances, complex interactions, inconsistent playfield detection, or severe occlusions (Sani 2023; Sabirin et al. 2015). Localization (Maglo et al. 2023) accuracy depends on image-world and real-world calibration, precise position estimation, and tracking precision.

Therefore, there are several OTS studies available focusing to address these limitations. Researchers have delved to improve detection (Sani 2023) and tracking (Herrmann et al. 2014; Martín and Martínez 2014) by particle swarm optimization (Manafifard et al. 2017; Manaafard et al. 2015), model field particles application (Baysal and Duygulu 2016), adaptive pattern recognition (Schlipsing et al. 2017), and automatic detection of events (Vidal-Codina et al. 2022). Others have used OTS to analyze player performance (Salvo et al. 2007; García-Aliaga et al. 2021), workload (Salvo et al. 2009; Castellano et al. 2014) and worst-case scenarios analysis (Novak et al. 2021), tactical analysis (Ponce-Bordón et al. 2021; Lorenzo-Martinez et al. 2020), and injury risk. Additionally, studies have assessed OTS to other tracking systems by assessing agreement with GPS in official match such as in Polish Ekstraklasa professional league (Makar et al. 2023), English Premier League (Ellens et al. 2022; Taberner et al. 2020) and Spanish professional clubs (Pons et al. 2021; Felipe et al. 2019). There are also studies conducted using broadcast cameras (Wei-Lwun et al. 2013; Maćkowiak 2013). OTS also have been used for other sports such as handball (Fleureau et al. 2020). Advancements in processing methods and tracking algorithms have significantly improved the accuracy and reliability of OTS data, making them an invaluable tool for a wide range of research and practical applications.

This paper aims to review and analyze published validity and reliability studies of optical tracking systems (OTS) in football, focusing on those that employed standardized test protocols on prescribed circuits. We prioritize studies from commercially available and verified OTS, using rigorous test protocols, and have published their data. A valid test protocol should include at least a test subject, a predefined circuit, and a criterion for comparison. The subject performs a series of prescribed movements on the circuit, and the measured OTS data is compared with the criterion values. Our review will specifically discuss the validity and reliability results. We exclude studies that relied on actual match data such as (Makar et al. 2023; Ellens et al. 2022) and those that used general video analysis software (Beato et al. 2016). To identify relevant studies, we conducted a comprehensive literature search using two online databases: Web of Science (WOS) and Scopus. These databases offer a vast collection of high-quality academic manuscripts. Based on our search criteria, we found and analyzed eight articles. This paper is organized as follows: Sect. 2 provides an overview of the published validity and reliability OTS studies. Sections 3 and 4 discuss the results of these studies, focusing on validity and reliability, respectively. Section 5 reviews the scales of measurement used, and Sect. 6 concludes with a summary and outlines future research directions.

2 OTS Studies Published to Date

Castellano et al. (Castellano et al. 2014) noted in their review that studies incorporating validity and reliability assessments represent high-quality research especially on prescribed circuits. However, only eight such studies were identified among OTS related publications, highlighting the scarcity of this type of research. Surprisingly, given the critical importance of system accuracy in engineering, such studies remain limited. Table 1 summarizes the validity and reliability of published OTS studies to date. We reviewed the work of seven researchers from eight publications, six of which are from sports related journal articles and two from PLOS ONE. The earliest study was published in 2006 by Valter et al. (Valter et al. 2006) and the most recent in 2022 by Aughey et al. (Aughey et al. 2022). Only three of the eight publications included reliability testing. In total, five OTS systems were tested, with Prozone®,

Table 1 Validity and reliability of OTS studies published to date

Author	Validity	Reliability	Journal	References
R. J. Aughey	Yes	No	Sports Engineering	Aughey et al. (2022)
D. Linke	Yes	No	PLOS ONE	Linke et al. (2020)
D. Linke	Yes	No	PLOS ONE	Linke et al. (2018)
J. Mara	Yes	Yes	Int. J of Comp. Science in Sports	Mara et al. (2017)
M. Buchheit	Yes	Yes	J of Sports Sciences	Buchheit et al. (2014)
M. Siegle	Yes	No	J of Sports Sciences	Siegle et al. (2013)
A. Redwood-Brown	Yes	No	Int. J of Sports Medicine	Redwood-Brown et al. (2012)
D. S. Valter	Yes	Yes	Int. J of Perf. Analysis in Sports	Valter et al. (2006)

SportVU and VisionKIT being studied twice, while TRACAB and Venatrack™ once each.

These studies were used to achieve several objectives, including validating OTS against motion capture (Aughey et al. 2022), comparing different versions of OTS (Linke et al. 2020), validating OTS against other EPTS (Linke et al. 2018), evaluating a new OTS (Mara et al. 2017), analyzing automated performance (Redwood-Brown et al. 2012), providing calibration equations for data interchangeability (Buchheit et al. 2014), analyzing performance at different speeds (Valter et al. 2006), and assessing positional accuracy (Scott et al. 2016). Typically, OTS data were compared against other tracking systems, such as GPS, LPS, timing gates, stopwatches, motion capture, laser devices, true data, or even other OTS. While GPS,

LPS, OTS, and motion capture can provide a range of data, including minimum, peak and instantaneous values, timing gates and stopwatches are limited to mean data. Motion capture systems (Bampouras and Thomas 2022) are limited to small area therefore usually not feasible for full sized pitch (Aughey et al. 2022). True data is determined using measuring tape, trundle wheel, theodolite, or etc. to measure position and distance. Several kinematic variables used for data validity and reliability assessments, with position, distance, speed, and acceleration being the most common. Interestingly, three publications relied solely on a single parameter for their analysis.

3 Results of Validity

Validity is crucial for any system, as it ensures accurate measurements that represent the actual kinematics of player activities on the pitch. It is a measure of accuracy, often expressed in terms of difference or error, correlation, and range of deviation or confidence. Table 2 describes the validity measures adopted by OTS studies published to date. The difference or error was quantified using metrics such as root mean square deviation (RMSD) (Aughey et al. 2022), root mean square error (RMSE) (Linke et al. 2020, 2018; Siegle et al. 2013), mean difference (Aughey et al. 2022; Linke et al. 2020; Mara et al. 2017; Redwood-Brown et al. 2012), percentage of difference (Mara et al. 2017), typical error (TEE) (Buchheit et al. 2014), and mean bias (Buchheit et al. 2014). Correlation between systems was measured using the Pearson correlation coefficient (r) (Redwood-Brown et al. 2012; Buchheit et al. 2014), and coefficient of determination, R-squared (Valter et al. 2006). Uncertainties were often presented as standard deviations (SD) (Linke et al. 2020, 2018; Mara et al. 2017), confidence intervals (CI) (Redwood-Brown et al. 2012; Buchheit et al. 2014), confidence limits (CL) (Buchheit et al. 2014), percentiles (PCTL) (Linke et al. 2020; Siegle et al. 2013), and coefficient of variations (CV) (Mara et al. 2017).

Table 2 Validity measures of OTS studies published to date

References	Parameter	Error measurement	Test
Aughey et al. (2022)	Pos; Vel	Mean ± SD; RMSD	–
Linke et al. (2020)	Pos; Vel; Acc	RMSE; Mean; 5th and 95th PCTL	T-test; p-value; Cohen's d effect sizes (ES)
	TD; DV; PS; PA; Dcn	Mean ± SD	% Diff.; p-value; Cohen's d effect sizes (ES)
Linke et al. (2018)	D; Vel; Acc	RMSE: Median ± SD	ANOVA: p-value; Cohen's d effect sizes (ES); Sig. Diff
	DV; PS	RMSE; RMSE%	ANOVA: p-value; Cohen's d effect sizes (ES)
Mara et al. (2017)	DV	Mean ± SD; Range; % Diff.; SD as CV%	–
Buchheit et al. (2014)	TS; PS; PA	Mean bias (90% CL); TEE (90% CL); r (90% CL)	–
	DV	Standardized bias (90% CI)	–
Siegle et al. (2013)	Pos	Mean RMSE; Max; 95th PCTL	Three-factor ANOVA; T-test
Redwood-Brown et al. (2012)	MS	Mean Diff. (LOA); r (95% CI)	–
Valter et al. (2006)	MS	R-squared	–

* Pos = Position; Vel = Velocity; Acc = Acceleration; TD = Total distance; DV = Distance velocity; PS = Peak speed; PA = Peak acceleration; Dcn = Deceleration; D = Distance; TS = Time spent; MS = Mean Speed; SD = Standard deviation; RMSD = Root mean square deviation; RMSE = Root mean square error; PCTL = Percentile; % Diff = Percent Difference; CV = Coefficient of variation; CL = Confidence limit; TEE = Typical error of the estimate

3.1 Location of Runs

OTS accuracy has been found to have significantly influenced by the spatial location. Regarding distance relative to camera position, lower validity is observed when the circuit is farther away than the camera. This is also true for vertical direction circuit in comparison to horizontal circuit (Siegle et al. 2013). Similar findings were reported by Linke et al. (2018), who observed larger errors in position validity. Additionally, OTS have been shown to overestimate maximal speed during vertical shuttle runs. These findings highlight the importance of considering both the spatial location of players and the direction of their movements when using OTS. By understanding these factors, researchers and practitioners can optimize camera placement and data analysis techniques to improve the accuracy and reliability of OTS measurements. Also, LPS have been found to be more valid than OTS in X, Y-axes measurement and was not influenced by the locations of runs. LPS also has more spatial accuracy than the OTS (Siegle et al. 2013) in which in Linke et al. (2018), LPS has reported smallest errors in spatial accuracy followed by OTS and GPS respectively. Also, RMSE data indicated LPS has the highest validity followed by GPS and OTS in which GPS and OTS showed comparable results in terms of speed accuracy but deviated when measuring distance velocity.

To improve OTS accuracy, it is recommended to strategically place cameras to minimize the distance between players and the cameras. This can be achieved by using multiple cameras with overlapping fields of view. Additionally, calibrating the cameras accurately is essential, especially when players are moving in the vertical direction. Regular calibration and maintenance of the OTS system can help mitigate these issues. Furthermore, considering the limitations of OTS in measuring vertical movements, it is important to supplement OTS data with other tracking systems, such as wearable devices, to obtain a more comprehensive understanding of player movements. Combining data from multiple sources can help to compensate for the limitations of individual systems and improve the overall accuracy of performance analysis.

3.2 Speeds and Intensities

Researchers typically perform trials with inclusion of low-speed, moderate-speed, and high-speed movements. To assess validity, researchers have measured average speeds, maximal speeds, and distance velocities. Comparing these parameters provides valuable information. OTS generally exhibit high validity at low-speed running but although still highly accurate, a relatively lower validity was reported at high-speed running (Valter et al. 2006). In (Redwood-Brown et al. 2012), coefficient of variations increased with shuttle speed, reaching their highest when shuttling at maximal speeds. However, overall, OTS demonstrates good validity across all range of speeds. Similarly, OTS in Siegle et al. (2013) found that RMSE increased from low to maximal speeds for both horizontal and vertical movements, with more significant RMSE for vertical movements. Meanwhile, in Buchheit et al. (2014), concluded that slightly-to-moderately greater results at high-speed running. Across all intensities, the OTS moderately overestimate the distance velocities and maximal speeds although generally acceptable. The RMSE in Linke et al. (2018) indicated larger residuals as the speed increased. In contrast, (Linke et al. 2020; Mara et al. 2017) found varying results, with larger and smaller mean deviations at low and medium speeds, respectively. For maximal speeds, OTS recorded trivial differences.

Most publications demonstrate acceptable validity across different speeds with increasing errors from low-speed to high-speed running. Distance velocity accuracy can be poor, especially at high speeds. Test protocols often include various movement patterns on known distance circuits. OTS validity can also be assessed based on movement complexity. Straight runs, runs with turns or curves, and more complex layouts can test OTS's basic ability to measure validity, simulate match situations, and estimate ecological validity. Activities with 45^0 and 90^0 turns often exhibit lower validity, as indicated by low R-squared, larger coefficient of variations, and poor-to-moderate results (Valter et al. 2006; Mara et al. 2017; Redwood-Brown et al. 2012; Buchheit et al. 2014) Z-turns and ball dribbling also present challenges (Redwood-Brown et al. 2012; Buchheit et al. 2014). Study in Siegle et al. (2013) found low validity for runs with momentary stops and 180^0 turns. Overall, OTS have been found to have lower validity with more complicated movements.

4 Results of Reliability

Reliability, a measure of precision, refers to the consistency of data obtained from repeated measurements under similar conditions. It can be categorized into intra-reliability and inter-reliability. Intra-reliability assesses the consistency of measurements made by a single rater, while inter-reliability evaluates the agreement between two or more raters. Reliability studies are crucial for ensuring consistent measurement across repeated trials.

Reliability data is typically presented in terms of difference, correlation, and range of confidence between repeated measurements. The difference can be quantified using metrics like typical error and total error. Intraclass correlation coefficient (ICC) is commonly used to assess the correlation between repeated measurements, while uncertainties can be expressed as coefficient of variations (CV) and confidence interval (CI).

The reviewed studies in Table 3 primarily focus on intra-reliability of OTS (Valter et al. 2006; Mara et al. 2017). These results can also be compared with reliability data from other systems, as worked in Buchheit et al. (2014). OTS reliability tends to be lower when measuring complex runs and distance velocities. Studies by Valter et al. (2006) and (Buchheit et al. 2014) reported relatively low ICC and CV values for runs with 90^0 turns and Z-turns, except for (Mara et al. 2017) which did not observe

Table 3 Reliability measures of OTS studies published to date

References	Parameter	Error measurement	Repetition
Mara et al. (2017)	DV	TEM; TE; ICC; TEM as CV%	2x
Buchheit et al. (2014)	PS;PA;HSD	TEM: CV%; ICC (90% CI)	2x
Valter et al. (2006)	MS	TEM (95% CL); TE (LOA); ICC; TEM as CV%	–

* DV = Distance velocity; PS = Peak speed; PA = Peak acceleration; HSD = High-speed distance; MS = Mean Speed; TEM = Typical error of measurement; TE = Total error; ICC = Intraclass correlation coefficient; CV = Coefficient of Variation; CI = Confidence interval; CL = Confidence limit; LOA = Limit of agreement

a significant difference between straight and complex runs. Additionally, a low ICC was found for high-

speed distance velocities in Buchheit et al. (2014).High intra-operator test–retest reliability indicates minimal human error and excellent system reproducibility. In (Linke et al. 2018) and (Buchheit et al. 2014), OTS measurements of mean speed against GPS and LPS demonstrated high reliability, suggesting these systems are consistent in measuring mean speed. However, all three studies exhibited low reliability when measuring complex runs and high-speed distance velocities.

5 Scale of Measurement

Regarding the scale of measurement, authors have used various magnitudes to interpret validity and reliability results. These scales can be based on existing literature, previous studies, or even adapted from different fields. A scale might be suitable for a specific parameter or for assessing all parameters in a study. In some cases, the same scale can be used for both validity and reliability assessments (Scott et al. 2016).

This review found scales of measurement declared in three publications (Linke et al. 2020, 2018; Buchheit et al. 2014). For magnitude of errors, standardized differences, typical error, and typical error of the estimate can be categorized as small (>0.2), moderate (>0.6), large (>1.2), and very large (>2) (Buchheit et al. 2014), following the method suggested by Hopkins et al. (2009). In (Linke et al. 2020, 2018), mean value differences were quantified into effect sizes (ES) and classified as trivial (0–0.19), small (0.20–0.49), moderate (0.50–0.79), and large (>0.80). Correlations can be interpreted as trivial ($\leq$0.1), small (>0.7–0.3), moderate (>0.3–0.5), large (>0.5–0.7), very large (>0.7–0.9), and almost perfect (>0.9–1.0) (Bampouras and Thomas 2022). For test–retest measurements, ICC can be assessed as very low (< 0.20), low (0.20–0.50), moderate (0.50–0.75), high (0.75–0.90), very high (0.90–0.99), and extremely high (> 0.99), based on unpublished work by WG Hopkins (Buchheit et al. 2014). Uncertainty can be measured as possible (25–75%), likely (75—95%), very likely (95–99%), and almost certain (>99%). Statistical inference regarding the population should be reported clearly, including the observed value's position within the uncertainty limit (Buchheit et al. 2014). Some publications did not define the scale of measurement. Authors used terms like 'acceptable,' 'small,' 'large,' 'strong,' 'high,' and 'more accurate' without providing specific numerical values. While there are no standard recommendations for these magnitudes in this field (Aughey et al. 2022), clear reporting is essential for accurate interpretation and comparison of results (Table 4).

6 Summary and Future Direction

While OTS studies have focused primarily on validity, there is a need for more research on reliability. Both validity and reliability are influenced by spatial location, vertical or horizontal movements, speed variations, trial complexity, and the specific parameters observed. No studies exist for monocular camera OTS, making it an intriguing area for future research. Given its low cost and simplicity, a single camera setup could be explored. Investigating OTS performance from higher camera positions and greater distances from the pitch would also contribute to this field. These research directions can expand our understanding of OTS and their potential applications in football analysis.

Table 4 Results summary of OTS studies published to date

No	Findings	Results	References
1	OTS validity is significantly influenced by the spatial location	*Distance:* Larger error for the circuit located far (RMSE; median ± SD = 0.92 ± 0.2) away in comparison to the circuit located near (RMSE; median ± SD = 0.60 ± 0.18) to the camera	Linke et al. (2018)
		Position: Larger error for the horizontal circuit located far (RMSE: mean ± 95^{th}PCTL = 0.33 ± 0.59) away in comparison to the circuit located near (RMSE: mean ± 95^{th}PCTL = 0.18 ± 0.41) to the camera. Similar results were found for vertical circuit located far (RMSE: mean ± 95^{th}PCTL = 1.72 ± 3.15) and near (RMSE: mean ± 95^{th}PCTL = 0.78 ± 1.71) to the camera	Siegle et al. (2013)
2	Larger errors in the vertical direction movements in which players move towards and away from the camera	*Speed:* OTS have found to overestimate the peak speed in the vertical direction movement (RMSE: mean = 0.71)	Linke et al. (2018)
		Position: Vertical direction circuit (RMSE: mean ± 95^{th}PCTL = 1.72 ± 3.15 and mean ± 95^{th}PCTL = 0.78 ± 1.71) recorded a relatively larger error in comparison to the horizontal circuit (RMSE: mean ± 95^{th}PCTL = 0.33 ± 0.59 and mean ± 95^{th}PCTL = 0.18 ± 0.41) despite the location from the camera	Siegle et al. (2013)
3	Players move at higher speed produces more OTS error	*Position:* In straight runs, the high-speed runs (RMSE: mean ± 95^{th}PCTL = 0.82 ± 2.61) has the largest error than the medium-speed (RMSE: mean ± 95^{th}PCTL = 0.67 ± 1.88) and low-speed (RMSE: mean ± 95^{th}PCTL = 0.72 ± 2.14) runs	Siegle et al. (2013)
		Speed: The errors in mean speed was at the highest at maximal-speed runs (Mean difference (LOA) = 0.64 (0.47)) than 19 km.h^{-1}-speed (Mean difference (LOA) = 0.27 (0.19)), 14 km.h^{-1}-speed (Mean difference (LOA) = 0.17 (0.13)), 11 km.h^{-1}-speed (Mean difference (LOA) = 0.10 (0.04)), and 7 km.h^{-1}-speed runs (Mean difference (LOA) = 0.07 (0.08))	Redwood-Brown et al. (2012)

(continued)

Table 4 (continued)

No	Findings	Results	References
		Speed: In the test protocol, the elevated-speed (RMSE: mean = 19.16 m) has the largest distance velocities error than the medium-speed (RMSE: mean = 15.52 m), and the low-speed (RMSE: mean = 11.24 m). Also, in the shuttle runs, the elevated-speed (RMSE: mean = 73.71 m) was larger than the medium-speed (RMSE: mean = 51.50), and the low-speed (RMSE: mean = 13.69 m)	Linke et al. (2018)
		Speed: In the curved runs, larger RMSE has been measured at high-speed (RMSE = 0.09) than the low-speed runs (RMSE = 0.03)	Linke et al. (2020)
4	Larger errors in the more complex trials in comparison to the simpler trials	*Position:* In straight circuit, the run-stop-turn-run (RMSE: mean ± 95^{th}PCTL = 0.77 ± 2.66) trials have the largest error than the run-stop-run (RMSE: mean ± 95^{th}PCTL = 0.74 ± 2.32) and straight run (RMSE: mean ± 95^{th}PCTL = 0.68 ± 2.21)	Siegle et al. (2013)
		Speed: The abrupt 20 m with a 90° turn at 10 m trials (R^2 = 0.9216) has the largest mean speed error than the 15 m straight (R^2 = 0.9501), 50 m with a 45° turn at 20 m (R^2 = 0.9998) and 60 m straight (R^2 = 1) trials	Valter et al. (2006)
		Speed: The zig-zag trials (Mean difference (LOA) = 0.50 (0.45)) have the largest mean speed error than the 50 m with 45° turn at 30 m trials (Mean difference (LOA) = 0.32 (0.50)), 20 m with 90° turn at 10 m (Mean difference (LOA) = 0.30 (1.18)), and straight (Mean difference (LOA) = 0.26 (0.47)) trials	Redwood-Brown et al. (2012)
		Distance: The 20 m with 90° turn at 10 m (Difference: mean ± SD = 0.48 ± 0.99) and 30 m with 45° turn at 20 m (Difference: mean ± SD = 0.46 ± 0.89) have larger error than the 30 m straight (Difference: mean ± SD = 0.31 ± 1.02) and 15 m straight (Difference: mean ± SD = 0.09 ± 0.71) runs	Mara et al. (2017)
5	OTS produces very large error in measuring the distance	*Distance:* OTS has large distance velocities error at 7.2 $km.h^{-1}$ (+47 m) and 14.4 $km.h^{-1}$ (+50 m)	Buchheit et al. (2014)
		Distance: Large distance velocities error recorded were at elevated-speed (RMSE = 73.71 m) and moderate-speed (RMSE = 51.50 m) during shuttle test. Similarly, large error in the total distance for the test protocol (RMSE = 3.59 m), shuttle test (RMSE = 19.77 m) and small sided games (RMSE = 4.60 m)	Linke et al. (2018)

(continued)

Table 4 (continued)

No	Findings	Results	References
		Distance: The distance velocities were large at low-speed (Difference: mean ± SD = 19.41 ± 13.24 m), sprinting (Difference: mean ± SD = 9.13 ± 25.58 m) and high-speed (Difference: mean ± SD = 8.94 ± 9.49 m)	Linke et al. (2020)
6	OTS are less reliable in the more complex trials	*Distance:* Larger distance velocities error in the 20 m with 90° turn at 10 m (TEE = 0.35 m) trials than the 30 m with 45° turn at 20 m (TEE = 0.31 m) and 15 m straight (TEE = 0.25 m)	Mara et al. (2017)
		Speed: Larger mean speed error in the 20 m with 90° turn at 10 m (TEE (90%CL) = 0.23 (0.58–0.15)) trials than the 50 m with 45° turn at 20 m (TEE (90%CL) = 0.07 (0.11–0.05), 60 m straight (TEE (90%CL) = 0.04 (0.06–0.03), and 15 m straight (TEE (90%CL) = 0.01 (0.04–0.01)	Valter et al. (2006)

References

Aughey RJ, Ball K, Robertson SJ, Duthie GM, Serpiello FR, Evans N, Spencer B, Ellens S, Cust E, Haycraft J, Billingham J (2022) Comparison of a computer vision system against three-dimensional motion capture for tracking football movements in a stadium environment. Sports Eng 25(1):2. https://doi.org/10.1007/s12283-021-00365-y

Bampouras TM, Thomas NM (2022) Validation of a LiDAR-based player tracking system during football-specific tasks. Sports Eng 25(1):1–6. https://doi.org/10.1007/S12283-022-00372-7/TABLES/1

Baysal S, Duygulu P (2016) Sentioscope: a soccer player tracking system using model field particles. IEEE Trans Circuits Syst Video Technol 26(7):1350–1362. https://doi.org/10.1109/TCSVT.2015.2455713

Beato M, Bartolini D, Ghia G, Zamparo P (2016) Accuracy of a 10 Hz GPS unit in measuring shuttle velocity performed at different speeds and distances (5–20 M). J Hum Kinet 54(1):15–22. https://doi.org/10.1515/hukin-2016-0031

Brosnan RJ, Watson G, Stuart W, Twentyman C, Kitic CM, Schmidt M (2022) The validity, reliability, and agreement of global positioning system units—can we compare research and applied data? J Strength Cond Res 36(12):3330–3338. https://doi.org/10.1519/JSC.0000000000004139

Buchheit M, Allen A, Poon TK, Modonutti M, Gregson W, Di Salvo V (2014) Integrating different tracking systems in football: multiple camera semi-automatic system, local position measurement and GPS technologies. J Sports Sci 32(20):1844–1857. https://doi.org/10.1080/02640414.2014.942687

Castellano J, Alvarez-Pastor D, Bradley PS (2014) Evaluation of research using computerised tracking systems (amisco® and prozone®) to analyse physical performance in elite soccer: a systematic review. https://doi.org/10.1007/s40279-014-0144-3

Di Salvo V, Baron R, Tschan H, Calderon Montero FJ, Bachl N, Pigozzi F (2007) Performance characteristics according to playing position in elite soccer. Int J Sports Med 28(3). https://doi.org/10.1055/s-2006-924294

Di Salvo V, Gregson W, Atkinson G, Tordoff P, Drust B (2009) Analysis of high intensity activity in premier league soccer. Int J Sports Med 30(3):205–212. https://doi.org/10.1055/s-0028-1105950

Ellens S, Hodges D, McCullagh S, Malone JJ, Varley MC (2022) Interchangeability of player movement variables from different athlete tracking systems in professional soccer. Sci Med Football 6(1):1–6. https://doi.org/10.1080/24733938.2021.1879393

Felipe JL, Garcia-Unanue J, Viejo-Romero D, Navandar A, Sánchez-Sánchez J (2019) Validation of a video-based performance analysis system (Mediacoach®) to analyze the physical demands during matches in LaLiga. Sensors (Switzerland) 19(19). https://doi.org/10.3390/s19194113

Fleureau A, Lacome M, Buchheit M, Couturier A, Rabita G (2020) Validity of an ultra-wideband local positioning system to assess specific movements in handball. Biol Sport 37(4):351–357. https://doi.org/10.5114/BIOLSPORT.2020.96850

García-Aliaga A, Marquina M, Cordón-Carmona A, Sillero-Quintana M, de la Rubia A, Vielcazat SJ, Garrosa FN, Román IR (2021) Comparative analysis of soccer performance intensity of the pre–post-lockdown covid-19 in laliga™. Int J Environ Res Public Health 18(7). https://doi.org/10.3390/ijerph18073685

Guitart M, Casals M, Casamichana D, Cortés J, Valle FX, McCall A, Cos F, Rodas G (2022) Use of GPS to measure external load and estimate the incidence of muscle injuries in men's football: a novel descriptive study. PLoS ONE 17(2):e0263494. https://doi.org/10.1371/JOURNAL.PONE.0263494

Herrmann M, Hoernig M, Radig B (2014) Online multi-player tracking in monocular soccer videos. AASRI Procedia 8:30–37. https://doi.org/10.1016/j.aasri.2014.08.006

Hopkins WG, Marshall SW, Batterham AM, Hanin J (2009) Progressive statistics for studies in sports medicine and exercise science. Med Sci Sports Exerc 41(1):3–12. https://doi.org/10.1249/MSS.0B013E31818CB278

Hoppe MW, Baumgart C, Polglaze T, Freiwald J (2018) Validity and reliability of GPS and LPS for measuring distances covered and sprint mechanical properties in team sports. PLoS ONE 13(2):e0192708. https://doi.org/10.1371/journal.pone.0192708

Imran Sainan K, Mohamad M, Mohamed Z, Bahari Saari S, Faiz Mat M, Syuhadah Khusaini N, Khusaini NS (2018) Athletes tracking using homography method: a preliminary study. Int J Eng Technol 7(4.27):6. https://doi.org/10.14419/ijet.v7i4.27.22427

Karungaru S, Tanioka H, Matsuura K (2022) Soccer players real location determination using perspective transformation. 2022 Joint 12th international conference on soft computing and intelligent systems and 23rd international symposium on advanced intelligent systems, SCIS and ISIS 2022, pp 1–4, 2022. https://doi.org/10.1109/SCISISIS55246.2022.10001939

Linke D, Link D, Lames M (2018) Validation of electronic performance and tracking systems EPTS under field conditions. PLoS ONE 13(7):e0199519. https://doi.org/10.1371/journal.pone.0199519

Linke D, Link D, Lames M (2020) Football-specific validity of TRACAB's optical video tracking systems. PLoS ONE 15(3):e0230179. https://doi.org/10.1371/journal.pone.0230179

Lorenzo-Martinez M, Kalén A, Rey E, López-Del Campo R, Resta R, Lago-Peñas C (2020) Do elite soccer players cover less distance when their team spent more time in possession of the ball?. Sci Med Football 5(4):310–316. https://doi.org/10.1080/24733938.2020.1853211

Lu W-L, Ting J-A, Little JJ, Murphy KP (2013) Learning to track and identify players from broadcast sports videos. IEEE Trans Pattern Anal Mach Intell 35(7):1704–1716. https://doi.org/10.1109/TPAMI.2012.242

Maćkowiak S (2013) Segmentation of football video broadcast. Int J Electron Telecommun 59(1):75–84. https://doi.org/10.2478/eletel-2013-0009

Maglo A, Orcesi A, Denize J, Pham QC (2023) Individual locating of soccer players from a single moving view. Sensors 23(18):7938. https://doi.org/10.3390/s23187938

Makar P, Silva AF, Oliveira R, Janusiak M, Parus P, Smoter M, Clemente FM (2023) Assessing the agreement between a global navigation satellite system and an optical-tracking system for measuring total, high-speed running, and sprint distances in official soccer matches. Sci Prog 106(3):00368504231187501. https://doi.org/10.1177/00368504231187501

Manaafard M, Ebadi H, Moghaddam HA (2015) Discrete particle swarm optimization for player trajectory extraction in soccer broadcast videos. Scientia Iranica D 22(3):1031–1044

Manafifard M (2024) A review on camera calibration in soccer videos. Multimedia Tools and Applications 83(6):18427–18458. https://doi.org/10.1007/s11042-023-16145-8
Manafifard M, Ebadi H, Moghaddam HA (2017) Multi-player detection in soccer broadcast videos using a blob-guided particle swarm optimization method. Multimedia Tools Appl 76(10):12251–12280. https://doi.org/10.1007/s11042-016-3625-6
Mara J, Morgan S, Pumpa K, Thompson K (2017) The accuracy and reliability of a new optical player tracking system for measuring displacement of soccer players. Int J Comput Sci Sport 16(3):175–184. https://doi.org/10.1515/ijcss-2017-0013
Martín R, Martínez JM (2014) A semi-supervised system for players detection and tracking in multi-camera soccer videos. Multimedia Tools Appl 73(3):1617–1642. https://doi.org/10.1007/s11042-013-1659-6
Novak AR, Impellizzeri FM, Trivedi A, Coutts AJ, McCall A (2021) Analysis of the worst-case scenarios in an elite football team: towards a better understanding and application. J Sports Sci 39(16):1850–1859. https://doi.org/10.1080/02640414.2021.1902138
Pino-Ortega J, Oliva-Lozano JM, Gantois P, Nakamura FY, Rico-González M (2021) Comparison of the validity and reliability of local positioning systems against other tracking technologies in team sport: a systematic review. Proc Inst Mech Eng Part P: J Sports Eng Technol 236(2):73–82. https://doi.org/10.1177/1754337120988236
Ponce-Bordón JC, Díaz-García J, López-gajardo MA, Lobo-Triviño D, Del Campo RL, Resta R, García-Calvo T (2021) The influence of time winning and time losing on position-specific match physical demands in the top one Spanish soccer league. Sensors 21(20):6843. https://doi.org/10.3390/S21206843
Pons E, García-Calvo T, Cos F, Resta R, Blanco H, López del Campo R, Díaz-García J, Pulido-González JJ (2021) Integrating video tracking and GPS to quantify accelerations and decelerations in elite soccer. Scientific Reports 11(1:1–10. https://doi.org/10.1038/s41598-021-97903-2
Redwood-Brown A, Cranton W, Sunderland C (2012) Validation of a real-time video analysis system for soccer. Int J Sports Med 33(8):635–640. https://doi.org/10.1055/s-0032-1306326
Robertson S, Duthie GM, Ball K, Spencer B, Serpiello FR, Haycraft J, Evans N, Billingham J, Aughey RJ (2023) Challenges and considerations in determining the quality of electronic performance & tracking systems for team sports. Frontiers Sports Active Living 53. https://doi.org/10.3389/fspor.2023.1266522
Sabirin H, Sankoh H, Naito S (2015) Automatic soccer player tracking in single camera with robust occlusion handling using attribute matching. IEICE Trans Inf Syst E98D(8):1580–1588. https://doi.org/10.1587/transinf.2014EDP7313
Sani MR (2023) Multi object tracking in soccer video focusing on occlusion detection and resolving. Multimedia Tools Appl 82(23):35913–35947. https://doi.org/10.1007/s11042-023-14798-z
Schlipsing M, Salmen J, Tschentscher M, Igel C (2017) Adaptive pattern recognition in real-time video-based soccer analysis. J Real-Time Image Proc 13(2):345–361. https://doi.org/10.1007/s11554-014-0406-1
Scott MTU, Scott TJ, Kelly VG (2016) The validity and reliability of global positioning systems in team sport. J Strength Cond Res 30(5):1470–1490. https://doi.org/10.1519/JSC.0000000000001221
Siegle M, Stevens T, Lames M (2013) Design of an accuracy study for position detection in football. J Sports Sci 31(2):166–172. https://doi.org/10.1080/02640414.2012.723131
Taberner M, O'Keefe J, Flower D, Phillips J, Close G, Cohen DD, Richter C, Carling C (2020) Interchangeability of position tracking technologies; can we merge the data? Sci Med Football 4(1):76–81. https://doi.org/10.1080/24733938.2019.1634279
Valter DS, Adam C, Barry M, Marco C (2006) Validation of Prozone ®: a new video-based performance analysis system. Int J Perform Anal Sport 6(1):108–119. https://doi.org/10.1080/24748668.2006.11868359
Vidal-Codina F, Evans N, · Bahaeddine, Fakir E, Billingham J (2022) Automatic event detection in football using tracking data 25:18. https://doi.org/10.1007/s12283-022-00381-6

Wu C, Zhao X, Jin G (2022) Application of intelligent analysis technology of football video based on online target tracking algorithm of motion characteristics in football training. Comput Intell Neurosci 2022(1):4739712. https://doi.org/10.1155/2022/4739712

Evaluating the Impact of Acceptance and Commitment Therapy Combined with Structured Exercise (ACT-EX) on Weight Loss in Overweight and Obese Young Adults: A Randomized Controlled Trial

Patricia Pawa Pitil and Siti Raudzah Ghazali

Abstract The rate of obesity has risen in Malaysia, particularly among university students, due to the COVID-19 pandemic and the implementation of the Movement Control Order. While Acceptance and Commitment Therapy (ACT) has been proven effective in addressing weight-related issues, the combined effects of ACT with a structured exercise program remain unclear. This randomized controlled trial aimed to compare the effectiveness of two ACT interventions for overweight and obese young adults: ACT alone and ACT combined with structured exercise (ACT-EX). The interventions were designed to improve body mass index (BMI) and address weight-related challenges. A total of 102 participants, aged 18 to 25 with a BMI above 23.0 kg/m^2, were randomly assigned to the ACT or ACT-EX group. The ACT group took part in six weekly 1.5-h group sessions, while the ACT-EX group participated in the same six ACT sessions, plus three 1.5-h structured exercise sessions per week over six weeks. Data were collected at baseline, post-intervention, and during a 12-week follow-up. The results indicated that the ACT-EX group experienced a more significant reduction in BMI and improved weight-related flexibility, although there were no significant differences between the two groups. These findings suggest that combining structured exercise with ACT can enhance psychological flexibility and assist in weight management.

Keywords Acceptance and commitment therapy · Overweight and obesity · Weight-related difficulties · Structured exercise · Young adults

P. P. Pitil (✉) · S. R. Ghazali
Department of Psychology, Faculty of Medicine and Health Sciences, Universiti Malaysia Sarawak, Kota Samarahan, Malaysia
e-mail: patriciapawa@uitm.edu.my

P. P. Pitil
Faculty of Sports Science and Recreation, Universiti Teknologi MARA Sarawak Branch, Kota Samarahan, Malaysia

M. H. A. Hassan et al. (eds.), *Proceedings of the 3rd International Conference on Innovation and Technology in Sports*, Lecture Notes in Bioengineering,
https://doi.org/10.1007/978-981-95-0804-4_13

1 Introduction

Obesity, characterized by an excessive accumulation of body fat, is a growing global health concern. Its prevalence has been steadily increasing, making it the eleventh most common health issue worldwide (World Health Organization 2017). In particular, the rates of overweight and obesity have risen significantly in many regions, with profound impacts on public health. In the Asian context, overweight is defined as having a body mass index (BMI) between 23.0 kg/m^2 and 27.4 kg/m^2, while obesity is classified as a BMI between 27.5 kg/m^2 and 34.9 kg/m^2 (Zainudin et al. 2011). In 2015, the percentage of adults who were overweight or obese (OW/OB) reached 30% and 17.7%, respectively (NHMS 2015). This trend has continued, with the most recent National Health and Morbidity Survey (NHMS) 2023 report showing that 54.4% of Malaysian adults are either overweight or obese (Institute for Public Health 2023). These figures highlight the urgent need for effective solutions to address the rising global burden of overweight and obesity.

Many university students encounter difficulties in maintaining a healthy body weight during the transition to early adulthood. Allegedly, 38.5% of university students in Malaysia are obese or overweight (Che Wan Jasimah et al. 2019). Another recent study discovered that 22.2% and 16.9% of them were obese, respectively (Pitil and Ghazali 2022). In other words, the fact that at least four out of ten university students are overweight or obese indicates a concerning trend. Throughout their lifetimes, these young adults, who typically ranged in age from 18 to 25, encountered a variety of challenges. It is imperative that they manage the numerous stressors they encounter in a constructive manner. One stressor is the problem of body weight. The COVID-19-induced confinement resulted in a sedentary lifestyle among the population, which led to weight gain and an energy imbalance (Rodríguez-Larrad et al. 2021).

Although most exercise professionals assert that regular exercise and a nutritious diet can resolve the obesity epidemic, the trend continues to rise year after year. Despite efforts to follow the latest exercise trends, and modified eating behaviors, the issue remains unresolved. This persistent challenge suggests that it may be the right time to emphasize the psychological aspect of this demographic, focusing on their behavior in relation to body weight.

Acceptance and Commitment Therapy (ACT) is a behavioral approach that encourages weight management by addressing a person's tendency to avoid unpleasant thoughts and feelings (Hayes et al. 1999). This third-generation cognitive-behavioral approach incorporates mindfulness, commitment, acceptance, and behavior modification techniques to enhance psychological flexibility, reduce negative thoughts, and embrace hard feelings while staying true to value-based actions (Hayes et al. 2006). A growing body of research has demonstrated that ACT is effective in addressing obesity-related stigma and overcoming weight management challenges (Levin et al. 2018; Palmeira et al. 2018; Petersen et al. 2021).

While numerous studies have identified the factors contributing to obesity in Malaysia, these have predominantly focused on epidemiological analyses, creating a research gap. Although psychological approaches to weight management are widely accepted in other countries, they are still relatively new in Malaysia (Curry et al. 2018). It is well-established that weight loss is closely linked to high levels of physical activity (Johnson et al. 2021), yet the combination of behavioral therapy and exercise remains unexplored. To address this gap, the ACT-EX was developed. This program integrates six key processes designed to improve weight-related difficulties (WRD): acceptance, cognitive defusion, mindfulness, self-as-context, values, and committed action.

This randomized controlled trial (RCT) aim was to assess the effectiveness of ACT in BMI and WRD changes, both with (ACT-EX) and without the addition of a structured exercise program (ACT). We hypothesized that ACT-EX would lead to a greater reduction in BMI compared to ACT alone. Furthermore, it was anticipated that ACT-EX would have a more significant effect on changes in WRD than ACT on its own.

2 Methodology

We selected participants during the COVID-19 pandemic using purposive sampling from two public institutions in Sarawak, Malaysia. This non-probability sampling method was employed to ensure that the target population included individuals who could provide valuable and specific insights for the research (Black 2019). To be taken into account, the participant had to fulfill the following requirements: (a) have a BMI of overweight to obesity (overweight = 23.0–27.4 kg/m^2, obese = 27.5 kg/m^2 and above); (b) be free of orthopedic injuries and chronic conditions (diabetes, hypertension); and (c) have reliable internet connectivity for all intervention video-conference sessions. On the other hand, the following criteria were used to exclude participants: (a) they had a history of orthopedic injuries; (b) they were currently participating in a supervised exercise program; or (c) they were currently taking medication. The study recruited 102 participants who satisfied the inclusion and exclusion criteria (Table 1).

Power calculations were utilized to determine the sample size. According to power estimates, 43 participants per condition are needed to detect a difference between the three-point measures (baseline, week six and week 12) (assuming a drop-out of 20% using the G* Power calculation, an alpha of 0.05, statistical power, and a 1-beta of 0.80). Consequently, 86 individuals in all are recommended. Nevertheless, N = 102 individuals were recruited and randomly allocated to either the ACT-EX (n = 50) or ACT (n = 52) groups. Table 1 displays the baseline demographic data for the intervention and control groups after randomization.

Design and Procedures

This was an online, parallel group, randomized controlled trial conducted in Malaysia between March 2021 and July 2021. The researchers used a simple randomization protocol to assign 102 participants to the ACT or ACT-EX conditions following

Table 1 Demographic characteristics by group at baseline

Demographics	ACT ($n = 52$)	ACT-EX ($n = 50$)
	Mean (SD)	
BMI (kg/m^2)	30.23 (4.83)	30.02 (4.63)
Age (years)	20.69 (1.65)	21.32 (1.20)
Sex	*n* (%)	
Male	19 (36.5)	11 (22)
Female	33 (63.5)	39 (78)
Race		
Malay	15 (28.8)	18 (36)
Bumiputera Sarawak	29 (55.8)	14 (28)
Bumiputera Sabah	8 (15.4)	4 (8)
Chinese	8 (15.4)	10 (20)
Indian	–	4 (8)
Study level		
Pre-university	3 (5.8)	–
Diploma	25 (48.1)	3 (6)
Bachelor's degree	24 (46.2)	47 (94)

baseline assessment. The intervention lasted for six weeks. For the ACT intervention, groups met once a week, and for the ACT-EX intervention, they met three times a week for six weeks. From week six until the follow-up at week 12, there was no intervention contact. Figure 1 details the flow of participants throughout the trial.

The ethical clearance was obtained from the Medical Ethics Committees of Universiti Malaysia Sarawak's Faculty of Medicine and Health Sciences. The study was conducted in compliance with the Declaration of Helsinki (UNIMAS/NC-21.02/03–02 Jld.4 (02)), the Malaysian Good Clinical Practice Guidelines, and the ICH Good Clinical Practice Guidelines. Those who fulfilled the inclusion requirements were given a permission form and the Physical Activity Readiness Questionnaire (for the ACT-EX group) as soon as they agreed to participate. They complete the questionnaire, which includes the Acceptance and Action Questionnaire for Weight-Related Difficulties Revised—18 (AAQW-R18), demographic data, and self-reported height and weight (Pitil and Ghazali 2021).

Measures

During the first screening interview, demographic data was gathered, including the participants' race, sex, age, and level of education. Additionally, participants self-reported their height and body weight. The self-reported weight and height of participants were used to calculate their BMI. This was done in order to classify their weight. The nearest kilogram (kg) and centimeter (cm) were used to record the weight and height, respectively. Participants were instructed to weigh themselves on bathroom scales at home, and wherever feasible, these readings were utilized to record their body weight. Weight in kilograms divided by height in meters squared was used to determine BMI (Zainudin et al. 2011).

The AAQW-R18 is an 18-item questionnaire that measures weight-related challenges by examining thoughts, emotions, and physical sensations in the context of experiential avoidance. It uses a 7-point rating; 1 denotes "never true" or "not at all believable," while 7 denotes "always true" or "completely believable." Higher scores indicate a greater avoidance of experiences related to weight. Test–retest reliability, validity, and internal consistency of the measure have all been demonstrated to be strong (Pitil and Ghazali 2021). Cronbach's α in the current study was 0.68 at baseline, but it improved to 0.85 after the intervention and 0.88 at the follow-up.

The ACT intervention program was conducted as a group-based treatment via videoconferencing. Participants took part in weekly 1.5-h sessions over six weeks. The program aimed to foster psychological flexibility by promoting openness, awareness, and active engagement. The content of the ACT sessions was adapted from previous ACT interventions focused on weight loss (Butryn et al. 2011; Tapper et al. 2009). Key elements of the program included (a) identifying values to boost motivation, (b) cognitive defusion to distinguish thoughts from behaviors related to food and exercise, (c) acceptance techniques for managing negative emotions, (d) mindfulness practices, and (e) commitment to values that support weight loss. Participants were provided with theoretical insights into the ACT model and were encouraged to take notes during the sessions. Homework assignments were also incorporated into the program. The content was tailored to address the participants' personally identified values to encourage their engagement (refer to Table 2 for details on the ACT session content).

Throughout a six-week period, participants in the ACT-EX program participated in three virtual videoconference exercise sessions in addition to weekly ACT sessions. A minimum of 24 h was designated between exercise sessions to ensure adequate recovery and reduce the risk of injury. The American College of Sports Medicine's guidelines were adhered to during each 60-min exercise session, which consisted of moderate-intensity routines. Furthermore, to support weight loss, individuals who are overweight or obese should aim for a minimum of 150 min of moderate-intensity exercise each week (World Health Organization 2017). A total body exercise was implemented during each session (refer to Table 3). Exercise intensity was measured as a Rate of Perceived Exertion (RPE) between 12 (fairly moderate) and 15 (intense), or 65% to 85% of the maximal heart rate (HRmax). The HRmax formula was employed to determine the objective heart rate: 220—age. The participants, aged 18 to 25, had an HRmax ranging from 195 to 202 beats per minute (bpm). The exercise intensity percentage was applied to the HRmax range to determine the target heart rate.

Participants were instructed to either monitor their heart rate using a smartwatch or assess their perceived exertion with the Borg scale during each exercise session. The exercise routines included bodyweight exercises targeting key muscle groups, such as push-ups, squats, lunges, and crunches. Each session began with a warm-up and concluded with a cool-down. The group-based format and virtual videoconference sessions were key factors in ensuring the intervention achieved 100% attendance.

Table 2 Content of ACT sessions

Session/ component	Aims	Key metaphors and exercises	Application to weight loss
1. Values	– To identify personal values – To determine whether weight loss would support such values	The choice point Values assessment rating	Boost motivation to achieve weight loss
2. Cognitive defusion	– To learn to distance from distressing thoughts about bodyweight and strengthen their commitment to health-related values	Push papers	To recognize when thoughts might undermine diet and exercise plans To help disrupt the connection between food-related thoughts and behaviors
3. Acceptance/ willingness	– To increase willingness to experience aversive internal states, the distress that associated with physical activity (e.g.; fatigue, physical discomfort, urges to stop) and diet-related feelings (e.g.; cravings)	Breathing exercise	To assist participants in tolerating negative emotions (such as those related to physical exercise and diet) instead of relying solely on control or avoidance strategies
4. Mindfulness	To develop a sense of self that can enable thoughts and feelings to flow without attachment To enable the participant to stay in non-judgmental contact with psychological and environmental events as they occur	Observational self-exercise Be where you are exercise	To support the use of defusion and acceptance strategies
5. Committed action	To become aware of the importance of committing to values To identify goals consistent with the participant's values	Goals, barriers and action exercise Living in accordance with values exercise	To assist participants in setting and achieving weight loss goals To promote long-term adherence to weight loss and maintenance strategies
6. Review	To review participants' experience over the intervening weeks and help clarify any difficulties they may have had To remind participants of key concept	Question and answer session Selection of exercises and metaphors from previous sessions	Same as before

Table 3 ACT-EX sessions

Session	Component
1	ACT session 1– Values (1.5 hour)
2 3 4	Exercise (1 hour) • Warm-up (5 minutes) -Jumping jacks -High knees • Major muscle groups exercises: -Squats -Lunges -Push up -Crunches -Plank } 3 Sets (20 seconds per exercise) * 3 - 5 minutes rest between sets • Cooling down (5 minutes)
5	ACT session 2 – Cognitive Defusion (1.5 hour)
6 7 8	Exercise (1 hour) • Warm-up (5 minutes) -Jumping jacks -High knees • Major muscle groups exercises: -Squats - Lateral Lunges -Push up -Leg raises -Side Plank (Left – Right) } 3 Sets (20 seconds per exercise) * 3 - 5 minutes rest between sets • Cooling down (5 minutes)
9	ACT session 3 – Acceptance/Willingness (1.5 hour)
10 11 12	Exercise (1 hour) • Warm-up (5 minutes) -Jumping jacks -High knees • Major muscle groups exercises: -Squats - Lateral Lunges -Push up -Alternate hand-feet touches - Side Plank (Left – Right) } 3 Sets (30 seconds per exercise) * 3 - 5 minutes rest between sets • Cooling down (5 minutes)
13	ACT session 4 – Mindfulness (1.5 hour)

(continued)

Table 3 (continued)

14 15 16	Exercise (1 hour) • Warm-up (5 minutes) -Jumping jacks -High knees • Major muscle groups exercises: -Squats - Lateral Lunges -Push up twist - Leg raises -Plank [3 Sets (30 seconds per exercise) * 3 - 5 minutes rest between sets] • Cooling down (5 minutes)
17	ACT session 5 – Committed Action (1.5 hours)
18 19 20	Exercise (1 hour) • Warm-up (5 minutes) -Jumping jacks -High knees • Major muscle groups exercises: -Squats - Lateral Lunges -Push up twist - Leg raises -Plank [3 Sets (12-15 repetitions per exercise) * 3 - 5 minutes rest between sets] • Cooling down (5 minutes)
21	ACT session 6 – Review (1.5 hours)
22	Exercise (1 hour) • Warm-up (5 minutes) -Jumping jacks -High knees • Major muscle groups exercises: -Squats - Single-leg lunges (left/right) -Push up twist - Leg raises -Plank [3 Sets (12 – 15 repetitions per exercise) * 3 - 5 minutes rest between sets] • Cooling down (5 minutes)

Data Analysis

The data analysis was conducted using IBM SPSS Statistics version 26. Descriptive statistics, such as frequencies, percentages, means, and standard deviations, have categorized the demographic and outcome variables. At baseline, post-intervention, and follow-up, the effects of the treatments (ACT and ACT-EX) on outcome measures, especially BMI and WRD, were evaluated using a repeated measures analysis of variance (ANOVA) (3×2). The partial eta squared (ηp^2) was computed to express the effect sizes for the ANOVA results. Effect sizes of 0.01, 0.06, and 0.14 were used to define small, moderate, and large effects, respectively. Cohen's d was computed using the estimated values in order to evaluate effect sizes from baseline to follow-up (week 12) and from baseline to post-intervention (week six). Within-group effect sizes were categorized as large (1.1), moderate (0.8), and small (0.5) (Cohen 1988). For all statistical analyses, a significance level of 0.05 ($p < 0.05$) was set.

3 Results

BMI Changes

There was no significant interaction between time and groups for BMI in the 3 × 2 repeated measures ANOVA [F (2, 99) = 2.705, p = 0.423, and $\eta p^2 = 0.05$, which indicate a small effect size]. A significant main effect for time was observed [F (2, 99) = 37.85, p = 0.001, $\eta p^2 = 0.43$], with both the ACT and ACT-EX groups observing a decrease in BMI over the time periods, indicating a large effect size. There was no significant main effect between the two intervention groups, as indicated by the following: [F (1, 100) = 0.646, $\eta p^2 = 0.06$], indicating that there was no difference in the effectiveness of the two interventions.

Figure 2 shows that both ACT and ACT-EX were successful in improving BMI. Overall, both groups experienced a decline in their BMI scores, with a significant slope observed in week 6. The reduction was more pronounced in the ACT-EX group from week six to week 12 compared to the ACT group. Concurrently, ACT sustained its increase in week 12. From baseline to post-intervention and from post-intervention to the Week 12 follow-up, the within-group effect sizes for ACT (ds = −0.37 and −0.09) and ACT-EX (ds = −0.35 and −0.13) were both small. At post-intervention and follow-up, the between-group effect sizes for ACT and ACT-EX were likewise small (d = 0.02 and 0.25, respectively) (Table 4).

WRD Changes

The 3 × 2 repeated measures ANOVA showed that time and group had a significant effect on WRD [F (2, 99) = 14.38, p = 0.001, $\eta p^2 = 0.25$, showing a large effect size). t was found that time had a significant main effect [F (2, 99) = 79.06, p = 0.001, $\eta p^2 = 0.62$], which means that both the ACT and ACT-EX groups had a decrease in WRD over time, and the effect size was big. The comparison of the two intervention groups yielded a significant main effect, [F (1, 100) = 46.43, $\eta p^2 = 0.32$], indicating a difference in the effectiveness of the interventions (Table 4).

Figure 3 shows that WRD was successfully decreased by both ACT and ACT-EX. In all groups, the WRD score decreased with time, with week six showing the largest drop. Compared to the ACT group, the ACT-EX group showed a more substantial decrease from week six to week twelve. ACT maintained its reduction in Week 12. ACT within-group effect sizes (ds = −0.74 and 0.06) showed small effects at follow-up and moderate effects from baseline to post-intervention. On the other hand, ACT-EX showed modest effects at follow-up and considerable effects from baseline to post-intervention (ds = −2.18 and −0.60). With values of d = 1.4 and 1.43, respectively, the effect sizes across groups for ACT and ACT-EX were significant at post-intervention and follow-up (Table 5).

Table 4 Analyses from baseline to post-intervention and baseline to Week 12 follow-up on study outcome measures ($N = 102$)

	ACT ($n = 52$)			ACT-EX ($n = 50$)				
	Baseline	Week 6	Week 12	Baseline	Week 6	Week 12	p (group x time)	P (time)
BMI (kg/m^2)	30.1 ± 4.83	28.3 ± 5.36	28.8 ± 4.39	30.0 ± 4.63	28.4 ± 5.36	27.7 ± 4.71	0.423	0.001*
WRD	81.60 ± 15.99	69.84 ± 8.68	70.35 ± 11.82	80.96 ± 10.69	57.66 ± 7.25	53.46 ± 9.11	0.001*	0.001*

Note BMI: Body mass index; WRD: Weight-related difficulties
*Significant at 0.05 ($p < 0.05$)

Table 5 Within-group and between-group effect size for outcome measures (Cohen's *d*)

Variable	Group	Within-group *ds*		Between-group *d*		
		Baseline to post-intervention	Post-intervention to follow-up	Group	Post-intervention	Follow-up
BMI (kg/m^2)	ACT	−0.37	−0.09	ACT—ACT-EX	0.02	0.25
	ACT-EX	−0.35	−0.13			
WRD	ACT	−0.74	0.06	ACT—ACT-EX	1.4	1.43
	ACT-EX	−2.18	−0.60			

Note ACT: Acceptance and commitment therapy group, ACT-EX: Acceptance and commitment therapy and structured exercise group, BMI: Body mass index, WRD: Weight-related difficulties

Positive *ds* indicate increases in score over time and negative *d's* indicate decreases

4 Discussion

The aim of this RCT was to examine the effectiveness of ACT and ACT-EX interventions in reducing BMI and WRD among OW/OB university students during the COVID-19 lockdown. Both the ACT and ACT-EX groups showed substantial reductions in BMI, with further improvements observed during the follow-up phase. The BMI reduction in the ACT-EX group was notably larger during the follow-up period. Additionally, the effectiveness of the ACT intervention appeared to decrease over time. As shown in Fig. 2, the mean BMI graphs for both groups across the three time periods reflect this trend. Although a significant difference in BMI was observed between ACT and ACT-EX (Table 3), both interventions demonstrated similar effectiveness in reducing BMI.

Our study's findings are consistent with the evidence that exercise interventions can improve BMI and facilitate weight loss (Bellicha et al. 2021; Cox 2017). Exercise was helpful in maintaining sustainable weight loss, as indicated by the follow-up data for the ACT-EX group. More specifically, the ACT-EX group experienced a 7.7% decrease in BMI, while the ACT group experienced a 4.3% decrease. The weight loss program was intended to promote ongoing improvements in BMI following its conclusion. Despite the fact that it is challenging to sustain weight loss following an intervention (Hall and Kahan 2018), this study demonstrates that it is feasible. The effect of the intervention was long-lasting, as the BMI continued to consistently decline even after the follow-up was administered 1.5 months after the conclusion of the intervention. This suggests that participants may continue to lose weight in the future. ACT-EX results in a greater reduction in BMI ACT alone, which is consistent with the primary hypothesis. Overall, the findings of this study suggest that ACT-EX significantly improves BMI in individuals who are OW/OB, both in the short and long term.

The congruence of healthy habits with fundamental personal values and commitment is emphasized by ACT, which serves as a motivational anchor during challenging times to assist individuals in maintaining healthy habits (Forman and Butryn 2015; Lillis and Kendra 2014). Structured exercise was incorporated into the ACT regimen (ACT-EX) to enhance weight management. Participants in the ACT-EX group likely continued their routine of exercise beyond the three weekly sessions that were part of the six-week intervention, which included weekly ACT sessions. It is likely that the development of a habit was the reason why individuals maintained their exercise routine post-intervention. This continuous physical activity contributed to sustained weight loss. Additionally, it was observed that ACT briefly increased physical activity levels (Butryn et al. 2011, 2018; Wang et al. 2020). It is recommended that participating in up to 200 min of physical activity per week can aid in managing body weight, as evidenced by the findings of the present study (Donnelly et al. 2009).

Although the literature extensively investigates the application of ACT for body weight issues, the results of incorporating a structured exercise program remain inconclusive. ACT-EX's efficacy in reducing WRD among OW/OB individuals was demonstrated by this research. From baseline to post-intervention, as well as from baseline to follow-up, the intervention exhibited a significant effect, as indicated by the mean plots (see Fig. 3). The ACT group continued to make progress from the post-intervention to the follow-up. ACT has been shown to be effective in addressing WRD in previous studies (Petersen et al. 2021; Failed 2018). This research is unique in that it incorporates structured exercise into the ACT intervention. The exercise component likely contributed to the significant outcomes observed. The efficacy of ACT in addressing experiential avoidance associated with weight concerns may be improved by the potential for exercise to increase self-esteem and body image (Zamani Sani et al. 2016). This implies that the most effective ACT intervention for reducing WRD in OW/OB university students includes an exercise component. Therefore, we verified the second hypothesis: ACT-EX had greater effects on WRD alterations than ACT alone.

Conclusion

The effectiveness of ACT interventions in promoting weight loss and improving weight-related flexibility among OW/OB university students was demonstrated in this study. The substantial weight loss that was observed as a result of ACT interventions contributes to the current body of literature regarding the prospective advantages of using behavioral strategies in weight management programs. In addition, the efficacy of the combined interventions was improved by the integration of a structured exercise program with ACT, resulting in more substantial weight loss outcomes for this population. This effect appears to be sustainable. These findings are valuable for addressing body weight challenges within this group, as they are the first RCT to compare ACT and ACT-EX in examining changes in WRD among OW/OB university students. Improvements in BMI and weight-related flexibility indicate that this demographic is less inclined to engage in avoidance behaviors and is more accepting of their internal experiences and weight. In conclusion, this study underscores the potential of integrating ACT with a structured exercise regimen to enhance psychological flexibility in relation to weight.

Acknowledgements We would like to express our gratitude to Universiti Malaysia Sarawak the Faculty of Medicine and Health Sciences, and the Medical Ethics Committees for their ethical approval. We also extend our thanks to the university students who participated in this research.

References

Bellicha A et al (2021) Effect of exercise training on weight loss, body composition changes, and weight maintenance in adults with overweight or obesity: an overview of 12 systematic reviews and 149 studies. Obesity Rev 22(Suppl 4): e13256. [Online]. Available: https://doi.org/10.1111/obr.13256

Black K (2019) Business statistics: for contemporary decision making. Wiley, Hoboken, NJ

Butryn ML, Forman E, Hoffman K, Shaw J, Juarascio A (2011) A pilot study of acceptance and commitment therapy for promotion of physical activity. J Phys Act Health 8:516–522. [Online]. Available: https://doi.org/10.1123/jpah.8.4.516

Butryn ML, Kerrigan S, Arigo D, Raggio G, Forman EM (2018) Pilot test of an acceptance-based behavioral intervention to promote physical activity during weight loss maintenance. Behav Med 44:77–87. [Online]. Available: https://doi.org/10.1080/08964289.2016.1170663

Che Wan Jasimah WMR, Hashem SJ, Ayed RAA, Mohd Istajib M, Mohd Zufri M, Nor Aishah A (2019) Analysis of obesity among Malaysian university students: a combination study with the application of Bayesian structural equation modelling and Pearson correlation. Int J Environ Res Public Health 16(3):492. [Online]. Available: https://doi.org/10.3390/ijerph16030492

Cohen J (1988) Statistical power analysis for the behavioral sciences, 2nd ed. Routledge. https://doi.org/10.4324/9780203771587

Cox CE (2017) Role of physical activity for weight loss and weight maintenance. Diabetes Spectrum 30(3):157–160. [Online]. Available: https://doi.org/10.2337/ds17-0013

Curry SJ, Krist AH, Owens DK, Barry MJ, Caughey AB, Davidson KW, Doubeni CA, Epling JW, Grossman DC, Kemper AR (2018) Behavioral weight loss interventions to prevent obesity-related morbidity and mortality in adults: US preventive services task force recommendation statement. JAMA 320(11):1163–1171

Donnelly JE, Blair SN, Jakicic JM, Rankin JW, Smith BK (2009) American college of sports medicine position stand: appropriate physical activity intervention strategies for weight loss and prevention of weight regain for adults. Med Sci Sports Exercise 41(2):459–471. [Online]. Available: https://doi.org/10.1249/MSS.0b013e3181949333

Forman EM, Butryn ML (2015) A new look at the science of weight control: how acceptance and commitment strategies can address the challenge of self-regulation. Appetite 84:171–180. [Online]. Available: https://doi.org/10.1016/j.appet.2014.10.004

Hall KD, Kahan S (2018) Maintenance of lost weight and long-term management of obesity. Med Clinics North America 102(1):183–197. [Online]. Available: https://doi.org/10.1016/j.mcna.2017.08.012

Hayes SC, Strosahl K, Wilson KG (1999) Acceptance and commitment therapy: an experiential approach to behavior change. The Guilford Press, New York, NY, USA

Hayes SC, Luoma JB, Bond FW, Masuda A, Lillis J (2006) Acceptance and commitment therapy: model, processes and outcomes. Behav Res Ther 44:1–25

Institute for Public Health (2015) National health and morbidity survey 2015 (NHMS 2015) Vol III: Non-communicable diseases, risk factors & other health problems

Institute for Public Health (2024) National health and morbidity survey (NHMS) 2023: Vol. I: NCDs – non-communicable diseases – main findings

Johnson NA, Sultana RN, Brown WJ, Bauman AE, Gill T (2021) Physical activity in the management of obesity in adults: a position statement from exercise and sport science Australia. J Sci Med Sport 24(12):1245–1254. [Online]. Available: https://doi.org/10.1016/j.jsams.2021.07.009

Levin ME, Potts S, Haeger J, Lillis J (2018) Delivering acceptance and commitment therapy for weight self-stigma through guided self-help: Results from an open pilot trial. Cogn Behav Pract 25(1):87–104

Lillis J, Kendra KE (2014) Acceptance and commitment Therapy for weight control: Model, evidence, and future directions. J Contextual Behav Sci 3(1):1–7. [Online]. Available: https://doi.org/10.1016/j.jcbs.2013.11.005

Palmeira L, Cunha M, Pinto-Gouveia J (2018) The weight of weight self-stigma in unhealthy eating behaviours: the mediator role of weight-related experiential avoidance. Eating Weight Disord 23(6):785–796. [Online]. Available: https://doi.org/10.1007/s40519-018-0540-z

Petersen JM, Durward C, Levin M (2021) Weight-related psychological inflexibility as a mediator between weight self-stigma and health-related outcomes. Bull Menninger Clin 85(3):316–330. [Online]. Available: https://doi.org/10.1521/bumc.2021.85.3.316

Pitil PP, Ghazali SRB (2021) Psychometric properties of Acceptance and Action Questionnaire (AAQ) for weight-related difficulties questionnaire among university students. Global Bus Manag Res 13(2)

Pitil PP, Ghazali SR (2022) Overweight and obesity: a study among university students in Sarawak, Malaysia. Int J Health Promot Educ, 1–13. [Online]. Available: https://doi.org/10.1080/14635240.2022.2040380

Potts SA (2018) Putting weight in context: acceptance and commitment therapy (ACT) guided self-help for weight self-stigma. Doctoral dissertation, Utah State University

Rodríguez-Larrad A et al (2021) Impact of COVID-19 confinement on physical activity and sedentary behaviour in Spanish university students: role of gender. Int J Environ Res Public Health 18(2):369. [Online]. Available: https://doi.org/10.3390/ijerph18020369

Tapper K, Shaw C, Ilsley J, Hill AJ, Bond FW, Moore L (2009) Exploratory randomised controlled trial of a mindfulness-based weight loss intervention for women. Appetite 52(2):396–404. [Online]. Available: https://doi.org/10.1016/j.appet.2008.11.012

Wang Y, Tarbox J, Chastain AN, Cameron MJ (2020) The effects of bilingual Acceptance and Commitment Training (ACT) on exercise in bilingual international university students. J Contextual Behav Sci 18:1–8, 2020. [Online]. Available: https://doi.org/10.1016/j.jcbs.2020.08.002

World Health Organization, "Obesity and overweight factsheet," 2017. [Online]. Available: https://www.who.int/news-room/fact-sheets/detail/obesity-and-overweight. Accessed Aug. 30, 2024

Zainudin S, Daud Z, Mohamad M, Boon ATT, Mohamed WMIW (2011) A summary of the Malaysian clinical practice guidelines on management of obesity 2004. J ASEAN Fed Endocrine Soc 26(2):101–101

Zamani Sani SH, Fathirezaie Z, Brand S, Pühse U, Holsboer-Trachsler E, Gerber M, Talepasand S (2016) Physical activity and self-esteem: testing direct and indirect relationships associated with psychological and physical mechanisms. Neuropsychiatr Dis Treat 12:2617–2625. https://doi.org/10.2147/ndt.S116811

The Need for Software in Assisting Coaches in Johor with Generating Periodization

Mohamed Rafi M. S. Solaiman, Asha Hasnimy Binti Mohd Hashim, Adjah Naqkiah Mazlan, and Nur Fatihah Raimi

Abstract Periodization is the process of designing and dividing training over a specific period to achieve peak performance for an individual. There are three distinct phases based on the targeted objectives: the preparation phase, the competition phase, and the transition phase. It is crucial for coaches to ensure that the training plan is suitable for reaching the athlete's peak performance level. However, developing periodization is challenging due to several issues, such as the vast amount of training information and the difficulty in accurately evaluating or viewing data. To address these challenges, a study was conducted to identify the need for software to assist coaches in creating periodization. This study employed a quantitative approach. Data was collected through a questionnaire on software needs for periodization, which was distributed to 30 respondents, all of whom are coaches from the state of Johor. The data obtained from the questionnaire was analyzed using the Statistical Package for the Social Sciences (SPSS). The findings from the questionnaire revealed an average value of 4.68, indicating that specialized software is highly needed to support coaches in developing periodization. In conclusion, coaches require software that can help them achieve peak athlete performance.

Keywords Periodization · Peak performance

A. H. B. M. Hashim · A. N. Mazlan · N. F. Raimi
Sport Innovation Technology Centre, Institute of Human Centered Engineering, Universiti Teknologi Malaysia, Johor Bahru, Johor, Malaysia

M. R. M. S. Solaiman (✉) · A. H. B. M. Hashim · A. N. Mazlan · N. F. Raimi
School of Education, Faculty of Social Sciences and Humanities, Universiti Teknologi Malaysia, Johor Bahru, Johor, Malaysia
e-mail: mohamedrafi@pis.edu.my

Politeknik Ibrahim Sultan, Pasir Gudang, Johor, Malaysia

M. H. A. Hassan et al. (eds.), *Proceedings of the 3rd International Conference on Innovation and Technology in Sports*, Lecture Notes in Bioengineering,
https://doi.org/10.1007/978-981-95-0804-4_14

1 Introduction

Modernization today heavily relies on technology to facilitate various processes in achieving planned objectives. The use of modern technology has become a cornerstone in the development of any country. The sports field is no exception in leveraging modern technology to foster creative, innovative, and excellent individuals. Technology is rapidly advancing and has become a key component in training planning. Training plans must be systematic and yield positive effects on the execution of training.

In sports, sophisticated equipment and facilities are used to make training more effective. Sports are not limited to training alone; they also encompass the use of technology, skills, management, and education. The application of technology in training is crucial for ensuring that athletes master sports science knowledge more easily. High-impact training effectiveness can be achieved through the use of modern technology.

The use of modern technology in planning annual or periodized training plans is essential to ensure smooth training processes. Periodization is known as the training plan designed to assess the effectiveness of training or the peak achievement of an individual's training. Periodization involves the management of training to reach peak performance at maximum levels (Tudor and Gregory 2009).

Periodization consists of three phases: the preparatory phase, the competitive phase, and the transition phase (Tudor and Gregory 2009). Success in a competition does not rely solely on the athlete; it also depends on the coach and team management. According to an interview with the Head of the Johor SUKMA 2020 Secretariat, apart from the athlete's capability to achieve success, the coach plays a crucial role in creating a training plan that ensures the training provided is suitable for reaching optimal levels.

A good relationship between the coach and the athlete is closely related to their performance. This relationship helps achieve success and consequently enhances the athlete's performance (Din and Awang 2015). Additionally, a coach's experience plays a significant role in crafting an effective planning strategy. Extensive experience builds trust between the coach and the athlete (Laios 2018). A coach's experience is influenced by three main aspects: planning the annual training program, training and competition experience, and coaching education experience (Santos et al. 2010).

Periodization is essential for coaches to achieve their athletes' peak performance. The structure of periodization greatly assists coaches in developing an effective training plan. This is based on the components involved in creating periodization. The structure or components of periodization include starting with the determination of planning phases: the preparatory phase, the competitive phase, and the transition phase (Tudor and Gregory 2009). This is followed by defining training objectives, seasons, goals, macrocycles, micro cycles, types of training, and factors that athletes must undergo within a training planning cycle.

In the field of sports medicine, software such as the BTrackS Balance Plate and Sport Balance software is used for diagnosing concussions among athletes (Goble et al. 2016). This indicates that technology is needed not only in sports medicine but also in assessing athlete performance over time to ensure the training provided is sufficient for achieving maximum results.

In 2018, the National Sports Institute (ISN) provided a Sports Performance Analysis System specifically for hockey in Malaysia to help the national men's hockey team win a gold medal at the 2018 Asian Games and qualify automatically for the Tokyo 2020 Olympics (Zatil 2018). This system included the 'Catapult Sport System' and 'Sportcode Performance Analysis Software V11' for the national men's hockey team. The Catapult Sport System is used to analyze athletes' movements during training and competition, while the Sportcode Performance Analysis Software V11 is used to study players' behavior, tactics, and techniques during training and competition through video recordings or live analysis.

2 Methodology

This study uses a quantitative approach to study the software needs in assisting Johor State coaches to produce sports training periodization. The factor that enabled the researcher to conduct the study effectively was the presence of detailed planning (Idris 2013). The methodology used involves collecting data through questionnaires and interviews. This study is used to obtain more comprehensive information related to the needs of periodization software among coaches.

This study is a descriptive study that aims to identify the needs of software in assisting coaches to produce training periodization. This study is carried out in the form of a survey using a questionnaire instrument that has been developed based on a literature review and research needs on periodization. If the population size is small, the researcher can use the entire population as the study sample (Mohd Najib Abdul Ghafar 2003). In this study, the researcher used the entire population (State Coaches of Johor) as the sample in various sports fields. While the study sample, the researcher has selected 30 coaches who have experience in preparing periodization. The sample selection criteria, the researcher chose in terms of coaches who have at least three years of experience in producing training plans as well as individuals who are directly involved in preparing athlete periodization.

The study instrument used in this research is a questionnaire that includes several main parts, namely part A related to demographic information and part B related to software requirements in periodization. This questionnaire uses a five-point Likert scale (1 = Strongly Disagree to 5 = Strongly Agree) to obtain coaches' perceptions of the need for periodization software.

This study was carried out in several main phases, namely questionnaire development, pilot testing, questionnaire distribution, data collection and data analysis. The questionnaire development was developed based on literature review and was refined with the help of experts in the field of sports. While pilot testing was carried out before the questionnaire was distributed, the pilot test was conducted on five coaches to ensure the validity and reliability of the instrument. The researcher distributed the questionnaire online and physically to the coaches involved in this study. The final phase is to collect and analyze data where the data obtained needs to be analyzed using statistical software to obtain the mean value, standard deviation and the relationship between the variables studied.

The researcher used SPSS software to obtain more accurate and systematic results. The researcher used SPSS software to obtain the mean value and standard deviation to analyze the coaches' perceptions of the need for periodization software. This systematic selection of methodology can help researchers provide a clear picture of the importance of software in producing more effective and systematic training periodization.

3 Results and Discussion

3.1 Results

The results from the training plan preparation workshop organized by the Johor State Sports Council reveal that coaches face difficulties in creating periodization using manual methods. Currently, coaches use Microsoft Excel and Periodization Schedules to prepare periodization. However, they find it challenging to evaluate the periodization created with these schedules due to the limited space for entering information. As a result, the periodization produced is often unclear, making it difficult to interpret the information accurately. Coaches require more time to prepare comprehensive periodization.

Table 1 shows the results of the survey conducted indicate that the average need for periodization software at a high level is 4.68. This specialized software can be used to analyze data easily, plan training more systematically according to a more organized training phase, and save time. Respondents strongly agree that specialized periodization software can help coaches plan appropriate training. Technology is an easy way to communicate among coaches and athletes because it facilitates the delivery of information. Through this method, the information to be conveyed or the planning to be implemented will reach the athletes easily and quickly.

3.2 Discussion

Periodization is important because it provides a systematic training plan that follows organized training phases according to specific needs, helping to avoid overtraining or injury. It is crucial for coaches to develop athletes who are capable of competing according to scheduled times (Tudor and Gregory 2009).

Additionally, data obtained from periodization is used to determine training levels and improve athletic performance. To achieve these objectives, coaches need to

Table 1 Mean construct the need for software to assist coaches in developing periodization

No	Item	Mean	SP
1	Periodization software can help me analyze data easily	4.60	0.621
2	Periodization software can help me design training more systematically	4.57	0.626
3	Periodization software can help me create training phases more orderly	4.70	0.466
4	Periodization software can help me develop training progressions clearly	4.77	0.430
5	Periodization software can help me determine training levels according to needs	4.77	0.430
6	Periodization software can help me assess athletes' abilities in training	4.70	0.466
7	Periodization software can help me evaluate athletes' performance	4.70	0.466
8	Periodization software can help me save time	4.60	0.621
9	Periodization software can help me plan appropriate training	4.83	0.379
10	Periodization software can help me enhance athletes' performance	4.57	0.504
	Overall Mean	**4.68**	**0.36**

provide training in physical, technical, tactical, and psychological aspects of the athletes (Ishak et al. 2012). Systematic and organized training helps coaches assess athletes' physical fitness and evaluate their performance (Ismail and Rajab 2010). Evaluations should not be limited to competitions or achievements but should also occur throughout training sessions. Coaches need to assess athletes' performance regularly to ensure that training aligns with the periodization phases.

Researchers have concluded that the majority of respondents agree that periodization effectively evaluates athlete performance. This was supported by interviews with the Head of the SUKMA Johor 2020 Secretariat, who stated that periodization enables coaches to assess athlete performance more accurately. Periodization is essential for monitoring training progress and setting training levels according to the athlete's capabilities in their respective sport.

Information technology greatly aids individuals in performing their tasks more easily, quickly, and effectively (Nejadirani and Behravesh 2011). This is evident as even less experienced coaches can develop training plans using technology. In today's technological world, coaches can easily and quickly learn or obtain information through applications or software. The use of technology now helps coaches analyse training, strengths, and weaknesses of opponents during competitions. Coaches have utilized technology in both training and competitions to ensure that the training provided to athletes is appropriate and can enhance their performance (Ismail et al. 2020).

4 Conclusion and Contributions

Software is a tool that can assist nearly everyone in making their tasks easier. Information technology greatly helps individuals perform their tasks more easily, quickly, and effectively. This is evident as even less experienced coaches can develop training plans using technology with ease. In today's technological world, coaches can quickly and easily learn or obtain information through applications or software. The use of technology now helps coaches analyse training, strengths, and weaknesses of opponents during competitions. Coaches have utilized technology in both training and competitions to ensure that the training provided to athletes is appropriate and can enhance their performance (Ismail et al. 2020).

Periodization is important because it provides a systematic training plan that follows organized training phases according to specific needs, helping to avoid over-training or injury. It is crucial for coaches to develop athletes who are capable of competing according to scheduled times (Tudor and Gregory 2009).

Acknowledgements Special thanks to Universiti Teknologi Malaysia, Skudai, Johor, Malaysia, Politeknik Ibrahim Sultan and Majlis Sukan Negeri Johor team members for your kind support.

References

Din A, Rashid SA, Awang MI (2015) Aspek pengurusan program latihan dan kesanya terhadap kepuasan atlet sekolah sukan. Universiti Utara Malaysia, Sintok, 06010 Bukit Kayu Hitam, Kedah. Malaysia

Ghafar MNA (2003) Reka Bentuk Tinjauan Soal Selidik Pendidikan, Penerbitan UTM, Skudai, Johor Bahru

Goble DJ, Manyak KA, Abdenour TE, Rauh MJ, Baweja HS (2016) An initial evaluation of the btracks balance plate and sports balance software for concussion diagnosis

Ishak A, Hashim HA, Jarwis MN, Ahmad H (2012) Modul Latihan dan Kecergasan. Universiti Pendidikan Sultan Idris, Perak

Ismail MH, Rajab MKNA (2010) Penilaian Tahap Prestasi Kecergasan Fizikal Berasaskan Kesihatan Bagi Atlet Bola Tampar SUKMA 2008. Universiti Teknologi Malaysia, Fakulti Pendidikan

Ismail FH, Karim ZA, Ghazali N (2020) Kompetensi guru sebagai jurulatih bola sepak dan kesannya terhadap prestasi pemain bola sepak Pusat Latihan Daerah Kuala Lumpur. Jurnal Sains Sukan Dan Pendidikan Jasmani 9(1):47–61

Laios A (2018) The educational system for training coaches in Greece. Int J Educ Manag 19(6):500–504

Nejadirani F, Rasouli R, Behravesh M (2011) The effect of applying information technology on efficiency of parks and green space organization: a case study. Middle-East J Scientific Res 10(2)

Noraini Idris (2013) Penyelidikan dalam Pendidikan. Edisi ke 2. McGraw-Hill Education

Santos S, Mesquita I, Graça A, Rosado A (2010) Coaches' perceptions of competence and acknowledgement of training needs related to professional competences. J Sports Sci Med 9(1):62

Tudor BO, Gregory HG (2009) Periodization theory and methodology of training. York University, United States, Human Kinetics

Zatil N (2018) ISN Serah Perisian Terkini Sistem analisis Prestasi Sukan kepada Konfederasi Hoki Malaysia. new.isn.gov.my

Does the Dynamics Strength Index Associate with the Swimming Performance in Age-Group Swimmers?

Radomyos Matjiur, Siwakorn Akkaneekool, Sakchai Pitakwong, Weerawat Limroongreungrat, Rodrigo Zacca, Karla de Jesus, and Phornpot Chainok

Abstract The Dynamic Strength Index (DSI) is highly valuable for assessing an athlete's training level in terms of neuromuscular strength and power. The normative data for DSI ratios and relationships to other performance (i.e., counter movement, sprint and swimming performance) have been examined. Since it is not known whether percentage DSI is well related with swimming performance particularly in age-group swimmers, we investigated the associations between DSI and the performance in 50-m sprint swimming across four different swimming techniques. Thus, 120 (67 boys and 53 girls) age-group swimmers (age: 12.0 ± 1.0 versus 11.9 ± 0.9 years old, height: 154.6 ± 12.3 versus 153.4 ± 11.2 cm, body mass: 46.7 ± 12.2 versus 45.7 ± 9.1 kg, maturity offset: -4.8 ± 1.2 versus -1.9 ± 1.1yrs) were recruited for this study. Inverse relationships between DSI and the performance of front crawl ($r = -0.55$, $p = 0.02$) and butterfly ($r = -0.53$, $p = 0.04$) were observed in girls and trend of inverse connection exhibited between DSI and 50-m time across all four techniques in girls. The DSI is useful in evaluating age-group sprinting performance and provide normative data for ballistic and isometric peak force measures.

R. Matjiur · S. Akkaneekool · P. Chainok (✉)
Faculty of Sport Science, Burapha University, Chonburi, Thailand
e-mail: phornpot@go.buu.ac.th

S. Pitakwong
Rattana Bundit University, Bangkok, Thailand

W. Limroongreungrat
College of Sport Science and Technology, Mahidol University, Nakhon Pathom, Thailand

R. Zacca
Research Center in Physical Activity, Health and Leisure (CIAFEL), Faculty of Sports, University of Porto (FADEUP), Porto, Portugal

Laboratory for Integrative and Translational Research in Population Health (ITR), Porto, Portugal

K. de Jesus
Human Performance Laboratory, Faculty of Physical Education and Physiotherapy, Federal University of Amazonas, Manaus, Brazil

M. H. A. Hassan et al. (eds.), *Proceedings of the 3rd International Conference on Innovation and Technology in Sports*, Lecture Notes in Bioengineering,
https://doi.org/10.1007/978-981-95-0804-4_15

Keywords Neuromuscular performance · Aquatics · Bio-banding

1 Introduction

Neuromuscular strength and swimming performance are two distinct yet interconnected aspects of physical fitness that have garnered significant attention in the realm of sports science and training perspective (Zacca et al. 2020a).The performance of swimming is influenced by the balance of powers between the force generated by the arm and leg movements and the resistance caused by the drag, or water resistance, experienced by the body while swimming (Nugent et al. 2019; Fone and Tillaar 2022; Carvalho et al. 2023). Thus, swimmers require a high level of mechanical power output and muscular strength to achieve optimal swimming performance (Fone and Tillaar 2022).

Swimmers who demonstrate high levels of strength are closely linked to improved performance in various aspects of swimming, such as swim block start, turning, and overall swim performance, especially in sprint swimming. Furthermore, the sprint swim performance is associated with maximum strength of both the upper and lower limbs, as well as specific markers of swim performance and neuromuscular function (Keiner et al. 2021).

The dynamic strength index (DSI), is a strength evaluation that is frequently used in the field of strength and conditioning. It can be utilized as a method for evaluating the lower-limbs strength attributes of an athlete and for guiding and evaluating future training interventions (Thomas et al. 2017). The DSI is calculated as the ratio between dynamic and maximal isometric strength, and provides insight into the ability to apply strength quickly and explosively (McMahon et al. 2017; Ong et al. 2024). When extrapolating DSI to the field of aquatic sports, it is observed that DSI is directly associated with sports-specific performance in surfers (Parsonage et al. 2018). The results showed that male surfers had significantly higher DSI values compared to females, and DSI was positively correlated with paddle power, which is an important component of surfing performance (Parsonage et al. 2018).

It is interesting to note that neuromuscular performance is likely to have a significant impact on swimming performance. However, the differences between the butterfly, backstroke, breaststroke, and front crawl in sprinting events, as well as differences between swimmers of the same age group, particularly between the ages of 11 and 14 years, when there are maturity-associated differences in size and function, are not well understood. Therefore, the purpose of this study was identify the associations between DSI and 50-m swimming performance across the four swimming techniques. Based on the findings of previous research in aquatic sport like surfing, it was hypothesized that statistically and practically significant differences in DSI and swimming performance would exist among swimming techniques.

2 Materials and Methods

2.1 *Subjects*

This study included 120 male (n = 63) and female (n = 57) age-group swimmers, specialized in short distance, provided written informed consent to participate in the study. Boys and girls characteristics were (respectively): 12.0 ± 1.0 versus 11.9 ± 0.9 years old, 154.6 ± 12.3 versus 153.4 ± 11.2 cm of height, 46.7 ± 12.2 versus 45.7 ± 9.1 kg of body mass, -4.8 ± 1.2 versus -1.9 ± 1.1yrs of maturity offset, 14.2 ± 0.5 versus13.8 ± 0.4 yrs of age of peak height velocity and 352.2 ± 94.5 versus 429.2 ± 76.6 of the World Aquatics point scoring. The inclusion criteria are included based on the following: (i) had a minimum of one years of experience at the regional or national competition, (ii) qualify to the national championships, (iii) free from current injury in the training period. The exclusion criterion is poor health status and physical condition with potential medical problems. The study conformed to the principles of the World Medical Association's Declaration of Helsinki at the local university (code number HS025/2567(C3)). The participants and their parents were provided with a comprehensive explanation of the testing processes before obtaining signed consent for participation.

2.2 *Design*

We conducted a cross-sectional study and tested age-group swimming performance and neuromuscular performance during the first macrocycle of a traditional three-peak preparation program.

2.3 *Methodology*

The experimental testing occurred in the last two week before the National Age-Group Championships (first macrocycle of the season). Participants were familiarized with the tests, which were conducted at the same time of the day to avoid systematic bias due to circadian variation (Zacca et al. 2020b). The official race time for the 50-m events at the National Age-Group Championships was obtained at the end of the first macrocycles. Age-group swimmers were instructed to avoid vigorous exercise the day before testing.

Measurements of anthropometry were evaluated in accordance with the standards established by the International Society for the Advancement of Kinanthropometry (ISAK). For the purpose of determining the standing height, body weight, and percentage of body fat, the bioelectrical impedance analysis (BIA) (Body Composition Analyzer: X-CONTACT 357S, Seoul, South Korea) was utilized.

Dynamic strength index (DSI): Before the start of testing, athletes performed a standardized 15-min warm-up routine consisting of light aerobic exercise and dynamic stretches for all major muscle groups and mobilization exercises. Age-group swimmer performed counter movement jump (CMJ) on the commercially available K-deltas dual force platforms (Kinvent Physio, Montpellier, France) at a sampling rate of 1000 Hz. The Kinvent Physio software (Kinvent Physio, Montpellier, France) was used to analyse and the obtained peak force (N) was used for further analysis. The isometric mid-thigh pull (IMTP) was conducted on the K-deltas dual force platforms 60 s after completing the counter movement jump (CMJ). Participants exerted maximum force by pulling the bar and driving their feet into the force platform for a duration of 5 s. Peak forces (N) was used for further analysis and DSI was calculated by dividing CMJ peak force by IMTP peak force (Thomas et al. 2017).

2.4 4Statistical Analysis

An a priori power analysis to determine sample size was completed using G*Power 3.1.9.2 software and showed that a sample size at least 112 to detect a small effect size 0.30 using a power of 0.90 and alpha of 0.05.The statistical analyses were performed using Jamovi (v2.3.2.0), which is a software package built on the R statistical language developed by the R Foundation for Statistical Computing. The assumption of normality was verified with the Shapiro–Wilk test. Mean and SD for descriptive analysis were obtained and reported for all studied variables. Pearson correlation coefficients were calculated to determine associations between DSI and 50-m time swimming performance across the four swimming techniques, with values interpreted as trivial (< 0.10), small (0.10–0.29), moderate (0.30–0.49), large (0.50–0.69), very large (0.70–0.89), and nearly perfect ($\geq$ 0.90) (Hopkins et al. 2009). Significance level alpha was established at 0.05.

3 Results

Statistical overview of the participants' DSI performance is presented in Table 1. The CMJ-PF and IMTP-PF correlation demonstrated excellent reliability and acceptable variability (boys: 0.59–0.92, girls: 0.62–0.99), as shown in Fig. 1. In boys, breaststroke and backstroke showed a trivial positive correlation with DSI ($r = 0.08$ and 0.07, $p > 0.05$), while front crawl exhibited a small positive correlation ($r = 0.26$, $p > 0.05$). On the other hand, butterfly had a small inverse correlation with DSI ($r = -0.26$, $p > 0.05$). Among girls, the backstroke showed a small inverse correlation ($r = -0.16$, $p > 0.05$) with DSI, while the breaststroke also exhibited a moderate inverse correlation ($r = -0.33$, $p > 0.05$) with DSI. Notably, the front crawl and butterfly strokes showed a large inverse correlation with DSI, with correlation coefficients of −0.55 and −0.53, respectively ($p < 0.05$) (Fig. 2).

Table 1 Descriptive statistics (mean ± SD) for dynamic strength index (DSI), counter movement peak force (CMJ-PF), isometric mid-thigh pull (IMTP-PF), and 50-m swimming performance of age-group swimmers

Variables	Girls (N = 57) MEAN ± SD	Boys (N = 63) MEAN ± SD	Total (N = 120) MEAN ± SD
DSI (%)	0.91 ± 0.18	0.95 ± 0.15	0.93 ± 0.17
CMJ-PF (N)	691.44 ± 456.69	759.07 ± 487.38	726.94 ± 472.29
IMTP-PF (N)	359.16 ± 239.23	508.93 ± 418.62	437.79 ± 351.96
50-m front-crawl time (s)	31.10 ± 1.83	29.38 ± 2.18	30.24 ± 2.16
50-m breaststroke time (s)	39.27 ± 2.99	38.53 ± 3.25	38.81 ± 3.12
50-m backstroke times (s)	35.67 ± 2.36	34.27 ± 2.96	35.04 ± 2.70
50-m butterfly time (s)	33.04 ± 1.19	31.17 ± 2.97	32.04 ± 2.47

4 Discussion

We evaluated the associations between the DSI and the performance in 50-m swimming across four different swimming techniques. Our main outcomes suggest that there is no link between DSI and a 50-m swimming performance in boys. Nevertheless, we observed an inverse relationship between DSI and the performance of front crawl and butterfly techniques in girls. Remarkably, a similar trend of inverse connection exhibited between DSI and 50-m swimming performance across all four techniques in girls. Therefore, our hypothesis was partly supported. This finding is important and emphasizes that, in a 50-m swim where explosive strength is crucial, an optimal DSI can serve as a key indicator of potential performance and inform training strategies to enhance sprint abilities, especially in age-group swimmers whose growth and maturation changes directly affect their swimming performance.

This is the first study to evaluate the relationship between neuromuscular performance by using DSI and 50-m swimming performance in age-group, particularly during pre-puberty and puberty (11–14 yrs) which is growth and maturation significantly influence 50-m swimming performance. During the ages of 11 to 14, the development of muscular strength and power varies significantly between boys and girls due to differences in the timing of puberty and hormonal changes (Zacca et al. 2020a).Our age-group swimmer's DSI values in both boys (0.95 ± 0.15) and girls (0.91 ± 0.18) were relatively high when compared with data from young Brazilian soccer players (0.76 ± 0.11) (Portella et al. 2024) and collegiate athlete (0.47- 0.51) (Suchomel et al. 2020).

From the perspective of propulsive efficiency in swimming, higher DSI typically indicates that swimmer's explosive strength is well-developed relative to their maximal strength, which can be beneficial in sports that require quickly cyclic of whole-body swimming (swimming with both upper- and lower-limb propulsive motions) and powerful propulsion to maximise the kinetic energy that directly contributes to the forward motion while swimming (Nugent et al. 2019; Fone and Tillaar 2022; Carvalho et al. 2023).

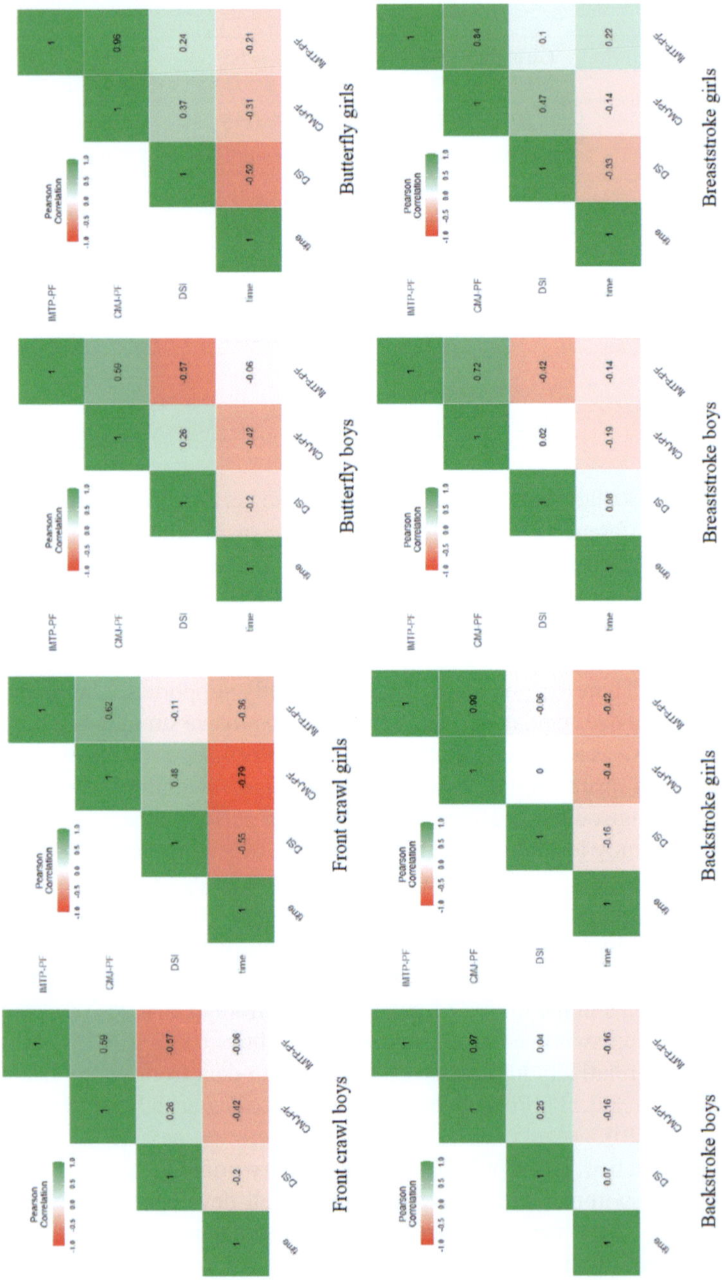

Fig. 1 Correlation among the component of DSI and 50-m swimming performance across the four techniques in boys and girls

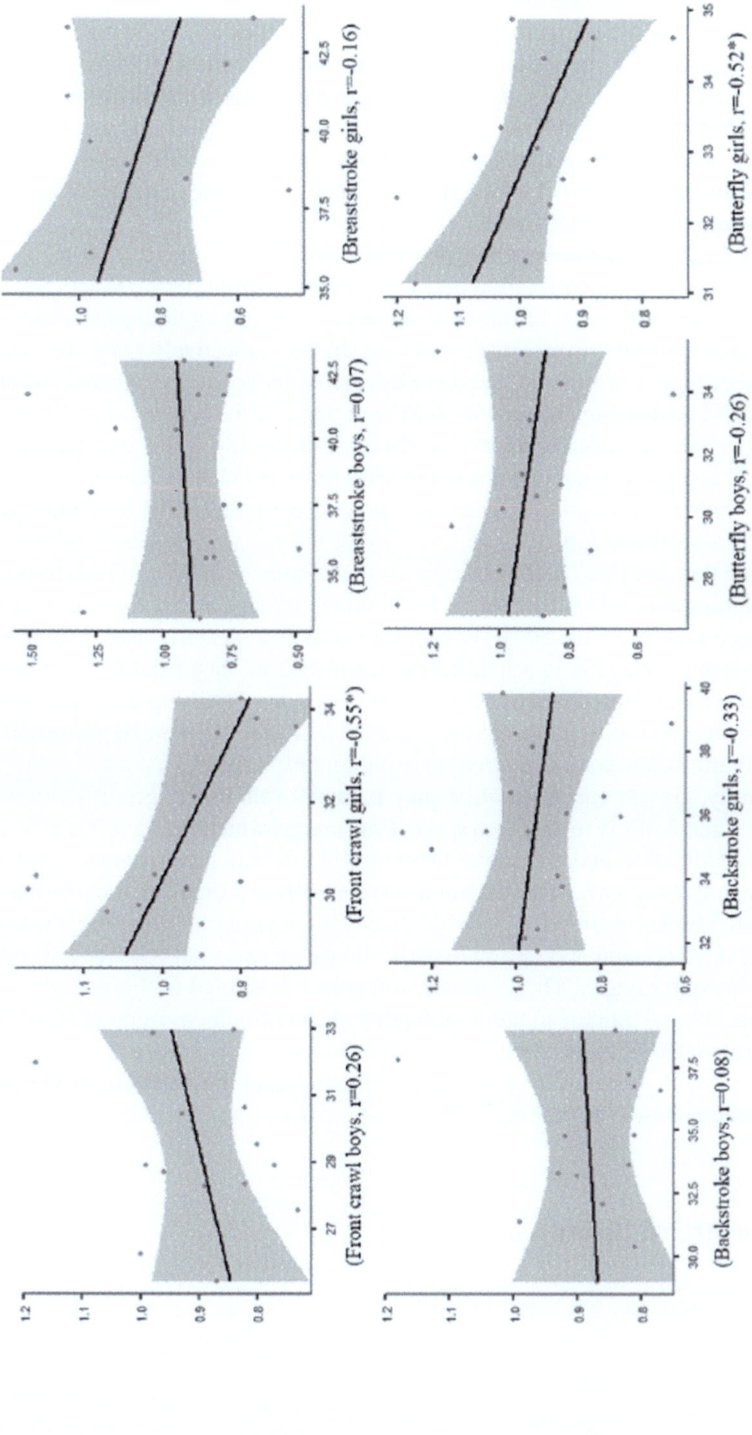

Fig. 2 Correlation between DSI and 50-m swimming performance across the four techniques

These findings are in line with previous research, which stated that upper and lower limb strength and power are important attribute to discriminate the performance levels in elite swimmers (Nugent et al. 2019; Carvalho et al. 2023). In regards to the correlation between the strength and power of the upper and lower limbs and sprint timings, Carvalho et al. (2023) discovered that sprint times were moderate-largely inversely correlated with upper and lower limb strength and power and greater levels of maximum strength were linked to longer stroke length in butterfly and front crawl strokes. Additionally, higher dry-land strength values were able to differentiate elite swimmers from non-elite swimmers.

Previous study on swimming force production, utilizing tethered force measurement, has shown a strong correlation between force and sprint performance. Specifically, the highest values of absolute maximum force production were observed in breaststroke, butterfly, front crawl, and backstroke, respectively (Morouço et al. 2014). Furthermore, it was noted that there were stronger connections between force parameters and swimming distance in short sprinting (50-m) compared to 100-m or even lower in 200-m. Taken together, the data demonstrating an inverse correlation between DSI and girls' front crawl and butterfly times provide evidence to the idea that young swimmers require a more specific and comprehensive neuromuscular performance assessment tool.

Following that, previous research suggests that a balanced DSI can help improve sprint swimming times. It is essential for swimmers to have an appropriate DSI for short-distance events like the 50-m swim since it affects their start, turns, and overall speed. For swimmers, DSI is a helpful indicator of how they establish a balance between power and explosiveness. Optimal DSI can drive training interventions to maximize sprint capabilities and serve as a crucial indicator of potential performance in a 50-m swim, when explosive strength is particularly critical.

Importantly, growth and maturation play a crucial role in age-group swimming performance, particularly in the 50-m sprint. Physical growth, hormonal changes, and improved coordination and technique all contribute to better performance. Understanding the impact of growth and maturation related to neuromuscular performance allows coaches to create effective, individualized training programs that support long-term athlete development and optimize performance as swimmers progress through their developmental stages. Therefore, the references presented in this study can be used for coaches and trainers to tailor strength and conditioning programs according to an athlete's stage of maturation. For instance, during early stages of maturation, training may focus more on developing coordination and motor skills, while later stages might emphasize enhancing explosive strength to improve DSI.

5 Disclosure of Interests.

There are no competing interests to report.

Acknowledgements We would like to acknowledge all the age-group swimmers, coaches and parents who voluntarily participated and provided impressive support throughout the data collection. We also extend our special thanks to the technicians of the Faculty of Sport Science, Burapha University for their time, collaboration and commitment to this study. Radomyos Matjiur and Phornpot Chainok are funded by the Faculty of Sport Science, Burapha University (005/2567). Rodrigo Zacca is supported by The Research Center in Physical Activity, Health and Leisure (CIAFEL), Faculty of Sport, University of Porto (FADEUP), which is part of the Laboratory for Integrative and Translational Research in Population Health (ITR); both are funded by the Fundação Para a Ciência e Tecnologia (FCT; grants UIDB/00617/2020 https://doi.org/https://doi.org/10.54499/UIDB/00617/2020; UIDP/00617/2020 https://doi.org/https://doi.org/10.54499/UIDP/00617/2020 and LA/P/0064/2020, respectively)

References

Carvalho DD, Monteiro AS, Fonseca P, Silva AJ, Vilas-Boas JP, Pyne DB, Fernandes RJ (2023) Swimming sprint performance depends on upper/lower limbs strength and swimmers level. J Sports Sci 41(8):747–757

Fone L, van den Tillaar R (2022) Effect of different types of strength training on swimming performance in competitive swimmers: a systematic review. Sports Med Open 8(1):19

Hopkins W, Marshall S, Batterham A, Hanin J (2009) Progressive statistics for studies in sports medicine and exercise science. Med Sci Sports Exerc 41(1):3

Keiner M, Wirth K, Fuhrmann S, Kunz M, Hartmann H, Haff GG (2021) The influence of upper- and lower-body maximum strength on swim block start, turn, and overall swim performance in sprint swimming. J Strength Cond Res 35(10):2839–2845

McMahon JJ, Jones PA, Dos' Santos T, Comfort P (2017) Influence of dynamic strength index on countermovement jump force, power, velocity, and displacement-time curves. Sports (Basel) 5(4):72

Morouço PG, Marinho DA, Keskinen KL, Badillo JJ, Marques MC (2014) Tethered swimming can be used to evaluate force contribution for short-distance swimming performance. J Strength Cond Res 28(11):3093–3099

Nugent F, Comyns T, Nevill A, Warrington GD (2019) The effects of low-volume, high-intensity training on performance parameters in competitive youth swimmers. Int J Sports Physiol Perform 14(2):203–208

Ong KY, Chua K, Lum D (2024) Relationship between dynamic strength index and early force production of isometric mid-thigh pull. J Sci Sport Exerc 1–7

Parsonage J, Secomb J, Dowse R, Ferrier B, Sheppard J, Nimphius S (2018) The assessment of isometric, dynamic, and sports-specific upper-body strength in male and female competitive surfers. Sports(Basel) 6(2):53

Portella DL, Jatene P, Da Silva AO, Dos Santos GS, Monteiro D, Teixeira JE, Forte P (2024) Comparison of force variables and dynamic strength index between age groups in elite young Brazilian football goalkeepers. Front Sports Act Living 6:1282214

Suchomel TJ, Sole CJ, Bellon CR, Stone MH (2020) Dynamic strength index: relationships with common performance variables and contextualization of training recommendations. J Hum Kinet 74(1):59–70

Thomas C, Dos' Santos T, Jones PA (2017) A comparison of dynamic strength index between team-sport athletes. Sports (Basel) 5(3):71

Zacca R, Azevedo R, Chainok P, Vilas-Boas JP, Castro FAdS, Pyne DB, Fernandes RJ (2020a) Monitoring age-group swimmers over a training macrocycle: energetics, technique, and anthropometrics. J Strength Cond Res 34(3):818–827

Zacca R, Azevedo R, JrVR R, Abraldes JA, Vilas-Boas JP, de Souza Castro FA, Fernandes RJ (2020b) Biophysical follow-up of age-group swimmers during a traditional three-peak preparation program. J Strength Cond Res 34(9):2585–2595

The Use of Technology Integration Software to Enhance Level of Movement Skills Based on Simpson's Psychomotor Domain

Adjah Naqkiah Mazlan, Jamalludin Harun, Zainal Abidin Zainuddin, and Nur Fatihah Raimi

Abstract With the rapid advancement of the increasingly technological world, educators are rightly encouraged to use technology in teaching, such as the use of multimedia. The advantages of multimedia are not only as facilitators but also as substitutes or supporters of the learning process being conducted. The field of sports is no exception, as the use of technology allows skills, techniques, and tactics to be clearly conveyed. The main purpose of this study is to observe the level of skill movement mastery based on Simpson's Psychomotor Domain Taxonomy and how software aids in the mastery of sports skills. The evaluation of the effectiveness of this software involves several instruments, such as observations, interviews, and questionnaires. The results of the analysis recorded the mean and standard deviation readings for observation findings as 4.61 ± 0.27 and for questionnaires as 4.52 ± 0.04, indicating that students' skill levels were at the third level, which is the guided response level, according to Simpson's psychomotor domain after using the software. Overall, the findings on how the software helps in mastering skill movements show a positive impact on the student learning process. In conclusion, the provided software has a positive effect on the level of mastery of students' skill movements.

Keywords Technology integration software · Level of movement skills · Psychomotor domain · Simpson's taxonomy

A. N. Mazlan (✉) · N. F. Raimi
Sport Innovation Technology Centre, Institute of Human Centered Engineering, Universiti Teknologi Malaysia, Bahru, Johor, Malaysia
e-mail: adjahnaqkiah@utm.my

A. N. Mazlan · J. Harun · Z. A. Zainuddin · N. F. Raimi
Fakulti Sains Pendidikan dan Teknologi, Universiti Teknologi Malaysia, Bahru, Johor, Malaysia

M. H. A. Hassan et al. (eds.), *Proceedings of the 3rd International Conference on Innovation and Technology in Sports*, Lecture Notes in Bioengineering,
https://doi.org/10.1007/978-981-95-0804-4_16

1 Introduction

In recent years, the use of technology integration software in education has expanded, offering new and innovative ways to enhance the learning experience across various disciplines, including physical education. Movement skills, which encompass fundamental motor skills such as running, jumping, throwing, and catching, are critical for the physical development of children and adolescents (Logan and Robinson 2020). These skills not only contribute to physical fitness but also support cognitive development, social interaction, and overall well-being (Barnett et al. 2022). As such, finding effective methods to teach and improve these skills is essential.

Through the keep changing era, the education system should go hand in hand with the progress of the world where the practice of the use of technology in the learning process should be compulsory (Adjah Naqkiah Mazlan 2014). Technology integration software provides unique opportunities to support and enhance the teaching of movement skills by offering interactive, personalized, and engaging learning experiences. For example, motion analysis software allows students to receive real-time feedback on their movements, enabling immediate correction and fostering an understanding of proper techniques (Vickery and Wulf 2021). Additionally, gamification and augmented reality applications can increase student motivation and participation by making physical education sessions more fun and rewarding (Granacher and Borde 2019).

Research has demonstrated that incorporating technology into physical education can significantly improve students' movement skills and overall engagement. A study found that students who used interactive video-based software for movement skills training showed notable improvements in their motor abilities compared to those who followed traditional methods (Garcia-Hermoso et al. 2021). Similarly, reported that augmented reality-based exercises could enhance both the motor skills and cognitive engagement of students, offering a more holistic approach to physical education (Hollis et al. 2023).

Despite the promising outcomes associated with technology integration, there are challenges to be addressed, such as ensuring equitable access to technological resources, providing adequate training for educators, and maintaining a balance between technology use and physical practice (Riso and Topolski 2020). It is crucial to understand how to effectively integrate technology into physical education to maximize its benefits without compromising the essence of physical activity (Adjah Naqkiah Mazlan 2014).

Thus, this study aims to investigate the effectiveness of various technology integration software in enhancing the movement skills of students. By examining the latest advancements in technology, including motion capture systems, interactive video analysis, and simulation applications, this study seeks to identify best practices for leveraging technology to improve physical education outcomes.

2 Methodology

In this study, a quantitative design is implemented first to obtain a preliminary overview of the matter. Subsequently, qualitative data is used to gain a deeper understanding of the issue. To identify the level of movement skill mastery and how multimedia learning software integrated with video analysis technology aids in mastering movement skills based on Simpson's psychomotor domain taxonomy, observation, interviews, and questionnaires were used.

The researcher used a post test-only design without involving a separate control group because it was anticipated that the respondents would be difficult to control, and it had been identified that they lacked proficiency in the skills to be tested, as determined through an initial survey conducted by the researcher. The respondents had no experience or knowledge of the required skills, leading the researcher to exclude pretest-post test testing, as it was assumed that there would be no interference from other factors.

The samples consists of students who were intensively involved in this study. A total of 30 students enrolled in the Gymnastics & Volleyball course participated in this study and used the developed learning software. This sample group was selected using a purposive sampling technique by choosing one section that was taking the relevant course. The purposive sampling method was chosen because the researcher did not want to disrupt the predetermined class or section assignments set by the faculty management and based on the agreement of the lecturers involved in teaching the subject.

Furthermore, the selection of samples to answer all the research questions through interviews was also based on the purposive sampling method. Respondents were selected voluntarily, consisting of nine students who were divided into three groups: active, moderately active, and less active in sports. Each of these student groups consisted of three respondents. However, the division into these three groups was not used as a measure in the conducted study.

3 Research Findings

For this research question, the researcher used quantitative data through observations and, as supporting data, also utilized qualitative data by conducting interviews. Additionally, to further support the findings, the researcher collected data from questionnaires. The findings are summarized below.

Table 1 Mean observation for the level and how the software assists in mastering movement skills based on Simpson's psychomotor domain taxonomy

Item	Level Simpson's psychomotor	Mean	SD
2	Perception	4.39	0.08
4			
6	Set	4.61	0.02
8			
10	Guided response	4.83	0.08
12			
14	Mechanism	4.50	0.26
16			
18	Complex overt response	4.56	0.02
20			
22	Adaptation	4.39	0.17
24			
26	Origination	4.22	0.13
28			
30	Overall software	4.22	0.83

3.1 Observation Findings on Students

Table 1 shows that the highest mean is at the third level, which is guided response, with a mean of 4.83 ± 0.08. The second highest is at the set level, recording a mean of 4.61 ± 0.02. This is followed by the mean for the complex overt response level at 4.56 ± 0.02, the mechanism level at 4.50 ± 0.26, and both the adaptation and perception levels, which recorded the same mean of 4.39 ± 0.08. The subsequent level is origination, with a mean of 4.22 ± 0.13. The mean reading for the overall items related to the software recorded a mean of 4.22 ± 0.83.

3.2 Interview Findings on Students

Table 2 shows overall, for basic movements, students agree and believe that this developed software can help foster their interest and generate creativity. Additionally, they state that this software indirectly encourages a variety of activities and makes their learning experiences more meaningful. Students also feel that this software helps them to adapt and perform movements perfectly. Moreover, they believe it facilitates learning, helps reduce risks, and indirectly adds to their knowledge. However, one student felt that this software did not help them in their learning.

Furthermore, the analysis of general movements shows that students agree the developed software helps provide a clearer picture and assists in identifying weaknesses as well as avoiding mistakes in their movements. Additionally, they state that

Table 2 The theme for the level and how the software assists in mastering movement skills based on Simpson's psychomotor domain taxonomy

Construct	Theme
Basic movement	Helps to foster interest and generate creativity (N = 5)
	Learning becomes more meaningful (N = 2)
	Encourages a variety of activities (N = 2)
	Helps to adapt and perform movements perfectly (N = 5)
	Facilitates learning and helps to reduce risk (N = 3)
	Summarizes and adds information (N = 2)
	Does not help in learning (N = 1)
General movement	Helps to provide a clearer picture (N = 4)
	Helps to identify weaknesses and avoid mistakes (N = 3)
	Helps to strengthen knowledge (N = 1)
	Easy to understand and clear (N = 6)
	Enhances and assists in mastering movements (N = 3)
	Interesting and exciting (N = 1)
Specific movement	Helps and allows freedom to explore learning (N = 5)
	Not bound by time and formal curriculum (N = 4)
	Easy, comfortable, and more open in learning (N = 2)
	Does not provide free opportunities (N = 1)
	Helps in improving learning (N = 4)
	Encourages responding to learning (N = 3)
	Helps develop cognitive abilities and enhance skills (N = 2)
	Helps detect weaknesses and serves as a reference material (N = 1)

this software indirectly supports their continuous learning and helps to strengthen their knowledge. They also believe that it enhances and aids in mastering movements, with some mentioning that the software is engaging and makes them excited.

The analysis of the transcripts for specific movements shows that 8 out of 9 students agree and believe that this developed software helps them freely explore learning movements. Additionally, they stated that students are not bound by time and formal curriculum, making the learning process easier and more open. However, one student felt that this software did not provide them with the freedom to experience the learning process independently. Overall, the students believe that this software helps improve their learning and encourages them to respond to learning. Furthermore, they feel that it helps develop cognitive abilities and enhance skills. Indirectly, the developed software also helps students identify their weaknesses, with some considering it a comprehensive reference material for their learning.

Table 3 Mean survey for the level and how the software assists in mastering movement skills based on Simpson's psychomotor domain taxonomy

Item	Level Simpson's psychomotor	Mean	SD
2	Perception	4.34	0.19
4			
6	Set	4.40	0.18
8			
10	Guided response	4.45	0.03
12			
14	Mechanism	4.39	0.12
16			
18	Complex overt response	4.40	0.00
20			
22	Adaptation	4.34	0.05
24			
26	Origination	4.30	0.04
28			
30	Overall SOFTWARE	4.50	0.57

3.3 Survey Findings on Students

Table 3 shows that the highest mean is at the third level, which is guided response, with a mean of 4.45 ± 0.03. The second highest is at the set and complex overt response levels, both recording a mean of 4.40 ± 0.18. This is followed by the mean for the mechanism level at 4.39 ± 0.12. The adaptation and perception levels recorded the same mean of 4.34 ± 0.05, and the next level is origination, with a mean of 4.30 ± 0.04. The overall mean reading for items related to the software recorded a mean of 4.50 ± 0.57.

4 Discussion

The researcher believes that the findings, which show the same sequence of levels, are due to the majority of students sharing similar interests and engaging in learning activities together. Observing the results from both design methods: observation and survey, the findings indicate that all levels are at the "agree" level, suggesting that students' achievement levels have been reached for all seven stages based on Simpson's Psychomotor Domain Taxonomy. This demonstrates that the software can help students achieve all levels of movement skills especially students have generally been able to reach the level of guided response.

This is supported by research conducted by Malik and Agarwal, which suggests that the process of knowledge acquisition becomes more efficient when students engage in learning using technology materials that are appropriate for their learning (Malik and Agarwal 2012). The use of technological materials also indirectly provides opportunities to enhance traditional "Chalk and Talk" teaching methods, making learning more flexible for students (Neo 2007). Recent studies further support these findings, emphasizing the role of interactive and adaptive learning environments in enhancing student engagement and performance. For example, Yilmaz and Keser highlight that technology-supported learning platforms significantly improve students' cognitive and psychomotor skills by providing interactive and personalized learning experiences (Yilmaz and Keser 2020). Similarly, Liu et al. found that the integration of multimedia resources in education promotes active learning and self-regulation among students, leading to better learning outcomes (Liu et al. 2021).

Additionally, the researcher believes that the effectiveness of this software is due to its ability to allow students to adapt through self-directed learning according to their individual preferences. The flexibility and adaptability provided by technology-enhanced learning tools enable students to engage with content more deeply, fostering both independent and collaborative learning experiences (Smith and Sivo 2020).

Next, from the interview results regarding basic movement skills, students believe that this developed software helps to foster their interest and generate creativity. Additionally, they state that this software indirectly encourages a variety of activities and makes their learning experience more meaningful. However, one student felt that this software did not assist them in their learning. According to the researcher, this may be because the student's statements suggest that they are an individual who better understands verbal explanations from instructors.

These findings align with recent studies that emphasize the importance of interactive and engaging learning tools in promoting student interest and creativity. The use of educational software that incorporates multimedia and interactive elements can significantly enhance student motivation and creative thinking skills (Wu et al. 2022). Diverse and meaningful learning experiences facilitated by technology can improve student engagement and satisfaction (Zhang et al. 2020). However, individual learning preferences vary, and some students may still prefer traditional verbal instruction over digital tools (Chen and Carliner 2021). This highlights the need for a balanced approach that combines technology-enhanced learning with traditional teaching methods to cater to different learning styles (Hwang and Lai 2020).

The responses regarding general movements show that all students believe this software is easy to understand and clear. Moreover, they also think that it enhances and aids in mastering movements, with some stating that the software is engaging and makes them excited. The researcher believes that the software helps students achieve positive outcomes across all movement levels because it is engaging and covers all aspects of learning. The software that is easy to understand and interact with can significantly enhance students' motivation and confidence in mastering new skills (Lin and Wu 2021). Additionally, engaging and interactive learning tools have been shown to improve the quality of learning experiences by making them more enjoyable and stimulating (Tan et al. 2020). This reinforces the importance of using technology that is not only informative but also engaging and adaptable to meet the diverse needs of learners (Huang and Yang 2023). Students using the software demonstrate a variety of methods in interpreting their opinions about the movements performed.

For specific movements, students believe that the developed software allows them to explore movement learning freely. Additionally, they stated that students are not bound by time or a formal curriculum, making the learning process easier and more open. Furthermore, they believe that it helps to develop cognitive abilities and enhance skills. Overall, the data triangulation findings (observation, interviews, questionnaires) prove that the software indeed helps students achieve all levels of movement skills. The researcher anticipates that this is because students use the software as a support tool in their learning process, which helps save time for students to master their skills.

5 Conclusion and Contributions

Recent studies support these findings, emphasizing the role of flexible and student-centered learning environments in enhancing skill acquisition and cognitive development. Educational software that allows for self-paced and exploratory learning positively impacts students' ability to master complex skills and improve their cognitive functions (Liu et al. 2023). When students are not restricted by formal time constraints and curriculum structures, they are more likely to engage deeply with the material and develop a better understanding of the content (Chen and Zhang 2022). This evidence underscores the importance of incorporating flexible learning tools in education to support student autonomy and skill mastery (Ng and Tan 2023). Therefore, the approach and learning methods used through this software show a positive impact on students' movement skills. The researcher found that the implications of this study can encourage students to learn according to their own preferences and access or search for the information they desire. With the developed software, it is also hoped that knowledge of movement skills can be fully mastered by those involved in the field of sports, and indirectly, it can help shape and produce a generation that progresses in line with environmental developments. In conclusion, the findings of this study demonstrate that this technology integrated software has a positive impact on all levels of movement skills based on Simpson's Psychomotor Domain Taxonomy.

Acknowledgements Special thanks to Universiti Teknologi Malaysia, Skudai, Johor, Malaysia for research grant scheme Q.J130000.2653.16J92 and Sports Science team members for your kind support.

References

Barnett LM, Salmon J, Hesketh KD (2022) The role of fundamental movement skills in the promotion of physical activity. J Phys Educ Sport 22(1):79–87

Chen S, Carliner S (2021) The impact of individual learning preferences on the effectiveness of technology-based instruction. Educ Tech Res Dev 69(1):203–220

Chen X, Zhang S (2022) Flexible learning environments and their effects on student engagement and understanding. J Learn Instruct 85:104176

Garcia-Hermoso A, Alonso-Martínez AM, Ramírez-Vélez R, Pérez-Sousa MA, Izquierdo M (2021) Effects of exercise on functional movement skills in young children: a systematic review and meta-analysis. Sports Med 51(4):737–749

Granacher U, Borde R, Hortobágyi T (2019) Effects of physical education on fundamental movement skills in children and adolescents: a systematic review. Int J Environ Res Publ Health 16(6):936

Hollis SJ, Coughlan E, Moran AP (2023) Augmented reality applications in physical education and their effect on motor skill development: a meta-analysis. J Phys Educ a Sport Sci 39(2):187–195

Huang YM, Yang SJH (2023) Designing adaptable learning technologies for diverse learning styles: insights and implications. J Educ Technol Dev Exchange 16(2):45–61

Hwang GJ, Lai CL (2020) Facilitating and bridging out-of-class and in-class learning: an interactive e-learning approach. Interact Learn Environ 28(5):570–583

Lin C, Wu C (2021) Impact of user-friendly educational software on student engagement and skill acquisition. J Educ Comput Res 59(1):65–83

Liu Y, Chen H, Wang L (2023) Impact of self-paced learning software on skill mastery and cognitive development. J Educ Technol Soc 26(1):14–28

Liu Y, Li H, Zhang H (2021) The effect of multimedia on learning effectiveness: evidence from a meta-analysis. Int J Educ Technol High Educ 18:25

Logan SW, Robinson LE, Getchell N (2020) The importance of fundamental motor skill development: a review of the literature. J Motor Learn Dev 8(1):30–45

Malik S, Agarwal A (2012) Use of multimedia as a new educational technology tool–a study. Int J Inform Educ Technol 2(5):468–471

Mazlan AN, Harun J, Zainuddin ZA (2014) Problems in learning movement skills among students in higher education. In: International conference on teaching and learning in computing and engineering (LaTiCE)

Neo M (2007) Learning with multimedia: engaging students in constructivist learning. Int J Instr Media 34(2):149–158

Ng S, Tan W (2023) Incorporating flexible learning tools in higher education: benefits and challenges. J Educ Res Pract 14(2):135–148

Riso EM, Topolski K, Kriemler S (2020) Barriers and facilitators to the implementation of physical education technology: a systematic review. Phys Educ Sport Pedag 25(4):376–393

Smith C, Sivo S (2020) The impact of technology-enhanced learning on student engagement and learning outcomes. J Educ Technol Res Dev 68(3):1457–1475

Tan SC, Lim KY, Koh E (2020) Fostering an engaging learning environment through interactive educational tools. Educ Technol Soc 23(3):99–110

Vickery WL, Wulf G, Lewthwaite R (2021) Feedback and motivation in physical education: Enhancing motor learning and performance with technology. Europ Phys Educ Rev 27(3):482–495

Wu H, Yu Y, Li S (2022) The impact of multimedia learning environments on student creativity and engagement. J Educ Technol Dev Exchange 15(2):30–46

Yilmaz RM, Keser H (2020) The impact of interactive learning environments on students' performance and engagement in learning activities. Educ Technol Soc 23(1):50–63

Zhang L, Zhao R, Lu Y (2020) Technology-enhanced learning experiences: effects on student engagement and learning satisfaction. Comput Educ 144:103701

A Meta-Analysis of Drone Technology in Sports

Fiderica Jeply, Diyana Zulaika binti Abdul Ghani, and Ahmad Nabil bin Md. Nasir

Abstract The utilization of drone technology in sports is experiencing significant growth. Accordingly, this meta-analysis was conducted to assess the advantages provided by drone technology within the sports domain. This study systematically reviewed and analyzed prior research from 2016 to 2024, focusing on various applications of drone technology in sports. Relevant studies were retrieved through academic databases including ScienceDirect and Scopus. The results demonstrate that drone technology offers substantial benefits across multiple areas, including athlete performance analysis, its application in football, broadcasting, position tracking, and ensuring security during sporting events. The predominant research design in the reviewed studies was experimental, aiming to verify the practical benefits of drone technology in sports empirically. In summary, the findings of this meta-analysis provide a foundational resource for future researchers seeking to explore and expand the use of drone technology in the sports field.

Keywords Drone technology · Drone in sports · Meta-analysis

F. Jeply (✉) · D. Z. A. Ghani · A. N. Md. Nasir
Faculty of Educational Sciences and Technology, Universiti Teknologi Malaysia, Skudai, Johor, Malaysia
e-mail: fiderica@graduate.utm.my

D. Z. A. Ghani
e-mail: diyanazulaika@utm.my

A. N. Md. Nasir
e-mail: ahmadnabil@utm.my

D. Z. A. Ghani
Sport Innovation and Technology Centre (SITC), Institute of Human Centered Engineering (iHumEn), Universiti Teknologi Malaysia, Johor, Malaysia

M. H. A. Hassan et al. (eds.), *Proceedings of the 3rd International Conference on Innovation and Technology in Sports*, Lecture Notes in Bioengineering,
https://doi.org/10.1007/978-981-95-0804-4_17

1 Introduction

The use of unmanned aerial vehicle (UAV) technology, also known as drones, has the potential to be a beneficial technology in the world of sports. Drones, which have become increasingly accessible to consumers, are being utilized in diverse domains such as precision agriculture, surveillance operations, film production, and in augmenting the viewing experience at sports events (Ayranci 2017). Drone technology is used professionally in sports for performance analysis, better refereeing, improving broadcast quality, aerial surveillance or crowd control and health and safety (Adorama 2025). This is attributable to the mobility and versatility of drone technology, which, unlike static cameras, allows for aerial perspectives of the playing field and the production of high-quality video content suitable for broadcasting, position tracking, and tactical analysis Russomanno et al. (2022). Although unmanned aerial vehicle (UAV) technology, or drones, shows promise for improving a variety of fields such as agriculture, filmmaking, and sports, it remains to be seen to what extent drones are currently being used in real-world athlete situations. Despite their potential for improving performance analysis, refereeing, broadcast quality, and other applications, systematic assessment of their impact and practical applications in sports is necessary. This study aims to fill this gap by assessing the extent to which drone technology is being integrated into sports performance, identifying its benefits, limitations, and potential areas for development.

2 Methodology

This meta-analysis entailed a systematic review, synthesis, and evaluation of scholarly studies examining the application of drone technology in sports. A comprehensive search strategy was employed, utilizing reputable academic databases such as ScienceDirect and Scopus. To ensure the inclusion of current and pertinent literature, the search was restricted to articles published between 2016 and 2024, using the keywords "drone in sports" and "unmanned aerial vehicle (UAV) in sports." Studies focusing on drone applications outside the realm of sports were excluded from this analysis. Following a rigorous screening process, 12 relevant articles were identified and classified into four principal domains of drone technology utilization in sports (Fig. 1).

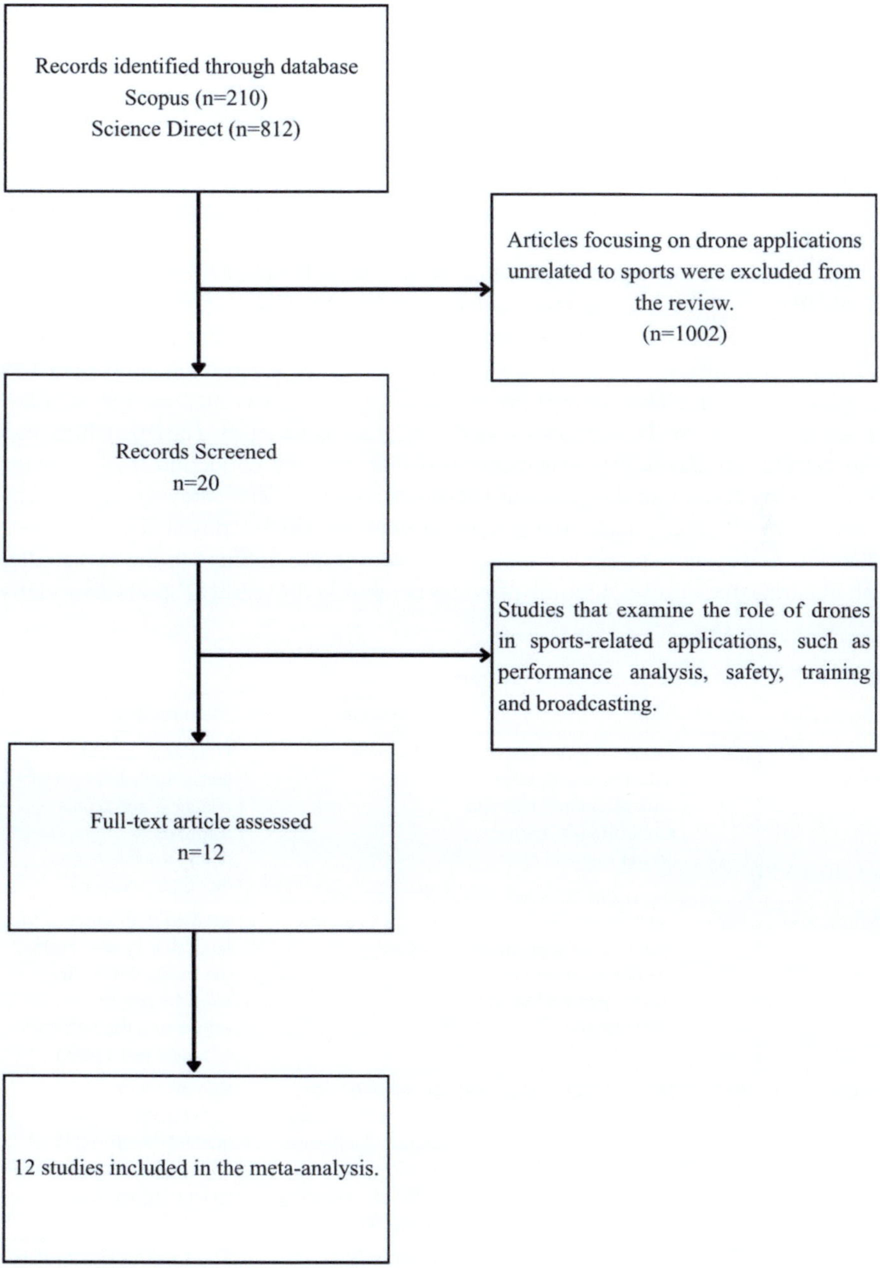

Fig. 1 Methodological flow

3 Findings and Discussions

3.1 Drone for Sports Performance Analysis

Drone technology has been found to be used in analyzing athlete performance. Four studies related to using drones to analyze athlete performance were analyzed (Table 1).

The use of drones in sports is proven to be beneficial for analyzing the performance of athletes. An experimental study carried out by Miura et al. (2021) using a bird-eye view provided in real-time via drone found that the error distance between the participants' position and the line significantly decreased when using the system than without the system. The findings prove that using drones can increase the accuracy of space perception. In addition, a study by Karungaru et al. (2019), which was also carried out physically, also confirmed that the data collection method using drones helps determine the speed of badminton athletes. The studies conducted by Olaniyan et al. (2023) and Mali (2020) did not use drones physically; they used interview and systematic review methods. However, the findings of the two studies still illustrate that the use of drones provides benefits in the world of sports, especially

Table 1 Drone for sports performance analysis

Researcher	Year	Title	Instrument	Main finding
Olaniyan et al.	2023	Enhancing on- pitch learning capabilities with data analytics and technologies in elite sports	Interview	Data analytics and technology help players, coaches, and teams improve decision-making and on-pitch practices
Miura et al.	2021	Bird's-eye view provided in real- time via drone improves spatial perception for team sports	Bird's-eye view system	Real-time drone feedback system from the bird's-eye view would work for improving the accuracy of space perception
Mali	2020	Modern technology and sports performance: an overview	Systematically searched through online databases: PubMed, google scholar, and google advance search	Recent sports technology developments enhance athletic performance and reduce injury risk
Karungaru et al.	2019	Ground sports strategy formulation and assistance technology development: player data acquisition from drone videos	Particle filter, template matching, HSV color space	Their method accurately captures and processes player data from drone videos

in performance analysis. The results of interviews in the study Olaniyan et al. (2023) found that using technology, such as drones, to analyze data or athlete performance increases the understanding of players, coaches and teams. A systematic review conducted by Mali (2020) found that recent developments in sports technology, such as drone technology, have created various products to enhance and increase athletic performance. In conclusion, the data obtained using drone technology provides a clearer view and makes it easier for sportsmen, whether coaches or athletes, to assess their level of performance from strategic and tactical aspects.

3.2 *Drone in Soccer*

Drone technology can also be used in football. Past studies are analyzed to conclude the contribution of drone technology in football (Table 2).

Table 2 Drone in soccer

Researcher	Year	Title	Instrument	Main finding
Scott et al.	2022	SoccerTrack: a dataset and tracking algorithm for soccer with fish-eye and drone videos	Multi-object tracking algorithms	SoccerTrack dataset enables tracking in soccer using both fish-eye and bird's-eye view videos
Russomanno et al.	2022	Drone-based position detection in sports-validation and applications	3D kinematic analysis like Vicon or Qualisys, Matlab	Drone-based position detection could serve as a promising alternative to existing radio and video-based electronic performance and tracking systems (EPTS)
Lee et al.	2021	Effect of image feedback by drones on elementary school students' satisfaction with after-school soccer classes	Questionnaire	Drones enhance students' satisfaction with physical education by providing unique visual feedback
Shahidul Islam	2020	Introducing drone technology to soccer coaching	Online and offline searches, electronic databases (PubMed, google scholar, google advanced search)	Drones provide valuable insights into team shape and strategy in soccer coaching, especially during practices and unofficial matches

Drone technology has emerged as a valuable tool in football for player tracking and performance analysis. Studies by Scott et al. (2022) and Russomanno et al. (2022) demonstrate the effectiveness of drones in monitoring player movements, with Scott et al. (2022) highlighting the SoccerTrack System's capacity to provide comprehensive field coverage through bird's-eye and fish-eye perspectives. Similarly, Russomanno et al. (2022) found drone-based position detection a viable alternative to conventional electronic performance and tracking systems (EPTS). These findings align with Lee et al. (2021), who emphasized the advantages of aerial perspectives in enhancing visual feedback, particularly for physical education students. Furthermore, Islam (2020) also identified the strategic benefits of drone footage in coaching, particularly for analyzing tactical formations and defensive structures. Overall, the integration of drone technology in football primarily enhances video-based performance analysis, contributing valuable insights for coaching and player development.

3.3 *Drone for Streaming, Broadcasting and Tracking*

Previous studies have found that drones can be used for streaming, broadcasting and tracking in the field of sports. Therefore, four studies related to the issue were analyzed (Table 3).

Table 3 Drone for streaming, broadcasting and tracking

Researcher	Year	Title	Instrument	Main finding
Mademlis et al.	2019	High-level multiple-UAV cinematography tools for covering outdoor events	Director's dashboard software	Cinematic tools for sports can also be applied in movies, newsgathering, and advertising
Koubaa and Qureshi	2018	DroneTrack: cloud-based real-time object tracking using unmanned aerial vehicles over the internet	DroneTrack a cloud-based system	Achieved 3.5-m tracking accuracy for slow-moving targets with UAVs
Wang et al.	2017	Networked drone cameras for sports streaming	A fog-networking based system architecture	Eight drones optimized to provide 94% coverage and 2 K video at 20 Mbps
Mademlis et al.	2017	Overview of drone cinematography for sports filming	UAV cinematography applications	Multiple drones expand possibilities for technical and industrial applications in cinematography

Drone technology has also been widely utilized in sports broadcasting and motion detection. Research conducted by Mademlis et al. (2019, 2018) examined the application of multiple drones in cinematography and sports filming, demonstrating that the use of multiple drones enables comprehensive sports coverage from various angles. Furthermore, Mademlis et al. (2019) highlighted that the extensive deployment of drones for recording extends beyond sports broadcasting, with potential applications in film production and advertising. These findings underscore the effectiveness and practicality of utilizing multiple drones in sports media coverage.

Additionally, Wang et al. (2017) found that drone-assisted sports broadcasting can achieve high video quality, reaching resolutions of up to 2 K, thereby enhancing audience satisfaction by providing clearer and more immersive viewing experiences. In a related study, Koubaa and Qureshi (2018) investigated the role of drones in motion detection, specifically evaluating the effectiveness of the DroneTrack system. Their findings indicated that the system achieved a detection accuracy of 3.5 m when tracking slow-moving targets.

Collectively, these studies highlight the significant contributions of drones to sports broadcasting and filming, emphasizing their broader applicability in various industries. Future research should further explore advancements in drone technology, particularly in integrating modern broadcasting and sensor systems, to enhance video quality and motion detection capabilities. Continued innovation in this area could lead to new applications and improvements in both sports and other fields.

3.4 Drone for Sports Safety

Safety in sports is very important. Some past studies have found that drone technology has the potential to be used for security surveillance and emergency first aid purposes. Four studies were analyzed to conclude whether drones should be used for sports safety or not (Table 4).

Drone technology is also applied in safety during sports. However, the studies above are system recommendations and have not yet been implemented in practice. Studies conducted by Al-Dosari et al. (2023b, 2023a) are focused on improving safety and security using drones. Both studies found that the use of the proposed drone in the security system increased awareness among safety and security professionals. However, the systems used are different, namely SitAw for mega sporting events and UAV-based supply chain management System (SCM) for FIFA mega sporting events.

The study of Dinesh Kumar and Jeeva (2017) and Bravo et al. (2016) is a proposed system using drones to help put out fires and send first aid to athletes with minor injuries. Dinesh Kumar and Jeeva (2017) propose a drone ambulance to deliver first aid kits and extinguishing balls to remote or difficult-to-access locations during emergencies, enhancing the speed and effectiveness of emergency responses. Similarly, Bravo et al. (2016) introduced an unmanned aerial assistance system (UAAS) to carry first aid kits (bandages, alcohol, cotton, gauze, water, and other emergency and

Table 4 Drone for sports safety

Researcher	Year	Title	Instrument	Main finding
Al-Dosari et al.	2023	Mega sporting event scenario analysis and drone camera surveillance impacts on command-and-control centre situational awareness for dynamic decision-making	Questionnaire, situational awareness system (SitAw)	Drone surveillance enhances situational awareness compared to fixed cameras for safety and security professionals
Al-Dosari et al.	2023	Security supply chain using UAVs: validation and development of a UAV-based model for Qatar's mega sporting events	Questionnaire, UAV-based supply chain management system (SCM)	Developed a reliable UAV-based cybersecurity framework for mega sporting events, with significant interest from IT professionals
Dinesh Kumar and Jeeva	2017	Drone ambulance for outdoor sports	GSM, GPS, and unmanned aerial assistance system	The drone ambulance delivers first aid kits and fire extinguishing balls to manage emergencies in remote and difficult-to-access locations
Bravo et al.	2016	First aid drone for outdoor sports activities	Unmanned aerial assistance system (UAAS)	The proposed drone system efficiently provides first aid for various injuries and emergencies during outdoor sports activities

survival products) and has real-time audio and video communications capabilities. Their system can address minor injuries, such as sprains and abrasions, and more severe conditions, such as dehydration and heat strokes. The UAAS represents a significant advancement in emergency medical support, offering timely assistance during sports activities.

The proposed system for security purposes and emergency first aid is seen to have a positive effect in the world of sports. However, the implementation of the proposed system needs to be done to assess whether the system should be applied or not. Therefore, it is hoped that future researchers will be able to realize the planned system proposal.

4 Conclusion

Overall, the studies show that drone technology provides various benefits in the world of sports. The main benefit that can be seen is from the aspect of analyzing the athlete's performance using drone video recording. Recording from the top view or bird's eye view gives athletes and coaches the opportunity to see the movement of each athlete more clearly which cannot be seen through video recording from the side view. Next, the use of drones is also able to give satisfaction to the audience due to the high quality and clearer broadcast video. Then, the proposed system for the purpose of surveillance and emergency first aid is also seen as useful, but the system needs to be tested to prove its effectiveness. Finally, drones are certainly suitable for use in the world of sports, especially sports that are performed outdoor. This is said to be so because of the various data that can be obtained through drone video recording such as speed and object detection. Future research holds considerable potential to build upon these findings. Further investigation into integrating advanced drone technologies and data analytics could yield more profound insights into athletic performance and enhance spectator engagement. Continued development and rigorous testing will be essential to optimize drone applications and fully realize their potential benefits within the sports industry.

Acknowledgements This work was supported by UTM Encouragement Research (UTMER) (Q.J130000.3853.31J68) from Universiti Teknologi Malaysia (UTM).

References

Adorama (2025) How professional sports are using drones. https://www.adorama.com/alc/professional-sports-using-drones/. Accessed 14 Feb 2025

Al-Dosari K, Hunaiti Z, Balachandran W (2023a) Mega sporting event scenario analysis and drone camera surveillance impacts on command-and-control centre situational awareness for dynamic decision-making. Safety 9(3). https://doi.org/10.3390/safety9030054

Al-Dosari K, Deif AM, Kucukvar M, Onat N, Fetais N (2023b) Security supply chain using UAVs: validation and development of a UAV-based model for Qatar's mega sporting events. Drones 7(9). https://doi.org/10.3390/drones7090555

Ayranci ZB (2017) Use of drones in sports broadcasting. Entertain Sports Law Spring 33(3):79–93

Bravo GC, Mendes L, Parra DM, Pereira AMDJ (2016) First aid drone for outdoor sports activities. IEEE, Vila Real, Portugal

Dinesh Kumar G, Jeeva B (2017) Drone ambulance for outdoor sports. Asian J Appl Sci Technol (AJAST) 1(5):44–49. www.ajast.net

Karungaru S, Matsuura K, Tanioka H, Wada T, Gotoda N (2019) Ground sports strategy formulation and assistance technology develpoment: player data acquisition from drone videos. In: Proceedings of 2019 8th International conference on industrial technology and management, ICITM 2019. pp 322–325. https://doi.org/10.1109/ICITM.2019.8710735

Koubaa A, Qureshi B (2018) DroneTrack: cloud-based real-time object tracking using unmanned aerial vehicles over the internet. IEEE Access 6:13810–13824. https://doi.org/10.1109/ACCESS.2018.2811762

Lee HS, Lee J, Kim HC, Shin JE (2021) Effect of image feedback by drones on elementary school students' satisfaction with after-school soccer classes. South African J Res Sport Phys Educ Recreat 43(2):69–84. https://www.researchgate.net/publication/355581761

Mademlis I et al (2019) High-level multiple-UAV cinematography tools for covering outdoor events. IEEE Trans Broadcast 65(3):627–635. https://doi.org/10.1109/TBC.2019.2892585

Mademlis I, Mygdalis V, Nikolaidis N (2018) Overview of drone cinematography for sports filming. In: European conference on visual media production. pp 3–4. https://doi.org/10.13140/RG.2.2.26966.96327

Mali NP (2020) Modern technology and sports performance: an overview. Int J Physiol 5(1), 212–216. www.journalofsports.com

Miura S, et al (2021) Bird's-eye view provided in real-time via drone improves spatial perception for team sports. https://doi.org/10.21203/rs.3.rs-148092/v1

Olaniyan O, Dehe B, Bamford D, Ward S (2023) Enhancing on-pitch learning capabilities with data analytics and technologies in elite sports. Eur Sport Manag Q. https://doi.org/10.1080/16184742.2023.2270565

Russomanno TG, Blauberger P, Kolbinger O, Lam H, Schmid M, Lames M (2022) Drone-based position detection in sports—validation and applications. Front Mater 13(March):1–11. https://doi.org/10.3389/fphys.2022.850512

Scott A, Uchida I, Onishi M, Kameda Y, Fukui K, Fujii K (2022) SoccerTrack: a dataset and tracking algorithm for soccer with fish-eye and drone videos. In: IEEE computer society conference on computer vision and pattern recognition workshops, vol. 2022, pp 3568–3578. https://doi.org/10.1109/CVPRW56347.2022.00401

Shahidul Islam M (2020) Introducing drone technology to soccer coaching. Int J Sports Sci Phys Educ 5(1):1. https://doi.org/10.11648/j.ijsspe.20200501.11

Wang X, Chowdhery A, Chiang M (2017) Networked drone cameras for sports streaming. In: Proceedings of the international conference on distributed computing systems (ICDCS). pp 308–318. https://doi.org/10.1109/ICDCS.2017.200

Enhancing Volunteer Management Using Preference Matching Algorithm

Suhizaz Sudin, Norfadila Mahrom, Priyadharishini Thurairaju, and Ahmad Faizal Annuar

Abstract To optimize volunteer scheduling and the management of sporting events, there are challenges that could be addressed by proposing a web-based solution. However, this kind of solution requires recognition of the critical role that the volunteers play in the success of these events. This project proposed a web-based volunteer scheduling system that is designed to streamline the allocation process of the volunteers by considering volunteers' preferences, skills, and experience. The core innovation of this project is the implementation of a Preference-based Matching approach that aligns volunteer roles with individual preferences and qualifications. To enhance overall satisfaction of the volunteers and operational efficiency, the system not only simplifies the scheduling process for event organizers, but it also ensures fair job distribution and maximizes volunteer engagement. To assess the system's effectiveness, key performance metrics such as match rate and average preference satisfaction were evaluated. This project aims to deliver a practical tool for sports event organizers, to optimize volunteer management while enhancing volunteer satisfaction and reducing setup time.

Keywords Preference matching · Volunteer scheduling · Sports scheduling · Sports event

---

S. Sudin · N. Mahrom (✉)
Advanced Computing, Centre of Excellence (AdvComp), Universiti Malaysia Perlis (UniMAP), Perlis, Malaysia
e-mail: norfadila@unimap.edu.my

Sports Engineering Research Centre, Centre of Excellence (SERC), Universiti Malaysia Perlis, Perlis, Malaysia

S. Sudin · N. Mahrom · P. Thurairaju
Faculty of Intelligent Computing, Universiti Malaysia Perlis, Perlis, Malaysia

A. F. Annuar
Faculty of Mechanical Engineering and Technology, Universiti Malaysia Perlis, Perlis, Malaysia

M. H. A. Hassan et al. (eds.), *Proceedings of the 3rd International Conference on Innovation and Technology in Sports*, Lecture Notes in Bioengineering,
https://doi.org/10.1007/978-981-95-0804-4_18

1 Introduction

It is known that sporting events not only offer entertainment, but also opportunities for community to engage and platforms for personal development. These events often rely heavily on volunteers who contribute their time, skills, and enthusiasm to ensure success (Dickson et al. 2022). However, due to the diversity of tasks and volunteer preferences, the organizers faced a significant challenge that required them to efficiently organize and schedule volunteers for various roles in the events.

In this project, a web-based system is developed to address the complexities of volunteer scheduling process. The system optimizes the allocation of volunteers based on their preferences and skills. The system aims to enhance the efficiency of sports event management by aligning volunteers with roles that match their interests and competencies. The system not only considers volunteers' preferences for specific tasks, but it also accounts for their individual capabilities and prior experience. This is to ensure a balanced and effective distribution of volunteers across various sporting events.

The primary objective of the system is to efficiently allocate volunteers with similar interests and expertise to the most suitable roles. It is supposed to be performed while ensuring that all events receive the necessary support from skilled volunteers. Hence, the system employs a Preference-Based Matching approach, to align volunteers with tasks they prefer and excel in. This approach ensures that volunteers are well-suited for their roles, as the scale and significance of sporting events continues to grow. The need for efficient volunteer management is increasingly evident nowadays. Thus, this project offers a practical solution to optimize volunteer scheduling with the aim to improve the operational efficiency of sports events, enhance volunteer satisfaction, and benefit the broader sports community.

2 Previous Works

Previous studies collectively emphasize the importance of incorporating preference-based matching in scheduling and allocation across various domains, such as healthcare, sports, event management, and assignment mechanisms. Several studies effectively highlight the benefits of considering individual preferences, improving satisfaction and operational outcomes (Agrawal et al. 2023; Guajardo and Jörnsten 2017). However, the repetition and overlap in discussing the significance of preference-based approaches across different contexts suggest a broader trend rather than offering distinct theoretical advancements. While some contribute innovative models for specific scenarios like event management and activity scheduling, they rely heavily on probabilistic methods without thoroughly addressing potential limitations in real-world applicability (Mancini et al. 2022; Aronshtam and Ilani 2022). Acharya et al. (2022) offer a robust framework that balances outcomes and preferences but could benefit from more practical validation, particularly in the context of computational feasibility and implementation challenges. Overall, while these studies advance the field by emphasizing the relevance of preferences, they sometimes fall short in critically examining the trade-offs and practical constraints inherent in their proposed models.

There were studies that offer insights into the management and experiences of volunteers at large-scale sporting events, yet they vary in their depth and practical implications. The "program management" model used at the London 2012 Olympic Games was critiqued, highlighting its inflexibility and negative impact on volunteer satisfaction (Holmes et al. 2018). Wilks (n.d) contributes to this discussion by applying the serious leisure framework to Olympic volunteering, offering a nuanced understanding of volunteer motivations and engagement. However, while the framework provides valuable theoretical insights, its practical recommendations for enhancing future volunteer programs could be more robust. However, these studies were complemented by focusing on volunteer satisfaction at the XVII Commonwealth Games (Rhoden et al. 2002). It revealed that while volunteers were generally satisfied with their roles, concerns about organization, communication, and training persisted. The study's reliance on quantitative data through satisfaction tables adds clarity but lacks the deeper qualitative analysis found in other works, limiting its ability to provide actionable recommendations. Overall, while these studies underscore the importance of volunteer management in mega-events, they often fall short in addressing the practical challenges and complexities of implementing more volunteer-friendly approaches in these high-pressure environments.

There were studies on volunteer scheduling systems to support the need for efficient, preference-based volunteer management at large-scale events. The studies present web-based systems that use distinct algorithms such as Round-Robin and First-Come, First-Serve (FCFS) to assign volunteers to tasks accurately and efficiently (Quan et al. 2024; Mahrom et al. 2024). These systems highlight the importance of considering volunteer preferences and skills, aligning with the issues on the challenges and strategies in volunteer management at mega-events like the Olympics. The contributions reinforce the need for flexible, volunteer-friendly scheduling approaches to enhance satisfaction and engagement.

3 Methodology

3.1 Specifications of the System

The web-based system of this project is tailored to the specific needs of sports administrators at Universiti Malaysia Perlis (UniMAP). It is targeted to facilitate the scheduling of volunteers for various sporting events. In this project, each sport is scheduled to have several sessions held on different days and at different venues. The venues considered in this project are Pusat Sukan Universiti Malaysia Perlis, Kompleks Belia Dan Sukan Negeri Perlis, and Dewan Serbaguna Kubang Gajah.

Within the system, various job scopes were defined to ensure comprehensive coverage of event needs. These include roles in crowd management, technical support, sports medicine, event planning, and sports event promotion. Each role was carefully outlined to match specific skill sets with the requirements of the events. This is to ensure that all aspects of the sporting events are effectively managed. The sports administrator plays a pivotal role in this system by leveraging the scheduler to assign volunteers to tasks based on the Preference-based Matching. An efficient allocation of manpower could be achieved by aligning the volunteer's roles with their individual preferences and skills. List of volunteers' skills considered in the system include computing, athletics, medical, event planner, and accountancy.

The system was developed based on the Agile Software Development Life Cycle (SDLC) model. It was chosen for its flexibility and responsiveness to the dynamic nature of sporting events (Ferreira 2023). This approach was particularly beneficial in addressing the evolving requirements of the system. Regular updates were necessary during the development process, to meet the specific demands of different sporting events. During the system design phase, considerable attention was given to ensure that the scheduler was scalable, secure, and efficient. The design phase involved creating a user-friendly interface that allows administrators to easily navigate and manage the scheduling process.

The backend of the system was developed using the Django framework. It was selected for its robustness and ease of integration with PostgreSQL (the database management system utilized in this project). The core of the system is the Preference-Based Matching algorithm, which was integrated into the scheduler to automate the allocation of volunteers to various tasks based on their skills and preferences (Acharya et al. 2022). The overall system architecture was refined through multiple Agile sprints, each targeting specific functionalities. They are such as user authentication, schedule creation, and volunteer allocation. These sprints allowed for iterative development and testing, ensuring that the system met the performance metrics and user requirements.

During operation, the system begins by allowing sports administrators to create and manage schedules for upcoming events. Volunteer details including their preferences and skills were obtained from UniMAP Sports Centre. The system's matching algorithm then processes this data to generate optimal schedules that align with the needs of the events and the preferences of the volunteers. The outcome is a set of schedules that balance the workload among volunteers while ensuring that the right skills are allocated to the appropriate tasks.

3.2 *Preference-Based Matching Algorithm*

The process of scheduling the volunteers based on their preference and skills is described as follows:

1. Initialize an empty list to store the schedule of volunteers.

2. Initialize total_assigned_volunteers, total_schedules, total_tasks, covered_tasks, preference_sum, and volunteers_with_preferences to 0.
3. Calculate total_tasks as the sum of all required tasks across sports.
4. While there are unassigned volunteers:
   (a) Sort volunteers based on their preferences and skills.
   (b) For each volunteer in sorted order:
      i. Check if the volunteer's preferred task is available and matches their skill.
      ii. If a match is found:
         - Assign the volunteer to the task.
         - Increment total_assigned_volunteers by 1.
         - Add the volunteer's assigned task to total_schedules.
         - Increment covered_tasks if the volunteer's skill matches the task requirement.
         - Update preference_sum with the preference index (0-based) of the assigned task.
         - If the volunteer's assigned task is in their list of preferences:
         - Increment volunteers_with_preferences by 1.

3.3 Performance Metrics

To evaluate the effectiveness of the scheduler, three key performance metrics were identified: match rate and average preference score.

1. Match Rate

This metric measures the percentage of volunteers who were assigned to one of their preferred tasks. A higher match rate indicates that the system effectively considers volunteer preferences in the allocation process.

$$\text{Match Rate} = \frac{\text{volunteers\_with\_preferences}}{\text{total\_assigned\_volunteers}} \times 100 \tag{1}$$

2. Average Preference Score

This score evaluates how well the assigned tasks match the volunteers' preferences, on average. It is calculated based on the ranking of assigned tasks within each volunteer's list of preferences. A lower average preference score indicates a better alignment between volunteer preferences and task assignments.

$$\text{Average Preference Score} = \frac{\text{preference sum}}{\text{total assigned volunteers}} \tag{2}$$

4 Volunteer Scheduler System

The home page of the Volunteer Scheduling System for sporting events at University Malaysia Perlis (UniMAP) is shown in Fig. 1. Meanwhile, Fig. 2 shows the admin dashboard of the system that serves as the central hub for administrators to manage volunteers, perform scheduling and view schedules. The sidebar provides links to key functionalities such as new volunteer registration, viewing volunteers, and running the schedule. The main content area displays a welcome message, an overview of the system, the current system status, and any recent activities. This user-friendly interface ensures efficient management and smooth operation of the volunteer scheduling process.

Figure 3 illustrates the sequence of steps involved in the volunteer scheduling process. The first step is to allocate volunteers to various sports events based on their task preferences and skills. Users are prompted to click "Allocate Volunteers" to initiate this process. Once the volunteers are allocated, users can then proceed to "Run the Scheduler," which finalizes the scheduling by considering the allocated volunteers and their assigned tasks. This two-step process ensures that the volunteers are optimally matched to tasks before the schedule is executed, enhancing the efficiency and effectiveness of volunteer deployment for the events.

A detailed list of volunteers who have been allocated based on their preferences and skills is shown in Fig. 4. Each row in the table corresponds to a volunteer and includes their name, matriculation number, phone number, task preference, and skill. The data highlights the matching of volunteers to tasks such as "Crowd Management" and "Sports Medicine," with associated skills like "Accounting," "Athletic,"

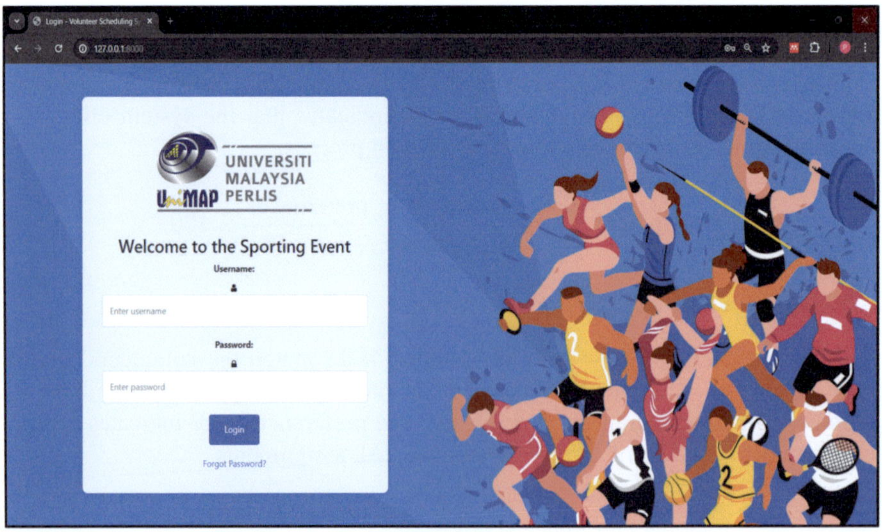

Fig. 1 Home page of the volunteer scheduler system

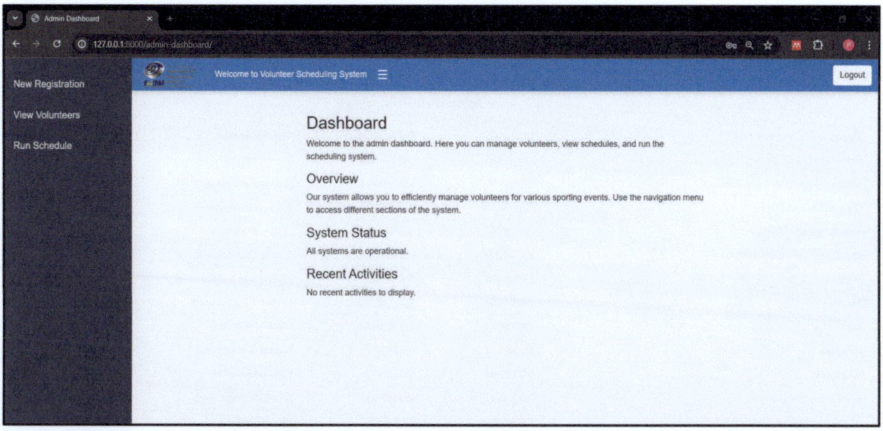

Fig. 2 Admin dashboard

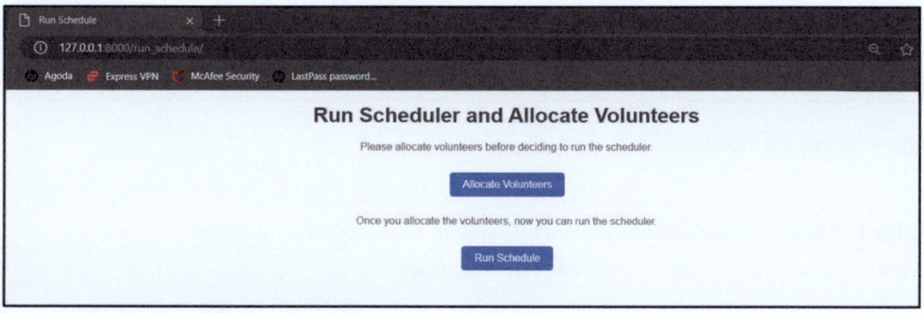

Fig. 3 Run scheduler and allocate volunteers

"Planner," and "Medical." This detailed view allows admin to easily see which volunteers have been assigned to each task, ensuring a clear and organized volunteer deployment for the events.

Each time the volunteer scheduler is run, the match rate and average preference score of the scheduled volunteers are measured. Table 1 presents the performance metrics for the volunteer allocation process, showcasing two key indicators: match rate and average preference score. For the 100 volunteers, the match rate is 81.63% reflecting that 81.63% of the volunteers were successfully assigned to tasks based on their preferences and skills. This high match rate highlights the system's efficiency in aligning volunteers with their preferred tasks. The average preference score is 1.3, indicating that most of the allocated volunteers were placed in their most preferred tasks, demonstrating the precision and effectiveness of the preference-based matching algorithm.

For the 300 volunteers, the match rate stands at 77.50%, meaning that only 77.50% of the volunteers were successfully assigned to tasks that matched their preferences and skills. This lower match rate suggests that as the number of volunteers increases,

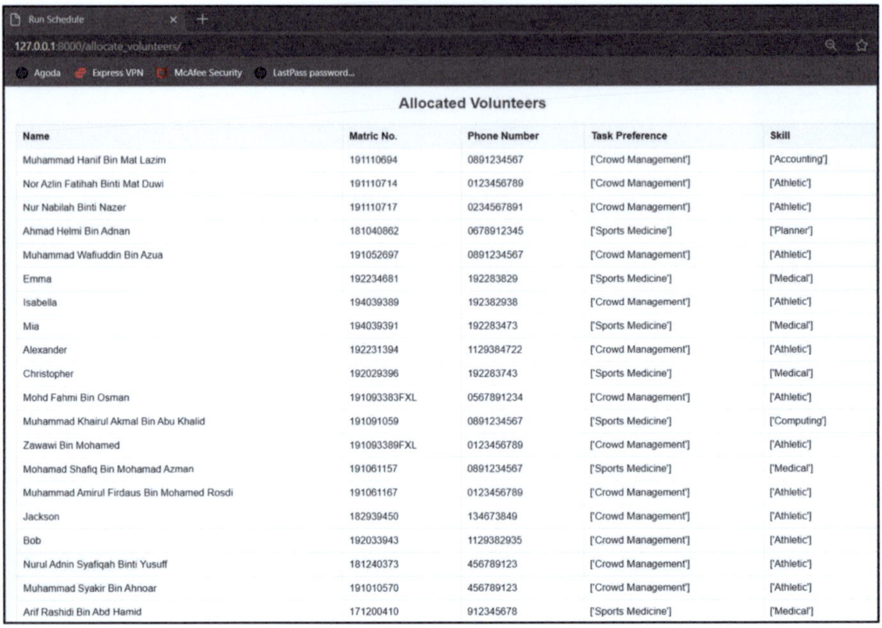

Run Schedule
127.0.0.1:8000/allocate_volunteers/
Agoda Express VPN McAfee Security LastPass password...

Allocated Volunteers

Name	Matric No.	Phone Number	Task Preference	Skill
Muhammad Hanif Bin Mat Lazim	191110694	0891234567	['Crowd Management']	['Accounting']
Nor Azlin Fatihah Binti Mat Duwi	191110714	0123456789	['Crowd Management']	['Athletic']
Nur Nabilah Binti Nazer	191110717	0234567891	['Crowd Management']	['Athletic']
Ahmad Helmi Bin Adnan	181040862	0678912345	['Sports Medicine']	['Planner']
Muhammad Wafiuddin Bin Azua	191052697	0891234567	['Crowd Management']	['Athletic']
Emma	192234681	192283829	['Sports Medicine']	['Medical']
Isabella	194039389	192382938	['Crowd Management']	['Athletic']
Mia	194039391	192283473	['Sports Medicine']	['Medical']
Alexander	192231394	1129384722	['Crowd Management']	['Athletic']
Christopher	192029396	192283743	['Sports Medicine']	['Medical']
Mohd Fahmi Bin Osman	191093383FXL	0567891234	['Crowd Management']	['Athletic']
Muhammad Khairul Akmal Bin Abu Khalid	191091059	0891234567	['Sports Medicine']	['Computing']
Zawawi Bin Mohamed	191093389FXL	0123456789	['Crowd Management']	['Athletic']
Mohamad Shafiq Bin Mohamad Azman	191061157	0891234567	['Sports Medicine']	['Medical']
Muhammad Amirul Firdaus Bin Mohamed Rosdi	191061167	0123456789	['Crowd Management']	['Athletic']
Jackson	182939450	134673849	['Crowd Management']	['Athletic']
Bob	192033943	1129382935	['Crowd Management']	['Athletic']
Nurul Adnin Syafiqah Binti Yusuff	181240373	456789123	['Crowd Management']	['Athletic']
Muhammad Syakir Bin Ahnoar	191010570	456789123	['Crowd Management']	['Athletic']
Arif Rashidi Bin Abd Hamid	171200410	912345678	['Sports Medicine']	['Medical']

Fig. 4 Allocated volunteers

Table 1 Match rate and average preference score of the scheduled volunteers

Performance metrics	Number of volunteers				
	100	200	300	400	500
Match rate (%)	81.63	80.00	77.50	83.00	87.93
Average preference score	1.3	1.5	2.2	1.8	1.5

the system faces challenges in aligning them with their preferred roles. Meanwhile, the average preference score is 2.2, signifying that a small number of volunteers who were allocated a task were placed in their most preferred role.

The table shows that as the number of volunteers increases, the system's performance in aligning tasks with volunteer preferences varies. The match rate, which indicates the percentage of volunteers assigned to their preferred tasks, generally improves with more volunteers, peaking at 87.93% with 500 volunteers. However, the average preference score, which measures how well tasks align with volunteer preferences (lower is better), fluctuates. It starts low at 1.3 for 100 volunteers, rises to 2.2 at 300 volunteers, and then improves again, returning to 1.5 at 500 volunteers. This suggests the system is more effective at aligning tasks with preferences when the volunteer count is either small or large, with a slight dip in performance at intermediate numbers.

The fluctuation in both the match rate and average preference score likely results from the complexity of balancing volunteer preferences and skills with the availability and distribution of tasks as the number of volunteers changes. When there are fewer volunteers (e.g., 100 or 200), it might be easier to closely match each volunteer to their top preferences, leading to lower average preference scores. However, as the number of volunteers increases to 300, the task allocation becomes more complex. There might be a higher competition for certain preferred tasks, making it harder to assign each volunteer to their top choices, which causes the average preference score to rise and the match rate to dip.

As the number of volunteers increases further (e.g., 400 and 500), the system might have more flexibility in task assignments due to a wider variety of tasks or better optimization strategies that can more effectively accommodate a large group. This can lead to an improved match rate and a reduction in the average preference score, as seen at 500 volunteers. Therefore, the fluctuation is likely caused by the dynamic interplay between volunteer preferences, task availability, and the system's ability to optimize assignments as the number of participants changes.

5 Conclusion

The developed system has successfully streamlined the process of assigning volunteers at Universiti Malaysia Perlis (UniMAP). By prioritizing volunteers' preferences and skills, the system not only improves volunteer satisfaction but also ensures that tasks are assigned efficiently, resulting in smoother event operations. The automated nature of the scheduling system significantly reduces the administrative workload for event organizers, allowing them to focus on more critical aspects of event management.

Furthermore, the system's ability to generate detailed performance metrics provides valuable insights into the effectiveness of the volunteer allocation process. These metrics enable continuous optimization and informed decision-making for future events. The successful deployment of this system at UniMAP demonstrates its potential as a scalable and efficient tool for managing volunteers, paving the way for more organized and effective sporting events in the future.

References

Acharya A, Bansak K, Hainmueller J (2022) Combining outcome-based and preference-based matching: a constrained priority mechanism. Polit Anal 30(1):89–112. https://doi.org/10.1017/pan.2020.48

Agrawal D, Pang G, Kumara S (2023) Preference based scheduling in a healthcare provider network. Eur J Oper Res 307(3):1318–1335. https://doi.org/10.1016/j.ejor.2022.09.027

Aronshtam L, Ilani H (2022) Bounds on the average and minimum attendance in preference-based activity scheduling. Discret Appl Math 306:114–119. https://doi.org/10.1016/j.dam.2021.09.024

Dickson TJ, Darcy S (2022) Surviving and thriving from the volunteer involvement in mega-sport events. Accessed on 7 Jan 2023 https://volunteeringstrategy.org.au/wp-content/uploads/2022/11/VRP_Surviving-and-Thriving-from-the-Volunteer-Involvement-in-Mega-Sport-Events.pdf

Ferreira A (2023) Deep dive: the software development life cycle (SDLC) in agile environments. https://medium.com/@argferreira1/deep-dive-the-software-development-life-cycle-sdlc-in-agile-environments-6b245bbea5ab

Guajardo M, Jörnsten K (2017) The stable tournament problem: matching sports schedules with preferences. Oper Res Lett 45(5):461–466. https://doi.org/10.1016/j.orl.2017.07.004

Holmes K, Nichols G, Ralston R (2018) It's a once-in-a-lifetime experience and opportunity- deal with it! volunteer perceptions of the management of the volunteer experience at the London 2012 Olympic Games. Event Manag 22(3):389–403. https://doi.org/10.3727/152599518X15252895715050

Mahrom N, Annuar AF, Sudin S, Quan CT (2024) Automated volunteer scheduler for sporting events using first come first serve approach. In: The proceedings of the 2nd international conference on innovation and technology in sports (ICITS 2023). https://doi.org/10.1007/978-981-97-3741-3_45

Mancini S, Triki C, Piya S (2022) Optimal selection of touristic packages based on user preferences during sports mega-events. Eur J Oper Res 302(3):819–830. https://doi.org/10.1016/j.ejor.2022.01.031

Quan CT, Mahrom N, Annuar AF, Sudin S, Raof RAA (2024) Optimizing volunteer scheduling for sporting events: a round-robin algorithm approach. In: The proceedings of the 2nd international conference on innovation and technology in sports (ICITS 2023). https://doi.org/10.1007/978-981-97-3741-3_31

Rhoden S, Ximeraki G, Ralston R (2002) 0161 247 2734 volunteering at the XVII commonwealth games

Wilks L (n.d.) The lived experience of London 2012 olympic and paralympic volunteers: a serious leisure perspective. http://www.tandfonline.com/eprint/7gNirHN3crAu77PP6Up6/full

EMG Analysis of Selected Lower Leg Muscles During Plantar Flexion in Various Sagittal Foot Positions

B. B. Shyamily, K. H. Shreaya, and Gerald Santhosh Dsouza

Abstract **(1) Background**: Plantar and dorsi flexion is a common movement in sports activities. The lower leg muscles namely Medial Gastrocnemius (MG), Lateral Gastrocnemius (LG), Soleus (SL) and Tibialis Anterior (TA) play an important role as agonists and antagonists in plantar and dorsi flexion. Plantar and dorsi flexion is a common movement in sports activities. These movements occur in sagittal plane such as on inclined, declined and flat surfaces as well as on frontal plane such as inversion, eversion and neutral positions. The objective of the study to investigate the relative activation of these four muscles, during different sagittal plane movements. The three frontal positions are combined for each of the sagittal positions. **(2) Methods**: Participants (N = 40) were required to undergo three trials of plantar flexion each on a flat surface and on 15 degrees raised wooden platform in a inclined as well as declined position. This was preceded by normalisation procedures for each of the muscles to find out the Maximum Voluntary Isometric Contraction. sEMG of the tested muscles were collected for all these movements. The percentage maximum voluntary isometric contraction (%MVIC) of MG, LG, SL and TA during plantar flexion in different sagittal positions was analysed using Analysis of Variance. **(3) Result**: There was significant statistical difference in muscle activation among MG, LG and SL in a combination of all the three sagittal foot variations at three inclined levels, although TA was found to be significantly different from the other three. MG showed the highest activation while soleus and LG were indicated similar activation levels. **(4) Conclusion**: This study indicates that the % MVIC of the tested muscles are highest when performed in an inclined plane and lowest in a declined plane. It is also found that MG is the most activated muscle among all the tested

B. B. Shyamily · G. S. Dsouza (✉)
Department of PG Studies and Research in Physical Education, Mangalore University, Mangalore, India
e-mail: gsansouza@mangaloreuniversity.ac.in

B. B. Shyamily
e-mail: shyamilybb3@gmail.com

K. H. Shreaya
Physical Education and Sports Science Division, Chanakya University, Bengaluru, India
e-mail: shreaya.kh@chanakyauniversity.edu.in

M. H. A. Hassan et al. (eds.), *Proceedings of the 3rd International Conference on Innovation and Technology in Sports*, Lecture Notes in Bioengineering,
https://doi.org/10.1007/978-981-95-0804-4_19

muscles. During the TA shows about 20–30% activation of MG as it is a dorsi flexor, the activation would have to arise during the eccentric phase. Strengthening of the lower leg muscles would have to be done primarily in inclined to derive greatest benefits, and eccentric movements could be combined to involve the TA also (BB S, Dsouza GS (2024) The contribution of lower leg muscles during vertical jump: an electromyographic study. Int J Phys Educ Sports Health 11(5):96–99. 10.22271/kheljournal.2024.v11.i5b.3496).

Keywords Muscles activation · Inclined · Declined · Plantar flexion · EMG · MVIC

1 Introduction

It's the world of contemporary science, and in sports these days, smart training is more important than rigorous training. Innovations like EMG have been crucial in assessing human capacity to achieve greatness in sports, contributing to the field of sports sciences' ongoing expansion. These innovations also aid in the prevention of injuries and aid in the process of recovery. Electromyography (EMG) is an electro-diagnostic medical method used to measure and record the electrical activity that skeletal muscles produce (Chen et al. 2020). An electromyograph, which creates a record known as an electromyogram, is used to perform EMG. The signals are then examined to look for patterns in muscle recruitment, the activation level, or medical anomalies. Analysing the biomechanics of animal or human movement is another application for the signals.

SEMG refers to surface electromyography and measures muscle activity in micro-volts. The measured muscle activity gives an indication as to which muscle is involved in the particular activity, thereby determining which muscle has to be relaxed and which muscle has to be fired in the proper sequence and amplitude. In addition to the usage of EMG techniques in training, they also provide information as to the status of an athlete's strength and conditioning or the benefits of a rehabilitation programme on the muscular system.

The Tibialis Anterior (TA) muscle is a crucial muscle in the knee and ankle joints and plays an important role in dorsiflexion. n essential element in the ankle and knee joints, plays an important role in bending the dorsiflexion of the foot and helps to keep constantly the axis of the ankle joint (Jeon 2020). The TA muscle is indispensable in a host of sports activities as almost all sports events required dorsiflexion of the ankle such as running, walking, kicking and so on. It is also essential for postural stability as its weakness might lead to limitations in running, walking and other functional activities (Jeon 2020). Soleus (SL) is situated opposite to the TA and is antagonist to it. Along with Gastrocnemius (GA), the SL produces about 93% of the force in plantarflexion (Lee et al. 2019; Schoenfeld et al. 2020). Just like the TA, SL also plays a crucial role in postural stability although TA offers more proprioception than the SL (Choi et al. 2018; Riemann et al. 2003).

Various studies point to the activation levels of Gastrocnemius, Soleus and Tibialis Anterior muscles while performing a variety of activities and for different populations. Studies show that the activation of tibialis anterior significantly decreases as inclination increases ($p < 0.05$). However, there was no significant change in activation of soleus, gastrocnemius medialis and gastrocnemius lateralis ($p > 0.05$), and no noticeable drifts were found (Malik et al. 2023a). Furthermore, there seemed to be negligible difference between all the muscles at ground level and inclination at the levels of 2 and 4.

Studies also indicate that the amplitude of gastrocnemius and soleus increased successively from a standing position to hopping and average values for gastrocnemius and soleus increased by 10.8 times and 6.3 times respectively. As demands for force and speed decreased, relative activation indicated a shift from soleus to gastrocnemius muscle. These findings indicate that these two functionally specialised muscles related to ankle extensor synergy, follow a task-specific preferential response (Moritani et al. 1991).

In another study, the Soleus, Gastrocnemius Medialis, Gastrocnemius Lateralis and Tibialis anterior were studied between males and females, while performing a standing heel raise during the concentric and eccentric phases. The results indicated that there was no significant difference between the baseline, concentric, and eccentric phases with respect to each of the muscles being studied. However, during the baseline and concentric phase, significant difference was observed in gastrocnemius medialis, between males and females. The data also indicates that the eccentric phase during the standing heel raise task is not gender based (Ugbolue et al. 2021).

While studying the activation patterns of Gastrocnemius and Soleus while performing various motor tasks, the EMG amplitude ratios of the soleus to the gastrocnemius was 1.16 in toe standing on one leg. Consequent to this, when necessities for force and speed were increased, the magnitude of relative activation shifted from soleus to the gastrocnemius muscle (Moritani et al. 1991). This would indicate that the soleus would activate significantly in static and tasks with lesser demands for force, while the gastrocnemius takes over in activities demanding more force and speed.

There are very few studies done on activation of Gastrocnemius, Soleus and Tibialis Anterior in different positions in sagittal plane. This study looked at the relationship between lower limb in electromyography (EMG) signals and activation of selected muscles (Soleus, Medial and Lateral Gastrocnemius and Tibialis anterior) while performing heel raises in combinations of Frontal positions (Neutral, Inversion, Eversion) and Sagittal Positions (Flat, Inclined, Declined).

Table 1 Demographic data of the participants

Characteristics	Males ($n = 25$)	Females ($n = 15$)
Age (years)	23.4 (1.5)	24.1 (1.5)
Body mass (kg)	68.4 (7.1)	53.7 (10.6)
Height (m)	1.72 (0.06)	1.59 (0.06)

2 Materials and Methods

2.1 Participants

A total of 40 healthy 20- to 25-year-old sportspersons enrolled at Mangalore University; India participated in this study [25 males: age 23.4 ± 1.5; height 1.72 ± 0.06 m; weight 68.4 ± 7.1 kg; Body Mass Index (BMI) 23.0 ± 2.8 kg/m^2; 15 females: age 24.1 ± 1.5 years; height 1.59 ± 0.06 m; weight 53.7 ± 10.6 kg; BMI 21.2 ± 3.7 kg/m^2] (Table 1).

The volunteers who agreed to participate in this study signed the Informed Consent Form and underwent an evaluation always performed by the same person. The subjects were explained the purpose of the study, and the procedures involved in the Study. Participants were screened for any conditions that may affect EMG readings such as skin conditions, injuries and were found to be free of the same. The study was approved by the university.

2.2 Instrumentation

The following tools were developed/used to collect data.

sEMG electrodes

Electromyography (EMG) was utilised to determine levels of muscle activity during heel raises in the combination of sagittal and frontal foot positions. The participants' limbs were prepared by shaving and cleaning the skin with alcohol wipes. Delsys Trigno Avanti EMG sensors, with double sided self-adhesive tapes (Delsys Inc. USA) were then placed on the mid portion of the Medial Gastrocnemius, Lateral Gastrocnemius, Soleus and Tibialis Anterior in accordance with SENIAM guidelines for electrode placement (Hermens et al. 2000) (Fig. 1).

Wooden box

A specially constructed wooden box with an inclination of 15° was used to rest the foot to perform heel raises in inclined and declined positions. White markings were drawn on the box to place the foot 15° laterally to assist in assuming eversion and inversion positions of the foot (Fig. 2).

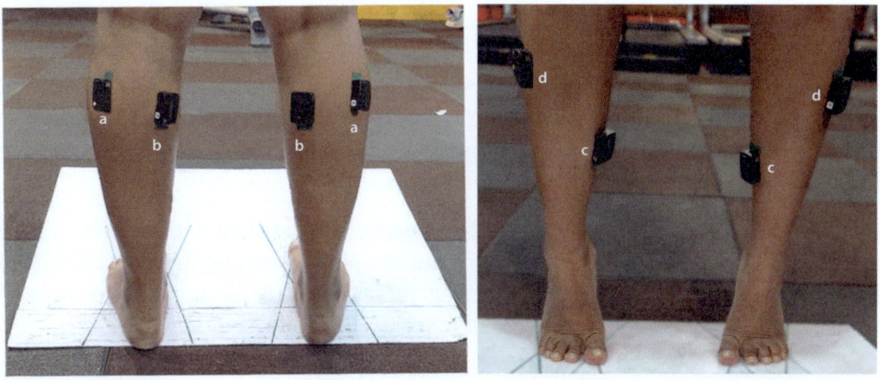

a - Lateral Gastrocnemius b- Medial Gastrocnemius c - Soleus d - Tibialis Anterior

Fig. 1 Sensor placement

Fig. 2 Wooden box

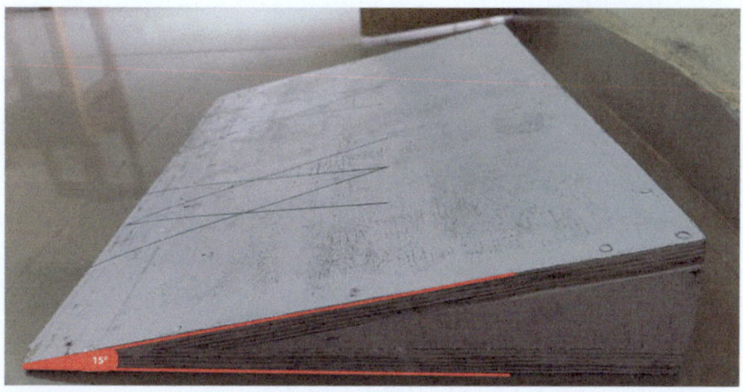

Software

The EMG signal was sampled at 500 Hz, amplified 1000, band pass filtered (20e450 Hz) and transmitted to the Labchart 8.0, software programme (Ad Instruments, USA).

2.3 *Experimental Protocol*

Muscle activation was normalised before acquiring the plantar flexion activation data.

Normalisation: Normalisation of muscle activation of each of the muscles was conducted by performing Maximal Voluntary Isometric Contractions (MVIC). Participants began with a warm up for 5 min, The subjects were then made to perform three trials of MVICs for each of the muscles followed by a rest period of 2 min in between each trial. The average of 3 trials was taken as the measure of MVIC for the respective muscle. The resistance was provided manually by the test assistants.

Medial and Lateral Gastrocnemius muscle: The MVIC of Gastrocnemius (lateral and medial) was performed by placing manual resistance against the ball of the feet, while the subject is sitting leaning back against the wall and the legs stretched out in front. conducted with the subject sitting with the back against the wall, with legs outstretched in the front. The subject is required to try to push to foot down using only the gastrocnemius muscle (BB and Dsouza 2024).

Soleus muscle: The subject is required to sit on a chair such that in the normal sitting position, the knees are flexed to 90^0 to isolate the soleus muscle. Resistance is given manually on the top of the thighs, MVIC is recorded while the subject tries to raise his heels only by contracting the soleus muscle. (BB and Dsouza, 2024).

Tibialis Anterior muscle: The MVIC of Tibialis Anterior (TA) was performed while assuming the same position as when recording the MVIC of the Gastrocnemius muscle. But in this instance the manual resistance was given at the top of feet (on the metatarsals). The subject was instructed to pull the toes upwards towards the tibia (dorsiflex) against the resistance (BB and Dsouza 2024).

Plantar flexion testing:

Immediately after MVIC testing, plantar flexion testing was carried out.

- The muscle activation was measured during heel raise while assuming the foot in three positions in the sagittal plane (flat, inclined and declined position, i,e, $-15°$, 0 degrees and $+ 15°$ to the horizontal), using the specially constructed box.
- In each of these sagittal positions plantar flexion was also performed while placing the foot three positions in the horizontal plane (neutral, inversion and eversion, i.e. 15° degrees on either side).
- Totally the positions assumed were—flat neutral, flat inversion, flat eversion, inclined neutral, inclined inversion, inclined eversion, declined neutral, declined inversion, declined eversion.
- The subject performed the heel raises against his/her own body weight for three trials in each of the positions, which were then averaged. The heel raises were performed in the following sequence. Two seconds raise the heel, hold for one second and two seconds lower the heel.
- The raw data is acquired in millivolts (mV) which is converted to Root Mean Square (RMS). All values are expressed as a percentage of MVIC for each of the muscles. The final activation values were computed as the average of the three trials.

3 Data and Analysis

The sEMG data acquired while performing MVC and plantar flexion in all the foot positions was recorded (Trigno Wireless System, Delsys, Natick, MA, United States) and analysed to quantify muscle activity (BB and Dsouza 2024). The data was converted into root-mean-square (RMS) technique (Chen et al. 2020) using

Labchart 8.0 software, by AD Instruments, USA. The data is shown in the form of graphs and tables to make a meaningful comparison between the muscle activation of each muscle during different repetitions. Descriptive statistics were calculated for selected EMG-dependent variables. ANOVA at a significance level of 0.05 was calculated to find the significance in difference, post hoc tests were carried out to find out the differences in individual muscles. The data was analyzed using advanced Excel and SPSS Software.

For computation of activation values, all three frontal positions (neutral, inversion and eversion) have been summed up for each of the three sagittal positions and data has been presented likewise i.e. muscle activations with reference to the three sagittal foot position variations namely, Flat, Inclined and Declined.

4 Results

4.1 Computation of Sample Size

Sample size for the respondents is estimated based on the statistical techniques that can be applied for the data analysis by considering power = 0.95 and effect size = 0.55 (moderate) by making use of a software G Power **t tests -ANOVA,** Means: Difference between two means (Fig. 3).

Analysis: A priori: compute required sample size

Input: Tail(s) = Two

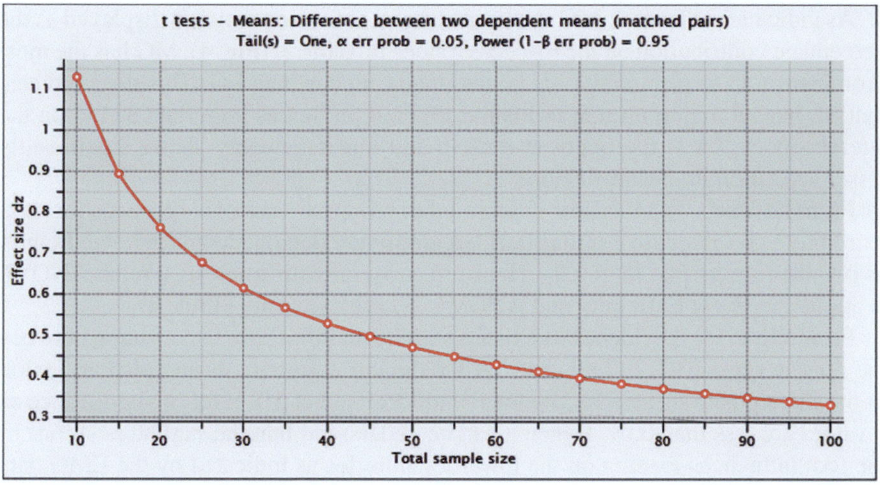

Fig. 3 Power test for deciding the sample size

	Effect size dz	= 0.55
	α err prob	= 0.05
	Power (1-β err prob)	= 0.95
Output:	Noncentrality parameter δ	= 3.3904277
	Critical t	= 1.6870936
	Df	= 39
	Total sample size	= 40
	Actual power	= 0.9537735

Hence the sample size is considered as 40.

4.2 *Muscle Activation of Each Muscle During Plantar Flexion in Each of the Sagittal Positions (Flat, Inclined and Declined) on Each Side*

Table 2 displays the descriptive data on the right lower leg muscles and left lower leg muscles. On both the sides, Medial Gastrocnemius shows the highest activation while Tibialis Anterior show the lowest activation, which is obvious as TA is the opposing muscle while performing plantar flexion. LG and SL are similar in their activation. At a glance one can ascertain that the variations in activation of all the four muscles are similar for both the right and left extremities.

As indicated in Table 2, MG shows the largest activation which is displayed as the percentage contribution on the respective sides in Table 3. (Fig. 4). MG has the most prominent role to play, while SL is marginally higher than LG. This does indicate that SL has an important role in Plantar Flexion, in fact as important as LG on the lateral aspect. TA is the opposite muscle and understandably shows significantly lesser activation in plantar flexion.

From Levene's Test it can be observed that groups are not homogeneous for right extremity as p values are less than 0.01 but groups are homogeneous for left extremity as p values are greater than 0.05. Hence to carry on with Analysis, Welch ANOVA is applied to Right Extremity and ANOVA to Left Extremity (Table 4).

From Table 5 it is evident that exists a significant difference in activation between the muscle types MG, LG, SL and TA for the left extremity during plantar flexion in the three positions i.e. Flat, Inclined and Declined at 1% level of significance as p values are less than 0.01. Hence it can be understood that the sagittal positions of the foot influences loading on the lower leg muscles as indicated by the EMG data (Table 6).

Table 2 Descriptive statistics of muscle activation of the right extremity in each of the sagittal positions

Position	Muscle	Mean	SD	Min	Max	Muscle	Mean	SD	Min	Max
Flat	RMG	94.30	20.70	58.11	165.20	LMG	100.84	31.24	58.11	165.20
	RLG	79.31	34.47	20.15	188.83	LLG	84.40	33.49	20.15	188.83
	RSL	85.95	23.18	47.29	236.50	LSL	77.50	32.89	47.29	236.50
	RTA	30.20	27.54	6.15	192.56	LTA	37.21	28.88	6.15	192.56
Inclined	RMG	94.15	20.45	62.60	175.87	LMG	96.60	30.95	44.60	233.83
	RLG	79.88	31.62	21.99	170.03	LLG	85.43	33.73	35.63	251.93
	RSL	94.78	19.61	49.52	139.21	LSL	86.72	29.65	41.80	220.62
	RTA	38.67	30.80	7.10	168.64	LTA	42.27	28.95	6.21	129.53
Declined	RMG	89.46	20.05	53.64	154.00	LMG	89.32	30.35	29.29	206.33
	RLG	79.62	33.16	16.06	191.54	LLG	77.71	31.27	22.79	214.36
	RSL	79.70	19.10	41.86	177.32	LSL	63.15	27.43	24.15	203.36
	RTA	25.93	19.50	7.13	111.26	LTA	31.94	28.052	6.09	126.91

Note Each of the sagittal feet positions are the sum of three frontal foot positions i.e. Neutral, Inversion and Eversion for 40 subjects each, making it 120 samples. The values are measured in RMS (mV) and are expressed in % of MVIC for each muscle

Table 3 Contribution of each lower leg muscle during heel raise in each of the sagittal foot positions

Position	Muscle	Right (% contribution)	Left (% contribution)
Flat	MG	32.5	33.6
	LG	27.4	28.1
	SL	29.7	25.8
	TA	10.4	12.4
Inclined	MG	30.6	31.1
	LG	26.0	27.5
	SL	30.8	27.9
	TA	12.6	13.6
Declined	MG	32.6	34.1
	LG	29.0	29.6
	SL	29.0	24.1
	TA	9.4	12.2

Note The data is collected for 120 samples and is expressed as a percentage of the sum of contraction of all the muscles in various sagittal foot positions on both sides

From Welch ANOVA it is evident that there exists a significant difference in % muscle activation between the muscle types Med Gast, Lat Gast, Soleus and Tib Ant for the Right Extremity in the three positions Flat, Inclined and Declined at 1% level of significance as p values are less than 0.01 (Table 7). Hence as indicated in the

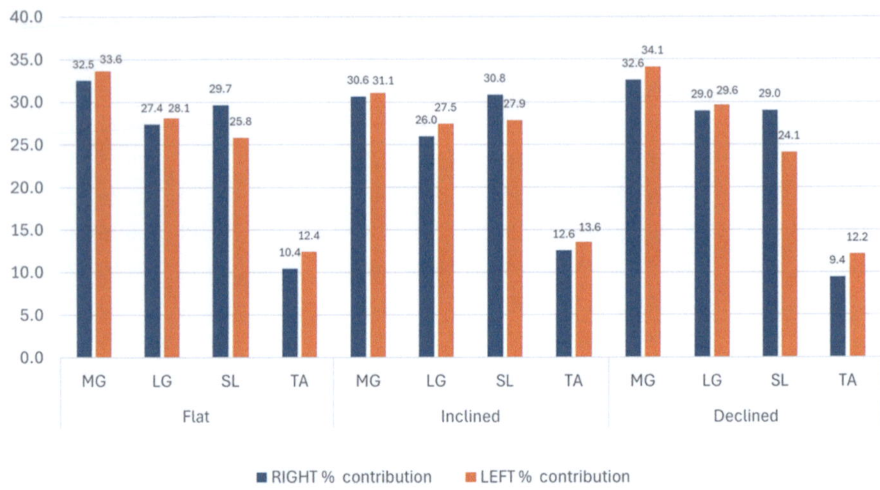

Fig. 4 Contribution of lower leg muscles during plantar flexion

Table 4 Test of homogeneity of variances

Position	Muscle type	Levene statistic	df1	df2	Sig
Flat	Right extremity	10.155	3	476	0.000
	Left extremity	0.440	3	476	0.725
Inclined	Right extremity	8.563	3	476	0.000
	Left extremity	0.049	3	476	0.985
Declined	Right extremity	10.227	3	476	0.000
	Left extremity	0.332	3	476	0.802

left extremity, the right extremity lower leg muscles also are affected by the sagittal positions of the foot.

Post Hoc Comparison

In order to find out whether any two or more-muscle type are homogeneous for the Right and Left Extremity on the basis of three positions Flat, Inclined and Declined Post Hoc Analysis is carried out which is revealed in the form of means plot.

Figure 5 A to F indicate that during plantar flexion on both right and left sides, in all the sagittal positions, there is no significant difference in activation between MG, LG and SL while all of them are significantly different from TA. In both the extremities, MG muscle works the hardest whereas the LG and SL are more or less equally activated. Obviously since TA is the opposite muscle it is least activated

Table 5 ANOVA table for significance in muscle activity between each lower leg muscle of Left Extremity in each of the Sagittal positions

Position		Sum of squares	df	Mean square	F	Sig
Flat	Between Groups	262,840	3	87,613.2	87.3	0.000**
	Within groups	477,708	476	1003.59		
	Total	740,548	479			
Inclined	Between groups	210,471	3	70,157	73.578	0.000**
	Within groups	453,871	476	953.51		
	Total	664,342	479			
Declined	Between groups	221,753	3	73,917.7	85.955	0.000**
	Within groups	409,340	476	859.958		
	Total	631,093	479			

**Significant at 1% level of significance

Table 6 Robust tests of equality of means statistics for significance in activity between each lower leg muscle of right extremity in each of the sagittal positions

Position	Muscle extremity		Statistic[a]	df1	df2	Sig
Flat	Right extremity	Weich	148.757	3	260.824	0.000**
Inclined	Right extremity	Weich	108.682	3	259.39	0.000**
Declined	Right extremity	Weich	249.627	3	261.086	0.000**

[a]Asymptotically F distributed
**Significant at 1% level of significance

Table 7 Descriptive statistics for sagittal foot positions of right and left extremity for sum of all the 4 muscles

Position	Extremity	N	Mean	Std. Deviation	Std. Error Mean
Flat	Right extremity	480	72.44	36.71	1.67588
	Left extremity	480	74.98	39.31	1.79468
Inclined	Right extremity	480	76.87	34.73	1.58557
	Left extremity	480	77.75	37.24	1.69984
Declined	Right extremity	480	68.68	34.42	1.57105
	Left extremity	480	65.53	36.29	1.65675

Note Each of the sagittal feet positions are the sum of three frontal foot positions i.e. Neutral, Inversion and Eversion for 40 subjects each, for the sum of 4 muscles adding to 480 samples. The values are measured in RMS (mV) and are expressed in % of MVIC for each muscle

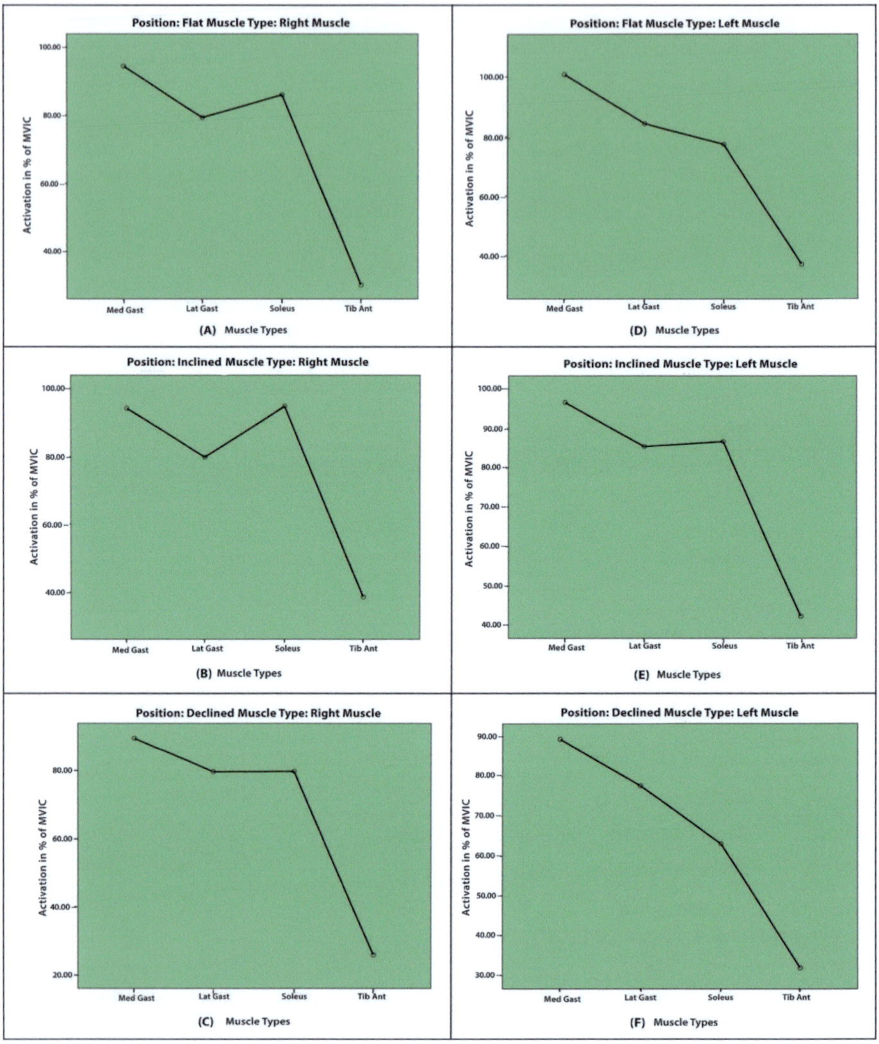

Fig. 5 Means plot for muscle activation during flat, inclined and declined positions

4.3 Muscle Activation of All Four Muscles Combined in Right and Left Extremity for Each of the Sagittal Positions

In this analysis the combined action of all the four muscles i.e. MG, LG, SL and TA on the left and right extremity separately during Plantar flexion in the flat, inclined and declined positions have been compared.

Table 8 Independent samples T test of significance for muscle activation of all four muscles combined in each of the sagittal positions

Position		Levene's test for equality of variances		t-test for equality of means		
		F	Sig	t	df	Sig. (2-tailed)
Flat	Equal variances assumed	0.044	0.834	−1.037	958	0.3
Inclined	Equal variances assumed	0.012	0.914	−0.381	958	0.703
Declined	Equal variances assumed	0	0.986	1.378	958	0.169

Table 7 indicates the descriptive statistics of the combined action of the four muscles on left and right side. It is evident from the data that muscle activity during inclined plantar flexion shows the highest activation and declined shows the lowest activation. This could be because of the greater range of movement, due to the muscles being in greater pre-contraction stretch in the inclined initial position when compared to the flat and declined positions. For the same reason, the flat position plantar flexion shows greater activity than the declined position.

Though the descriptive statistics show differences in activation in all the three positions, the activations are not significantly different from each other, which is evident from Table 8. From Independent sample t test t is evident that there exists no significant difference in % activation of lower leg muscles between Right and Left Extremities for three positions Flat, Inclined and Declined at 5% level of significance as p values are greater than 0.05'.

4.4 *Muscle Activation of Each Pair of Muscle on Right and Left Extremity Combined for Each of the Sagittal Positions*

In this section, activation of each pair of muscle on both sides is summed up and analysed for difference in activation in each of the sagittal foot positions.

Table 9 portrays the activation of each muscle on both sides combined for each of the foot positions. It is evident that MG is activated the most in all the three positions while LG and SL activate nearly equally. TA activates the least.

From the test for Homogeneity of Variances it is observed that all the significance values are greater than 0.05 which leads to the conclusion that groups are homogeneous and hence qualifies for further applying ANOVA.

From ANOVA it can be observed that there exists a significant difference in activation between muscle types Med Gast, Lat Gast, Soleus and Tib Ant for the three positions Flat, Inclined and Declined at 1% level of significance as p values are less than 0.01 (Table 10). Thus, it can be concluded that there is significant impact of various sagittal foot positions while performing plantar flexion on lower leg muscles.

Table 9 Descriptive statistics for combined activation of each muscle on both sides for each of the Sagittal positions

Position	Muscles	N	Mean	SD	Minimum	Maximum
Flat	MG	240	97.57	26.65	48.83	231.64
	LG	240	81.86	34.01	20.15	254.43
	SL	240	81.73	28.71	35.64	237.12
	TA	240	33.71	28.38	6.15	192.56
Inclined	MG	240	95.38	26.21	44.60	233.83
	LG	240	82.66	32.75	21.99	251.93
	SL	240	90.75	25.41	41.80	220.62
	TA	240	40.47	29.88	6.21	168.64
Declined	MG	240	89.40	25.68	29.29	206.33
	LG	240	78.67	32.18	16.06	214.36
	SL	240	71.43	25.01	24.15	203.36
	TA	240	28.94	24.29	6.09	126.91

Note Each of the sagittal feet positions are the sum of three frontal foot positions i.e. Neutral, Inversion and Eversion for 40 subjects each, for the sum of two muscles on either side making it 240 samples. The values are measured in RMS (mV) and are expressed in % of MVIC for each muscle

Table 10 ANOVA table for muscle activation combined on both sides for each of the Sagittal positions

Position		Sum of squares	df	Mean square	F	Sig
Flat	Between groups	552,063	3	184,021.03	210.489	0.000**
	Within groups	835,786	956	874.254		
	Total	1,387,849	959			
Inclined	Between groups	454,292	3	151,430.75	183.653	0.000**
	Within groups	788,267	956	824.547		
	Total	1,242,559	959			
Declined	Between groups	505,376	3	168,458.68	231.527	0.000**
	Within groups	695,583	956	727.597		
	Total	1,200,959	959			

Post Hoc Comparison

In order to find out whether any two or more-muscle type are homogeneous on the basis of three positions Flat, Inclined and Declined Post Hoc Analysis is carried out which is revealed in the form of means plot.

As far as muscle activation for all the foot positions is concerned (Fig. 6a–c) Med Gast, Lat Gast and Soleus forms one group who are at high level on an average whereas Tib Ant forms another group who is at low level. This is obvious as MG,

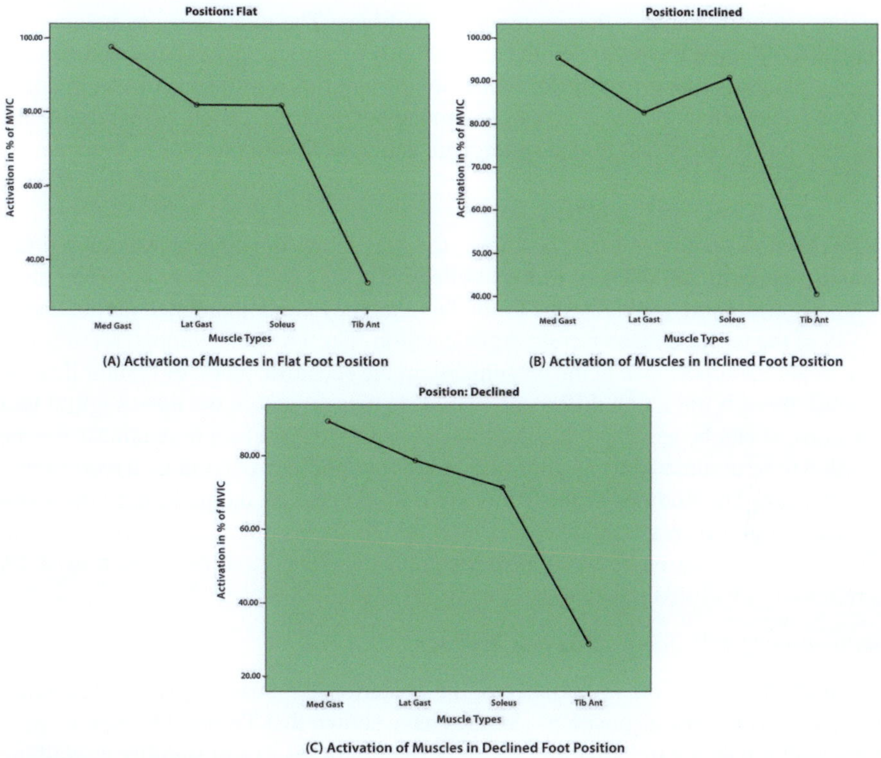

Figures 6 a–c Means plot of activation of muscles on both sides combined while performing plantar flexion in three sagittal positions

LG and SL are the main movers during plantar flexion and TA only comes to play during the eccentric phase of plantar flexion. But a point to note here is that, just as evidenced in previous analyses, MG has the highest activation, which goes to show that MG takes most of the load in activities involving heel raises. Soleus being located on the lateral side of the leg shows equal muscle activation to LG, since it is a purely plantar flexor. The activation of TA can be considered to occur mostly in the eccentric phase of plantar flexion. The post hoc analysis shows that the activation of LG and TA does not differ significantly, while it does so with MG and TA.

5 Discussion

Foot placement in the sagittal planes such as inclination and declination and frontal planes such as eversion and inversion, occur in various sports such as football, basketball, hockey etc. Similarly motor skills and sport skills are executed in a variety of challenging situations involving plantar flexion (pushing off from the toes) and dorsi flexion (front of the foot flexing towards the tibia). The main muscles involved in plantar flexion are the lower leg muscles namely Gastrocnemius lateral and medial, soleus, plantaris, tibialis posterior while tibialis anterior would be involved in dorsiflexion. The effect of these dynamic movements causes varied demands on the muscles individually and on the synergistic action of all the muscles as a whole.

Recruitment of muscles

In all the combinations of foot positions, the MG shows the highest activation, individually on both the sides as well as collectively. LG and SL show approximately similar recruitment, probably due to the fact that they are both aligned on the lateral aspect of the leg, and theirs is a synergistic action. The TA shows enough recruitment to indicate its importance in the presumably in the eccentric phase of Plantar flexion. Overall, there is not much difference in muscle recruitment in the three sagittal foot positions, it can be assumed that the muscles orient themselves in readiness for the position to be assumed during game play, and the total effect of combined recruitment is important. The findings of this study are corroborated by the findings in the study of Malik et al. (Torres-Banduc et al. 2021) which shows that there is no significant difference in activation between gastrocnemius and soleus and the activation of TA is minimal in inclined position.

Implications for Rehabilitation and training

The study highlights the importance of the posterior leg muscles as a whole (synergistic action) in various positions. Studies have shown that TA and SL are crucial in maintaining balance integrity and are important muscles to add stability in walking and running. As an application of training, it highlights the importance of the soleus muscle, which is a dedicated plantar flexor, and offers informative data as to its role in plantar flexing in relation to gastrocnemius muscle. The role of tibialis anterior in the totality of muscle contraction during plantar flexion, is also brought to light though it is a dorsiflexor. The data obtained from the study can also be used to evaluate to activation levels of the lower leg muscles in various athletes and assessments can be made as to the fitness of the lower leg muscles in relation activation levels. Any significant deviation from the percentage role of the muscles established in this study, could be looked into by prescribing training drills for the individual muscle in consideration.

Limitations of sEMG

Though enough care has been taken within the standard procedures for sEMG testing, it should be noted that sEMG is not without its shortcomings. sEMG or Kinesiological EMG is a functional tool as compared to a clinical or neurological EMG which is a diagnostic tool. Water, communication of signals and muscle movement under the electrodes might lead to cross-talk or might measure other activities. An accurate placement of the electrodes on the midpoint of the contracted muscle, might minimise cross-talk. It is also a challenge apply MVC techniques in dynamic situations. EMG does not represent force or torque. Further, thickness of the skin and patterns of subcutaneous layer under the skin might require more study (Malik et al. 2023b).

Another important factor limiting the wide use of these technique is represented by the limited access to the software needed for separating the motor unit action potentials in the sEMG (Malik et al. 2023b). It seems that this trend is now quickly (Clarys et al. 2010; Merletti et al. 2020; Farina et al. 2014) changing due to the uncontroversial advancement in sEMG technique of the last 10 years coupled to the increased power of the analytical tool specifically designed to decipher the sEMG signal (Vecchio et al. 2019). On the other hand, the lack of extensive dissemination of accurate information on sEMG among professionals continues to prevent implementation of sEMG in their evaluation protocols. sEMG equipment also needs to be feasible for the field usage.

While researchers underestimate sEMG, sports professionals are found to give more importance and credence to it (Feeney et al. 2018). Though opinions regarding the applicability of sEMG is varied among experts, on a broader scale, production of wearable high density electrodes allowing testing large muscle groups which are nearer to the surface show results reliable enough to draw conclusions and apply them to training and rehabilitation settings.

6 Conclusion

The above study was conducted with the purpose of assessing the muscle activity through EMG studies on the important lower limb muscles while performing plantar flexion or heel raises. The three sagittal positions namely flat, inclined and declined was considered for performing the heel raises. Additionally, in each of these positions the plantar flexion was also considered in the frontal movements namely, neutral, inverted and everted positions. The frontal positions were not considered individually, rather were combined and analysed for each of the sagittal foot positions separately.

The study concluded the following based on the findings:

1. Medial Gastrocnemius is the highest activator during plantar flexion.
2. Lateral Gastrocnemius and Soleus being on the same side of the foot i.e. lateral side, show marginal difference in activation.
3. Though it is evident that Tibialis Anterior is not a plantar flexor, it is a dorsi flexor, it also shows approximately 20–30% activation of the other muscles, which most probably occurs during the eccentric or lowering phase of plantar flexion.
4. The activation in the inclined position is the highest, since it has the highest pre-stretch before the actual execution of the heel raise, which increases the stored energy in the muscle. The activation is the lowest in the declined position, as the muscle is already in a semi contracted position while the neutral position exhibits average muscle activation percentage.

Though there are noticeable differences in muscle activation, the post hoc analysis indicates no significant difference between the actual plantar flexors namely, MG, LG and SL, while it is obvious that there would be significant difference between them and TA, as it is a dorsi flexor. This indicates that the three plantar flexors cannot be isolated while training, and they work in sync in activities involving plantar flexion.

More studies could be done on muscle activation of the lower limb muscles while separating the concentric phase and eccentric phase of plantar flexion. It would also be educative to analyse these muscles during plantar flexion in the knee flexed position, to assess the load borne by the Soleus muscle, as it is the main mover in this position. The combined role of quadriceps, hamstrings and the lower leg muscles could also be investigated.

References

BB S, and Dsouza GS (2024) The contribution of lower leg muscles during vertical jump: an electromyographic study. Int J Phys Educ Sports Health 11(5):96–99. https://doi.org/10.22271/kheljournal.2024.v11.i5b.3496

Chen C-K, Lin S-L, Wang T-C, Lin Y-J, Wu C-L (2020) Lower-limb electromyography signal analysis of distinct muscle fitness norms under graded exercise intensity. Electronics 9(12):12. https://doi.org/10.3390/electronics9122147

Choi T, Im CH, Kim SJ, Kim H, Lee JM (2018) Prediction method of walking speed at swing phase using soleus electromyogram signal at previous stance phase. Annu Int Conf IEEE Eng Med Biol Soc. 2018:2308–2311

Clarys JP, Scafoglieri A, Tresignie J, Reilly T, Von Roy P (2010) Critical appraisal and hazards of surface electromyography data acquisition in sport and exercise. Asian J Sports Med 1(2):69. https://doi.org/10.5812/ASJSM.34868

Del Vecchio A, Negro F, Holobar A, Casolo A, Folland JP, Felici F, Farina D (2019) You are as fast as your motor neurons: speed of recruitment and maximal discharge of motor neurons determine the maximal rate of force development in humans. J Physiol 597(9):2445–2456. https://doi.org/10.1113/JP277396

Farina D, Holobar A (2016) Characterization of human motor units from surface EMG decomposition. Proc IEEE 104(2):353–373. https://doi.org/10.1109/JPROC.2015.2498665

Farina D, Merletti R, Enoka RM (2014) The extraction of neural strategies from the surface EMG: an update. J Appl Physiol 117(11):1215–1230. https://doi.org/10.1152/JAPPLPHYSIOL.00162.2014

Feeney DF, Mani D, Enoka RM (2018) Variability in common synaptic input to motor neurons modulates both force steadiness and pegboard time in young and older adults. J Physiol 596(16):3793–3806. https://doi.org/10.1113/JP275658

Hermens HJ, Freriks B, Disselhorst-Klug C, Rau G (2000).Development of recommendations for SEMG sensors and sensor placement procedures. Seniam Guidel 10:361e374

Jeon I-C (2020) Comparison of the electromyographic activity of the tibialis anterior and isometric dorsiflexor strength during dorsiflexion according to toe postures in individuals with ankle dorsiflexor weakness. J Kor Phys Ther 32:233–237

Lee HS, Hong SB, Chin HN, Choi JL, Seon HC, Jeong DY (2019) Characteristics of muscle activity in the lower extremity during stepping over various obstacle. J Korean Soc Phys Med. 14:55–62

Malik RJ, Waris A, Gilani SO, Iqbal J, Kashif AS, Kamboh HM, Ghafoor MF (2023) The effects of surface inclination on gastrocnemius, soleus and tibialis anterior muscle activation during gait. J Back Musculoskelet Rehabil 36(1):181–186. https://doi.org/10.3233/BMR-210371. PMID: 35964168

Merletti R, Roatta S, Felici F, Del Vecchio A, Vecchio D (2020) Surface electromyography: what limits its use in exercise and sport physiology? 11:578504. https://doi.org/10.3389/fneur.2020.578504

Moritani T, Oddsson L, Thorstensson A (1991) Activation patterns of the soleus and gastrocnemius muscles during different motor tasks. J Electromyogr Kinesiol 1(2):81–88. https://doi.org/10.1016/1050-6411(91)90001-L. PMID: 20870497

Riemann BL, Myers JB, Lephart SM (2003) Comparison of the ankle, knee, hip, and trunk corrective action shown during single-leg stance on firm, foam, and multiaxial surfaces. Arch Phys Med Rehabil 84:90–95

Schoenfeld BJ, Vigotsky AD, Grgic J, Haun C, Contreras B, Delcastillo K et al (2020) Do the anatomical and physiological properties of a muscle determine its adaptive response to different loading protocols? Physiol Rep 8:e14427

Torres-Banduc M, Ramirez-Campillo R, Andrade DC, Calleja-González J, Nikolaidis PT, McMahon JJ, Comfort P (2021) Kinematic and neuromuscular measures of intensity during drop jumps in female volleyball players. Front Psychol 12:724070. https://doi.org/10.3389/fpsyg.2021.724070

Ugbolue UC, Yates EL, Ferguson K, Wearing SC, Gu Y, Lam W-K, Baker JS, Dutheil F, Sculthorpe NF, Dias T (2021) Electromyographic assessment of the lower leg muscles during concentric and eccentric phases of standing heel raise. Healthcare 9:465. https://doi.org/10.3390/healthcare9040465

Comparative Analysis of Heel Contact Force and Electromyographic Activity Among Different Nordic Hamstring Exercises

Tibeth Sukantameth, Jutharat Promson, Nawarat janpolsaen, Prisana phonhan, Sirinya Audonphon, Radomyos Matjiur, Siwakorn Akkaneekool, Rodrigo Zacca, and Phornpot Chainok

Abstract The Nordic hamstring exercise (NHE) is particularly effective in an enhancing performance and reducing the risk of hamstring injuries. Therefore, we examine and compare the changes in heel contact force and muscle activation between different variations of dynamic and static break points in collegiate soccer players. Twelve healthy collegiate soccer players (mean ± SD age 20.15 ± 1.1 yrs, height 171.8 ± 4.9 cm, and body mass 66.1 ± 6.1 kg) who had not experienced hamstring strain injuries participated in the study. A one-way repeated measures ANOVA was conducted to determine whether there was a statistically significant difference in heel contact force and electromyographic activity in difference variations of dynamic and static break points (40°, 50° and 60°). The NHE at the isometric 40° breaking point resulted in significantly higher peak force variables in both the dominant ($p < 0.001$; $\eta^2 = 0.39$) and non-dominant lower limbs ($p < 0.001$; $\eta^2 = 0.27$), as well as increased iEMG activity in the biceps femoris long head ($p < 0.001$; $\eta^2 = 0.22$), and semitendinosus ($p < 0.001$; $\eta^2 = 0.71$) were statistically significant higher than the other static break points angle and dynamic free fall. The static NHE offers several advantages, particularly in terms of hamstring strength, enhance muscular endurance which is beneficial for activities requiring sustained hamstring activation.

Keywords Nordic hamstring break point · Force · Muscle activation

T. Sukantameth · J. Promson · N. janpolsaen · P. phonhan · S. Audonphon · R. Matjiur · S. Akkaneekool · P. Chainok (✉)
Faculty of Sport Science, Burapha University, Chonburi, Thailand
e-mail: phornpot@go.buu.ac.th

R. Zacca
Research Center in Physical Activity, Health and Leisure (CIAFEL), Faculty of Sports, University of Porto (FADEUP), Porto, Portugal

Laboratory for Integrative and Translational Research in Population Health (ITR), Porto, Portugal

M. H. A. Hassan et al. (eds.), *Proceedings of the 3rd International Conference on Innovation and Technology in Sports*, Lecture Notes in Bioengineering,
https://doi.org/10.1007/978-981-95-0804-4_20

1 Introduction

The Nordic hamstring exercise (NHE) is a widely recognized eccentric exercise designed to enhance the strength of the hamstrings, specifically for the purpose of injury prevention (Vicente-Mampel et al. 2022). Coaches and strength and conditioning coach can use the most appropriate NHE variation based on the athlete's specific needs, whether for rehabilitation, injury prevention, or strength development (Sconce et al. 2015; Nishida et al. 2022). In addition, the NHE can be an effective tool in the rehabilitation process for those who are healing from hamstring injuries. It can assist in the reconstruction of strength and lower the likelihood of re-injury (Sconce et al. 2015; Nishida et al. 2022; Bautista et al. 2021).

In theory, different variations of the NHE have the potential to regulate the degree of muscular activation, particularly in the hamstrings and other supporting muscle groups (Vicente-Mampel et al. 2022; Bautista et al. 2021). This information is crucial for customizing exercise programs to meet the individual requirements of athletes, whether it is to improve performance, prevent injuries, or benefit in the recovery from an injury (Ribeiro-Alvares et al. 2021). Besides, electromyography (EMG) research on the Nordic hamstring exercise has consistently demonstrated its great efficacy in activating the hamstring muscles, especially during the eccentric phase (Soga et al. 2023; Suskens et al. 2023). Therefore, within the context of hamstring exercises, electromyography (EMG) can be utilized to ascertain the level of activation exhibited by each individual muscle within the hamstring group, including the biceps femoris, semitendinosus, and semimembranosus, during different movements (Nishida et al. 2022).

From the perspective of NHE performance, one important measure of performance to consider for the NHE is the knee angle at the moment of downward acceleration (Vicente-Mampel et al. 2022; Sconce et al. 2015; Soga et al. 2023). Commonly, this is referred to as the Nordic Hamstring break point. Following this stage, there is a decline in the activation of the hamstring muscles, accompanied by a fast elongation of the fascicles in the long head of the biceps femoris (Šarabon et al. 2019; Moreno-Pérez et al. 2020). An individual's strength, level of experience, and level of muscular training all play a role in determining their Nordic break-point angle. Notably, previous studies have found that the break-point angle in trained soccer players was around 50° (Soga et al. 2023), 31–37° in young men (Nishida et al. 2022), and 36–46° in male and female soccer players (Sconce et al. 2015), respectively.

It is remarkable that a higher break-point angle is linked to the eccentric flexors being stronger than the concentric extensors, which may potentially decrease the risk of hamstring injury. In addition, the electromyographic (EMG) activity in the hamstrings can vary significantly depending on the angle at which the static (isometric) and traditional NHE is performed and different angles place different levels of tension on the hamstrings, affecting muscle activation levels. Thus, it appears that further studies are required to clarify whether difference dynamic and static break points indicates the difference in heel contact force and muscle activation. Therefore, the present study examined and compare the changes in heel contact force and muscle activation between different variations of dynamic and static break points in collegiate soccer players. Based on the findings of previous research revealed that EMG activity before and after break point was difference and a weak relationship observed between break-point angle and the angle of peak torque, it was hypothesized that statistically and practically significant differences heel contact force (peak force and % of asymmetry) and muscle activation of biceps femoris long head, semitendinosus and erector spinae muscle would exist among difference variations of dynamic and static break points.

2 Materials and Methods

2.1 Subjects

Sample size was estimated using the data from a previous study (Vicente-Mampel et al. 2022).An a priori power analysis to determine sample size was completed using G*Power 3.1.9.2 (Heinrich-Heine-Universitat Dusseldorf, Dusseldorf, Germany; http://www.gpower.hhu.de/) software. Based on the Cohen d effect size of 0.64 for, it was estimated that at least twenty participants were necessary, with an alpha level of 0.05 and power ($1 - \beta$) of 0.80. Twenty male sub-elite university soccer players age: 20.15 ± 1.14 years old; body height: 171.82 ± 4.77 cm; body weight: 66.10 ± 6.06 kg; body mass index: 23.25 ± 3.11 $kg{\cdot}m^{-2}$) volunteered to participate in the study (Table 1). All players competed in the amateur level of the professional football competition league. Ethical approval for the study was obtained from the Ethical Advisory Committee at local university. Participants were recruited through personal connections and social media platforms. The inclusion criteria were as follows: (i) had a minimum of 5 years of training experience; (ii) all players participated in at least 95% of the pre-season training period; (iii) free from current injury in the six-weeks training period before testing. Exclusion criteria are as follows: (i) goalkeepers were excluded and; (ii) poor health status and physical condition with potential medical problems or a history of ankle, knee, or back pathology, and (iii) incomplete participation in the training and testing program.

2.2 Procedure

A repeated-measures design was used to analyse the effect of difference Nordic hamstring exercise (i.e., free fall vs. static at three difference break-point angle at 40°, 50° and 60°) on biceps femoris long head (BF), semitendinosus (ST) and erector spinae (ES) muscles EMG activity, heel contact force component (peak force and % of asymmetry) during NHE. The participants reported to the laboratory on two separate days with a week between visits. The participants were introduced to

Table 1 General characteristics of the participants

Variables	Mean + SD
Age (yrs)	20.15 ± 1.44
Weight (kg)	66.10 ± 6.06
Height (cm)	171.82 ± 4.77
Body mass index (kg·m^{-2})	23.25 ± 3.11
Isokinetic peak torque (Nm)	
– Dominant knee 60° flexion	196.07 ± 42.83
– Non-dominant knee 60° flexion	171.49 ± 41.31
– Dominant knee 60° extension	123.11 ± 31.34
– Non-dominant knee 60° extension	113.72 ± 27.32
– Dominant knee 180° flexion	136.49 ± 28.79
– Non-dominant knee 180° flexion	117.07 ± 25.62
– Dominant knee 180° extension	93.54 ± 38.25
– Non-dominant knee 180° extension	91.31 ± 32.08
– Trunk flexion 60°	193.96 ± 46.36
– Trunk extension 60°	187.06 ± 54.06
– Trunk flexion180°	160.72 ± 31.34
– Trunk extension 180°	152.84 ± 31.34

the tests, which were carried out at the same time of day to prevent any systematic bias caused by circadian variation (Zacca et al. 2020). The first visit consisted of measuring the individual's standing height, body weight, and body mass index. This was done using a bioelectrical impedance analysis (BIA) (Body Composition Analyzer: X-CONTACT 357S, Seoul, South Korea) was utilized. The knee and trunk flexion and extension peak torque were evaluated in isokinetic conditions at a speed of 60 deg·s^{-1} (slow velocity), 180 deg·s^{-1} (medium velocity). All isokinetic contractions were performed and carried out separately for both legs and trunk using Isoforce dynamometer (TUR GmbH, Berlin, Germany).The torque signals were digitized with a sampling frequency of 1000 Hz and analyzed using the Isoforce software (TUR GmbH, Berlin, Germany). Follow up the anthropometric and isokinetic evaluation, a familiarization session was conducted for all participants to practice NHE (3 sets of 3 repetitions). The purpose of this session was to ensure that participants understood and followed the correct technique.

On data collection day, surface electromyogram (sEMG) amplitudes were measured from biceps femoris long head, semitendinosus and erector spinae muscles in both lower extremities and both side of the trunk using BTSFREEEMG (BTS-FREEEMG1000; BTS Bioengineering, Milan, Italy) was set to a sampling rate of 1,000 Hz per channel, and the signals were band-pass filtered from 20 to 500 Hz. The study utilized active silver chloride surface electrodes (Dormo, Telic, S.A., Spain) equipped with preamplifiers (AD621BNZ; Analog Devices, Norwood, MA, USA), following the guidelines outlined in the European Recommendations for Surface

Electromyography (Hermens et al. 2000). Each participant was first required to perform three maximal voluntary isometric contractions (MVIC) for 5 s each, with a 5-min rest between contractions and verbal encouragement was given (Chainok et al. 2022). EMG signal processing, with raw EMG signals filtered using a fourth-order Butterworth band-pass filter (bandwidth 20–500 Hz), rectified and averaged to obtain the full-wave signals. The rectified EMG integration was calculated per unit of time (iEMG/T) for each turning phase to eliminate the phase duration effect and EMG signals were partitioned in 40 ms windows to find the maximal iEMG values (iEMGmax) for all studied muscles. iEMG/T was expressed as a percentage of iEMGmax to normalize the results (Chainok et al. 2022).

Following the completion of a dynamic warm-up, consisting of 5 min of dynamic stretching that mainly focuses on the muscles in the torso and thighs, each participant assumed a kneeling position on a specially designed NHE device (Kinvent Physio, Montpellier, France). Their ankles were then secured above the lateral malleolus using an ankle brace that was connected to a load cell. The load cell, connected to Kinvent Physio software (Kinvent Physio, Montpellier, France), was used to measure the vertical force applied to the ankle brace at a sampling rate of 1000 Hz. Kinvent software was used to analyse and generate all the absolute peak force for each lower extremity limb (N) and the percentage of asymmetry. The participants were asked to perform the NHE in a kneeling position over the padded board and arms across the chest. The participants were asked to perform the experiment protocol consisting of four variations of the NHE as follows: (1) isometric contraction at 60° of breaking point; (2) isometric contraction at 50° of breaking point; (3) isometric contraction at 40° of breaking point and, (4) dynamic free fall NHE (Fig. 1).

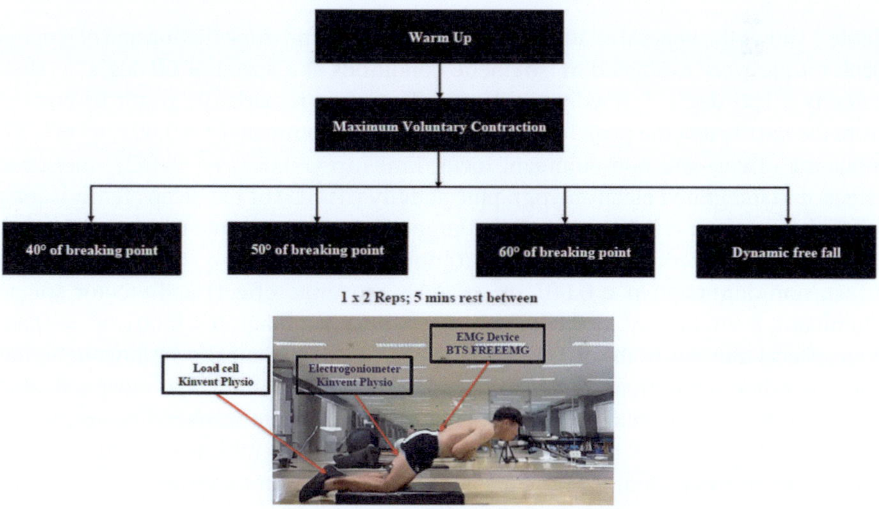

Fig. 1 Schematic representation of the research design and data collection

A single-axis electrogoniometer Kmove (Kinvent Physio, Montpellier, France) was attached to the lateral aspect of the dominant knee to monitor the knee joint angle during the NHE. The participants were instructed to maintain the isometric contraction for a duration of five seconds. If the participant was unable to sustain the desired angle, the trial was disregarded, and the subject was instructed to repeat the trials. For the dynamic free fall, they were instructed to lean forward "as slowly as possible." while maintaining a straight posture from the knee to the head. The average of the two NHE trials in each condition was utilized for subsequent analysis.

2.3 Statistical Analysis

Statistics were performed in SPSS for Windows version 24 (SPSS, Chicago, IL, USA), with the significance level being set at 0.05. The assumption of normality was verified with the Shapiro–Wilk test. A one-way repeated measures ANOVA was conducted to determine whether there was a statistically significant difference in an iEMG and heel contact force data. Eta squared (η^2) test was used to quantify the percentage of the variance explained by each covariate (effect size) and interpreted as follows: $0 < \eta^2 < 0.04$ without effect, $0.04 \leq \eta^2 \leq 0.24$ minimum, $0.25 \leq \eta^2 < 0.64$ moderate and $\eta^2 \geq 0.64$ large effect (Ferguson 2009) (Table 2).

3 Results

Table 1 shows the general characteristics and the knee and trunk flexion and extension peak torque were evaluated in isokinetic conditions at a speed of 60 deg$^\bullet$s^{-1} (slow velocity), 180 deg$^\bullet$s^{-1} (medium velocity).The most remarkable result to emerge from the data is that the peak force variables of both dominant ($p < 0.001$, $\eta^2 = 0.39$, moderate effect) and non-dominant lower limbs ($p < 0.001$, $\eta^2 = 0.27$, moderate effect) and integrated electromyographic activity (iEMG) of the biceps femoris long head (dominant; $p < 0.001$, $\eta^2 = 0.71$, large effect, non-dominant; $p < 0.002$, $\eta^2 = 0.26$, moderate effect), semitendinosus (dominant; $p < 0.004$, $\eta^2 = 0.22$, minimum effect, non-dominant; $p < 0.001$, $\eta^2 = 0.41$, moderate effect) and erector spinae (dominant; $p < 0.001$, $\eta^2 = 0.86$, large effect, non-dominant; $p < 0.001$, $\eta^2 = 0.86$, large effect) muscles of the 40° of breaking point were statistically significant higher than the other static break points angle and dynamic free fall. The most activated muscle of the 40° of breaking point was found in dominant semitendinosus muscle (121.15 ± 24.66) follow by non-dominant semitendinosus muscle (114.90 ± 14.98) and dominant biceps femoris long head (107.78 ± 7.95), respectively.

Table 2 The heel contact force and electromyographic activity in difference variations of dynamic and static break points during NHE variations

Variables	Breaking point angle				F-test	
	40° Mean ± SD (95% CI)	50° Mean ± SD (95% CI)	60° Mean ± SD (95% CI)	Free fall Mean ± SD (95% CI)	p	Partial η2
1. Breaking point angle (deg)						
Average breaking point	40.88 ± 0.28	50.21 ± 0.21	59.93 ± 0.14	44.13 ± 5.32		
2. Heel contact force (N)						
Dominant peak force	317.67 ± 19.04 (277.48, 357.86)	250.10 ± 17.59 (212.98,287.22)	201.65 ± 23.40 (152.27,251.03)	288.55 ± 20.43 (245.48,331.65)	0.001	0.39
Non-dominant peak force	292.59 ± 22.25 (245.64,339.54)	241.18 ± 23.84 (190.88,291.48)	201.65 ± 23.41 (152.27,251.04)	287.55 ± 21.43 (244.41,321.64)	0.001	0.27
% of asymmetry	11.33 ± 2.07 (6.96,15.71)	11.19 ± 2.17 (6.61,15.77)	15.41 ± 2.75 (9.61,21.21)	13.47 ± 2.45 (8.34,18.62)	0.148	0.12
3. The integrated electromyography (iEMG)						
Dominant BF	107.78 ± 7.95 (90.99,124.55)	99.92 ± 7.49 (6.61,15.77)	82.65 ± 6.37 (69.21,96.08)	64.02 ± 7.21 (48.79,79.24)	0.001	0.71
Non-dominant BF	98.17 ± 16.18 (64.03,132.32)	82.29 ± 7.64 (66.17,98.40)	69.63 ± 6.82 (55.24,84.02)	54.84 ± 5.11 (44.06,65.62)	0.002	0.26
Dominant ST	121.15 ± 24.66 (69.12,173.18)	103.35 ± 11.00 (80.14,126.56)	86.30 ± 9.57 (66.12,106.48)	59.80 ± 6.74 (45.59,74.01)	0.004	0.22
Non-dominant ST	114.90 ± 14.98 (83.29,146.51)	94.86 ± 9.35 (75.14,114.57)	80.67 ± 8.37 (63.02, 98.32)	64.64 ± 7.63 (48.54,80.73)	0.001	0.41
Dominant ES	92.41 ± 8.57 (74.33,110.49)	79.84 ± 8.60 (59.59,100.09)	59.18 ± 6.84 (44.76,73.61)	48.07 ± 7.91 (31.38,64.75)	0.001	0.86
Non-dominant ES	80.36 ± 6.65 (66.32,94.39)	73.23 ± 4.74 (63.23,83.24)	56.34 ± 5.16 (45.46,67.22)	40.60 ± 3.95 (32.26,48.93)	0.001	0.86

4 Discussion

The aim of this study was to analyze and compare the changes in heel contact force (peak force and % of asymmetry) and muscle activation of the biceps femoris long head, semitendinosus, and erector spinae muscles among various dynamic and static break points in collegiate soccer players. In overall, the NHE at the isometric 40° breaking point resulted in significantly higher peak force variables in both the dominant and non-dominant lower limbs, as well as increased iEMG activity in the biceps femoris long head, semitendinosus, and erector spinae muscles. These effects were moderate to large compared to other isometric breaking point angles and dynamic free fall. Interestingly, when it comes to muscle activation, the dominant semitendinosus and biceps femoris long head were revealed to be the most recruited muscles at the 40° breaking point.

The data provide evidence to support of our hypothesis, indicating statistically and practically significant variations in force characteristics and muscular activity as demonstrated by the integrated electromyography. Differences in dynamic and static break points result in changes in heel contact force, including peak force and percentage of asymmetry. Additionally, variations in muscular activation can be observed in the biceps femoris long head, semitendinosus, and erector spinae muscles. Specifically, according to our results may suggest that suitable static nordic hamstring exercise offers several advantages, particularly in terms of hamstring strength and injury prevention. In addition, static nordic hamstring exercise can serve as a progression tool for beginners before they move on to the more challenging dynamic version. It helps in building the foundational strength needed for more advanced exercises (Soga et al. 2023; Jeanguyot et al. 2023).

Considering breaking point angles from the dynamic free fall, our collegiate soccer players breaking point angles' values (44.13 ± 5.32) were lower relatively low when compared with data from trained soccer players (50°) (Soga et al. 2023), but higher than the value from young men (31–37°) (Nishida et al. 2022) and are in line with the value from male and female soccer players (36–46°) (Sconce et al. 2015), respectively. In regards to the optimum breaking point angle, Soga et al. (Soga et al. 2023) state that it is recommended to keep the breaking point angle of the Nordic Angle within 35° to effectively prevent recurrent hamstring injuries during NHE.

From a perspective of the neuromuscular activation, electromyographic (EMG) activity in the hamstrings can vary significantly depending on the angle at which the static (isometric) NHE is performed (Vicente-Mampel et al. 2022; Bautista et al. 2021). In addition, numerous studies on electromyography (EMG) have consistently shown that the NHE is highly effective in activating the hamstring muscles (i.e., biceps femoris long head and semitendinosus) particularly during the eccentric phase (Soga et al. 2023; Suskens et al. 2023). The present study found that the semitendinosus muscle is highly activated when the body is in a static position alongside isometric muscular contractions, While performing the NHE, the semitendinosus muscle is preferentially activated to control the descent of the body by resisting knee extension, ensuring that the movement is slow and controlled [Bourne et al., 2016].

Furthermore, if NHE is performed at a shallow knee flexion angle, it will increase the activity of the bicep femoris muscle (Hirose et al. 2021).

In conclusion, this study clarified that the NHE at the isometric 40° breaking point resulted in significantly higher peak force variables in both the dominant and non-dominant lower limbs, as well as increased iEMG activity in the semitendinosus biceps femoris and semitendinosus muscles are quite equally activated during NHE. The static NHE offers several advantages, particularly in terms of hamstring strength, enhance muscular endurance which is beneficial for activities requiring sustained hamstring activation, such as running or cycling. Therefore, it is important to prioritize the static version as it enables a more concentrated emphasis on correct form and muscle activation. Controlling the body's position makes it simpler to ensure precise targeting of the hamstrings without any compensatory movements. Finally, in order to enhance the strength of athletes' hamstrings, static Nordic Hamstring Exercise (NHE) can be utilized as a foundational exercise for beginners prior to transitioning to the more demanding dynamic variation. It aids in developing the fundamental strength required for more complex exercises.

Acknowledgements We would like to acknowledge all the subjects who voluntarily participated and provided impressive support throughout the data collection. We also extend our special thanks to the technicians of the Faculty of Sport Science, Burapha University for their time, collaboration and commitment to this study. Rodrigo Zacca is supported by The Research Center in Physical Activity, Health and Leisure (CIAFEL), Faculty of Sport, University of Porto (FADEUP), which is part of the Laboratory for Integrative and Translational Research in Population Health (ITR); both are funded by the Fundação Para a Ciência e Tecnologia (FCT; grants UIDB/00617/2020 https://doi.org/https://doi.org/10.54499/UIDB/00617/2020; UIDP/00617/2020 https://doi.org/https://doi.org/10.54499/UIDP/00617/2020 and LA/P/0064/2020, respectively)

Disclosure of Interests There are no competing interests to report.

References

Bautista IJ, Vicente-Mampel J, Baraja-Vegas L, Segarra V, Martín F, Van Hooren B (2021) The effects of the Nordic hamstring exercise on sprint performance and eccentric knee flexor strength: a systematic review and meta-analysis of intervention studies among team sport players. J Sci Med Sport 24(9):931–938

Chainok P, Lauer J, Gonçalves P, de Jesus K, Fernandes RJ, Vilas-Boas JP (2022) Backstroke-to-Breaststroke Turns Muscular Activity. A Study Conducted in Age Group Swimmers. J Sports Sci Med 21(3):402

Ferguson CJ (2009) Is psychological research really as good as medical research? Effect size comparisons between psychology and medicine. Rev Gen Psychol 13(2):130–136

Hirose N, Tsuruike M, Higashihara A (2021) Biceps femoris muscle is activated by performing Nordic hamstring exercise at a shallow knee flexion angle. J Sports Sci Med 20(2):275

Jeanguyot E, Salcinovic B, Johnson A, Van Dyk N, Whiteley R (2023) Eccentric hamstring strength in young athletes is best documented when normalised to body mass: a cross-sectional study with normative data of 590 athletes from different age categories. Biol Sport 40(4):1079–1095

Moreno-Pérez V, Méndez-Villanueva A, Soler A, Del Coso J, Courel-Ibáñez J (2020) No relationship between the Nordic hamstring and two different isometric strength tests to assess hamstring muscle strength in professional soccer players. Phys Ther Sport 46:97–103

Nishida S, Nakamura M, Kiyono R, Sato S, Yasaka K, Yoshida R, Nosaka K (2022) Relationship between Nordic hamstring strength and maximal voluntary eccentric, concentric and isometric knee flexion torque. PLoS ONE 17(2):e0264465

Ribeiro-Alvares JB, Oliveira GDS, De Lima-E-Silva FX, Baroni BM (2021) Eccentric knee flexor strength of professional football players with and without hamstring injury in the prior season. Eur J Sport Sci 21(1):131–139

Šarabon N, Marušič J, Marković G, Kozinc Ž (2019) Kinematic and electromyographic analysis of variations in Nordic hamstring exercise. PLoS ONE 14(10):e0223437

Sconce E, Jones P, Turner E, Comfort P, Graham-Smith P (2015) The validity of the Nordic hamstring lower for a field-based assessment of eccentric hamstring strength. J Sport Rehabil 24(1):13–20

Soga T, Yamaguchi S, Inami T, Saito H, Hakariya N, Nakaichi N, Laddawong T (2023) Hamstring activity before and after break-point angle calculated by smartphone application during the Nordic hamstring exercise. Int J Sports Phys Ther 18(6):1290–1298

Suskens JJ, Maas H, van Dieën JH, Kerkhoffs GM, Goedhart EA, Tol JL, Reurink G (2023) The effect of the Nordic hamstring exercise on muscle activity: a multichannel electromyography randomized controlled trial. J Appl Biomech 39(6):377–387

Vicente-Mampel J, Bautista IJ, Martín F, Maroto-Izquierdo S, Van Hooren B, Baraja-Vegas L (2022) Effects of ankle position during the Nordic Hamstring exercise on range of motion, heel contact force and hamstring muscle activation. Sports Biomech 1–13

Zacca R, Azevedo R, Ramos VR Jr, Abraldes JA, Vilas-Boas JP, de Souza Castro FA, Fernandes RJ (2020) Biophysical follow-up of age-group swimmers during a traditional three-peak preparation program. J Strength Cond Res 34(9):2585–2595

Sex Comparison of ACL Injury Risk Using a Clinic-Based Algorithm Among Volleyball Athletes UiTM

Raihana Sharir, Muhammad Hidayat Jamaludin, Ezhan Iqbal Zulkiflie, and Radin Rafeeuddin

Abstract This study evaluates sex differences in the probability of high knee load, a key predictor of anterior cruciate ligament (ACL) injury risk, using a clinic-based injury prediction algorithm among volleyball players at Universiti Teknologi MARA (UiTM, Shah Alam, Selangor). Twenty-two injury-free players (11 males, 11 females) participated in the study. The participants completed drop vertical jump tasks, as well as concentric knee extension and flexion tasks utilizing the isokinetic dynamometer on their dominant leg. Female players exhibited greater knee valgus motion and smaller knee flexion angles at landing compared to their male counterparts. However, both sexes showed no significant differences in the probability percentage of high knee load ($p = 0.78$). Particularly, most male players have longer tibia length, which may be contributing to the high probability. In contrast, female players tend to move their knees inwards during landing which may cause extra loading on the ACL during impact. The findings indicate that although differences in knee mechanics between sexes are present, they do not substantially impact the overall risk evaluation provided by the clinic-based algorithm. Implementing effective prevention strategies should prioritize improving landing mechanics and enhancing the strength of lower limb muscles to mitigate the risk of ACL injuries.

Keywords Volleyball · ACL · Field-based testing · Clinic-based algorithm · Injury screening · Sex differences

---

R. Sharir (✉) · M. H. Jamaludin · E. I. Zulkiflie · R. Rafeeuddin
Faculty of Sports Science and Recreation, Universiti Teknologi MARA, Selangor, Malaysia
e-mail: raihanasharir@uitm.edu.my

R. Sharir
Research Institute for Sport and Exercise Sciences, Liverpool John Moores University, Liverpool, UK

R. Sharir · R. Rafeeuddin
Sports and Exercise Biomechanics Special Interest Group, Universiti Teknologi MARA, Selangor, Malaysia

M. H. A. Hassan et al. (eds.), *Proceedings of the 3rd International Conference on Innovation and Technology in Sports*, Lecture Notes in Bioengineering,
https://doi.org/10.1007/978-981-95-0804-4_21

1 Introduction

In dynamic sports, injury to the anterior cruciate ligament (ACL) is one of the most frequent injuries to occur. This injury typically happens during movements such as sudden changes of direction, landing from a jump, and sudden stops or decelerations (Wetters et al. 2016). At the same time, studies have shown that females are four to six times more likely to sustain an injury compared to males (Hewett et al. 2005), indicating that females may have a higher risk of injury. Prospective studies have provided insights into the risk factors and mechanisms that could lead to ACL injuries. Therefore, proactive measures should be taken to reduce these risks before they arise, with screening being a crucial step in identifying high-risk athletes.

It is well-known that the gold standard for biomechanical laboratory-based ACL injury screening can be quite expensive. As a result, various field-based ACL injury screening methods have started to emerge (McLean et al. 2005; Myer et al. 2008, 2010; Padua et al. 2009). The knee abduction moment (KAM) has been recognized as a significant risk factor for non-contact ACL injuries (Hewett et al. 2005). The clinic-based algorithm was created to provide a systematic methodology to simplify and predict KAM (Myer et al. 2011a). The algorithm was invented through a combination of evaluations of the anthropometric measurements, lower-limb muscle imbalances and landing mechanics during the drop vertical jump to predict the probability of high KAM. A nomogram was also created to allow a simple estimation of high KAM.

Volleyball is a sport that demands the athletes to jump both for attacking (jump spike) and defending (jump block) during every match (Borras et al. 2011; Sattler et al. 2012). As a result, ACL injuries are common among volleyball players due to incorrect landing techniques after jumps (Briner and Gallo 2010). Additionally, volleyball players are less being observed compared to other dynamic sports (Leppänen et al. 2017; Mok et al. 2016). Therefore, the goal of this study is to compare the probability of high knee load between sexes using a clinic-based injury prediction algorithm among volleyball players at Universiti Teknologi MARA (UiTM).

2 Methods

The clinic-based algorithm (Myer et al. 2011a) was used to predict ACL injury risk by observing the probability of high knee load between sexes among 22 injury-free UiTM volleyball players (M: 21.18 ± 1.54 years, 172.68 ± 7.75 cm, 69.41 ± 11.58 kg; F: 22.27 ± 1.10 years, 155.55 ± 4.23 cm, 53.88 ± 6.80 kg). Participants were tested in a laboratory at UiTM. Anthropometric measurements were taken as required by the algorithm (Fig. 1).

Participants were required to perform a 10-min warm-up and dynamic stretching prior to testing. Knee extension and flexion (concentric/concentric muscle action) movements were assessed using an isokinetic dynamometer (Prima DOC, EasyTech, Italy) at a speed of 300 degrees for the leg. The quadriceps-to-hamstrings (QuadHam)

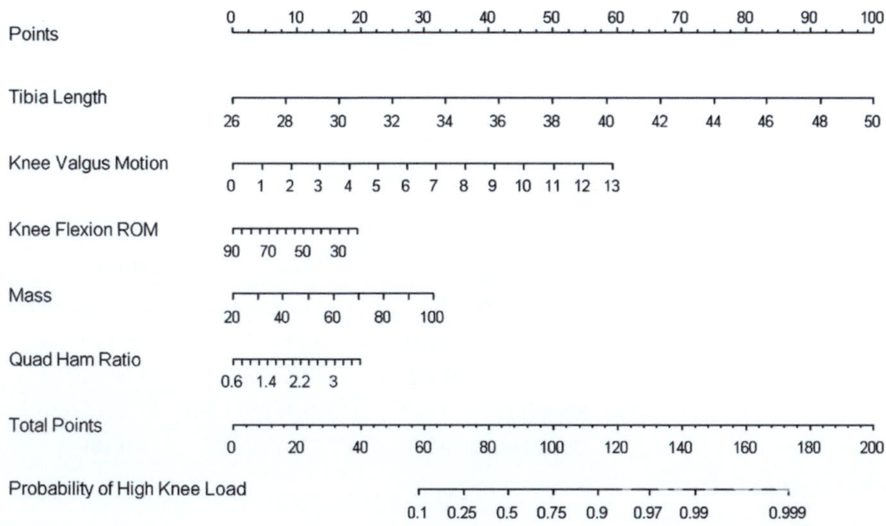

Fig. 1 Clinical-based injury prediction algorithm (Myer et al. 2010)

was calculated as the ratio of quadriceps to hamstrings peak isokinetic torque. Participants performed 3 sets of knee extensions and flexions, with 5 repetitions per set.

Two video cameras were positioned at a height of 1.2 m, perpendicular to each other in the frontal and sagittal planes, and 3.8 m away from the landing zone. Participants performed three trials of bilateral 30 cm DVJ. They needed to land past 50% of the length of their lower body (landing zone). Video analysis of the DVJ was conducted using Kinovea software (version 0.8.25) to measure the knee flexion angle and knee valgus motion. Shapiro–Wilk tests were used to test for normality. All five measurements were taken and summed as total points as per Fig. 1 to obtain the probability of high knee load. An independent T-test with a significance level set at $p < 0.05$ was used to compare the differences between sexes.

3 Results

This study aimed to compare the probability of high knee load between male and female volleyball players at UiTM. Female players exhibited higher valgus motion and smaller knee flexion angles during drop vertical jump (DVJ) landings compared to male players, who had longer tibia lengths and greater body masses. Both sexes displayed nearly identical Quadriceps-to-Hamstrings (QuadHam) ratio scores.

No significant differences were found in the probability percentage of high knee load between males and females ($p = 0.78$) (Table 1). However, there were visible differences in the scores for individual variables between the sexes (Table 2).

Table 1 Independent Sample T-test for the probability percentage of knee load

	Sex	N	Mean (SD)	Levene's test	t	df	p-value
Probability percentage of knee load	M	11	70.32 (25.51)	0.484	0.283	20	0.780
	F	11	67.46 (21.75)				

Table 2 Participant's scores (mean ± SD) for each variable

	Male	Female
Tibia Length (cm)	41.25 (±3.73)	36.91 (±2.91)
Knee Valgus Motion (cm)	3.04 (±1.26)	5.60 (±1.80)
Knee Flexion ROM (∘)	87.27 (±19.20)	75.64 (±27.32)
Mass (kg)	69.41 (±11.58)	53.88 (±6.80)
QuadHam Ratio	1.44 (±0.24)	± 0.18)

Although the mean tibia length was longer in males than females, the difference was not statistically significant.

4 Discussion

The findings suggest that an increase in tibia length, as observed in males, may be associated with increased knee abduction moment (KAM) (Myer et al. 2011b). Additionally, greater joint forces, which are more challenging to balance and dampen, occur when body mass increases with longer joint levers during high-velocity movements in the lower extremities (Anderson and Pandy 2001). Males have been found to adapt to neuromuscular spurts to meet the increased demands of growth and development, unlike females, who do not show similar neuromuscular adaptations during puberty, potentially leading to increased KAM (Hewett et al. 2005; Quatman et al. 2006).

Increased knee valgus motion during landing, as seen in female athletes, may also contribute to high KAM. This study's findings align with Chappell et al. (Chappell et al. 2002), who observed that females experience greater valgus forces during the landing phase of a vertical jump. Reduced neuromuscular control of the knee joint, particularly in females, lowers knee joint stiffness and raises the risk of ligament injury (Ford et al. 2003). However, some studies suggest that varus-valgus loading rarely occurs in isolation, and the ACL loading pattern differs when combined with transverse-plane knee loads (Shimokochi and Shultz 2008).

A small knee flexion angle during landing can lead to overloading of the knee. Males demonstrated higher knee flexion angles compared to female athletes. A higher knee flexion angle during landing is safer as it helps absorb forces during the drop jump task. Taylor et al. (Taylor et al. 2011) found that peak ACL strain occurs during pre-landing when knee flexion angles are at their lowest. Additionally, Laughlin et al. (Laughlin et al. 2011) reported that peak ACL force could be reduced by a soft landing, portrayed by increased knee flexion angles and reduced impact forces.

Finally, this study found only a minimal difference in QuadHam ratios between males and females. Previous research suggests that decreased hamstrings strength and QuadHam ratios in females may be linked to neuromuscular imbalances associated with maturation (Ahmad et al. 2006; Holm and Vøllestad 2008). Hewett et al. (Hewett et al. 2005) reported that while male athletes demonstrate significantly greater hamstrings peak torques with increasing maturity, female athletes maintain stable peak hamstrings torque throughout maturation. This study supports the findings of Hewett et al. (Hewett et al. 2008), which indicate that gender differences in isokinetic QuadHam ratios are not evident in post-pubertal athletes at slower angular velocities but become significant during sports activities.

5 Conclusion

Increased knee valgus motion with decreased knee flexion angle at landing, along with naturally longer tibia length and a larger QuadHam strength ratio, contribute to high knee load. Therefore, preventing ACL injuries is vital. Through this easy clinic-based algorithm technique, players at higher risk of injury can be easily identified. Training that focuses on strengthening the lower limb muscles and practicing safe landing techniques may reduce the probability of ACL injury.

References

Ahmad CS, Clark AM, Heilmann N, Schoeb JS, Gardner TR, Levine WN (2006) Effect of gender and maturity on quadriceps-to-hamstring strength ratio and anterior cruciate ligament laxity. Am J Sports Med 34(3):370–374

Anderson FC, Pandy MG (2001) Static and dynamic optimization solutions for gait are practically equivalent. J Biomech 34(2):153–161

Borras X, Balius X, Drobnic F, Galilea P (2011) Vertical jump assessment on volleyball: a follow-up of three training seasons. Rev Brasil Med Esporte 17(1):12–16

Briner WW, Gallo RA (2010) Injuries in volleyball. Curr Sports Med Rep 9(5):275–279

Chappell JD, Yu B, Kirkendall DT, Garrett WE Jr (2002) A comparison of knee kinetics between male and female recreational athletes in stop-jump tasks. Am J Sports Med 30(2):261–267

Ford KR, Myer GD, Hewett TE (2003) Valgus knee motion during landing in high school female and male basketball players. Med Sci Sports Exerc 35(10):1745–1750

Hewett T, Myer G, Ford K, Heidt R Jr, Colosimo A, McLean S, van den Bogert J et al (2005) Biomechanical measures of neuromuscular control and valgus loading of the knee predict anterior cruciate ligament injury risk in female athletes: a prospective study. Am J Sports Med 33(4):492–501. https://doi.org/10.1177/0363546504269591

Hewett TE, Myer GD, Zazulak BT (2008) Hamstrings to quadriceps peak torque ratios diverge between sexes with increasing maturation. J Sci Med Sport 11(5):395–403

Holm I, Vøllestad N (2008) Gender differences in isokinetic strength and muscular fatigue in adolescent soccer players. Pediatr Exerc Sci 20(1):13–26

Laughlin WA, Weinhandl JT, Kernozek TW, Cobb SC (2011) Effects of single-task versus dual-task paradigms on landing mechanics in females. J Sports Rehabil 20(2):250–265

Leppänen M, Pasanen K, Kujala UM, Vasankari T, Kannus P, Äyrämö S, Parkkari J (2017) Stiff landings are associated with increased ACL injury risk in young female basketball and floorball players. Am J Sports Med 45(2):386–393

McLean S, Walker K, Ford K, Myer G, Hewett T, van den Bogert J (2005) Evaluation of a two-dimensional analysis method as a screening and evaluation tool for anterior cruciate ligament injury. Br J Sports Med 39(6):355–362

Mok KM, Petushek E, Krosshaug T (2016) Reliability of knee biomechanics during a vertical drop jump in elite female athletes. Gait Post 46:1–6

Myer G, Ford K, Hewett T (2008) Tuck jump assessment for reducing anterior cruciate ligament injury risk. Athlet Ther Today J Sports Health Care Profess 13(5):39

Myer G, Ford K, Khoury J, Succop P, Hewett T (2010) Development and validation of a clinic-based prediction tool to identify female athletes at high risk for anterior cruciate ligament injury. Am J Sports Med 38(10):2025–2033

Myer G, Ford K, Hewett T (2011a) New method to identify athletes at high risk of ACL injury using clinic-based measurements and freeware computer analysis. Br J Sports Med 45(4):238–244

Myer G, Brent J, Ford K, Hewett T (2011b) A clinical model to estimate knee-abduction moment: a step toward identifying female athletes at high risk for ACL injury. J Sports Rehabil 20(4):415–426

Padua D, Marshall S, Boling M, Thigpen C, Garrett W Jr, Beutler A (2009) The landing error scoring system (LESS) is a valid and reliable clinical assessment tool for identifying individuals at high risk for anterior cruciate ligament injury. Am J Sports Med 37(10):1996–2002. https://doi.org/10.1177/0363546509343200

Quatman CE, Ford KR, Myer GD, Hewett TE (2006) Maturation leads to gender differences in landing force and vertical jump performance: a longitudinal study. Am J Sports Med 34(5):806–813

Sattler T, Sekulic D, Hadzic V, Uljevic O, Dervisevic E (2012) Vertical jumping tests in volleyball: reliability, validity, and playing-position specifics. J Strength Cond Res 26(6):1532–1538

Shimokochi Y, Shultz SJ (2008) Mechanisms of noncontact anterior cruciate ligament injury. J Athl Train 43(4):396–408

Taylor K, Ford KR, Nguyen AD, Terry LN, Miller A (2011) Biomechanical comparison of single-leg and double-leg stop-jump tasks in female basketball athletes: implications for ACL injury risk assessments. J Appl Biomech 27(4):340–351

Wetters N, Weber A, Wuerz T, Schub D, Mandelbaum B (2016) Mechanism of injury and risk factors for anterior cruciate ligament injury. Operat Tech Sports Med 24(1):2–6. https://doi.org/10.1053/j.otsm.2015.09.001

Effectiveness of Foam Rolling in Reducing Delayed Onset Muscle Soreness (DOMS)

Nur Wahida Amalin Muhammad Helmi Kannan, Nur Fatihah Raimi, Adjah Naqkiah Mazlan, and Ayu Suzailiana Muhamad

Abstract Foam rolling is a rehabilitation technique used to alleviate delayed onset muscle soreness (DOMS), which can impair performance following physical activity. This study aims to examine the effectiveness of foam rolling by comparing the Visual Analogue Scale (VAS), gastrocnemius muscle circumference, and Range of Motion (ROM) between a foam rolling group (FR) and a control group (C) after performing a DOMS protocol. Twenty subjects (10 males, 10 females, aged 22.75 ± 0.91 years) were divided into two groups randomly. The DOMS protocol involved 5 sets of 10 Counter Movement Jumps. Afterward, the FR group performed foam rolling for two minutes, while the control group received no treatment. Measurements were taken before the DOMS protocol and at 24, 48, and 72 h post-exercise. Statistical analysis (Independent t-test, SPSS) revealed a significant improvement in VAS after 24 h ($p < 0.01$) in the FR group, but no significant differences in gastrocnemius muscle circumference ($p > 0.05$) or ROM ($p > 0.05$). The results suggest that while foam rolling accelerates subjective recovery as indicated by VAS, it does not significantly impact muscle circumference or ROM. These findings highlight foam rolling's potential in reducing perceived pain but suggest limitations in objective recovery measures.

Keywords Foam Rolling · Delayed Onset Muscle Soreness (DOMS) · Visual Analog Scale (VAS) · Gastrocnemius Muscle Circumference · Range of Motion (ROM)

N. W. A. M. H. Kannan · N. F. Raimi (✉) · A. N. Mazlan
School of Education, Faculty of Social Science and Humanities (FSSH), Universiti Teknologi Malaysia, Skudai, Johor, Malaysia
e-mail: nurfatihah.raimi@utm.my

N. F. Raimi · A. N. Mazlan
Sports Innovation and Technology Centre (SITC), Institute of Human Centred Engineering (iHumEN), Universiti Teknologi Malaysia, Skudai, Johor, Malaysia

A. S. Muhamad
Exercise and Sports Science Programme, School of Health Sciences, Universiti Sains Malaysia, Health Campus, Kota Bharu, Malaysia

M. H. A. Hassan et al. (eds.), *Proceedings of the 3rd International Conference on Innovation and Technology in Sports*, Lecture Notes in Bioengineering,
https://doi.org/10.1007/978-981-95-0804-4_22

1 Introduction

Delayed-onset muscle Soreness (DOMS) is well-known as muscle pain or discomfort that occurs after engaging in unfamiliar exercises. Theodore Hough first introduced the term DOMS in 1902, describing it as discomfort that appears within the first 24 h after exercise, peaks between 24 to 72 h, and gradually subsides over 5 to 7 days (Clarkson et al. 1992).

DOMS symptoms include muscle discomfort, swelling in the affected limbs, and joint stiffness, which can temporarily reduce the range of motion. Individuals with DOMS often feel pain or discomfort when moving the affected muscles, especially during activities involving stretching or contraction. DOMS is also associated with localized swelling and inflammation as part of the body's natural healing process, which can lead to joint stiffness and limited range of motion (ROM) (Braun and Sforzo 2011; Hedges and Womack 2022; Bennett and Andrews 2023).

DOMS is common when athletes try new or intense activities, especially those involving eccentric movements that cause more muscle damage. Exercise length and intensity also contribute to the development of DOMS (Cheung et al. 2003). Susceptibility to DOMS varies among individuals.

Athletes experiencing DOMS may see a decline in performance, which can also impact their mental well-being (Figueiroa and Costa 2023). Various recovery methods, such as foam rolling, are used to manage DOMS. Consistency in foam rolling is important, as it can gradually reduce muscle soreness. The effectiveness of foam rolling also depends on the duration of treatment, with studies using anywhere from 30 s to several minutes per muscle group. According to earlier research, athletes frequently opt for foam rolling as a recovery technique (Figueiroa and Costa 2023; Pearcey et al. 2015). On the other hand, sports massage is another popular method. Passive recovery has been demonstrated in earlier research to improve performance, lessen tiredness, and produce better outcomes (Buchheit et al. 2009; Contro et al. 2016). Few studies, nonetheless, have compared various recovery techniques for treating DOMS, including sports massage, foam rolling, stretching, passive recovery and cryotherapy. Since foam rolling may be an alternate healing treatment, researchers have decided to investigate its efficacy in lowering DOMS.

2 Methodology

2.1 Sampling

Random sampling is the sample technique employed in the present study. The study included 20 subjects in total, 10 of whom are in the control group (C) and 10 of whom undertake the foam rolling (FR) testing procedure. The subjects consist of

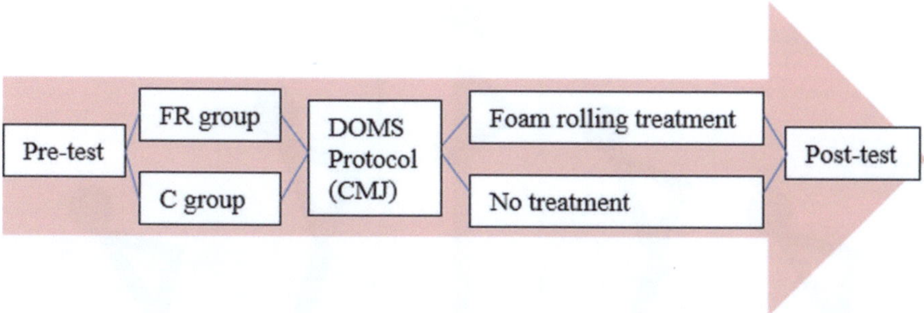

Fig. 1 Flow chart of the research procedure

male and female students from the School of Education at UTM, aged between 22 and 24 years, with no health issues or history of serious injury.

2.2 *Research Procedure*

An initial meeting was attended by these students, during which the background of the study, including a brief overview of the testing procedures and treatments, was explained. The flow chart of the research procedures was displayed in Fig. 1.

A measuring tape was used to measure the circumference of the gastrocnemius muscle, a universal goniometer was used to measure range of motion (ROM), and the Visual Analog Scale (VAS) was used to measure pain levels before and after the DOMS protocol. The patients completed five sets of Counter Movement Jumps (CMJ) with 10 repetitions each and 30 s of rest in between sets as part of the DOMS protocol after the pre-test. The researcher gave the FR group a foam rolling recovery treatment after the DOMS regimen was finished, but the C group received no treatment. After the treatment, the subjects' muscle discomfort levels were recorded using the VAS, the gastrocnemius muscle circumference was measured, and ROM was assessed using the goniometer. The data collected by the researcher were compiled, analyzed, and reported in detail.

2.3 *Testing Procedure*

The testing procedure includes the DOMS protocol, the foam rolling protocol, VAS (Visual Analog Scale), gastrocnemius muscle circumference measurement, and ROM (Range of Motion).

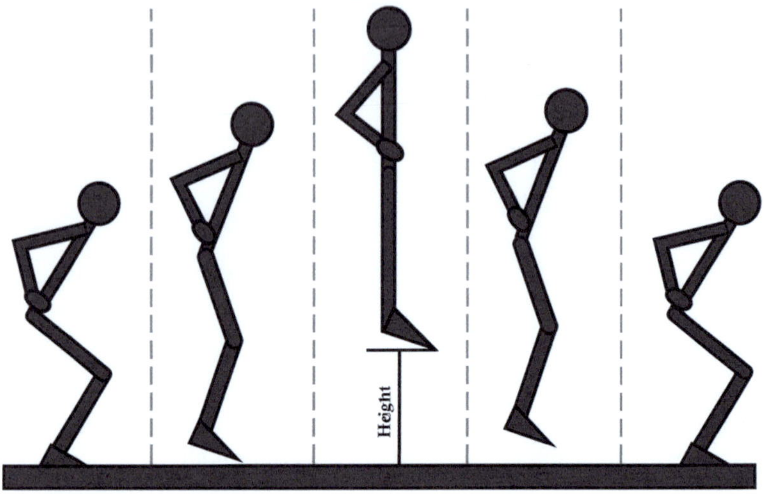

Fig. 2 Counter Movement Jump (CMJ)

2.3.1 Protocol of DOMS

The counter-movement jump (CMJ) techniques started from a standing position, they performed a counter movement by bending their knees and hips, and then jumped as high as possible vertically. The protocol began with the subjects standing with their feet shoulder-width apart and their hands on their hips. At the signal "start," the subjects performed the CMJ from a squat position and landed back in the same position. Each subject completed 5 sets of 10 repetitions, with 30 s of rest between sets. After completing the test, the subjects performed the foam rolling protocol. The CMJ as shown in Fig. 2.

2.3.2 Protocol of Foam Rolling

After completing the CMJ protocol, subjects in FR group received foam rolling treatment on the gastrocnemius muscle. The goal was to relieve muscle soreness caused by the previously performed CMJ activity. Subjects were instructed to lie on their backs with one leg elevated and placed on the foam roller. Comfortable pressure was applied to relax the muscle. Gradual rolling movements were performed from the heel to the calf for 2 min. After completing this, the foam roller was moved to the other leg, and the foam rolling process was repeated. During the process, subjects were asked to breathe normally and move their knee joints. After the foam rolling, subjects performed static stretching on the gastrocnemius muscle for 30 s, repeating it 3 times, through a pressure-free dorsiflexion movement. The aim of foam rolling was to improve blood circulation, relieve affected muscles, and reduce muscle soreness from previous physical activity. This was expected to expedite the recovery process.

2.3.3 Visual Analog Scale (VAS)

The subjects were given a Visual Analog Scale (VAS) measurement. The pain was measured using a device called the VAS, which is a 100 mm straight line with labels at either end that describe the pain experienced. Prior to the DOMS protocol, as well as 24, 48, and 72 h later, the subjects were given the VAS. The subjects were asked to mark the level of discomfort they felt using a permanent marker on the VAS scale. Measurements were taken after the vertical jump test was performed three times. The subjects marked their level of discomfort on a 100 mm continuous line, with '0' denoting no pain and '100' denoting severe pain.

2.3.4 Gastrocnemius Muscle Circumference

To assess swelling, the circumference of the gastrocnemius muscle was initially measured using a measuring tape. After the explosive strength testing protocol using CMJ was performed, the gastrocnemius muscle circumference was measured again to determine changes in muscle size. The subjects were asked to stand while the measurement was taken. The measurement location was determined as one inch below the calf fold. The muscle circumference was measured with the tape placed parallel to the muscle, ensuring it was neither too loose nor too tight. The measurement value was recorded with a permanent marker and kept for a 3-day trial period. Measurements were taken before the test, and at 24, 48, and 72 h afterward. The measurement was conducted thrice to acquire more precise and dependable data, and the mean value was utilized. This measurement aimed to determine changes in the gastrocnemius muscle volume before and after the CMJ testing protocol. Changes in muscle size indicated muscle stretching or contraction resulting from physical activity. The location for measuring the gastrocnemius muscle circumference is shown in Fig. 3.

2.3.5 Range of Motion (ROM)

A universal goniometer was used to measure the Range of Motion (ROM) of the gastrocnemius muscle in the subjects. The subjects were instructed to stand facing a solid support, such as a wall or door, with their feet about two feet away from the support. The leg being tested, which was the dominant leg, was positioned behind the opposite leg. The flat surface of the goniometer was placed on the posterior tibia as the zero-degree reference point. The goniometer was then positioned at the ankle so that the axis of movement was parallel to the tibia. The test began after the subject pushed the foot downward to its maximum extent. The angle values were read and recorded at both the lowest and highest positions. To ensure reliability, measurements were repeated three times with a rest period between each measurement. The measurement data were processed to obtain the average maximum ROM value. This measurement was taken before and after the intervention to assess changes in muscle flexibility.

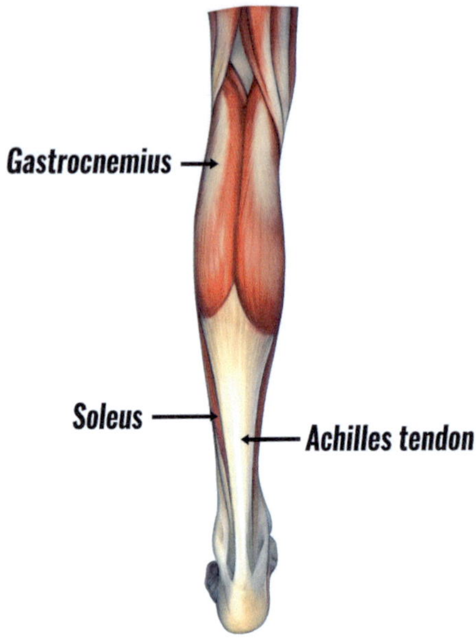

Fig. 3 Location of gastrocnemius muscle circumference measurement

3 Data Analysis

Range of Motion (ROM), gastrocnemius muscle circumference, and Visual Analog Scale (VAS) data were analyzed before to, 24 h, 48 h, and 72 h following the DOMS protocol for both the foam rolling (FR) and control (C) groups. At each time point prior to, 24 h following, 48 h following, and 72 h following the DOMS protocol, these three parameters were compared. In order to determine whether the two groups differ significantly, the analysis employed an independent t-test.

3.1 Comparison of Visual Analog Scale (VAS) Scores Between the Foam Rolling (FR) Group and the Control (C) Group Before, 24 h, 48 h, and 72 h After the DOMS Protocol

Based on Table 1, there were significant differences in the Visual Analog Scale (VAS) scores between the FR group and the C group recorded at 24 h ($t = -3.45$, $p = 0.003$), 48 h ($t = -5.906$, $p = 0.001$), and 72 h ($t = -4.669$, $p = 0.001$) after the protocol.

Table 1 Independent t-test for comparing VAS between the FR group and the C group

Visual Analog Scale (VAS)				
		Mean ± Standard deviation	t	Value P
After the DOMS protocol	FR	0 ± 0		
	C	0 ± 0		
After 24 hours	FR	29 ± 12.87	−3.45	0.003
	C	54 ± 18.97		
After 48 hours	FR	17 ± 11.60	−5.906	0.001
	C	50 ± 13.33		
After 72 hours	FR	8 ± 6.33	−4.669	0.001
	C	35 ± 17.16		

3.2 Comparison of Gastrocnemius Muscle Circumference Between the Foam Rolling (FR) Group and the Control (C) Group Before, 24 Hours, 48 Hours, and 72 Hours After the DOMS Protocol

Based on Table 2, there were no significant differences in the gastrocnemius muscle circumference between the FR group and the C group recorded before ($t = -4.68$, $p = 0.645$), at 24 h ($t = -1.413$, $p = 0.175$), at 48 h ($t = -1.948$, $p = 0.067$), and at 72 h ($t = -1.721$, $p = 0.102$).

Table 2 Independent t-test for comparing gastrocnemius muscle circumference between the FR group and the C group

Gastrocnemius muscle circumference				
		Mean ± Standard deviation	t	Value P
After the DOMS protocol	FR	31.35 ± 5.20		0.645
	C	32.20 ± 2.44	−4.68	
After 24 hours	FR	30.83 ± 5.03	−1.413	0.175
	C	33.30 ± 2.31		
After 48 hours	FR	30.10 ± 4.90	−1.948	0.067
	C	33.40 ± 2.16		
After 72 hours	FR	29.95 ± 5.08	−1.721	0.102
	C	32.95 ± 2.14		

Table 3 Independent t-test for comparing Range of Motion (ROM) between the FR group and the C group

Range of Motion (ROM)				
		Mean ± Standard deviation	t	Value P
After the DOMS protocol	FR	53.50 ± 10.06		0.140
	C	63.30 ± 17.35	−1.545	
After 24 hours	FR	56.70 ± 10.87	−0.505	0.620
	C	59.70 ± 15.33		
after 48 hours	FR	58.80 ± 11.11	0.322	0.751
	C	56.90 ± 14.96		
After 72 hours	FR	29.95 ± 5.08	−1.721	0.102
	C	32.95 ± 2.14		

3.3 Comparison of Range of Motion (ROM) Between the Foam Rolling (FR) Group and the Control (C) Group Before, 24 Hours, 48 Hours, and 72 Hours After the DOMS Protocol

Based on Table 3, there were no significant differences in the gastrocnemius muscle circumference between the FR group and the C group recorded before ($t = -1.545$, $p = 0.140$), at 24 h ($t = -0.505$, $p = 0.620$), at 48 h ($t = 0.322$, $p = 0.751$), and at 72 h ($t = 0.887$, $p = 0.387$).

4 Discussions

This study evaluated the effects of foam rolling on recovery from Delayed Onset Muscle Soreness (DOMS) by comparing its impact on Visual Analog Scale (VAS) scores, gastrocnemius muscle circumference, and Range of Motion (ROM) between the foam rolling (FR) and control (C) groups at various intervals post-protocol.

4.1 Findings of the Foam Rolling Group and the Control Group on the Comparison of Visual Analog Scale (VAS) Before, 24 Hours, 48 Hours, and 72 Hours After the DOMS Protocol

The results indicated that foam rolling was effective in reducing pain associated with DOMS. Although both groups experienced an increase in VAS scores after 24 h, a significantly lower increase was observed in the FR group compared to the C group ($p < 0.01$). This clearly indicated that foam rolling aided in the recovery from DOMS. According to the previous study, the VAS was used to assess individuals' perceptions of their pain before and after treatment, which was important for determining the effectiveness of the intervention (Hawker et al. 2011). This trend continued, with a more significant decrease in pain observed in the FR group over 72 h ($p < 0.01$), suggesting that foam rolling enhanced recovery by reducing DOMS-related discomfort. Philip (2010) noted that DOMS typically resolves within seven days without specific treatment (Philips 2010).

4.2 Findings of the Foam Rolling Group and the Control Group on the Comparison of Gastrocnemius Muscle Circumference Before, 24 Hours, 48 Hours, and 72 Hours After the DOMS Protocol.

Initially, no significant difference was found between the foam rolling (FR) group and the control (C) group before the DOMS protocol ($p > 0.05$). After treatment, the FR group showed a decrease in muscle circumference at various time points (24, 48, and 72 h), while the C group initially showed an increase followed by a decrease. Despite these trends, statistical analysis using independent t-tests revealed no significant differences in muscle circumference between the FR and C groups at any time point (24 h: $p > 0.05$, 48 h: $p > 0.05$, 72 h: $p > 0.05$). This suggested that while foam rolling might influence muscle circumference, these changes were not significant within the short-term observation period. According to Peake et al. (2017), short-term changes in muscle circumference post-exercise are often minimal. Significant hypertrophy or changes in muscle size typically require a longer duration to become detectable (Peake et al. 2017). Thus, while foam rolling may impact muscle recovery, its effects on muscle circumference were not immediately significant and may require longer-term studies to fully understand its impact.

4.2 Findings of the Foam Rolling Group and the Control Group on the Comparison of Range of Motion (ROM) Before, 24 Hours, 48 Hours, and 72 Hours After the DOMS Protocol

The average ROM in the foam rolling (FR) group showed an improvement, while a decline was experienced by the control (C) group from before the DOMS protocol to 72 h post-protocol. Before the intervention, no significant difference in ROM was found between the two groups ($p > 0.05$), indicating similar initial conditions. This was consistent with Cheung et al. (2003), who noted that DOMS can reduce ROM (Hedges and Womack 2022). Post-protocol, after 24 h, an increase in ROM was exhibited by the FR group compared to the C group, which experienced a decrease. Despite this trend, independent t-tests indicated no significant differences between the FR and C groups at 24 h ($p > 0.05$). Similar trends were observed after 48 h (FR group increased, C group decreased) and 72 h (FR group continued to improve while

C group remained decreased), with no significant differences detected at these time points (48 h: $p > 0.05$; 72 h: $p = 0.05$).

The findings suggested that foam rolling might support recovery from DOMS, as evidenced by the increasing ROM in the FR group over time. Behm et al. (2019) highlighted that while intense exercise can impair ROM due to micro-injuries, adequate recovery time is essential (Behm and Wilke 2019). Additionally, the CMJ protocol used in this study might not have effectively induced DOMS, as supported by Braun and Dutto (2003), who successfully used Downhill Running to induce DOMS (Braun and Dutto 2003). Thus, while foam rolling appeared beneficial in promoting ROM recovery, the effectiveness of the DOMS protocol could have influenced the observed outcomes.

5 Conclusions

This research was conducted due to the importance of identifying the effectiveness of foam rolling in the recovery process following physical activity. The information was intended to be applied and practiced by coaches and athletes and to serve as a precursor to future studies using this foam rolling method. The study presented could still be improved to ensure more accurate and detailed results. The recommendations provided were intended to serve as a guideline for future research.

Acknowledgements Special thanks to all the participants in this research.

References

Behm DG, Wilke J (2019) Do self-myofascial release devices release myofascia? Rolling mechanisms: a narrative review. Sports Med 49(8):1173–1181. Springer International Publishing. https://doi.org/10.1007/s40279-019-01149-y

Bennett HJ, Andrews JJ (2023) The impact of exercise-induced muscle damage on joint stiffness and range of motion: a systematic review. Sports Med 53(1):45–59

Braun W, Sforzo G (2011) Delayed onset muscle soreness (DOMS). American College of Sport Medicine. www.acsm.org

Braun WA, Dutto DJ (2003) The effects of a single bout of downhill running and ensuing delayed onset of muscle soreness on running economy performed 48 h later. Eur J Appl Physiol 90(1–2):29–34. https://doi.org/10.1007/s00421-003-0857-8

Buchheit M, Cormie P, Abbiss CR, Ahmaidi S, Nosaka KK, Laursen PB (2009) Muscle deoxygenation during repeated sprint running: effect of active vs. passive recovery. Int J Sports Med 30(6):418–425. https://doi.org/10.1055/s-0028-1105933

Cheung K, Hume PA, Maxwell L (2003) Delayed onset muscle soreness treatment strategies and performance factors. Sports Med 33(2)

Clarkson PM, Nosaka K, Braun B (1992) Muscle function after exercise-induced muscle damage and rapid adaptation. Med Sci Sports Exerc 24

Contro V, Pieretta E, Patrizia P (2016) Delayed onset muscle soreness (DOMS) management: present state of the art. Trends Sport Sci 3(23):121–127

Figueiroa J, Costa T (2023) The effects of foam rolling on recovery and performance in athletes with delayed onset muscle soreness: a systematic review and meta-analysis. J Strength Cond Res 37(6):1491–1502

Hawker GA, Mian S, Kendall RJ, Bosse MJ, Dworkin RH (2011) Measures of adult pain. J Pain Res 4:1–6. https://doi.org/10.2147/JPR.S12973

Hedges RD, Womack CJ (2022) Delayed onset muscle soreness: an updated review. J Sports Sci Med 21(4):517–526

Peake JM, Neubauer XO, Della Gatta PA, Nosaka XK (2017) Recovery from exercise muscle damage and inflammation during recovery from exercise. J Appl Physiol 122:559–570. https://doi.org/10.1152/japplphysiol.00971.2016.-Unac

Pearcey GEP, Bradbury-Squires DJ, Kawamoto J-E, Drinkwater EJ, Behm DG, Button DC (2015) Foam rolling for delayed-onset muscle soreness and recovery of dynamic performance measures. J Athl Train 50(1):5–13

Philips SM (2010) The role of exercise in the management of delayed onset muscle soreness. J Sports Sci 28(6):1–8

Effects of Hand–Eye Coordination Training on Accuracy and Reaction Time Among E-Sports Players

Mohamad Aiman Hakimi Mamat, Nur Fatihah Raimi, and Adjah Naqkiah Mazlan

Abstract The growing popularity of electronic sports, or e-sports, emphasizes the necessity of looking into the variables influencing player performance. This study investigates how e-sports players' accuracy and reaction times are affected by hand–eye coordination training. This study is to evaluate if the training may improve hand–eye coordination, which is essential for playing video games. Twenty participants were randomly assigned to two groups; the experimental group (HEC) ($n = 10$), which received training in hand–eye coordination techniques such wall ball bounce, ball dribbling, and ball juggling, and the control group (C) ($n = 10$), which received no training. Before and after the training, AimLab was used to evaluate the accuracy and reaction time of both groups. According to the results, participants who participated in the hand–eye coordination training had significant improvements in their reaction times when compared to the control group. However, there was no noticeable rise in accuracy as a result of the training. The experimental and control groups' reaction times differed significantly ($p < 0.05$), according to an independent t-test. According to the study's findings, hand–eye coordination training improves reaction time but has no noticeable impact on e-sport accuracy.

Keywords Hand–eye coordination training · Reaction time · Accuracy · E-sport · Aimlab

M. A. H. Mamat · N. F. Raimi (✉) · A. N. Mazlan
School of Education, Faculty of Social Science and Humanities (FSSH), Universiti Teknologi Malaysia, Skudai, Johor, Malaysia
e-mail: nurfatihah.raimi@utm.my

N. F. Raimi · A. N. Mazlan
Sports Innovation and Technology Centre (SITC), Institute of Human Centred Engineering (iHumEN), Universiti Teknologi Malaysia, Skudai, Johor, Malaysia

M. H. A. Hassan et al. (eds.), *Proceedings of the 3rd International Conference on Innovation and Technology in Sports*, Lecture Notes in Bioengineering,
https://doi.org/10.1007/978-981-95-0804-4_23

1 Introduction

Competitive gaming has been redefined by the explosive growth of electronic sports, or e-sports, which has brought attention to several variables that affect player success. Understanding the significance of both cognitive and physical abilities, such as accuracy, hand–eye coordination, and response time, is becoming increasingly essential as the demand for excellence in e-sports rises. E-sports have only lately started to be examined through this lens, in contrast to traditional sports where physical training has long been researched to maximize performance.

The capacity to precisely complete tasks by coordinating hand movements with visual information is known as hand–eye coordination (Ammann et al. 2016). This ability is essential in e-sports since players must constantly interpret visual cues and convert them into accurate motor movements, frequently under time constraints. Accuracy, defined as the degree to which a player's actions align with intended outcomes, is another critical factor in e-sports (Schmidt and Lee 2019). Meanwhile, reaction time is the speed at which a player responds to a given stimulus (Welford 1980). These three elements interact dynamically, impacting overall performance in high-stakes gaming environments.

AimLab is a specialized tool made to measure and enhance e-sports players' accuracy and reaction times. It offers a variety of exercises and tests that mimic actual game situations, enabling players to sharpen their abilities in a monitored setting. Researchers can test a player's accuracy in striking targets and response time to stimuli objectively by utilizing AimLab. Because of this, AimLab is a perfect tool for assessing how well hand–eye coordination programs work. This study investigates the effects of hand–eye coordination training on the accuracy and reaction times of e-sports players. Despite the similarities between e-sports and traditional sports, there is currently not much study on physical training designed for e-sports.

In e-sports, the combination of precision, reaction time, and hand–eye coordination is essential. Players that have good hand–eye coordination are able to move accurately, which has a direct effect on their ability to hit targets or carry out strategy. The speed at which these movements can be started in response to in-game events depends on reaction time. Players with faster response times, for instance, may interpret visual information more quickly and move their hands precisely in fast-paced games, which improves accuracy and performance overall (Liu and Li 2020; Huang and Wei 2018).

This study utilises AimLab as a measurement instrument to investigate how hand–eye coordination training affects players' accuracy and response time in the context of the tactical and cognitive demands of e-sports. Through this investigation, the study aims to expand knowledge of the physical elements that improve performance in the quickly changing e-sports industry.

2 Methodology

2.1 *Sampling*

A total of 20 participants were selected for this study. These individuals had previous experience playing video games, either first-person shooters (FPS) or multiplayer online battle arenas (MOBA) involving PC gaming. The participants were fourth-year students from Universiti Teknologi Malaysia who were still actively playing video games for no less than 3 months. Participants who had played video games but had been inactive for 3 months or more were not selected as participants. They were divided into two groups: the experimental group and the control group. The experimental group underwent controlled hand–eye coordination training, while the control group followed their usual routine without any specific hand–eye coordination training.

2.2 *Research Procedure*

Upon signing consent form, all participants were required to undergo a familiarization followed by a pre-test of accuracy and reaction time measured by Aimlab application in a computer prepared by the researchers at the data collection location. After the pre-test, 20 participants were randomly allocated into two groups, either HEC ($n = 10$) or C group ($n = 10$). The participants in HEC group were required to undergo a four-weeks hand–eye coordination training meanwhile the participants in C group acted as control for four weeks. Upon completion of the four weeks, they were required to undergo a post-test of accuracy and reaction time measured by Aimlab application as in the pre-test. Mean differences of the accuracy and reaction time scores between pre-test and post-test within both groups were compared using independent t-test analysis in SPSS. Figure 1 displays the flow chart of the research procedures.

2.3 *Protocol of Accuracy and Reaction Time Testing Using Aimlab*

The study began with the researcher collecting test results from the athletes using the AimLab application, as shown in Fig. 2. The athletes were asked to complete a test in the app one day before starting their training. This test was designed to measure how fast and accurately they responded to visual stimuli in an e-sports setting. These results helped the researcher understand the athletes' performance levels before the training began.

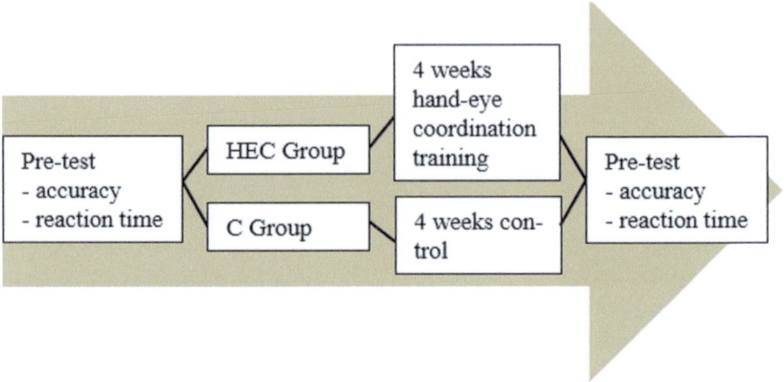

Fig. 1 Flow chart of the research procedure

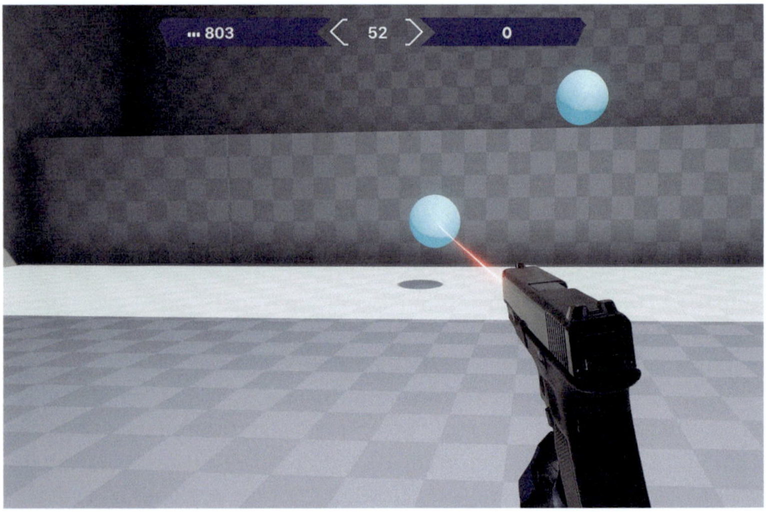

Fig. 2 AimLab reaction time and accuracy test

The athletes took the AimLab test again after the training sessions to see if there were any improvements. This allowed the researcher to determine how much the hand–eye coordination training affected their reaction time and accuracy.

During the test, participants had to shoot at blue balls shown in the AimLab app. The test required them to use precise mouse movements guided by their hands. The test lasted for one minute, challenging the athletes to react quickly and accurately. If they missed a shot, points were deducted to ensure accuracy standards were maintained. For this study, the researcher focused on the scores related to reaction time and accuracy. The scores for reaction time and accuracy is shown as in Fig. 3.

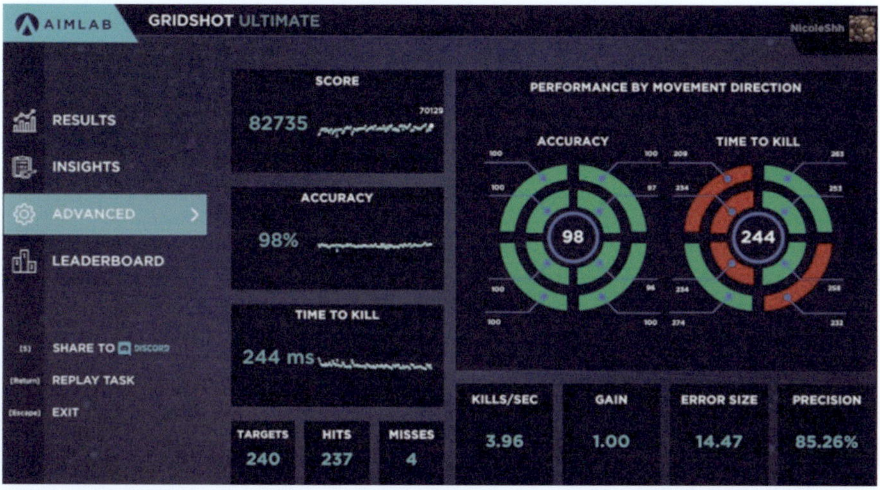

Fig. 3 Full score of the reaction time and accuracy test as displayed in AimLab

2.4 *Protocol of Hand–eye Coordination Training*

The researcher provided three types of exercises to the experimental group to follow for 4 weeks. Each week, the participants attended three training sessions conducted in the afternoon. The three exercises, named Wall Ball Bounce, Ball Dribbling, and Ball Juggling, were performed on the same day. The intensity of the exercises was progressively increased each week to ensure improvement in participants' performance.

Wall Ball Bounce. The researcher provided 10 tennis balls and a wall for exercise. Participants were instructed to stand 2 m from the wall and throw the tennis ball towards it. They were required to catch the ball with the opposite hand using an upward palm grip. If the ball was missed, participants simply retrieved it from the ground and continued the exercise. In the first two weeks, participants performed the exercise 10 times per set. For weeks 3 and 4, the number of sets was increased, and participants were required to throw and catch the ball using a downward palm grip.

Ball Dribbling. The researcher provided three basketballs for the exercise. Participants stood still with their feet shoulder-width apart and bounced the ball on the court with one hand. The bounce was done at waist level to the side. They had 30 s to complete each set. If the ball went far, they simply picked it up from nearby. In the first two weeks, participants bounced the ball 25 times per set for 2 sets using one hand, and then switched to the other hand. In weeks 3 and 4, the number of sets increased. Participants stood with their feet wider apart, and knees bent, bouncing the ball from one hand to the other in front of them, completing 25 bounces per set for 2 sets.

Ball Juggling. The researcher provided a ping pong bat and ball. Participants stood still and tossed the ball into the air, making sure it went above shoulder level.

They did 20 tosses per set, switching hands between sets. This was done for 2 sets in weeks 1 and 2. In weeks 3 and 4, they increased the exercise to 20 tosses for 4 sets, again switching hands between sets.

3 Data Analysis

Mean differences between pre-test and post-test for accuracy and reaction time scores in both groups were compared using an independent t-test analysis. P value for the significant level is below than 0.05.

3.1 Comparison of Accuracy in Pre-Test and Post-Test Between Groups

Table 1 shows that the mean and standard deviation before the intervention for C group are 88.30 ± 2.37, while for the HEC group, they are 93.15 ± 2.12. After the intervention, the mean and standard deviation for the C group are 91.54 ± 2.31, and for the HEC group, they are 92.66 ± 1.82. The p-value in the post-test, which is greater than 0.05 ($p = 0.268$), indicates that there is no significant difference in the mean post-test accuracy between the C and HEC groups. The HEC group (which underwent hand–eye coordination training) demonstrated a higher mean accuracy (92.66) compared to the control group (91.54). Although there was a difference in pre-test accuracy for both groups, the HEC group showed a decrease in mean accuracy in the post-test compared to their higher pre-test scores. Conversely, the C group exhibited an improvement in accuracy scores, despite not undergoing any training. Therefore, hand–eye coordination training did not have a significant effect on accuracy among e-sports players.

Table 1 Comparison of accuracy in pre-test and post-test between groups

Test	Group		Mean ± SD	t	P value
Pre-test	HEC		93.15 ± 2.23	−4.57	<0.001
	C		88.30 ± 2.50		
Post-test	HEC		92.66 ± 1.91	−1.14	0.268
	C		91.54 ± 2.43		

Table 2 Comparison of reaction time in pre-test and post-test between groups

Test	Group		Mean ± SD	t	P value
Pre-test	HEC		935.6 ± 101.6	3.86	0.001
	C		1181.9 ± 174.3		
Post-test	HEC		698.9 ± 36.2	8.10	<0.001
	C		1039.9 ± 128.1		

3.2 Comparison of Reaction Time in Pre-Test and Post-Test Between Groups

Table 2 shows that the mean and standard deviation before the intervention for the control group are 1181.9 ± 174.3, while for the experimental group, they are 935.6 ± 101.6. After the intervention, the mean and standard deviation for the control group are 1039.9 ± 128.1, and for the experimental group, they are 698.9 ± 36.2. The p-value in the post-test, which is significantly less than 0.05 ($p < 0.001$), indicates that there is a significant difference in the average post-test reaction time between the control and experimental groups. The experimental group exhibited a higher average reaction time (698.9 ms) compared to the control group (1039.9 ms). There is a significant difference in the post-test reaction time for both groups. This is because the reaction time for the experimental group is higher than that of the control group in the post-test, following the completion of the 4-week hand–eye coordination training.

4 Discussions

4.1 Accuracy Before and After Hand–Eye Coordination Training

Both the control and experimental groups showed different results in accuracy. The experimental group had a small drop in average accuracy, suggesting that the hand–eye coordination training might not have improved accuracy overall. The high standard deviation means there was a lot of variation in how different people responded to the training, with some improving and others not.

In contrast, most people in the control group improved their accuracy after the test. According to Nuri and Shadmehr (2013), athletes who make better decisions often have faster reaction times, which helps them prepare their movements better (Nuri and Shadmehr 2013). This was seen in the control group, where higher reaction times were linked to better accuracy. Meanwhile, the experimental group, with faster reaction times, tended to have lower accuracy. So, there was a noticeable change in accuracy performance among e-sports players before and after the hand–eye coordination training.

4.2 Reaction Time Before and After Hand–Eye Coordination Training

The research demonstrates that the control and experimental groups exhibited differences in results. A significant increase in reaction time was observed in the experimental group following hand–eye coordination training in comparison to the control group. The control group, which did not receive any training, also exhibited an improvement; however, the change was less significant. The control group experienced an average reduction in reaction time of −142 ms, while the experimental group experienced an average reduction of −236.7 ms.

Hand–eye coordination training has a beneficial impact on the reaction times of athletes, as per Kuzu and Çetin (2017) (Kuzu and Çetin 2017). This implies that such training can effectively enhance the reaction times of e-sports participants. The experimental participant with the most significant improvement demonstrated a reduction in reaction time from 1111 to 737 ms, which is a significant change of −374 ms. a quicker reaction time is indicated by a decrease in milliseconds (ms). Consequently, the reaction time of e-sports participants was significantly enhanced both prior to and following the hand–eye coordination training.

4.3 Effects of Hand–Eye Coordination Training on Accuracy and Reaction Time

The reaction time of the HEC group demonstrated a significant improvement following the hand–eye coordination training. This is in opposition to the C group, which also experienced improvement, though to a lesser extent. The reaction time performance of the HEC group significantly improved in both the pre-test and post-test, whereas the C group only demonstrated minor improvements.

Additionally, the accuracy assessments demonstrated significant differences between the two groups. The C group, which did not receive the training, demonstrated a significant increase in accuracy, with the majority of participants demonstrating improvement. In contrast, the HEC group demonstrated a decrease in accuracy for certain participants, despite an improvement in reaction time.

This implies that hand–eye coordination training tends to improve reaction time more than accuracy. In contrast, participants who did not participate in the training and exhibited slower reaction times tended to enhance their accuracy. This is supported by Thomson and Watt (2009), who discovered that athletes frequently make decisions at the expense of accuracy in order to move quickly (Thomson and Watt 2009), and by Causer and Williams (2013), who discovered that a higher stimulus intensity reduces reaction time and enhances accuracy (Causer and Williams 2013).

The results of this study are consistent with the existing literature, which emphasizes the beneficial effects of physical hand–eye coordination training on reaction time, particularly in the context of e-sports.

Initially, Bediou and Pichon (2018) conducted a thorough meta-analysis of a variety of training programs, including those that focused on physical hand–eye coordination. Their research offers compelling evidence that physical coordination training significantly enhances reaction times. This meta-analysis illustrates that enhancements in hand–eye coordination, which are accomplished through physical exercise, result in quicker reaction times (Bediou and Pichon 2018). This discovery corroborates our research, which indicates that players' capacity to react to in-game events is improved by physical hand–eye coordination training.

In order to further solidify this concept, Yang and Zhang (2020) conducted a detailed analysis of the impact of physical hand–eye coordination training on the reaction times of e-sports participants. Their research revealed that participants' reaction times significantly improved as a result of this training (Yang and Zhang 2020). The results of this research are directly supported by our study, suggesting that reaction time can be improved through targeted physical training, a critical factor in e-sports performance. The findings indicate that participating in physical activities that are intended to enhance hand–eye coordination can result in more rapid and accurate responses during gameplay.

Furthermore, the research conducted by Hsu and Lu (2019) emphasizes the more extensive implications of physical hand–eye coordination training. The impact of physical coordination training on both cognitive and motor abilities, including reaction time, is underscored by their research. The researchers discovered that physical training not only enhances hand–eye coordination but also improves overall cognitive and motor functions, which are essential for effective performance in competitive environments (Hsu and Lu 2019). Our conclusion that physical hand–eye coordination training is advantageous for enhancing reaction time in e-sports contexts is further substantiated by this broader perspective.

In conclusion, the assertion that physical hand–eye coordination training can substantially improve reaction time is collectively supported by the evidence from these studies. This improvement is essential for e-sports participants, who depend on the ability to execute rapid and precise movements in order to succeed in competitive environments. By incorporating physical coordination exercises into training regimens, athletes can enhance their reaction times, which in turn enhances their performance in e-sports.

This discussion integrates evidence from the cited studies to substantiate the argument that physical hand–eye coordination training enhances reaction time, making it important and specific to the e-sports industry.

5 Conclusions

In summary, the findings of this study indicate that hand–eye coordination training significantly improves reaction time. Moreover, incorporating such training into e-sports regimens has the potential to enhance player performance, particularly under competitive conditions. However, further research is needed to explore the extent to which accuracy can be improved through these training methods. Ultimately, these insights can serve as a foundation for developing more effective e-sports training programs, optimizing player performance and competitive outcomes.

Acknowledgements Special thanks to all the participants in this research.

References

Ammann R, Bertucco M, Omlin X (2016) The effect of hand-eye coordination training on reaction time and performance in a simulated driving task. Hum Factors Ergon Manuf Serv Ind 26(1):18–28. https://doi.org/10.1002/hfm.20520

Bediou B, Pichon A (2018) Improving cognitive and motor performance through video game training: a meta-analysis. J Cogn Train Ther

Causer J, Williams AM (2013) The effect of stimulus intensity on reaction time and accuracy in sports. J Sports Sci 31(7):755–764. https://doi.org/10.1080/02640414.2012.746730

Hsu CH, Lu YH (2019) Hand-eye coordination training and its impact on cognitive and motor skills in competitive gaming. J Sport Sci Med

Huang X, Wei Z (2018) The role of reaction time and hand-eye coordination in competitive gaming. In: Proceedings of the international conference on competitive gaming, pp 20–30

Kuzu M, Çetin E (2017) The effects of hand-eye coordination training on athletes' reaction time and performance. J Sport Sci Med 16(4): 521–527. https://doi.org/10.1177/1558925017740840

Liu T, Li L (2020) Effects of hand-eye coordination on performance in e-sports: a review. In: Proceedings of the conference on e-sports research, pp 1–10

Nuri A, Shadmehr R (2013) Decision-making and the speed of motor responses in athletes. J Neurophysiol 109(7):2050–2058. https://doi.org/10.1152/jn.00770.2012

Schmidt RA, Lee TD (2019) Motor control and learning: a behavioral emphasis, 6th edn. Human Kinetics

Thomson D, Watt R (2009) Decision-making in athletes: the impact of speed and accuracy. J Sport Sci 27(3):323–332. https://doi.org/10.1080/02640410802508705

Welford AT (1980) Reaction time. Academic Press

Yang H, Zhang X (2020) Effects of physical hand-eye coordination training on reaction time in e-sports players. J E-Sport Perform.

Correlation Between Leg Muscle Power, Flexibility and Shooting Precision in Blind Football: A Study of Malaysian Athletes

Sohaila Ishak and Hosni Hasan

Abstract This study investigates the relationship between leg muscle power alongside flexibility and shooting accuracy performance among Malaysian blind football athletes while filling a gap in sports science research that studies adaptive sports for visually impaired participants. The demanding sport of blind football requires an advanced understanding of performance factors that extend past standard physical fitness measurements. Method: Fifteen male blind athletes aged 18 to 39 participated in this quantitative correlational study. The study utilized three specific assessments adopted from three specific evaluations: the standing broad jump test for measuring leg muscle power, the modified sit and reach test for flexibility assessment, and the modified target goal test for shooting accuracy measurement. The analysis used Spearman's rank correlation coefficient to determine how well variables related to each other. Results: The results showed that shooting accuracy did not correlate with leg muscle power ($r = 0.299$, $p = 0.280$) or flexibility ($r = 0.278$, $p = 0.315$). Conclusion: The obtained study outcomes contradict established beliefs about how physical fitness measures of muscle power and flexibility help determine blind football performance. The research findings demonstrated auditory and spatial awareness as potential factors that enhance performance outcomes. The research supports developing specialized training programs focusing on sensory and spatial abilities to boost athletic field performance. A refined understanding of adaptive training requirements for visually impaired athletes emerges from this study to guide future research and practical sports science applications.

Keywords Visual impaired athletes · Leg muscle power · Flexibility · Accuracy shooting goals

S. Ishak · H. Hasan (✉)
Faculty of Sports Science and Recreation, Universiti Teknologi MARA (UiTM) Shah Alam, Shah Alam, Selangor Darul Ehsan, Malaysia
e-mail: hosnihasan@uitm.edu.my

M. H. A. Hassan et al. (eds.), *Proceedings of the 3rd International Conference on Innovation and Technology in Sports*, Lecture Notes in Bioengineering,
https://doi.org/10.1007/978-981-95-0804-4_24

1 Introduction

The universally popular sport of football welcomes all athletes with disabilities and explicitly accommodates sports participation for people with vision impairments. Blind football and goalball, alongside other Paralympic sports, received official recognition during the 1976 Summer Paralympics in Toronto as competitive sports for visually impaired athletes. The International Blind Sports Federation (IBSA) officially regulated blind football after its origins in Spain during the 1920s and its subsequent development in Brazil throughout the 1960s. The sport of blind football started in Malaysia in 2009 and brought its first victories during the 2015 ASEAN Para Games championship. High-performance achievements face ongoing difficulties because of scarce resources, insufficient facilities, and weak support structures.

Blind football teams played with five members who belong to B1 (totally blind) and B2 (partially impaired) categories while using a sighted goalkeeper and goal guide (Rules And 2022). The International Blind Sports Association (IBSA) creates and regularly updates detailed rules that enhance fairness and competition standards. Blind football requires outstanding performance from athletes combining physical and sensory abilities (Oliver et al. 2024). To excel at blind football, athletes need exact timing, correct body movements, and adequate leg muscle strength (Gamonales Puerto et al. 2018; Wijaya 2020). Previous studies demonstrated that explosive leg muscle power relates to scoring effectiveness (Almizan et al. 2020; Soniawan 2021) and flexibility maintains body position while enhancing motion efficiency, which affects shooting accuracy (França et al. 2021; Zul et al. 2022). The intricate nature of the sport demands immediate research into how physical qualities, including leg muscle power and flexibility, affect shooting performance. Although traditional training regimens often emphasize general physical fitness, blind football's unique demands require a more tailored approach (Ishak et al. 2024).

The current research gap received attention through an investigation that measured the role of leg muscle power and flexibility in Malaysian blind football athletes' shooting accuracy. By addressing these needs, this research seeks to advance the field of adaptive sports, offering practical recommendations for coaches and practitioners working with visually impaired athletes and contributing to the broader sports science literature.

2 Methodology

2.1 Participants and Sampling

This research analysis targeted the population of Malaysian blind football athletes. All eligible athletes were included using a total sampling method to ensure comprehensive representation. The study selected fifteen male athletes aged between 18 and 39 to meet its research requirements for analyzing leg muscle power, flexibility, and shooting accuracy (Ishak et al. 2024). All prospective athletes are required to be classified into the designated categories, either B1 (totally blind) or B2 (partially blind), according to the standards set by the International Blind Sports Association (IBSA) (Ishak et al. 2024; Mann and Ravensbergen 2018; Rules And 2022) and capable of providing informed consent. The study excluded athletes who suffered from physical injuries or had recently undergone surgery, sighted athletes, and female athletes to guarantee consistent physical characteristics throughout the study. The study obtained approval from the UiTM Ethics Committee through letter 600-FSR (PT.5/1/PG/750).

2.2 Material and Method

The research utilized three specifically created tools to measure essential physical attributes for accurate shooting performance among national blind football athletes. The standing broad jump test served as the initial instrument for measuring leg muscle power through athlete' forward jump distances, which revealed their explosive leg muscle strength. The modified sit and reach test evaluated hamstring and lower back flexibility as the second measuring tool. This modified sit and reach test was adapted from the standard protocol by considering different arm lengths and trunk dimensions to deliver precise flexibility measurements for different body types. The modified target goal test functioned as the third instrument to determine accuracy levels in shooting performance. During the test, the goal was divided into five labeled areas containing five labeled sections: A, B, C, D, and E (Fig. 1), with point values set at one.

The goal measurement had 3.62 m of width and 2.22 m of height, with 0.8 m by 1.1 m defined for each segment. The goal guide used auditory cues to help athletes stand 8 m from the goal while shooting at specified zones. The goal-labeled areas received successful shots that were recorded for accuracy calculation by dividing successful shots by total attempts.

The researcher instructed athletes to perform proper warm-up and cool-down routines throughout trials to ensure their safety and as injury prevention. Before beginning the testing process, a researcher demonstrated each protocol to verify their understanding of procedures and expected results. The athletes received modified testing procedures that enabled them to experience equipment and setup physically before formal testing began. Athletes gained complete process comprehension through hands-on experience, which led to dependable data collection and an encouraging atmosphere that supported peak performance (Fig. 2).

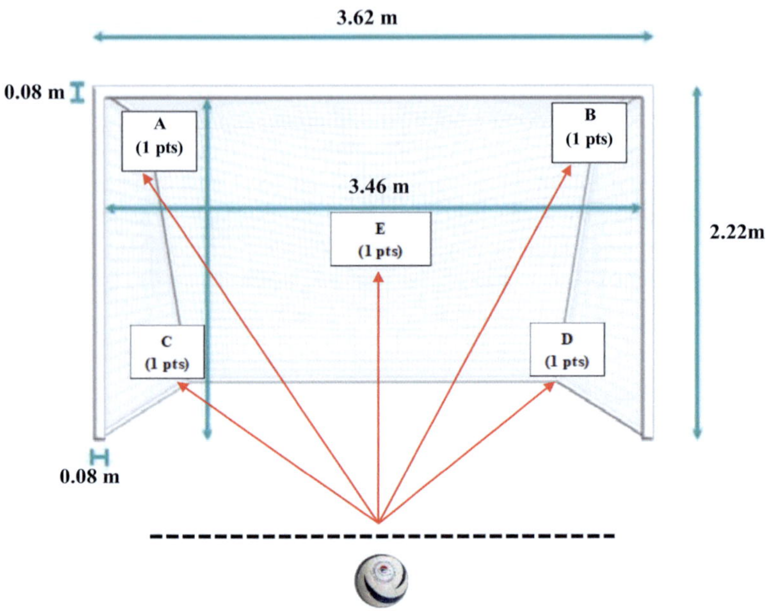

Fig. 1 Scoring for modified target goals test

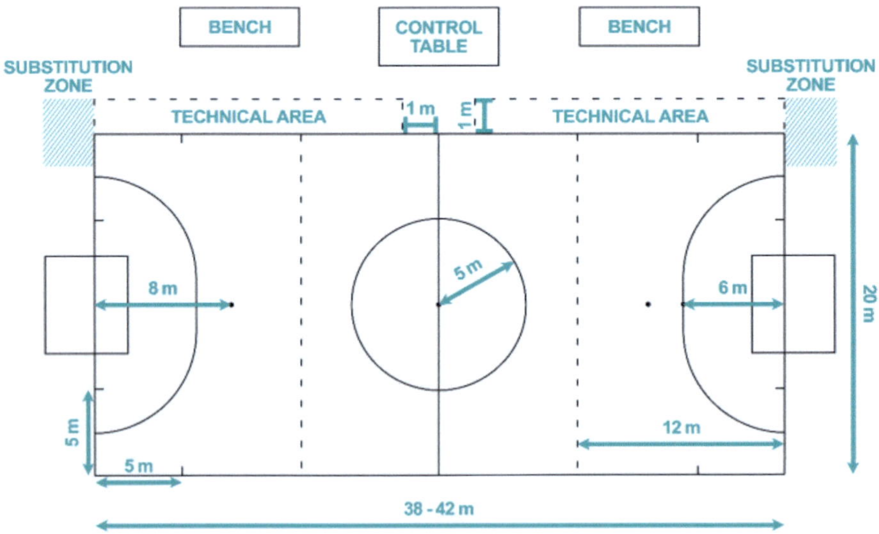

Fig. 2 Pitch measurement

2.3 Data Analysis

Experimental data analysis took place using the Statistical Package for the Social Sciences (SPSS) version 29. The research used descriptive and inferential statistics to analyzed the relationships between leg muscle power, flexibility, and shooting goal accuracy. Spearman's rank correlation coefficient assessed variable correlation in this analysis.

The researchers chose this non-parametric method because it successfully examines ordinal data and detects monotonic relationships without requiring normal distribution assumptions. The researchers performed initial tests on the data to confirm that it satisfied the requirements for statistical analysis before beginning hypothesis testing. Analyses proceeded at a significance threshold of $p \leq 0.05$ to establish statistical significance for observed correlations. A thorough analysis of how leg muscle power and flexibility affect shooting accuracy in blind football athletes was enabled through this analytical method.

3 Result

3.1 Descriptive Statistic

Table 1 displays descriptive statistical data that reveals important information about the physical characteristics of national blind football athletes. The standard standing broad jump measure for leg muscle power showed 1.993 m as the mean result and a standard deviation of 0.2282. The participants showed similar levels of explosive leg strength based on this outcome. The modified sit and reach test showed an average flexibility score of 33.200 cm, while participants displayed a standard deviation 5.2762.

Multiple flexibility levels existed throughout the sample group, as indicated by the observed variability. The mean shooting accuracy score was 3.67, while its standard deviation reached 0.488. The score demonstrated moderate performance stability across athletes while showing variable results between participants.

Table 1 The mean and s of age, height, and weight of national blind football athletes

	N	Mean	Std. deviation
Leg muscle power	15	1.993	0.2282
Flexibility	15	33.200	5.2762
Accuracy shooting goal	15	3.67	0.488
Valid N (listwise)	15		

Table 2 Correlation between leg muscle power and accuracy shooting goals

	Correlation coefficient, r	Sig. value (p-value)
Leg muscle power	0.299	0.280

Note * $p > 0.05$

3.2 Leg Muscle Power Versus Accuracy Shooting Goals

Spearman's rank correlation analysis determined the relationship between leg muscle power and goal-shooting accuracy. Table 2 shows a correlation coefficient 0.299 with a significance value of 0.280 ($p > 0.05$). The analyzed data showed a weak positive relationship between shooting accuracy and leg muscle power.

The data showed that athletes with stronger leg muscles tended to shoot more accurately, but the statistical analysis confirmed that the trend was insignificant. The current study demonstrates that variations in leg muscle power did not cause meaningful changes in shooting accuracy among the participants. Both random chance effects and genuine relationships could justify the observed association; thus, its statistical significance remained absent.

The results rejected the hypothesis that leg muscle power significantly affects shooting accuracy. The study results showed that goal-shooting accuracy works through multiple layers. Studies revealed that factors such as coordination, proper body control technique, and proprioception may matter more than leg muscle power for visually impaired athletes.

3.3 Flexibility Versus Accuracy Shooting Goals

The statistical analysis presented in Table 3 showed a correlation coefficient of ($r = 0.278$) with ($p = 0.315$) indicating ($p > 0.05$). The analysis established that flexibility showed a weak positive relationship with shooting accuracy since more flexible athletes performed marginally better during shooting tests.

The calculated significance value ($p = 0.315$) showed that this relationship was not statistically significant. The research results indicated that the relationship between flexibility and shooting accuracy could be explained by random factors instead of substantial underlying effects. The data failed to support the hypothesis that flexibility substantially impacts shooting accuracy. The data shows flexibility helps athletic

Table 3 Correlation between flexibility and accuracy shooting goals

	Correlation coefficient, r	Sig. value (p-value)
Flexibility	0.278	0.315

Note * $p > 0.05$

performance but does not seem crucial for accurate shooting among visually impaired football athletes.

3.4 Leg Muscle Power and Flexibility Versus Accuracy Shooting Goals

This study employed Spearman correlation analysis to investigate the relationships between leg muscle power and flexibility and shooting goal accuracy among blind football athletes. Table 4 results showed that leg muscle power demonstrated a correlation coefficient of ($r = 0.299$) and a significance value of ($p = 0.280$), while flexibility displayed a correlation coefficient of ($r = 0.278$) with a significance value of ($p = 0.315$). Statistical analysis showed that the significance values above 0.05 ruled out any significant relationship between shooting accuracy and flexibility and leg muscle power measurements.

However, the study showed a weak negative relationship between leg muscular power and flexibility ($r = -0.175$) with a significance level of ($p = 0.533$). The research data showed that leg muscle power and flexibility demonstrated no significant relationship among these populations.

The results revealed a weak relationship between the independent variables of leg muscle power and flexibility and the dependent variable of shooting accuracy. Specifically, this weak positive correlation indicated that leg muscle power and flexibility had no statistical significance regarding shooting accuracy. The athletes examined in the study showed no significant link between their goal-shooting accuracy and leg muscle power and flexibility variations. Consequently, the research findings suggest that leg muscle power and flexibility do not affect shooting accuracy.

Table 4 Correlation between leg muscle power, flexibility, and accuracy shooting goals

			Leg muscle power	Flexibility	Accuracy shooting goal
Spearman's rho	Leg muscle power	Correlation Coefficient	1.000	−0.175	0.299
		Sig. (2-tailed)	.	0.533	0.280
		N	15	15	15
	Flexibility	Correlation Coefficient	−0.175	1.000	0.278
		Sig. (2-tailed)	0.533	.	0.315
		N	15	15	15
	Accuracy shooting goal	Correlation Coefficient	0.299	0.278	1.000
		Sig. (2-tailed)	0.280	0.315	.
		N	15	15	15

4 Discussions

4.1 Leg Muscle Versus Accuracy Shooting Goals

Research showed that national blind football athletes experienced no significant relationship between their leg muscle strength and their ability to score goals. The research demonstrated that elevated leg muscle power levels failed to improve shooting accuracy consistently. Moreover, the researchers also found similar results to Katis et al. (2013) and Nurseto et al. (2019), who established that powerful kicks matter but are not the sole factor for successful goal-scoring. The researcher discovered that additional elements beyond muscle explosiveness remained fundamental for achieving accurate shooting results. The expert confirmed explosive leg strength served as a vital fitness component yet failed to function as the exclusive predictor of shooting results.

In contrast, other studies have highlighted the significance of leg muscle power in enhancing athletic performance. For instance, Deprez et al. (2015) emphasized the crucial role of leg power in achieving high-performance levels, and similarly Bahenský et al. (2020) observed that explosive leg strength was vital for optimal jumping performance in soccer. Furthermore, Bell et al. (2010) and Gardasevic et al. (2017) further supported the view that leg power significantly influences kicking strength and overall performance. Soniawan (2021) also found that leg muscle strength directly impacts shooting accuracy. All these studies collectively suggested a strong connection between leg muscle power and various aspects of football performance.

However, accuracy in football shooting is a multifaceted skill beyond mere physical strength. Katis et al. (2013) highlighted that the efficiency of a kick relied on both its force and precision, suggesting that achieved shooting accuracy involved more than just leg strength; other factors, for example, technique, timing, coordination, and spatial awareness, are also critical for successful goal-scoring. The differences between the findings of this study and those of previous research may be attributed to the unique characteristics of blind football compared to sighted football (Ponchillia et al. 2002). The physical and perceptual demands of visually impaired athletes differ significantly from those of sighted competitors. Blind athletes rely heavily on auditory signals and spatial awareness (Mackenzie and Cushion 2013), which may influence how leg muscle strength affects their performance. Consequently, the specific requirements of blind football could mean that muscle strength plays a different performance role than mainstream football.

Therefore, this study's lack of a significant correlation may reflect the distinctive attributes and requirements of blind football and differences in study participants and methodologies. The complex nature of sports performance, with its interplay of various factors, likely impacts shooting accuracy in ways that extend beyond leg muscle strength alone.

4.2 Flexibility Versus Accuracy Shooting Goals

The study found no significant relationship between flexibility and shooting goal accuracy among visually impaired football athletes. This result may be explained by the athletes' reliance on strategies specifically adapted to accommodate their sensory limitations. Unlike sighted athletes, visually impaired athletes may not have required the same degree of flexibility, as they optimized their performance through aural signals and tactile feedback from the ball and ground. In support of this Nascimento et al. (2022) explored the impact of vision impairment on postural control in visually challenged football athletes and found that they exhibited comparable postural control to sighted athletes when deprived of visual input. This suggested that visually impaired athletes, although lacking vision, managed to maintain balance and posture effectively through non-visual means, further indicating that flexibility may not have been as critical to their performance as it is for sighted athletes.

The equivalent postural control outcomes observed between visually impaired and sighted athletes demonstrated that visual information deprivation played a more significant role than improved compensatory mechanisms in postural control. The visually impaired athletes in this research might have depended on alternative movement methods because they needed flexibility and extended muscles for accurate shooting but not postural control. The athletes perhaps reduced their flexibility in kicking because other sensory inputs became more important for guiding their movements.

The postural control results from visually impaired athletes showed significant divergence compared to sighted competitors who kept their eyes open. Postural control results from visually impaired athletes showed lower performance levels, highlighting the fundamental importance of visual information in maintaining balance and posture. Schwesig et al. (2011) demonstrated that proprioception and vestibular feedback alone did not compensate for the lack of visual sensory input. A heightened sense development in alternative senses among visually impaired individuals did not lead to postural control abilities matching those of sighted athletes (Ponchillia et al. 2002).

The study revealed how visual information plays a vital role in maintaining postural control yet demonstrated how visually impaired athletes adapt their movements to handle their sport-specific challenges. These modifications do not fully address the lack of visual input, revealing specific performance issues visually impaired athletes face in competitive situations. The research indicates that alternative approaches do not fully substitute the performance advantages that vision offers in sports activities.

4.3 *Leg Muscle Power and Flexibility Versus Accuracy Shooting Goals*

The findings of this study revealed that national blind football athletes showed no correlation between their leg muscle power and flexibility and their ability to hit the goal. The findings opposed previous studies by Burhaein et al. (2020) and Soniawan (2021), which stressed that explosive leg muscle power and flexibility determine shooting accuracy in football athletes. The analysis by Budiharjo et al. (2018) and Hidayati et al. (2023) revealed that football athletes experienced better kicking after their leg muscle power and flexibility improved.

According to Aminudin et al. (2020) and Arjunnaja et al. (2022), research showed better muscle strength of the legs, enhanced hip flexibility, dynamic balance, and improved football shooting accuracy. The research team demonstrated that physical elements are critical for enhancing goal-scoring performance. The researchers indicated that strong leg muscles, motion balance, and flexible hips determine how precise goal-shooting techniques become.

The distinct characteristics of blind football explain why no significant connection emerged between leg muscle power flexibility and shooting accuracy in this research. The distinctive characteristics of blind football, including sensory input dependence and technical variations with specialized training protocols, probably reshape how leg muscle power and flexibility influence the sport compared to sighted football. The study findings indicated that blind athletes develop alternative strategies to overcome visual limitations, which results in performance factors that differ from those of sighted athletes.

5 Conclusions and Recommendations

This study revealed distinct insight into the relationship between leg muscle power and flexibility and their relationship to shooting accuracy among national blind football athletes, thus contradicting standard assumptions from sighted football. The research showed that leg muscle power alongside flexibility did not affect shooting accuracy across blind football athletes. Blind football demands athletes to depend on auditory feedback and spatial understanding because these elements form the basis for optimal performance. Visual-impaired athletes modify their playing techniques by developing timing skills, coordination, and spatial awareness capabilities that utilize their non-visual senses of hearing and touch. Postural control adaptations between sighted and blind football players help explain the lack of measurable impact that flexibility training had on shooting accuracy.

Research in blind football performance excellence must proceed by studying how visually impaired athletes respectively use auditory and tactile perceptions to optimize their capabilities. A comparison between visual and non-visual football athletes would explain how the lack of visual feedback affects posture control and balance capabilities. Research on diverse training methods used by blind athletes will deliver essential information about enhancing technique and coordination since these elements directly determine shooting accuracy. Analyzing skill and adaptation development through time in longitudinal research would advance our understanding of athlete advancement. The development of training programs for blind football athletes would benefit from studies that investigate psychological aspects such as spatial cognition and cognitive mapping. Future research combining these areas will substantially improve training methods that produce better outcomes for visually impaired athletes.

References

Almizan MY, Desman MA, Ilham M (2020) Explosive power exercises models on shooting ability on football extracurricular students at SMAN 2 Lubuk Basuk Agam

Aminudin A, Sugiyanto S, Liskustyawati H (2020) Contribution leg muscle strength, dynamic balance and hip joint flexibility to the accuracy of football shooting. Bp Int Res Crit Linguist Educ (BirLE) J 3(2):912–918. https://doi.org/10.33258/birle.v3i2.985

Arjunnaja A, Asnawi S, Pamungkas AT, Letran MN (2022) Analisis gerakan tendangan instep pada atlet sepakbola dalam peningkatan performa. Sepakbola 2(1):18. https://doi.org/10.33292/sepakbola.v2i1.148

Bahenský P, Fry A, Novak D (2020) Explosive leg power in elite youth soccer players: relationships with performance in jumping and sprinting. Int J Environ Res Public Health 17(6):2145

Bell W, Dutton J, Ian D (2010) Leg power and functional performance in elite soccer athletes: A positional analysis. J Sports Sci Med 9(1):32–37. https://www.jssm.org/

Budiharjo B, Florentinus TS, Sugiharto M, Soegiyanto KS (2018) The effectiveness of power of leg muscles and flexibility of leg muscles to improve lofted kick method. Adv Soc Sci Educ HumIties Res (ASSEHR) 247

Burhaein E, Ibrahim BK, Pavlovic R (2020) The relationship of limb muscle power, balance, and coordination with instep shooting ability: a correlation study in under-18 football athletes. Int J Hum Mov Sport Sci 8(5):265–270. https://doi.org/10.13189/saj.2020.080515

Deprez D, Fransen J, Boone J, Lenoir M, Philippaerts R, Vaeyens R (2015) Characteristics of high-level youth soccer players: variation by playing position. J Sports Sci 33(3):243–254

França C, Gouveia ÉR, Caldeira R, Marques A, Martins J, Lopes H, Costa M, Ihle A (2021) Predictors of explosive leg power in elite football players. J Phys Educ Sport 21(6):3599–3605. https://doi.org/10.7752/jpes.2021.06486

Gamonales Puerto JM, Muñoz Jiménez J, León Guzmán K, Ibáñez Godoy SJ (2018) Efficacy of shots on goal in football for the visually impaired. Int J Perform Anal Sport 18(3):393–409. https://doi.org/10.1080/24748668.2018.1475194

Gardasevic J, Bjelica D, Vasiljevic I (2017) Muscle strength and power differences among football players and its impact on kicking strength. J Phys Educ Sport 17(4):2181–2185

Hidayati AN, Sulaiman S, Hartono M (2023) Contribution of limb muscle power, flexibility of limb muscles and self-confidence to the accuracy of futsal shooting. JUARA : Jurnal Olahraga 8(1):100–110. https://doi.org/10.33222/juara.v8i1.2595

Ishak S, Hasan H, Rum AZM (2024) A comparative assessment of static balance among malaysian blind football athletes. Lecture Notes in Bioengineering, 23-31. https://doi.org/10.1007/978-981-97-3741-3_3

Katis A, Giannadakis E, Kannas T, Amiridis I, Kellis E, Lees A (2013) Mechanisms that influence accuracy of the soccer kick. J Electromyogr Kinesiol 23(1):125–131. https://doi.org/10.1016/j.jelekin.2012.08.020

Mackenzie R, Cushion C (2013) Performance analysis in football: a critical review and implications for future research. J Sports Sci 31(6):639–676. https://doi.org/10.1080/02640414.2012.746720

Mann DL, Ravensbergen HJC (2018) International paralympic committee (IPC) and International blind sports federation (IBSA) joint position stand on the sport-specific classification of athletes with vision impairment. Sports Med 48(9):2011–2023. https://doi.org/10.1007/s40279-018-0949-6

Nascimento MM, Silva PRO, Felício LR, Mainenti MRM, Miranda HL, Paz GA, Lemos T, de Sá Ferreira A, Vigário PS (2022) Postural control in football players with vision impairment: Effect of sports adaptation or visual input restriction? Motriz. Revista de Educacao Fisica, 28. https://doi.org/10.1590/S1980-657420220010821

Nurseto F, Suwarli S, Cahyadi A (2019) The effect of balance and strength of muscle leg on the accuracy of kicking a ball towards the goal in the student football university of lampung. Qual Sport 5(4):13–20. https://doi.org/10.12775/qs.2019.020

Oliver JL, Ramachandran AK, Singh U, Ramirez-Campillo R, Lloyd RS (2024) The effects of strength, plyometric and combined training on strength, power and speed characteristics in high-level, highly trained male youth soccer players: a systematic review and meta-analysis. Sports Med 54(3):623–643. Springer Science and Business Media Deutschland GmbH. https://doi.org/10.1007/s40279-023-01944-8

Ponchillia PE, Strause B, Ponchillia SV (2002) Athletes with visual impairments: attributes and sports participation. J vis Impair Blind 96(4):267–272. https://doi.org/10.1177/0145482x0209600408

Rules And G (2022) International Blind Sports Federation. IBSA goalball rules and regulations, 2022–2024

Schwesig R, Goldich Y, Hahn A, Müller A, Kohen-Raz R, Kluttig A, Morad Y (2011) Postural control in subjects with visual impairment. Eur J Ophthalmol 21(3):303–309. https://doi.org/10.5301/EJO.2010.5504

Soniawan V (2021) A contribution of leg muscle explosion power and flexibility to football shooting accuracy

Wijaya DH (2020) The relationship of leg muscle explosive power and students straddle style results at Universitas Riau. https://doi.org/10.1088/1742

Zul RH, Nasuka N, Priyono B (2022) The effect of training methods and power on shooting accuracy of porma football school players in Kudus Regency. J Phys Educ Sports 11(3):315–321. https://journal.unnes.ac.id/sju/index.php/jpes

A Study on the Muscle Performance of Drivers During Long Versus Short-Cornered Course Using Electromyography (EMG)

Muhammad Izzdihar Mahadzir, Mohammad Azzeim Mat Jusoh, Ahmad Khushairy Makhtar, and Sukarnur Che Abdullah

Abstract Driving posture is an important ergonomic criterion for ensuring the comfort of the driver. Long periods of driving can cause significant strain on body posture, especially in moving vehicles due to external factors such as motion and vibration. Incorrect driving posture can also lead to discomfort in the shoulders and arms. The purpose of this study is to monitor the activation of the biceps brachii muscle during driving in two distinct postures and two driving styles (short-corner vs. long-corner). The two steering wheel positions used in the experiment were the stretched position (posture A) and the lean-back position (posture B). Five males with two years of driving experience participated in the experiment. The results indicate that participants exhibited higher muscle activation in posture B. This posture showed a high value of RMS, suggesting that individuals use more muscle when in this position. Further research is needed to better understand how long-distance driving affects the activation of specific muscles, such as the biceps brachii. Additionally, more strategies are required to reduce the risk of skeletal and muscular problems in drivers. These results are especially beneficial for academicians and designers in the automotive sector, providing guidelines to develop an ideal sitting posture for all users.

Keyword Surface Electromyography (sEMG) · Biceps brachii · Driving posture

1 Introduction

Driving for extended periods can lead to neuromuscular fatigue and discomfort, especially in the biceps brachii muscle, when there is little opportunity to change driving positions. In Malaysia, consumers rely on cars for daily transportation since

M. I. Mahadzir · M. A. M. Jusoh (✉) · A. K. Makhtar · S. C. Abdullah
School of Mechanical Engineering, College of Engineering, Universiti Teknologi MARA (UiTM) Shah Alam, Shah Alam, Selangor, Malaysia
e-mail: m_azzeim@uitm.edu.my

M. H. A. Hassan et al. (eds.), *Proceedings of the 3rd International Conference on Innovation and Technology in Sports*, Lecture Notes in Bioengineering,
https://doi.org/10.1007/978-981-95-0804-4_25

public transport is not as efficient as in other countries. When a person is driving, ergonomics is the most important factor to consider, as it directly impacts health and safety (Gao et al. 2022). Incorrect sitting posture while driving can result in severe muscle fatigue, particularly in the biceps brachii. The risk of developing this fatigue is influenced by several ergonomic factors in the driving environment. From a designer's perspective, it is critical to address these ergonomic elements to improve driver comfort and reduce the likelihood of musculoskeletal problems, such as lower back pain (Mohamad et al. 2010).

The level of comfort in the driving environment plays a crucial role in preventing fatigue, which is a significant factor contributing to microsleep episodes. Uncomfortable seating, poor ergonomic design, and a lack of support can lead to driver fatigue, increasing the likelihood of microsleep. As a result, modern vehicles are often equipped with adjustable seats, lumbar support features, and improved ergonomic design to promote a more comfortable and supportive driving experience.

Ensuring comfort while driving is important for several reasons, encompassing both the well-being of the driver and overall road safety. Prioritizing comfort is not only about individual well-being; it also contributes to safer roads and a more enjoyable driving experience. Modern vehicle design, incorporating ergonomic features and advanced technologies, reflects an industry-wide commitment to enhancing driver comfort and safety (Lecocq et al. 2020). By addressing both physical and mental aspects of comfort, the driving experience becomes more enjoyable, ensuring that drivers are better equipped to navigate the road safely and with greater satisfaction.

The aim of this study is to identify the muscle activation of the biceps brachii when a driver is speeding in two different conditions: a long-cornered course and a short-cornered course, with two different body postures, referred to as Posture A and Posture B. By identifying the highest muscle activation, we can determine the best sitting posture for drivers. Additionally, an electromyography (EMG) device will be used to assess muscle fatigue, helping us identify the most suitable driving position (Kyung and Nussbaum 2009).

2 Methodology

2.1 Participants

In this experiment, five male drivers with more than two years of driving experience and in good health participated. The participants were instructed to have enough rest before the experiment began. The average age of the participants ranged from 22 to 28 years. Informed consent was obtained from each participant before the experiment began. Criteria of the test subject is shown in Table 1.

The experiment was conducted in the Ergonomic Laboratory at Universiti Teknologi MARA (UiTM), where the driving simulator prototype was located (Refer

Table 1 Characteristics of the participants

Gender	Male
Age	22–28 years old
Medical history	• Active in healthy lifestyle • Have no critical injury issue

Fig. 1). The laboratory maintained an ambient temperature of 21.7 °C and humidity of 58%. The driving simulator was fully equipped with standard automotive interfaces, including a steering wheel, a car seat with a seatbelt, a brake pedal, and a gas pedal. This setting was chosen because it minimizes disturbances such as heat, noise, and vibration, providing a controlled environment for the experiment. The steering wheel is fixed in position, whereas the seat position and inclination are adjustable. The raw data from the EMG signals will be processed using a software called ProEMG. Figure 1 shows the setup for the sEMG devices.

Preparing the skin for electrode placement on the biceps brachii for electromyography (EMG) involves several steps to ensure accurate and reliable readings. First, clean the skin thoroughly with soap and water to remove any dirt, oils, and lotions. Then, use an alcohol wipe to ensure the area is completely clean. Allow the skin to air dry completely to avoid any alcohol residue, which can affect adhesion and conductivity. Next, gently exfoliate the skin using an abrasive gel or pad to remove the top layer of dead skin cells, which helps reduce skin impedance. If there is excessive hair on the biceps, trim or shave it carefully to improve electrode contact, being mindful to avoid causing skin irritation or cuts. Ensure that the skin is intact and free from any cuts, rashes, or irritation before proceeding with electrode placement (Simonetti et al. 2022).

For the electrode placement on the biceps brachii, follow the specific guidelines for positioning. Place the electrodes over the muscle belly of the biceps, ensuring they are aligned with the muscle fibers. Press the electrodes firmly onto the skin to ensure good adhesion and reduce movement artifacts; use medical tape if necessary to secure the electrodes further. Once the electrodes are in place, use the EMG machine to check the impedance of the electrode–skin interface, aiming for an impedance

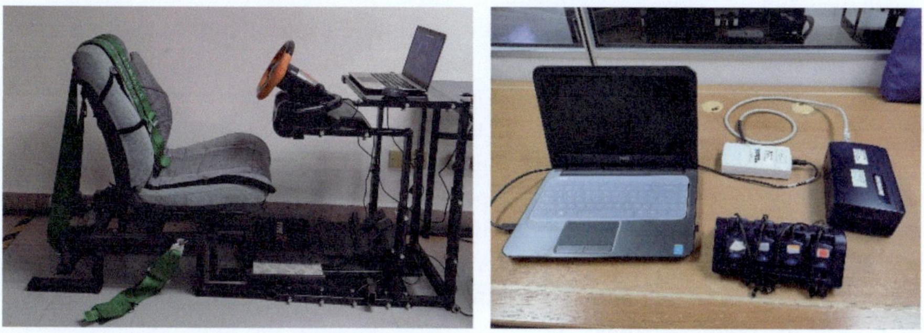

Fig. 1 Driving simulator and EMG set-up

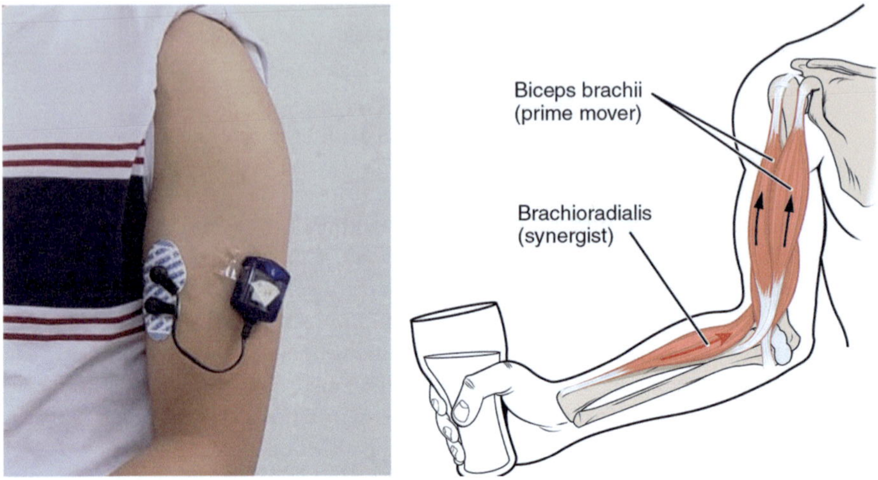

Fig. 2 Electrode placement for biceps brachii

below 10 kΩ (Soltani et al. 2024). If the impedance is too high, re-clean the skin and adjust the electrodes as needed. Figure 2 shows the electrode placement on specified muscle.

3 Experimental Procedure

The experiment was conducted in the Ergonomic Lab of the School of Mechanical Engineering at Universiti Teknologi MARA (UiTM) Shah Alam. Before the experiment began, all participants were briefed on the proper procedures. It was also important for the participants to have enough rest and not feel tired before the experiment. Following the regulations was crucial to avoid any disturbances during the experiment.

The Maximum Voluntary Contraction (MVC) for the biceps brachii refers to the highest amount of force the muscle can generate during a single maximal effort against resistance. Participants underwent an exercise involving lifting a dumbbell to maximize muscle engagement. Three trials of MVC were performed, each lasting 7 s, with 1 min of rest between each set to minimize muscle fatigue (Refer Fig. 3).

Posture A requires participants to place their hands close to the steering wheel, while in Posture B, their hands should be positioned further away. The experiment will begin with participants moving the steering wheel based on a 2-min video played for the long-cornered course in Posture A. After a 2-min rest, participants will continue with the short-cornered course in Posture A. The same procedure will be repeated for Posture B. Figure 4 shows the comparison between the long-cornered versus

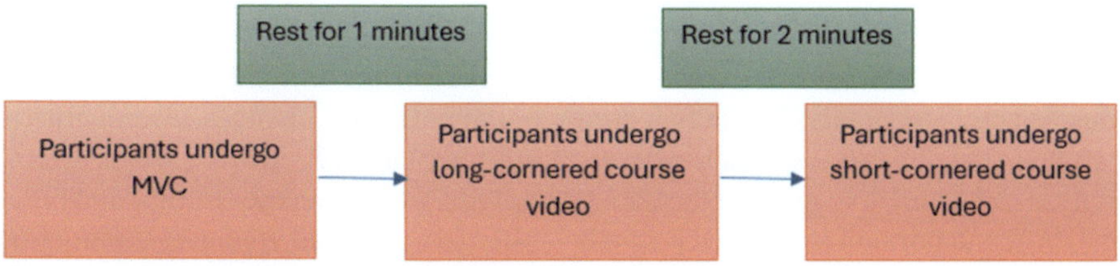

Fig. 3 Activity of the participants

short-cornered course map. Whereas Table 2 describes the criteria between the two specified postures.

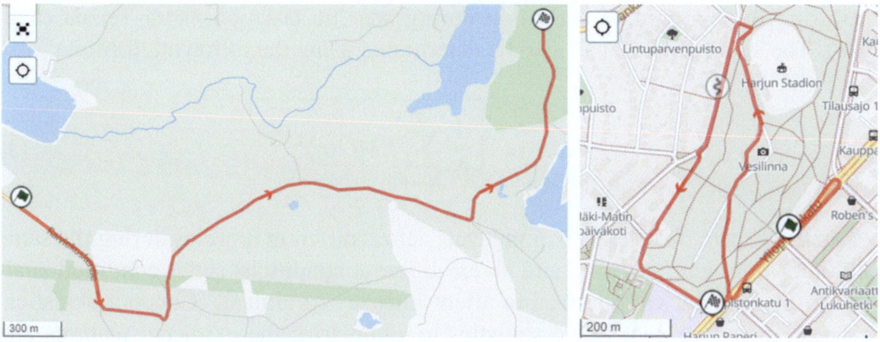

Fig. 4 Long-cornered course map versus short-cornered course map

Table 2 Posture position when using the driving simulator

Posture A	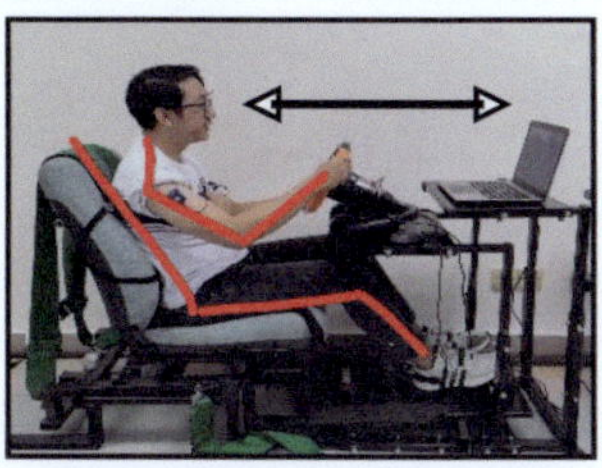	• Participants were instructed to sit in comfortable driving positions • The seat back angle should be 110° from the seat base • The distance between the steering wheel and the chest is 0.25 m
Posture B		• Participants were instructed to sit slightly farther from the steering wheel • The seat back angle should be 135° from the seat base • The distance between the steering wheel and the chest is 0.40 m

3.1 Data Collection Process

Muscle activation analysis using electromyography (EMG) involves several critical steps to accurately assess muscle activity. A software called ProEMG 2.0 was used to analyze the data. First, prepare the data by applying filters to remove noise and baseline drift, and normalize the EMG signals—often to a maximal voluntary contraction (MVC) value—for comparison. Segment the EMG signal into relevant phases or time periods corresponding to specific tasks or movements, using event markers to define the start and end of each phase (Avdan et al. 2024).

Next, feature extraction is performed, where the root mean square (RMS) of the EMG signal is calculated to quantify muscle activation, and frequency domain analysis is conducted to understand muscle fatigue and activation patterns. Timing analysis includes determining the onset and offset times of muscle activation and measuring the duration of activation during specific tasks (Amador-Tejada et al. 2024). The value of RMS can also be determined using the following formula:

$$RMS = \sqrt{\frac{1}{N}\sum_{N=1}^{N} EMG(n)^2} \tag{1}$$

Next, compare muscle activation levels across different muscles during the same task (inter-muscular analysis) and under different conditions (inter-condition analysis) to assess how various factors affect muscle recruitment. Conduct statistical analyses, including descriptive statistics to summarize the extracted features and inferential statistics to identify significant differences or relationships in the data. Finally, interpret the results in the context of the specific task or movement, leading to clinical or performance insights that can inform interventions, rehabilitation strategies, or athletic performance enhancements. This comprehensive approach ensures a thorough understanding of muscle function and coordination through EMG analysis (Mira et al. 2021).

4 Results and Discussion

4.1 RMS Signal

Converting a raw EMG signal to its root mean square (RMS) value involves several steps to effectively quantify muscle activity. First, the raw EMG signal is typically pre-processed by applying a band-pass filter to remove noise and artifacts, ensuring the signal is centered around zero. In the ProEMG software, the RMS value was automatically calculated (Amador-Tejada et al. 2024). Figure 5 shows a sample of the RMS signal from one of the participants.

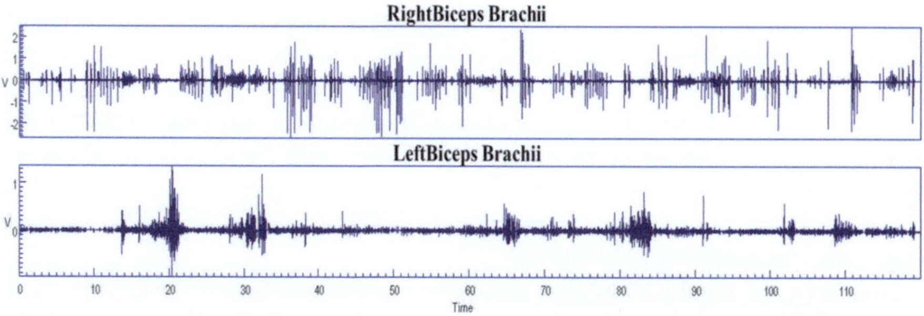

Fig. 5 Sample of the RMS signal data for participants 3 (voltage versus time in seconds)

The average RMS value for maximum voluntary contraction (MVC) shows that the RMS values differ between the right and left biceps brachii (Refer Fig. 6). The data indicates that the RMS value was higher for the dominant hand used by the participant. Four out of five participants were right-handed, while one participant was left-handed. Performing an MVC before using EMG is essential for normalizing the data. Raw EMG signals can vary greatly between individuals and muscles due to differences in electrode placement, skin conductivity, and muscle properties. Normalizing EMG data to a percentage of MVC allows for standardized measurements, reducing variability and improving the reliability of comparisons across different sessions, muscles, and individuals.

MVC serves as a benchmark for evaluating muscle strength and endurance. It helps determine a muscle's capacity to generate force under maximal effort, providing a baseline for assessing muscle performance and fatigue over time. This is crucial for tracking changes in muscle function, monitoring recovery, and evaluating the effectiveness of training and rehabilitation programs (Avdan et al. 2024).

The RMS values for long-cornered and short-cornered postures (Posture A and Posture B) were also considered. As mentioned earlier, each participant had a higher

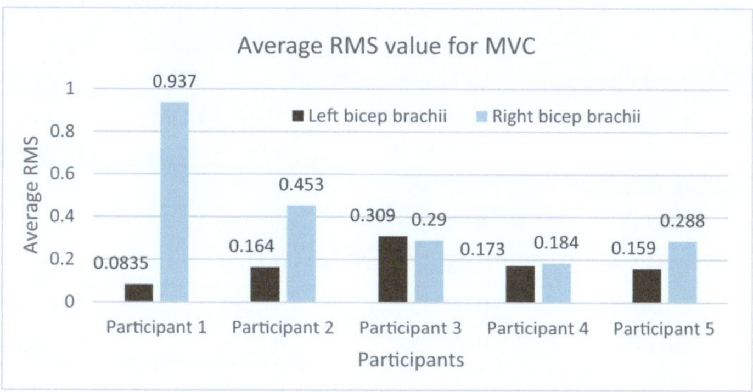

Fig. 6 RMS value for MVC

RMS value for their dominant hand. The RMS value offers insights into the signal's power and is particularly useful because it is less sensitive to outliers compared to other amplitude measures, making it a reliable indicator of overall muscle activation.

From Table 3, the data above shows that each participant had a higher RMS value for Posture B (far from the steering wheel) when navigating the long-cornered course, indicating that they used more muscle in Posture B. For Participant 1, the RMS value reached 0.31785 V in Posture B, compared to 0.2251 V in Posture A. This suggests higher muscle activation in Posture B.

Similarly, the results show that muscle activation was higher during the short-cornered course in Posture B. The figure above demonstrates that Posture B (far from the steering wheel) resulted in greater muscle activity compared to Posture A (close to the steering wheel). All five participants showed higher muscular activation in Posture B, which aligns with the expected results.

Participant 1 had the highest RMS value for Posture B compared to the other participants, with a value of 0.31785 V for the long-cornered course and 0.41285 V for the short-cornered course. These results may be influenced by the participant's characteristics, such as greater muscle mass, which could generate more electrical activity in the muscle. In contrast, Participant 4 had the lowest RMS value, possibly due to technical issues or lower muscle mass. A lack of sufficient rest may also have contributed to the lower RMS value compared to the other participants. Visuals is shown in Fig. 7.

Overall, the RMS values for the short-cornered course were higher for all participants because this course involves more movement compared to the long-cornered course. The biceps brachii muscle exhibited the lowest RMS value in the posture closest to the steering wheel. This suggests that, to reduce muscle activity, particularly in the biceps area, a driver should not sit too far from the steering wheel. Discomfort experienced while steering was found to be correlated with muscle activity. As a result, a closer position to the steering wheel is preferable to a farther position. Muscle activation can be influenced by several factors, such as steering motion and hand positioning on the wheel. Proper hand positioning can also help reduce muscle soreness.

Table 3 RMS value for long cornered and short cornered

Participants	Long cornered posture A	Long cornered posture B	Short cornered posture A	Short cornered posture B
1	0.2251	0.31785	0.29055	0.41285
2	0.1774	0.1866	0.2273	0.239
3	0.07765	0.0855	0.1105	0.1218
4	0.03085	0.03315	0.0291	0.03785
5	0.1146	0.16355	0.1519	0.2585

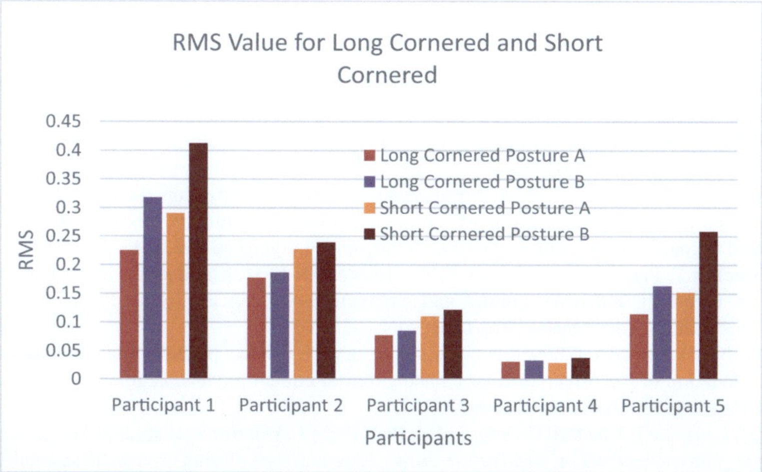

Fig. 7 RMS value for long cornered and short cornered

5 Conclusion

With the help of ProEMG software, all the necessary data was obtained, and the main objectives of the experiment were achieved, which was to determine muscle activation for both postures across different driving courses. The results show that sitting far from the steering wheel increases muscle activity, potentially leading to muscle fatigue in drivers. It is crucial to focus on comfort while driving to prevent accidents. This experiment also highlights the importance of sitting comfortably to avoid excessive muscle use. For more accurate results, further research should be conducted, as this experiment was performed in a simulator rather than in real-life conditions. Including more subjects is essential for improving the accuracy and reliability of the results. Also, consideration of female test subjects is also expected. A larger sample size increases statistical power, reducing the risk of overlooking significant differences or effects. It also ensures that findings are more generalizable, accounting for individual variability and minimizing bias. Overall, more participants lead to more dependable and widely applicable conclusions.

As a recommendation, the automotive industry should prioritize designing car seats with ergonomics in mind. To enhance seat comfort and functionality, seats should feature adjustable lumbar support, multi-way adjustments for height, angle, and depth, and reinforced side bolsters for proper posture. High-quality, breathable, and durable materials, such as memory foam or gel inserts, should be used to reduce pressure points (Sunagawa et al. 2023). Additionally, it is important for individuals to get enough rest to prevent muscle fatigue while driving. It is also recommended to study other muscles involved in driving, so drivers can be aware of the areas they need to focus on while driving.

References

Amador-Tejada A, McGillivray JE, Kumbhare DA, Noseworthy MD (2024) Denoising of the gradient artifact present in simultaneous studies of muscle blood oxygen level dependent (BOLD) signal and electromyography (EMG). Magn Reson Imaging 111:179–185. https://doi.org/10.1016/J.MRI.2024.05.004

Avdan G, Chen CB, Onal S (2024) An alternative EMG normalization method: heterogeneous recurrence quantification analysis of isometric maximum voluntary contraction movements. Biomed Signal Process Control 93:106219. https://doi.org/10.1016/J.BSPC.2024.106219

Gao Y, Fischer T, Paternoster S, Kaiser R, Paternoster FK (2022) Evaluating lower body driving posture regarding gas pedal control and emergency braking: a pilot study. Int J Ind Ergon 91:103357. https://doi.org/10.1016/J.ERGON.2022.103357

Kyung G, Nussbaum MA (2009) Specifying comfortable driving postures for ergonomic design and evaluation of the driver workspace using digital human models. Ergonomics 52:939–953. https://doi.org/10.1080/00140130902763552

Lecocq M, Lantoine P, Bougard C, Allègre JM, Bauvineau L, Bourdin C et al (2020) Neuromuscular fatigue profiles depends on seat feature during long duration driving on a static simulator. Appl Ergon 87. https://doi.org/10.1016/j.apergo.2020.103118

Mira RM, Molinari Tosatti L, Sacco M, Scano A (2021) Detailed characterization of physiological EMG activations and directional tuning of upper-limb and trunk muscles in point-to-point reaching movements. Curr Res Physiol 4:60–72. https://doi.org/10.1016/j.crphys.2021.02.005

Mohamad D, Deros BM, Wahab DA, Daruis DDI (2010) Integration of comfort into a driver's car seat design using image analysis

Simonetti D, Koopman B, Sartori M (2022) Automated estimation of ankle muscle EMG envelopes and resulting plantar-dorsi flexion torque from 64 garment-embedded electrodes uniformly distributed around the human leg. J Electromyogr Kinesiol 67:102701. https://doi.org/10.1016/J.JELEKIN.2022.102701

Soltani M, Shafiei M, Ariabod A, Farhani F, Abbasi A (2024) Acute effects of blood flow restriction on EMG activity of arm muscles during karate tsuki strike in young men. J Bodyw Mov Ther 40:482–485. https://doi.org/10.1016/J.JBMT.2024.05.005

Sunagawa M, Shikii S ichi, Beck A, Kek KJ, Yoshioka M (2023) Analysis of the effect of thermal comfort on driver drowsiness progress with predicted mean vote: an experiment using real highway driving conditions. Transp Res Part F Traffic Psychol Behav 94:517–27. https://doi.org/10.1016/j.trf.2023.03.009

Early Phase Development of Training Efficiency Application for Seni Silat Cekak Malaysia Using EEG Signal

Aznor Hanah Abdul Halim, Rohani Farook, Fakhrul Mukmin Mansor, M. F. Mohd Fairuz, Siti Azura Abuzar, A. R. Mohamad Rizal, and Suhizaz Sudin

Abstract This paper uses the Emotiv Epoc to analyze the EEG signal of Seni Silat Cekak Malaysia (SSCM) practitioners towards the development of application in determining training efficiency. SSCM students currently lack a method for measuring Silat Cekak Malaysia Repetition training efficiency and rely on normal instructor-led training without knowing if their training achieves efficiency. In this project, the methodology for finding the efficiency using EEG signal and Standard of Procedure (SOP) in data collection was explored. SOP in collecting data was created and enrolled in the collection of data for sampling and explained with the result obtained. Further, the data will be analyzed using the AI technique to find the optimum technique to be used in recognizing the Silat Cekak Malaysia Training Efficiency.

Keywords Training efficiency application · EEG signal · Seni Silat Cekak Malaysia

A. H. Abdul Halim (✉) · R. Farook · F. M. Mansor · M. F. Mohd Fairuz · S. A. Abuzar · A. R. Mohamad Rizal · S. Sudin
Faculty of Electronic Engineering and Technology (FKTEN), Universiti Malaysia Perlis, Arau, Malaysia
e-mail: aznor@unimap.edu.my

A. H. Abdul Halim · R. Farook · M. F. Mohd Fairuz · S. A. Abuzar · A. R. Mohamad Rizal · S. Sudin
Sports Engineering Research Center (SERC), Universiti Malaysia Perlis, Arau, Malaysia

A. H. Abdul Halim · R. Farook · F. M. Mansor
Persatuan Seni Silat Cekak Malaysia (PSSCM), Kuala Lumpur, Malaysia

M. H. A. Hassan et al. (eds.), *Proceedings of the 3rd International Conference on Innovation and Technology in Sports*, Lecture Notes in Bioengineering,
https://doi.org/10.1007/978-981-95-0804-4_26

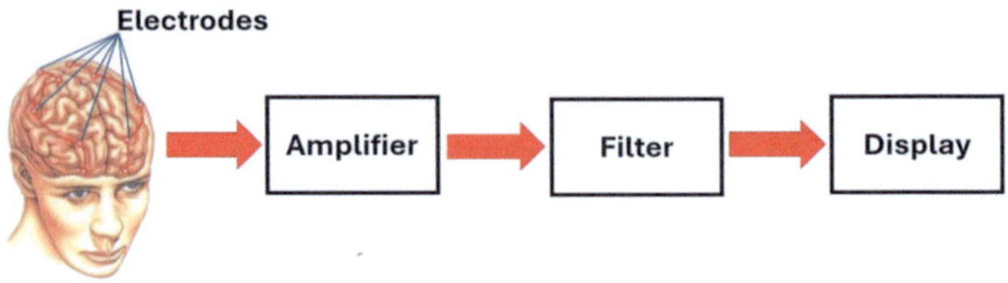

Fig. 1 The flow of electrical signals from the brain is being amplified, filtered and displayed

1 Introduction

1.1 EEG Signal

EEG, or electroencephalography, is a non-invasive method for measuring electrical brain activity by placing electrodes on the scalp and using computer software to capture synchronized neural firing. This provides insights into cognitive processes and states such as attention, arousal, and motor function. EEG detects and measures electrical activity resulting from the flow of ionic currents in the neurons of the brain. The procedure involves placing electrodes along the scalp to detect electrical signals that originate from the myelin-covered pathways of the cranial nerve axes.

The discovery of electrical phenomena in the brains of animals was made by Richard Caton in 1875, while Hans Berger recorded the first human EEG in 1925. One advantage of EEG is its non-invasiveness, as it does not require the implantation of equipment into the human body. However, this also means that the electrical signals conducted from the brain to the electrodes are relatively weak due to several layers, such as skin and skull, between them. To address this, the acquired signals are amplified and filtered by a computer to remove noise. The resulting brainwaves are then displayed on a computer screen as shown in Fig. 1.

In muscle training, a study has shown that EEG provides an excellent tool with sensitivity to changes in the muscles associated with fatigue, enabling real-time monitoring as well (Rahman et al. 2019). Recent advancements have allowed researchers to explore the neural correlates of strength training, thereby enabling mental outcome ability change capacity constructions based on individualized cortical motor potentials associated with maximal isometric contractions (Rahman et al. 2019).

1.2 Seni Silat Cekak Malaysia

Seni Silat Cekak Malaysia (SSCM) is practiced under an association named Persatuan Seni Silat Cekak Malaysia (PSSCM). This Malay martial arts have 13 branches (Persatuan Seni Silat Cekak Malaysia 2022, 2024) SSCM experts offer PSSCM,

Fig. 2 *Latihan Tangan* example (left) and *Latihan Kaki* example (right)

which has a comprehensive syllabus, to most universities in Malaysia. It is called *Tenaga Pengajar.* SSCM have 21 *buah* and there are four level named *buah asas*, *buah jatuh*, *buah potong* and *tamat* (Halim et al. 2016; Mustapha et al. 2014a).

In the practical martial arts class, students engage in hand and leg exercises to improve coordination, concentration, and balance. They also practice fend-off techniques, such as Kaedah A (Mustapha et al. 2014a), B, C, and D, to improve muscle memory, timing, and execution. The final phase, *Latihan Buah Silat*, involves predetermined movements involving fend-off techniques and defensive strategies, enhancing precision and timing in complex combinations and transitions. Figure 2 displays the "Latihan Tangan" and "Latihan Kaki" examples.

1.3 Motivation

There is a lack of systems or applications developed for SSCM that measure training efficiency based on EEG signals. Most SSCM research analyses, such as those conducted by Mustapha et al. (Mustapha et al. 2014a, 2014b), focus on hand movements. Currently, SSCM students can only improve their efficiency through traditional training with an instructor, without knowing whether their training achieves the desired efficiency. By using EEG analysis, the goal is to determine the level of efficiency based on training repetitions, as the signals can provide insights into the cognitive and physiological effects of training. Therefore, this paper presents an EEG signal standard of procedure for the experiment on SSCM training repetitions and GUI design for the application, as an initial step towards developing an SSCM application.

2 Method

2.1 EEG Signal Using EMOTIV EPOC

Emotiv EPOC is a wireless headset developed by Australian electronics company Emotiv Systems, capable of receiving EEG signals from the human scalp, with sixteen electrodes for accurate positioning (EMOTIV 2024). The Emotiv Epoc EEG headset uses the international 10–20 electrode location system. This system includes electrode names such as AF3, F7, F3, FC5, T7, P7, O1, O2, P8, T8, FC6, F4, F8, and AF4. These electrode locations correspond to specific brain regions based on the 10–20 system. In this project, the electrode locations provided by the Emotiv Epoc headset can be used to capture EEG signals during Seni Silat Cekak Malaysia training repetitions. The selection of electrode locations should be based on brain regions involved in motor control, attention, and cognitive processing that are relevant to the project goal. Based on the international 10–20 electrode system commonly used in EEG studies, the following electrode locations will be used in the project:

- F3 and F4: These electrodes are positioned over the prefrontal cortex and are associated with attention, decision-making, and cognitive processes.
- FC5 and FC6: These electrodes are placed over the prefrontal and central regions and are related to motor control and movement planning.
- T7 and T8: These electrodes cover the temporal regions and are involved in auditory processing, memory, and language comprehension.
- P7 and P8: These electrodes are positioned over the parietal lobes and are associated with visuospatial processing and attention.
- O1 and O2: These electrodes are placed over the occipital lobes and are involved in visual processing.

Considering that Seni Silat Cekak Malaysia involves physical movements, attention, and cognitive processes, selecting electrode locations that cover the frontal, central, temporal, parietal, and occipital regions would provide a comprehensive assessment of brain activity during the training repetitions.

2.2 Experimental Procedure

During the pre-experiment, the subjects are briefly informed about the basic information related to the study and the procedure of each task as shown in Fig. 3. Participants in a study on Seni Silat Cekak Malaysia's training repetitions are required to follow standard operating procedures (SOP) to minimize external influences and artifacts in the EEG signals and are given an informed consent form to confirm their participation. The flowchart in Fig. 3 shows the summarization of the experimental protocol flow. After the consent letter is signed, the EEG device is set up. The subject was required to wear the Emotiv EPOC with 14 electrodes as shown in Fig. 4.

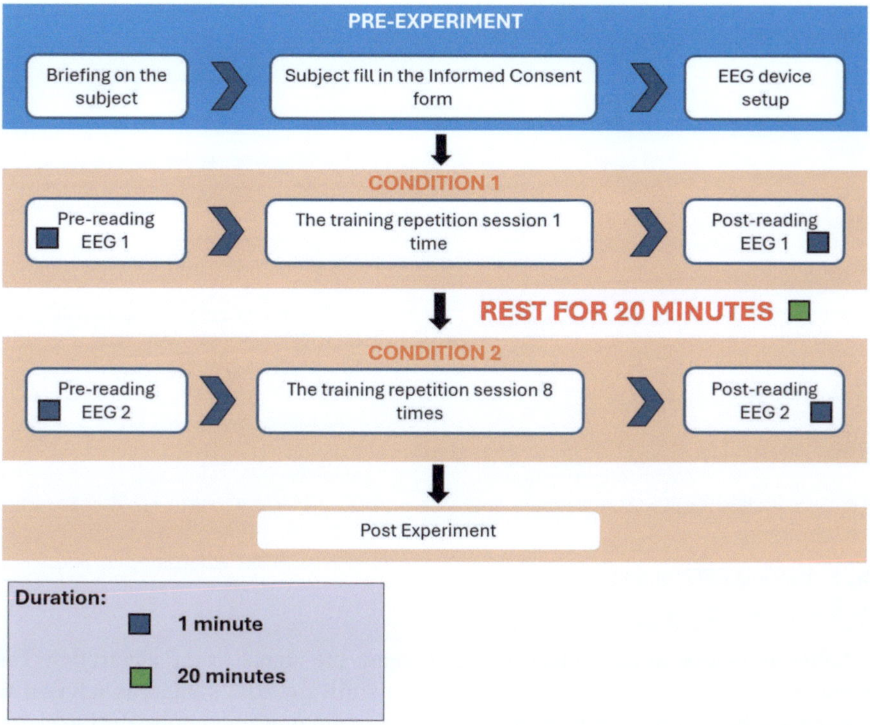

Fig. 3 SOP of the experimental process

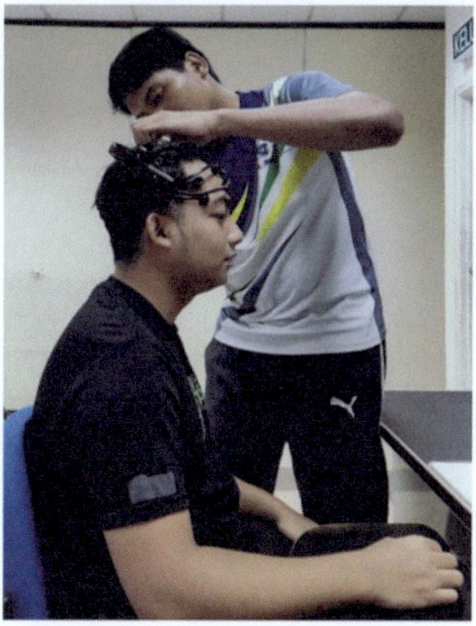

Fig. 4 Subject wear the device

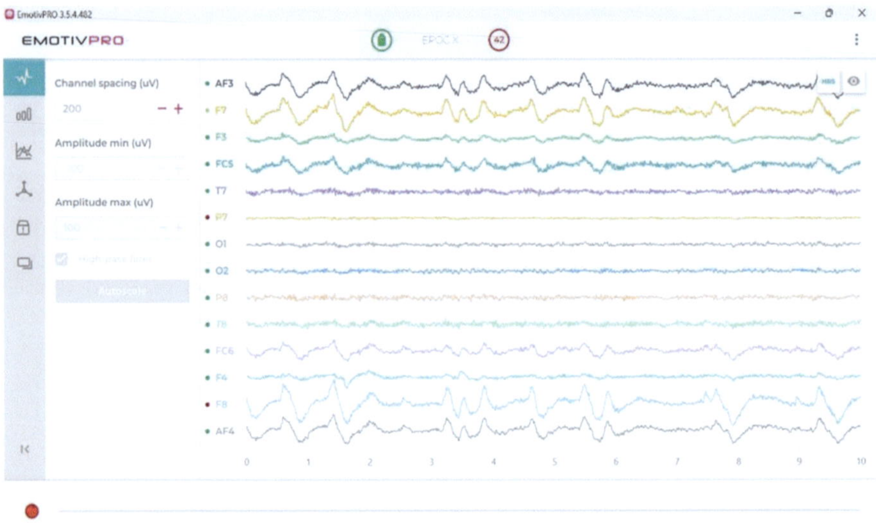

Fig. 5 Baseline EEG for Subject 1

Saline is placed in each electrode to increase the signal to the electrodes. The signal is displayed in the Emotiv Pro system software. The subject is advised to reduce their body movements or not to speak to avoid unwanted noise during signal recording. The protocol begins with a pre-experiment session in which subjects wear the Emotiv Epoc devices and are calibrated with the Emotiv PRO software to obtain the baseline EEG signal for one minute, as shown in Fig. 5.

2.3 Application Development

In application development, Android Studio was used to create software solutions, specifically mobile applications. Android Studio is a user-friendly interface for creating Android applications. Functionalities such as data comparison can be included during the final development phase. Firebase was used as a backend solution to store and access comparison data in real time. Firebase provides a real-time database that keeps data up to date and accessible across several devices. This project can construct strong applications with seamless data comparison and secure storage by integrating the advantages of Android Studio and Firebase.

3 Result

The experiment is divided into two conditions, namely Condition 1 and Condition 2, as listed in Fig. 3. In the first condition, participants performed a single repetition of Seni Silat Cekak Malaysia movements for data sets, collecting data immediately after without rest as shown in Fig. 4. This approach allows comparative analysis of multiple repetitions and single repetitions on brain activity, with each condition having a 5-min rest period.

In the second condition, the participants perform eight repetitions of Seni Silat Cekak Malaysia movements, collecting data immediately after each set to study brain activity in real-time. This process is repeated for three sets, providing a comprehensive data set for analysis. At the end of the sessions, the evaluation form was given to the subject. Figure 5 depicts the EEG baseline for one of the subjects following the process from SOP that lined up for the experiment. Figures 6 and 7 are the filtered alpha and beta bands for Condition 1 and Condition 2 as explained for the subject.

Later, the data collection phase will be conducted for 100 subjects following the SOP engraved. The raw EEG signals obtained from Seni Silat Cekak Malaysia training repetitions will undergo signal preprocessing to improve the quality of the data.

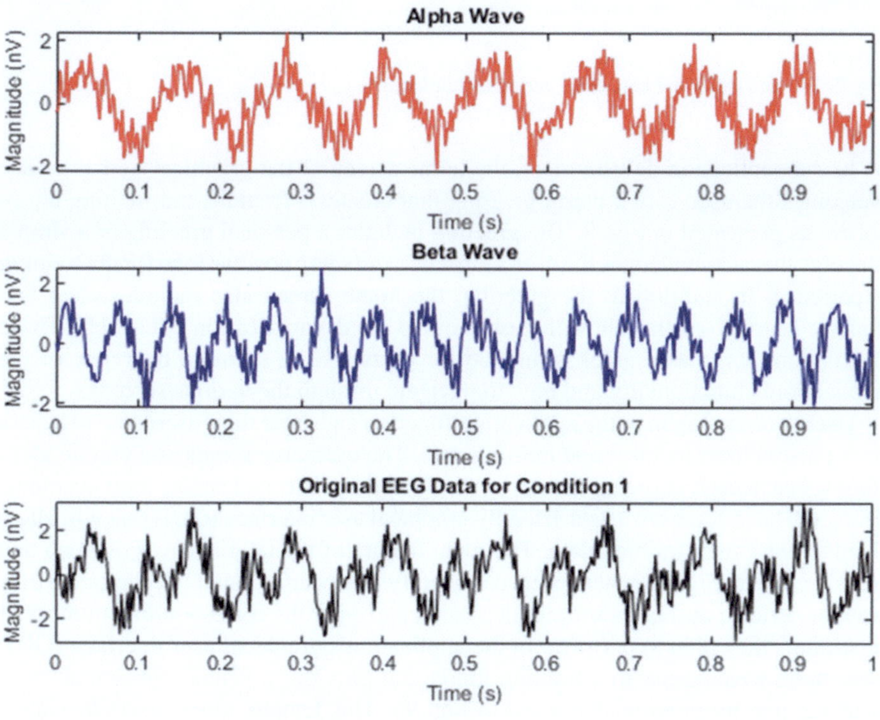

Fig. 6 Filtered alpha and beta band during Condition 1

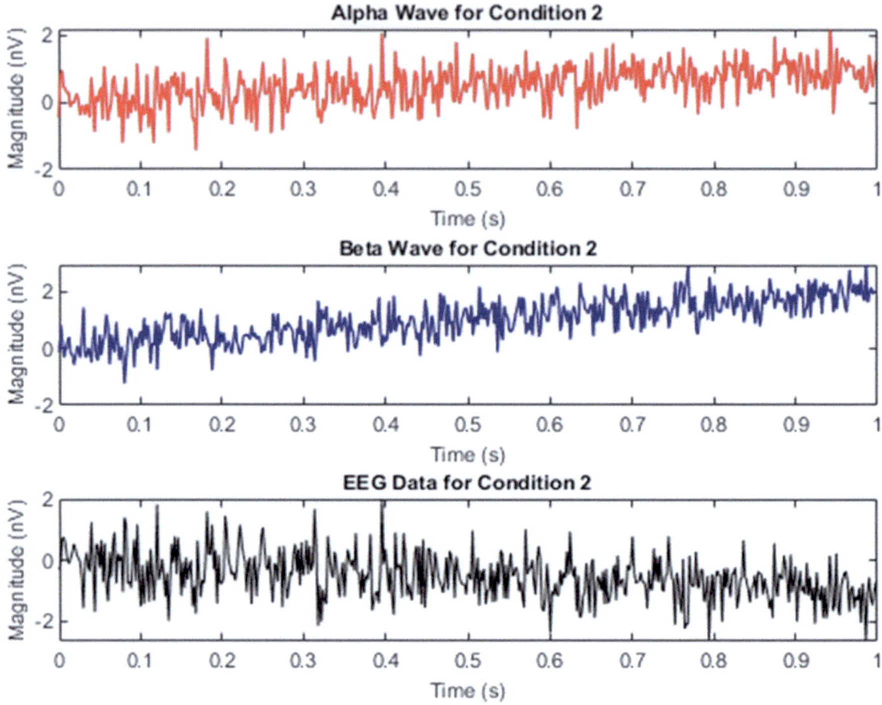

Fig. 7 Filtered alpha and beta band during Condition 2

In the application development, the home screen of the graphical user interface welcomes the user with a warm greeting that creates a friendly and inviting atmosphere as presented in Fig. 8. The greeting includes a personal greeting or a simple message that acknowledges the user's presence and sets a positive tone for the training experience. In addition to the greeting, the home screen also includes clear and concise instructions that show the user how to use the application effectively. These instructions serve as a quick reference for users, giving them an overview of the application's functionality and how to navigate through the various screens.

Users need to log in to the application function and enter their username to access their personalized features and training data. The username serves as a unique identifier for each user, allowing them to securely authenticate and access their accounts. users will have access to a user-friendly graphical user interface (GUI) that will allow them to enter their training data. The main feature of the GUI is the "Fill Buah and Reps" screen in Fig. 9a, where users can specify the specific "buah" (techniques) they want to perform during their training session and enter the corresponding number of repetitions they plan to perform. In the application's graphical user interface (GUI) also, users have access to a training graph that provides a visual representation of their training progress as displayed in Fig. 9b. This feature allows users to view a

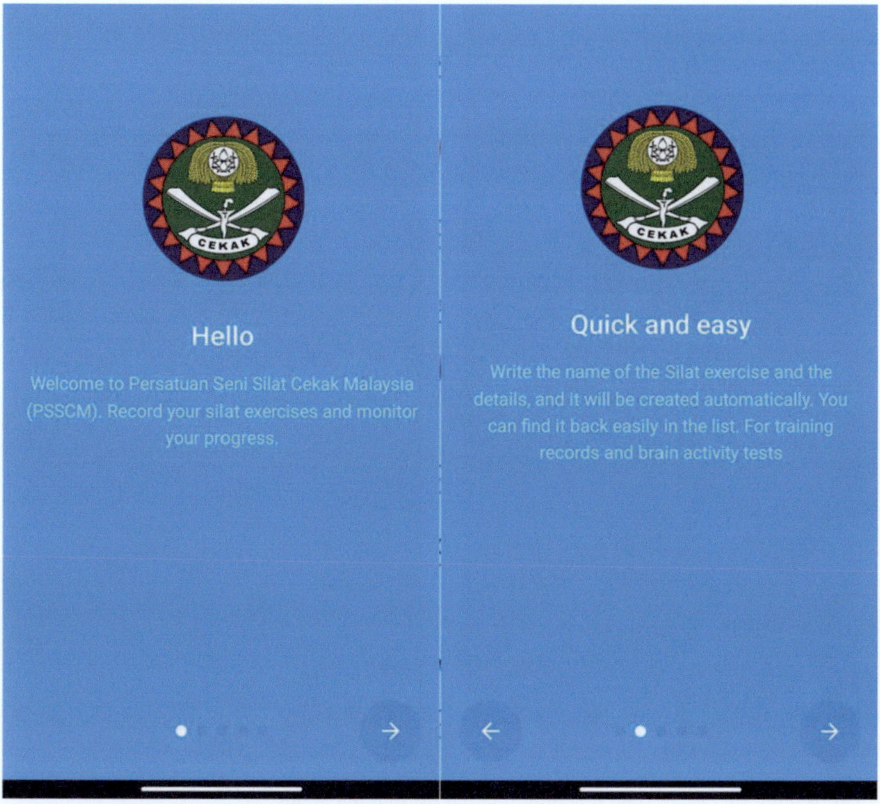

Fig. 8 Home screen GUI for SSCM application

graph representing the recorded EEG signals and relevant metrics related to their training sessions.

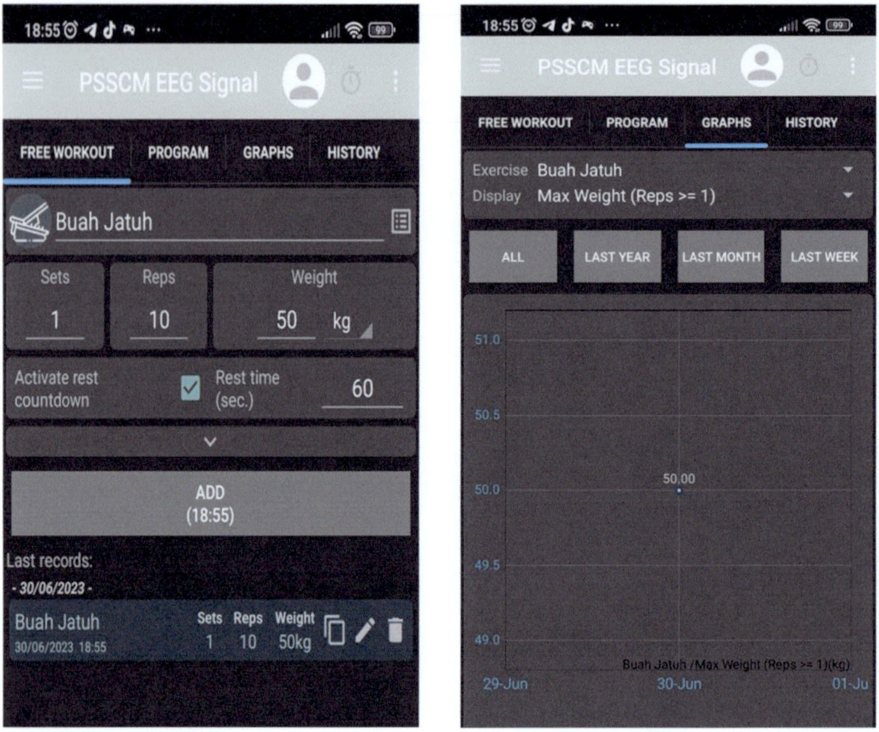

(a) (b)

Fig. 9 **a** Input for "Buah" and repetition; **b** GUI for training graph

4 Conclusion

In conclusion, this paper has discussed the SOP for Experimental setup for data collection to ultimately be used in the development of the SSCM Application—for training efficiency. The SOP is explained and the sample data from the process is shown and discussed. The GUI of the SSCM Apps is also being presented which allow users to check their training progress. Therefore, the established SOP and the SSCM Application contribute to an objective and systematic method for assessing and enhancing Silat training efficiency.

Acknowledgements The authors gratefully acknowledge the Persatuan Seni Silat Cekak Malaysia (PSSCM) and Center of Excellence—Sports Engineering Research Center (SERC), Universiti Malaysia Perlis for research funds and equipment to further facilitate this study.

References

EMOTIV, EPOC X—14 channel wireless EEG headset. https://www.emotiv.com/products/epoc-x. Accessed 30 Sep 2024

Halim AHA, Ahmad N, Farook R, Syuhada A, Omar N, Ali H (2016) The 11th ISDEV international islamic development management conference (11th IDMAC 2016). In: The 11th ISDEV international islamic development management conference, pp 1–12

Persatuan Seni Silat Cekak Malaysia, "CAWANGAN NEGERI." https://www.silatcekak.org.my/cawangan/. Accessed 3 Sep 2024

Persatuan Seni Silat Cekak Malaysia, Mesyuarat Agung Perwakilan Persatuan Seni Silat Cekak Malaysia kali ke-50—25 Jun 2022. 2022

Mustapha G, Zakaria M, Wan Sulaiman WR, Mahmud J (2014) The mechanical aspects of martial arts: total time of execution and kinematics of Kaedah A. In: Proceedings of the international colloquium on sports science, exercise, engineering and technology 2014 (ICoSSEET 2014), vol 2014, no ICoSSEET 2014, pp 3–12. https://doi.org/10.1007/978-981-287-107-7

Mustapha G, Wan Sulaiman WR, Abdul Majeed APP, Mohd Yahya NH, Mahmud J (2014) Kinematics and efficacy analysis of the Seni Silat Cekak Malaysia (Kaedah A). Appl Mech Mater 680(October):267–270. https://doi.org/10.4028/www.scientific.net/AMM.680.267

Rahman M, Karwowski W, Fafrowicz M, Hancock PA (2019) Neuroergonomics applications of electroencephalography in physical activities: a systematic review. Front Hum Neurosci 13(June):1–21. https://doi.org/10.3389/fnhum.2019.00182

Real-Time Monitoring and Alert System for Hikers Using IoT Technology

Muhammad Shazwan Baderol, Siti Azura Abuzar, Dayang Khadijah Hamzah, Aznor Hanah Abdul Halim, Rohani Farook, and Suhizaz Sudin

Abstract Real-time monitoring and alert system are a technological framework designed to continuously observe, track, and analyze data or activities in real-time, triggering alerts when certain predefined conditions or thresholds are met. It has been applied to in sport activities such as hiking to improve hiker's safety via peer-to-peer connection technology. Therefore, a project-based research is developed to monitor or track the hiker's location and also update their condition. Common causes of accidents include altitude sickness, missing persons, and unintended falls. This project is used ESP32 and Arduino Uno as microcontrollers and LoRa as a peer-to-peer connection. A heartbeat and temperature sensor are used to assess the hiker's condition. This project also makes use of the modernize library, which makes it suitable for usage with any microcontroller. The system design to reduces the search area while preserving up-to-date information on hiking conditions. The data can be presented to the receiver, who can then provide information about the victim's locations.

Keywords Real-time monitoring · Alert system · Hiking safety · Microcontroller · Sensor · Location tracking

1 Introduction

Hiking is a popular outdoor activity that is useful to a healthy lifestyle due to the quantity of mountains. Because this is a demanding exercise, the hikers face numerous dangers. When such circumstances are encountered, rescuers will take urgent action

M. S. Baderol · S. A. Abuzar (✉) · D. K. Hamzah · A. H. A. Halim · R. Farook · S. Sudin
Faculty of Electronic Engineering & Technology (FKTEN), Universiti Malaysia Perlis, Arau, Malaysia
e-mail: azuraabuzar@unimap.edu.my

S. A. Abuzar · D. K. Hamzah · A. H. A. Halim · R. Farook · S. Sudin
Sports Engineering Research Center (SERC), Universiti Malaysia Perlis, Arau, Malaysia

M. H. A. Hassan et al. (eds.), *Proceedings of the 3rd International Conference on Innovation and Technology in Sports*, Lecture Notes in Bioengineering,
https://doi.org/10.1007/978-981-95-0804-4_27

to begin a search and rescue operation. However, if the rescuers are unable to determine the victim's last known location, it will be difficult to locate them. The problem with current system shows lack of monitoring system and low credibility in allocation system using internet because of thick mountain canopy (Aditya et al. 2019). Proposed the Real-Time Monitoring System technology that can constantly update hikers' location while monitoring hiker heart rate and surrounding air pressure can be beneficial. The system consists of two main components which is transmitter and receiver to control and monitor sensor to determine user condition to alert rescuer when necessary.

The information produced by this real-time monitoring system for hiking activities is particularly vital for rescue operations. Hikers must always wear the given module during the hiking process until they reach the checkpoint. The system will continue to upload hiker information to the cloud server. Rescuers can retrieve data from the server during a search and rescue operation. Rescuers can narrow down the search area by identifying the hiker's most recent position.

2 Methodology

Wireless interfaces protocols are creation of a network of interconnected sensors in the current stage of the Internet of Things (IoT) depends heavily on long-range communication. A Wireless Sensor Network (WSN) can be created by connecting the sensors in a network to provide data on the physical conditions of the technology, such as location or temperature. Today, WSN are utilized for a variety of purposes, including security and surveillance, vehicle monitoring, and precision agriculture. A network structure is necessary for sending data within the network, there are two options, centralized networks or decentralized networks, each with unique properties (https://www.semtech.com/lora).

Machine-to-machine (M2M) communication is possible through a variety of wireless interfaces. Each one is distinguished by the information it transmits, coverage, along with form of spectrum control. LoRa is a wireless technology that provides long-distance, low-power, and secure data transport for M2M and IoT applications. The origins of LoRa, a spread spectrum modulation technique, can be traced back to chirp spread spectrum (CSS) technology (http://www.sghoslya.com/p/lora-is-chirp-spread-spectrum.html). LoRa allows link between sensors, machines, gateways, and other devices.

The prototype's major goal is to build a low-cost, long-range peer-to-peer network. Since the system does not require high bandwidth, bandwidth and range are the system's least and most crucial features, respectively. These elements were compared in a variety of connected wireless interfaces in order to find an interface that would work well for the project.

Certain networks are set up to route data through a centralized server. For instance, consider a LoRaWAN network with two suitable devices and a downlink-capable gateway. These two gadgets are each in need of data from the other. In order for one

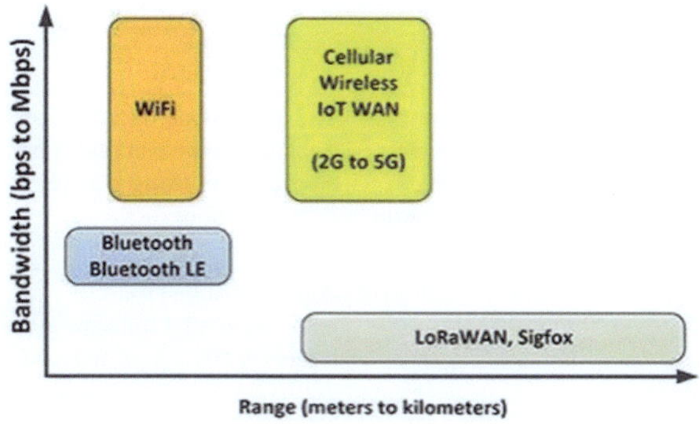

Fig. 1 Protocol comparison use in wireless interfaces (https://www.digikey.com/en/articles/welcome-wireless-menagerie-rf-bands-rotocol-choices-embedded-developers-part-1)

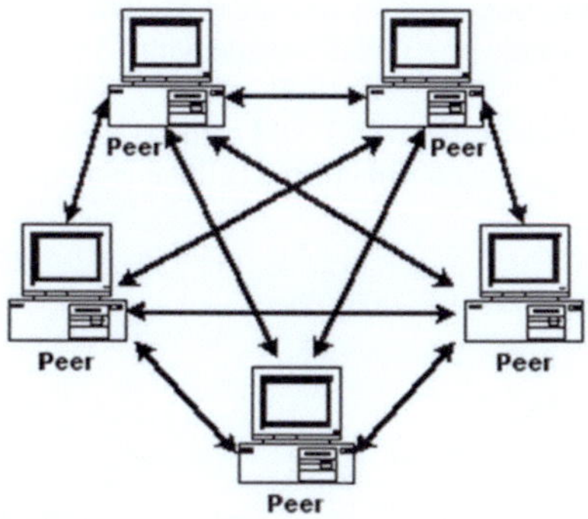

Fig. 2 Peer-to-peer network

node to receive data from another, data must be transferred to the gateway and then from the gateway to the end node. Without an authoritative server, end nodes can send and receive information. This type of implementation is known as peer-to-peer communication as shown in Fig. 2.

2.1 Related Research Project

Research related	Research related	Research related
Smartphone GPS locations of student movements (Doyle-Baker 2021)	Using GPS data to monitor student behavior and measure Student engagement in the campus	– Mobile phone GPS data

(continued)

(continued)

Research related	Research related	Research related
Design and implementation of vehicle navigation system in urban environments using internet of things (Godavarthi et al. 2017)	To continuously monitors and track the vehicle at real-time environment using web page in smart phone	– GPS/GPRS/GSM module – Raspberry Pi B+ board – Temperature sensor (DS18B20)
IoT for smart transportation system (Ahmed 2017)	A monitoring system to navigate school transportation based on concept of IoT and 4G/LTE cellular network service	– Raspberry Pi – 4G shield – DIV antenna – GPS antenna
Design and implementation of vehicle tracking system using GPS/GSM/GPRS technology and smartphone application (Lee et al. 2014)	Develop a vehicle tracking system to track exact location using GPS/GSM/GPRS and Smartphone	– Arduino Atmega38 – GPS module (EM-406) – GPRS module with SM5100B – AT&T SIM card(50 MB) – Quad-band cellular antenna

2.2 *System Block Diagram*

The are two design stage to accomplish the aim of the project where each had a transceiver and a microprocessor with a variety of functions (https://how2electronics.com/interfacing-sx1278-lora-module-with-arduino/). LoRa module integrate with versatile microcontroller known as ESP32 which used as a radio communication which has low voltage and open spectrum. Listed are the requirements must be met consisting:

- Located GPS coordinates
- Measure blood pressure
- Send all measured data via LoRa transmitter to receiver.

Figure 3 show the system made up of two modules, each of which had a transmitter and receiver with a variety of functions. The ESP32 board with a LoRa and GPS module was used to set up the single module as the receiver's microprocessor. To accomplish mobility and replicate the processing unit of an autonomous machine, micro-controller with equivalent modules, such as the Arduino Uno and Raspberry Pi used to achieve the same outcomes.

In the software part, it was built for both modules to communicate GPS coordinates with one another. An open-source code is used where it's compatible to allows LoRa packages to be sent between nodes while maintaining the same shield. Along with sending the delivery, it also determines the necessary measures for precise Geo-location. Both modules are designed to operate inverted from one another and are

Fig. 3 The block diagram of transmitter and receiver

each executing a straightforward loop that processes data. Figure 4 describes the loop in a flowchart diagram.

2.3 Project Design

Figure 5a, b show the circuit connection of ESP32's with low power consumption, integrated Wi-Fi and equipped with Bluetooth functionalities has compatibility with a wide range of sensors assures the ongoing versatility of this project, keeping its functionality while bringing new options for innovation and development in the Internet of Things (IoT) sphere.

3 Results and Discussion

Initial prototype as in Fig. 6, represent an initial version of the project and serve as electronic functional applications.

This project monitoring and alert system demonstrates a proof of concept for sending GPS coordinates using P2P LoRa connectivity while also keeping track of hiker performance. The system prototype that was created met the requirements for a P2P network structure, which made it feasible to monitor the hiker. The prototype capable of communicating with a similar machine via software, and this communication could be assessed. Table 1 shows the preliminary result before expedition start while Table 2 shows the result when the expedition arrived at the top of Bukit Lagi, Perlis. Based on data collection, the prototype able to communicate and generate the required data for the monitoring purposes.

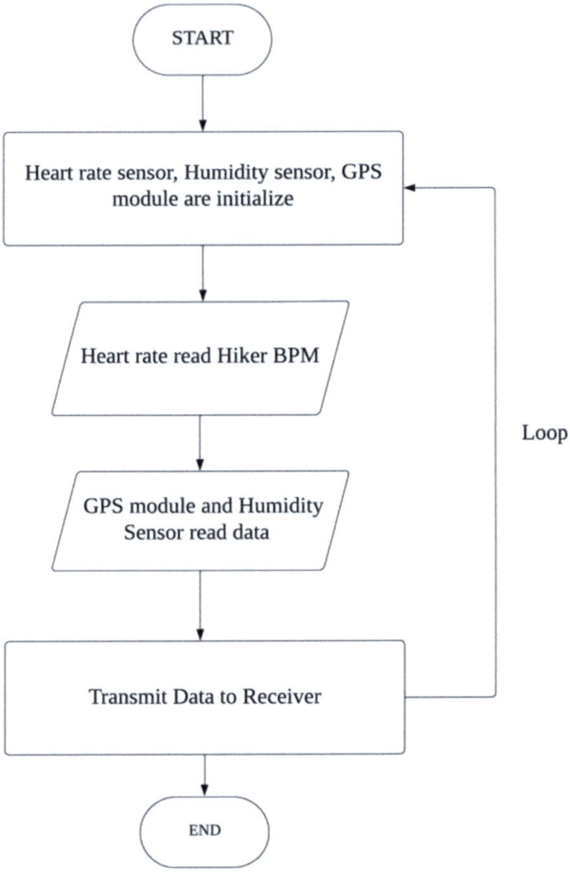

Fig. 4 General overview when all loops is activates

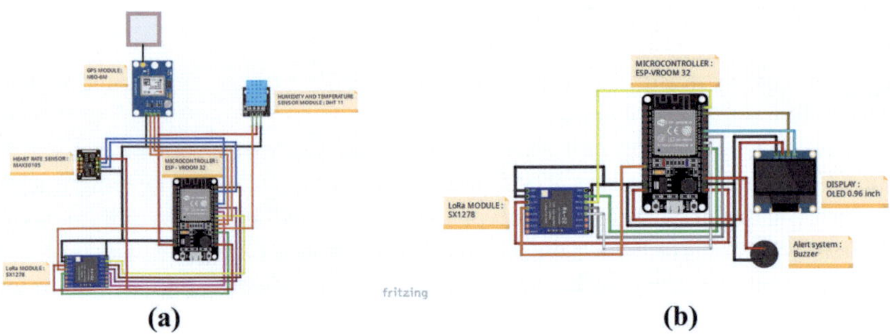

Fig. 5 **a** Circuit diagram for transmitter. **b** Circuit diagram for receiver

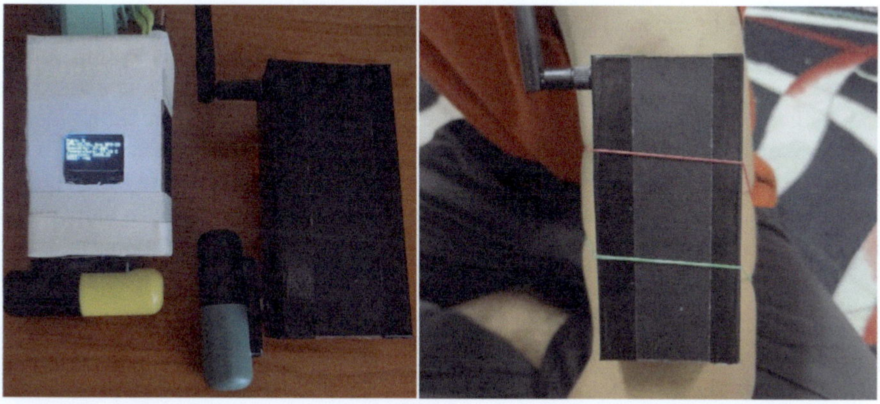

Fig. 6 Prototype real-time monitoring and alert system for hikers using iot technology

Table 1 Preliminary data reading before hiking

Packet	BPM	Avg. BPM	Humidity (%)	Temperature (°C)	Latitude	Longitude
1	85	82	66	27	6.43278	100.206867
2	83	79	65	28	6.432342	100.206845
3	84	82	68	28	6.432244	100.206784
4	85	82	68	27	6.432453	100.206791
5	87	84	69	27	6.432567	100.206821

Table 2 Data reading generated from the prototype

Packet	BPM	Avg. BPM	Humidity (%)	Temperature (°C)	Latitude	Longitude
1	110	95	70	27	6.43819	100.20701
2	105	92	68	28	6.438262	100.20698
3	112	98	72	30	6.438294	100.20662
4	106	95	65	31	6.438267	100.20676
5	108	94	70	29	6.43823	100.206835

4 Conclusion

In this Prototype, a LoRa based monitoring system with IoT was created. A safety system that communicates the user's current position during emergency conditions was built by monitoring the heart rate and temperature. This technology allows rescuers to be aware of the user's condition. The rescue method can be made more efficient by using Geo-location to narrow the search region and starting the search early when the user's data is aberrant. Although the proposed model is designed for the monitoring and early rescue of hikers, it can also be used for a variety of other applications such as monitoring miners, workers, patients, and other hazardous environments and dangerous activities where it would be useful in detecting abnormalities early and providing adequate care. Furthermore, since LoRa is low power and longer range, compared to other wireless data communication technologies, LoRa is versatile in applications. This proposed approach demonstrates that in the event of a critical situation, geographical data is reliably transmitted. In future developments, the LoRa range can be expanded further by employing various modulation algorithms, code rates, spread spectrum, and hardware, such as a directional antenna.

References

Aditya V, Deva Prasaad V, Faheem Kamaludeen M, Gophika T (2019) Systematic review of mountaineers using LORA. IJRAR Int J Res Anal Rev. http://ijrar.com/upload_issue/ijrar_issue_20543234.pdf

Ahmed AY (2017) Internet of things for a smart transportation system". University of Baghdad, Iraq, Bachelor of Engineering

Doyle-Baker PK (2021) Smartphone GPS locations of students' movement to and from campus. ISPRS Int J Geo Inf. https://www.mdpi.com/2220-9964/10/8/517

Godavarthi B, Nalajala P, Ganapuram V (2017) Design and implementation of vehicle navigation system in urban environments using internet of things (IoT). IOP Conf Ser Mater Sci Eng 225 (2017). https://doi.org/10.1088/1757-899X/225/1/012262/pdf

https://how2electronics.com/interfacing-sx1278-lora-module-with-arduino/—Sx1278 LoRa module interface using Arduino Uno

https://www.semtech.com/lora—Design resource technology of LoRa

Lee S, Tewolde G, Kwon J (2014) Design and implementation of vehicle tracking system using GPS/GSM/GPRS technology and smartphone application. IEEE World Forum Internet Things

LoRa is a chirp spread spectrum modulation based wireless networking standard. http://www.sghoslya.com/p/lora-is-chirp-spread-spectrum.html

Welcome to the Wireless Menagerie: RF Bands and Protocol Choices for Embedded Developers. https://www.digikey.com/en/articles/welcome-wireless-menagerie-rf-bands-protocol-choices-embedded-developers-part-1

Evaluating the Hydrodynamic Performance of Macon, Big, and Fat Blades During the Drive Phase in Rowing

Ab Aziz Mohd Yusof, Hazim Sharudin, Haszeme Abu Kasim, Ainaa Maya Munirah Ismail, Ahmad Naufal Adnan, and Kamariah Md Isa

Abstract The performance of the rowing blade significantly influences propulsion efficiency in rowing. This study aims to investigate the fluid flow on the shapes of the existing blades and the corresponding hydrodynamic forces generated in the drive phase at the beginning of the competition. The study used computational fluid dynamics (CFD) with Star CCM+ software. The simulation was based on established experimental data, with boundary conditions reflecting real rowing tank conditions. Results revealed distinct variations in fluid flow patterns and force distributions across the blade surfaces. The highest hydrodynamic force happens to the Fat blade due to the largest blade's surface area, resulting in enhanced performance compared to the other two blade designs. The analysis demonstrated a correlation between blade aspect ratio and hydrodynamic efficiency, indicating that blades with higher aspect ratios tended to generate greater thrust. In conclusion, the study highlights the importance of optimizing blade shape for improved rowing efficiency. These findings can inform future designs of rowing blades, contributing to advancements in competitive rowing technology and athlete performance.

Keywords Rowing · Blade design · Hydrodynamic force

1 Introduction

Understanding the hydrodynamic force generated and fluid characteristics around the moving blade is an informative knowledge that can enhance rowing performance. The studies of the oar blade propulsive mechanisms mostly use simplified mechanisms due to uncontrolled conditions and difficulties in capturing the reading (Sliasas and Tullis 2009, 2010). The blade dynamic motion is replaced by flowing the water to hit

A. A. Mohd Yusof (✉) · H. Sharudin · H. Abu Kasim · A. M. M. Ismail · A. N. Adnan · K. M. Isa
Faculty of Mechanical Engineering, Universiti Teknologi MARA Johor Branch, Pasir Gudang Campus, Masai, Malaysia
e-mail: abaziz86@uitm.edu.my

M. H. A. Hassan et al. (eds.), *Proceedings of the 3rd International Conference on Innovation and Technology in Sports*, Lecture Notes in Bioengineering,
https://doi.org/10.1007/978-981-95-0804-4_28

the blade at a specific position or rotating the blade at a constant speed, which neglects the rower stroke. In addition, Macon and Big Blades's fluid dynamics characteristics were investigated previously, where Big Blade increased hydrodynamic efficiency by 2% compared to Macon Blade under quasi-static conditions. However, no studies have been reported on the Fat Blade (Caplan and Gardner 2007a).

The Macon, Big, and Fat blades are popular rowing blade designs, each offering distinct characteristics tailored to different rowing styles and conditions (Caplan and Gardner 2007a, b; Nolte 2009, 2023). The Macon blade is characterized by its narrow shape and moderate surface area, promoting efficient entry and exit from the water. This design allows smoother transitions and reduced drag, making it suitable for sculling and lighter conditions. The Big blade, in contrast, features a broader profile and larger surface area, generating increased force and propulsion (Nieuwburg et al. 2023). This design provides powerful strokes and is favoured in heavy conditions where maximizing thrust is crucial. However, it may result in higher drag during recovery. The Fat blade combines elements of both designs, boasting a wide and thick profile that enhances stability and power during the stroke. This blade is particularly effective for athletes seeking maximum thrust while maintaining control and balance. The choice between these blades depends on the rower's style, water conditions, and performance goals, offering unique advantages to optimize efficiency and speed on the water.

Appropriate boundary conditions are essential to study the hydrodynamics of a rotating blade. One of the important elements to be considered is the no-slip condition, which is typically implemented at the blade surface, where the fluid velocity must be equal to the rotation speed of the blade. This condition is critical for properly modelling the interaction between the blade and the fluid in its vicinity. The water flow around the blade is represented by the realistic conditions of fluid moving inside the simulation domain, represented by the moving blade and boat speed (Nieuwburg et al. 2023). Previous studies have demonstrated that understanding these dynamics is crucial for optimizing blade performance, as variations in flow patterns can significantly affect hydrodynamic forces and efficiency (Ahmed et al. 2013; Coppel et al. 2010). Investigating blade performance under these conditions is vital for improving design parameters, enhancing propulsion efficiency, and ensuring stability in various operational scenarios, ultimately leading to advancements in rowing technology and competitive performance.

Despite the importance of dynamic conditions, limited studies have focused on blade performance under actual stroke conditions. This paper's main objective is to investigate the fluid flow on the shapes of the existing blades and the corresponding hydrodynamic forces generated in the drive phase at the beginning of the competition. The start phase is critical, as rowers must achieve precise blade placement and apply substantial power to accelerate the boat effectively.

2 Method

The study employed a computational simulation approach, utilizing Star CCM+ software to analyze the hydrodynamic forces involved in rowing propulsion under unsteady, incompressible, turbulent flow conditions. A numerical simulation was performed to investigate the fluid flow characteristics and force generated by the blade at the beginning of a rowing competition. The associated hydrodynamic forces were measured and recorded. After completing the simulation, the results were validated, and a detailed analysis was conducted on the generated forces. The present study was a continuity of the previous study of Yusof et al. (2018). In the previously mentioned study, only the Macon blade was explored with fluid flow around the blade. The study used experimental and computational methods validated with an 8.3 per cent error.

2.1 *Blade Models*

The moving blade model consisted of the fluid region (background domain) and the blade subdomain. The Chimera method was used, based on the overlapping mesh method, to simulate the motion of the blade (Houzeaux et al. 2014). The solution was achieved by overlapping the blade's secondary mesh grid with the surrounding fluid's background mesh grid. Tri-linear interpolation of the dependent variables at their boundaries facilitated communication between the two mesh grids.

The Macon blade model was used to validate computational work, as illustrated in Fig. 1a, the details of which can be found in Yusof et al. (2018). After the computational simulation was validated, the Big and Fat blades (Fig. 1b, c) were analyzed using computational simulation. All the blades were submerged in the water with a submerged angle, θs of 10 degrees relative to the water-free surface.

The blade details are provided in Table 1. The Big and Fat blade's blade surface area was more significant than the Macon blade, with 36% and 50% surface higher, respectively. Surface area changes impacted the aspect ratio, a key determinant in hydrodynamic performance. The blade aspect ratio is the blade width divided by the blade area, which represents the design's performance towards hydrodynamic performance (Binning and Fulton 2011; DeVoria and Ringuette 2012).

2.2 *Boundary Conditions*

The CFD simulation utilized data from experimental studies, with boundary conditions designed to replicate the environment of rowing tank tests. No-slip walls defined most of the domain surfaces. The upper boundary of the fluid domain was treated as a symmetry plane to save computations. Water and air properties were assigned

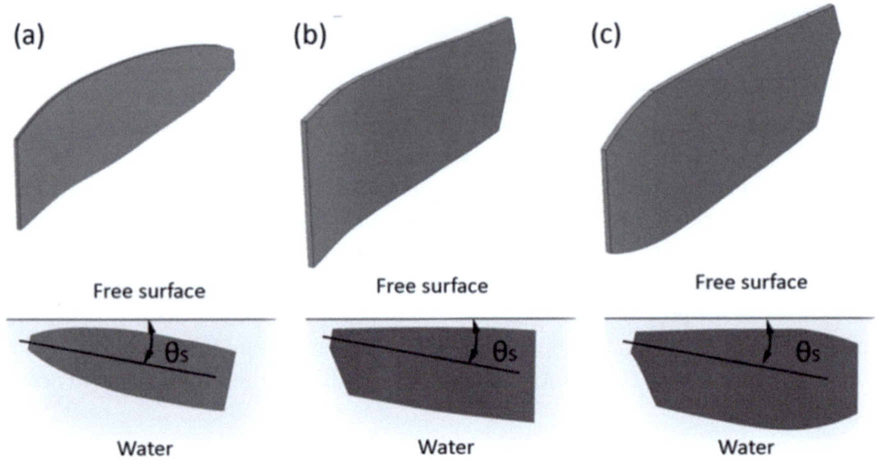

Fig. 1 Blade model on which CFD analysis was performed at a submerged angle, θs **a** macon blade situated at the submerged position **b** big blade situated at the submerged position **c** fat blade situated at the submerged position

Table 1 Dimension of the investigated blades

Blade type	Length (m)	Width of tip (m)	Area (m^2)	Aspect ratio
Macon blade	0.57	0.17	0.0910	0.317
Big blade	0.55	0.26	0.1242	0.544
Fat blade	0.57	0.28	0.1371	0.572

using a volume fraction approach, dividing the domain into two regions to represent heavier and lighter fluids. A multiphase fluid technique modelled the interaction between the two fluid fractions at the free surface. The simulations employed a k - ω turbulence model. This turbulence model was selected to consider the near-wall turbulence effect as the transitional regime between fully laminar and fully turbulent boundary flow happening around the blade. This study treated the fluid as incompressible, with water having a density of 997.561 kg/m^3 and a dynamic viscosity of 8.8871 $\times$ 10^{-4} kg/(ms) (Phillips et al. 2015). A density of 1.18415 kg/m^3 and a dynamic viscosity of 1.85508 $\times$ 10^{-5} kg/(ms) were used for air.

2.3 Boundary Conditions

As depicted in Fig. 2, the blade was scheduled to perform rotational and translation movements that replicated the oar's angular and the boat's translation velocity. Using a superposition method, this motion was defined in a Cartesian coordinate system. 1 represents the normal force, 2 is the shear force, 3 is the total hydrodynamic force,

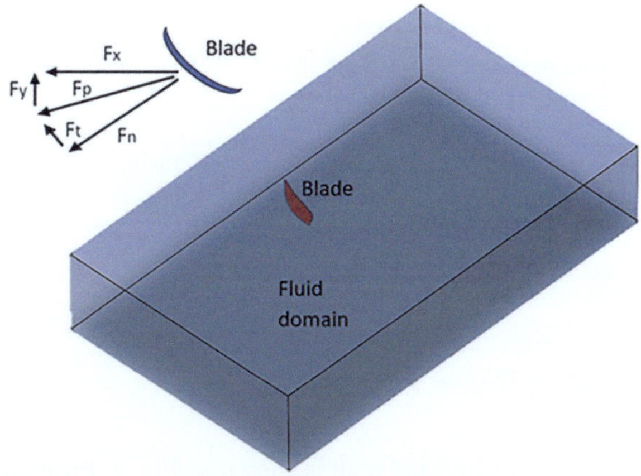

Fig. 2 The study's CFD model for the study which includes the fluid domain and rotational domain of blade models

4 is the force in the x direction, and 5 is the force in the y direction. These forces are analyzed to assess and compare the performance of the blade.

Three-dimensional hexahedral elements discretized the fluid domain. The mesh scale was increased to correctly capture the boundary layer, hydrostatic pressure, and water surface effects near the blade and the middle of the fluid domain. This was accomplished using a volumetric control setting, with mesh elements 50% smaller than the maximum element size assigned.

Various mesh convergence studies were performed across a range of mesh element numbers to identify the optimal mesh element count for analysis. The fluid domain was split into two sub-domains (including the rotating domain, which contained the blade model and the background domain. Mesh convergence was evaluated using usual plot trends and average percentage error. The final mesh size for CFD convergence was chosen to be 4,367,030 elements. Total pressure is employed as the reference to determine the error percentage of the investigation based on the convergence analysis.

Higher mesh densities generally increased total pressure until convergence reached 4 million mesh elements. The percentage error was calculated to determine the optimum number of elements in comparing the total pressure obtained with that of the four-million-element model. A mesh size of 868,853 elements with the lowest percentage error was selected for further analysis (see Table 2). Additionally, the pressure contour for the 868,853-element mesh closely resembled that of the four million-element model, with the red-coloured high-pressure zones consistently distributed near the blade tips, unlike the other mesh configurations.

Table 2 Result of mesh convergent study

The mesh size of the rotation part (m)	The mesh size of the background part (m)	No. of mesh element	Computational time (Hour)	Percentage error
0.09	0.20	79896	3.59	6.75
0.06	0.10	361484	17.89	4.79
0.04	0.08	868853	25.00	1.05
0.02	0.06	4367030	131.20	Reference

3 Result and Discussion

The blade's projected path is shown in Fig. 3. From the catch to the finish position, the overall drive angle was 61.7 degrees. The blade moved initially slowly during the drive phase, but after 0.2 s, its speed was greater, and it achieved a steady state velocity of 0.5 s, which was then maintained for the rest of the drive phase. As there was no blade slip in the drive, the blade's path did not cross at the same point. No path crossing happened because the speed of the blade was greater than the speed of the boat, resulting in a positive relative speed with active hydrodynamic forces generated.

Figure 4a, b show the instantaneous total pressure contour on the blade. In the first 0.4 s, the pressure was evenly distributed across the blade surface. As the relative speed increased, the maximum pressure shifted toward the tip and the lower side of the blade. This pressure distribution remained stable until 1.3 s of the drive, after which it moved toward the centre as the drive phase ended.

The blade shape is very influential on the driving phase pressure contours. The maximum total pressure on the front of the blade took place at the peak force at 0.6 s, which was 3325 Pa, 3303 Pa, and 3423 Pa from Macon, Big, and Fat blades, respectively, as depicted in Fig. 4b. During the next phase of the drive, the angular

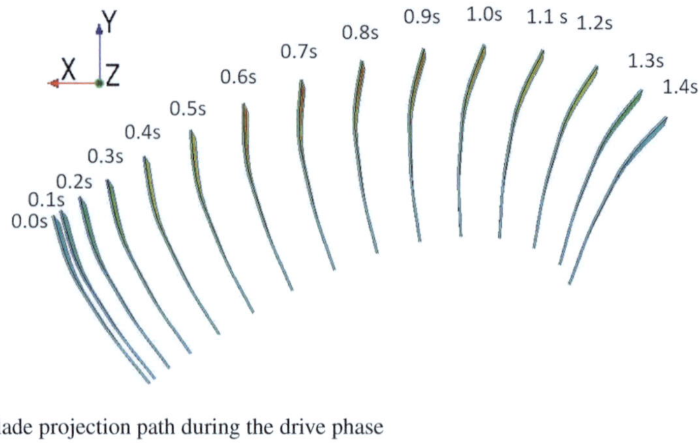

Fig. 3 Blade projection path during the drive phase

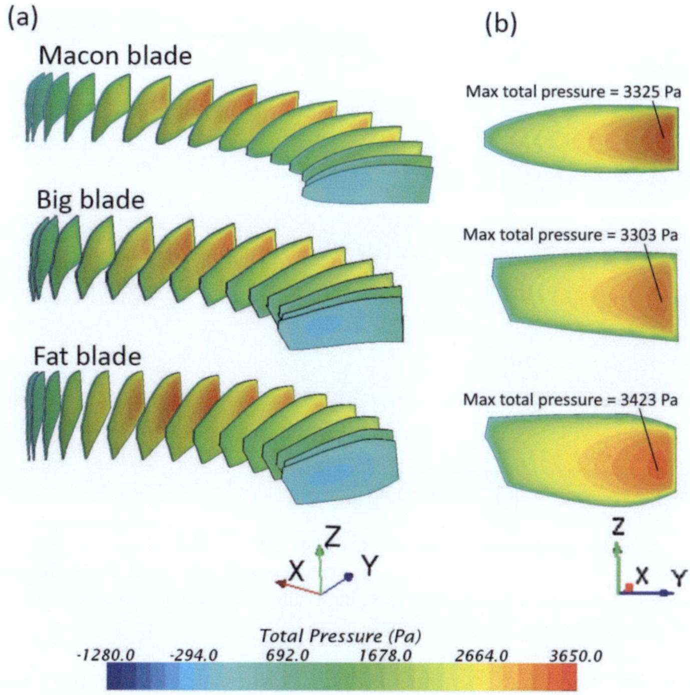

Fig. 4 Total pressure contour of the blade during the drive phase: **a** balde projection of the blade at 1.4 s of drive phase on the macon, big and fat blades; and **b** pressure contour at peak force, occurring at 0.6 s into the drive phase

velocity of the oar decreased. At the same time, the speed of the boat increased, resulting in an opposing force on the blade. This reversal of force direction resulted from water striking the blade from the opposite direction.

The force on the blade is equal to the pressure over the area of the blade, and the peak pressure correlates with the blade's centre, where the force is applied (as indicated in the pressure distribution plot). In the first part of the stroke, the flow was uniform over the blade, resulting in a relatively even distribution of fluid momentum over the blade's surface and, hence, a relatively uniform distribution of hydrodynamic pressure. The hydrodynamic force at this time was at the blade midspan. According to Novais et al. (2012), hydrodynamic drag acting on a body at depth has increasing magnitude up to a maximum depth of approximately 0.75 m, beyond which the magnitude of drag decreases (Novais et al. 2012).

However, when the submerged blade was at a depth of less than 0.7 m, the effect of hydrodynamic drag was significant (Fuss et al. 2016). The drag effect caused by the blade's depth started to dominate during the drive phases as the forward speed increased. This moved the location of maximum pressure downwards, closer to the lower edge and tip of the blade. At this point, the force was concentrated on the

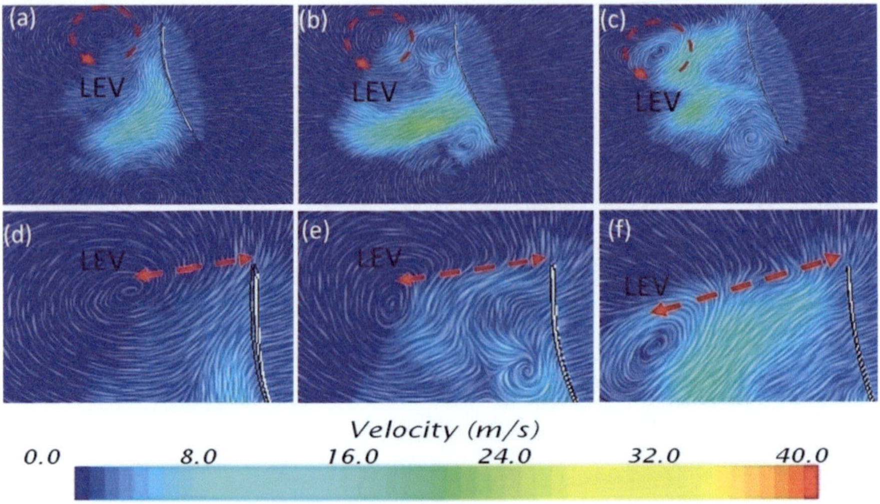

Fig. 5 LEV circulation around the blade at 0.6 s from the top view of the simulation model. **a–c** LEV circulation of macon blade, big blade, and fat blade, respectively. **d–f** The distance between LEV circulation and the tip of the macon blade, big blade, and fat blade, respectively

lower side of the blade close to the tip. This shifting pressure pattern occurred in both symmetrical and asymmetrical blade models.

Fluid flow contour around the blade at 0.6 s of the drive. A leading-edge vortex (LEV) is formed at the blade's edge in this stage. Fluid was circulated most vigorously around the Fat blade, followed by the Big and Macon blades. The leading-edge vortex produced during the stroke can be seen at the same time-step in Fig. 5, where the leading-edge vortex location is highlighted in red. These measurements led to a maximum distance of the LEV from the blade tip of $x = 0.35$ m for the Fat blade, $x = 0.27$ m for the Big blade, and $x = 0.22$ m for the Macon blade.

The leading-edge vortex (LEV) is a ring-shaped vortex that develops near the blade and has been demonstrated to boost drag-aided propulsion (Kim and Gharib 2011). Early in the stroke, LEVs form at the tip and along all edges of the blade. However, as speed increases, the top and bottom edges dissipate due to wake formation, leaving the vortex at the blade tip to dominate the propulsive force, as illustrated in Fig. 5. As the catch angle increased (see Fig. 3), the LEV shifted outward from the blade tip (see Fig. 5d–f), driven by the spanwise flow across the blade. According to Kim et al. (2011), the spanwise flow interacts with the LEV, pushing it further away from the blade tip. This effect becomes more pronounced with larger aspect ratio blades, which generate more spanwise flow and push the LEV farther out, as the larger tip area provides more surface for fluid interaction.

The generated LEV plays a significant role in influencing energy efficiency and boat speed. As the blade moves causes, flow separation at the leading edges enhances the hydrodynamic force as it reduces energy wasted to the drag induced. Leveraging these advantages allows the rower to maximize power transfer to the boat, which then increases the boat's speed as less energy is wasted.

Figure 6a shows the normal force generated by the Macon, Big, and Fat blades. While the overall pattern of the normal force is similar for all three blades, the magnitude differs. The normal force peaks of 0.6 s were 231 N for the Macon blade, 307 N for the Big blade, and 347 N for the Fat blade. These differences arise from the distinct shapes of the blades, affecting fluid movement and the forces applied per blade.

CFD analysis was used to obtain the shear force on the blade surface. As shown in Fig. 6b, the peak shear force appeared between 0.6 and 0.8 s for each type of blade. Toward the end of the drive, the shear force increased due to the adverse pressure effects. The magnitude of the shear force was from 0.1 to 0.17 N. This was 0.045% smaller than the normal force. Although the shear force magnitude was small, it plays an important role as a component of the hydrodynamic forces and contributes to the formation of LEV.

LEV vortices form due to the shear layer, which arises from the shear force (Yusof et al. 2018). Significant fluid tangential velocity (dv/dy) is experienced in the shear layer. As the Reynolds number increases, the shear layer's thickness also increases. During the drive phase, the Reynolds number increased from 0 to 1.04×10^6, with the shear force illustrated in Fig. 6b. Changes in the shear layer's size, driven by the increase in the oar's angular speed, affect the strength of the LEV circulation. The Fat blade produced a higher shear force than the other blade types because its larger aspect ratio increases the inner area enclosed by the vortex, generating stronger circulation along the blade edge (Phillips et al. 2015; Hord and Lian 2016).

The Macon blade generated the lowest hydrodynamic force, followed by the Big and Fat blades, as shown in Fig. 6c. The blade's relative speed was low early in the stroke, resulting in smaller hydrodynamic forces. As the relative speed increased, so did the hydrodynamic force, which rapidly increased after 0.2 s as the oar's angular speed accelerated. The effect of blade design on hydrodynamic force becomes more apparent at the start of the stroke, especially as the difference in speed between the oar's angular speed and the boat's translation speed increases. The Fat blade showed a 51% increase in maximum force compared to the Macon blade, while the Big blade showed a 33% increase. The hydrodynamic force decreased sharply by 1.2 s into the drive phase, reaching zero at 1.4 s as the stroke entered the finish phase. It is important to note that the normal force dominated the propulsive mechanics, so the magnitude and pattern of the hydrodynamic force closely followed that of the normal force.

The forces x, F_x, and Force y, F_y during the drive are shown in Fig. 7a, b. F_x denotes the force acting parallel to the boat's direction, while F_y acts perpendicular to it. During the drive phase, F_x rapidly increased as the blade moved from the catch position to approximately 17.5 degrees at 0.6 s. As the boat received the necessary force to accelerate, F_x gradually decreased until 1.2 s. From 1.2 to 1.4 s, F_X dropped rapidly to zero. Overall, the magnitude of F_x was higher than that of F_y, consistent with findings by Leroyer et al. (2008).

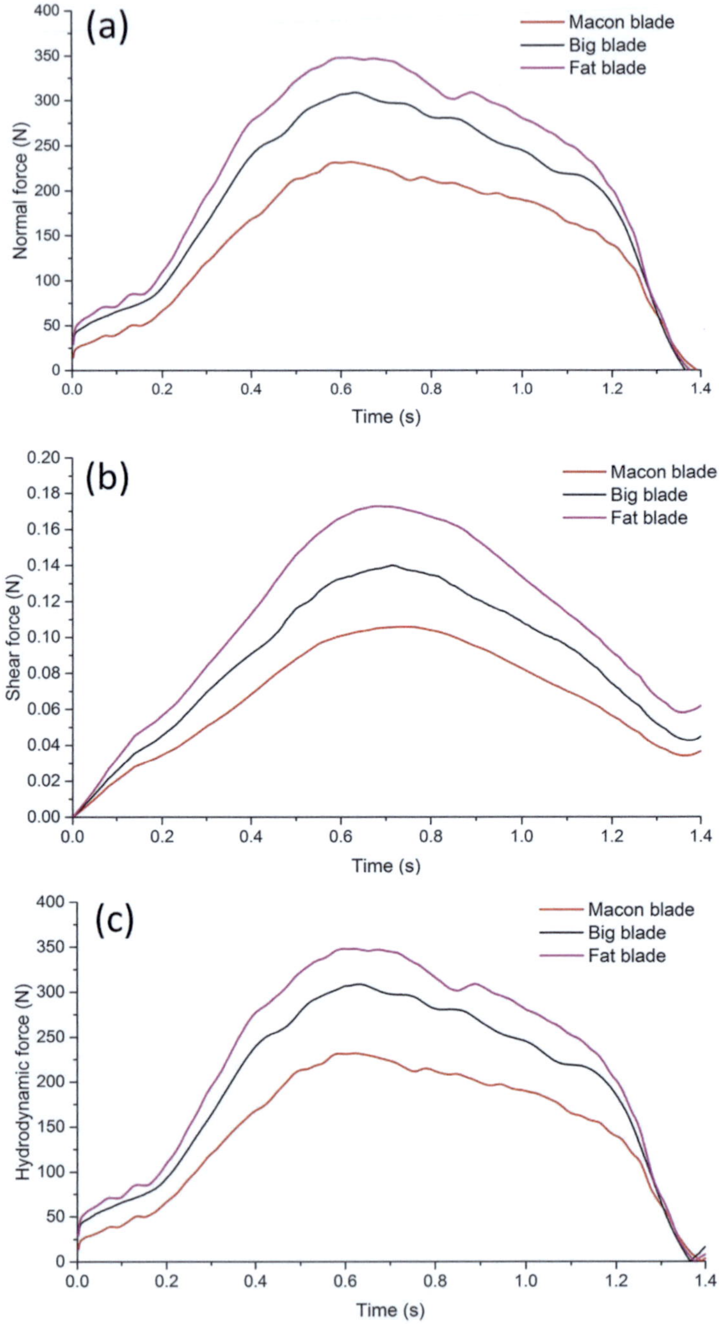

Fig. 6 Graph showing the force components on the blade: **a** normal force generated by macon, big and fat blade, **b** the shear force on macon, big and fat blade, and **c** hydrodynamic force on macon, big and fat blade

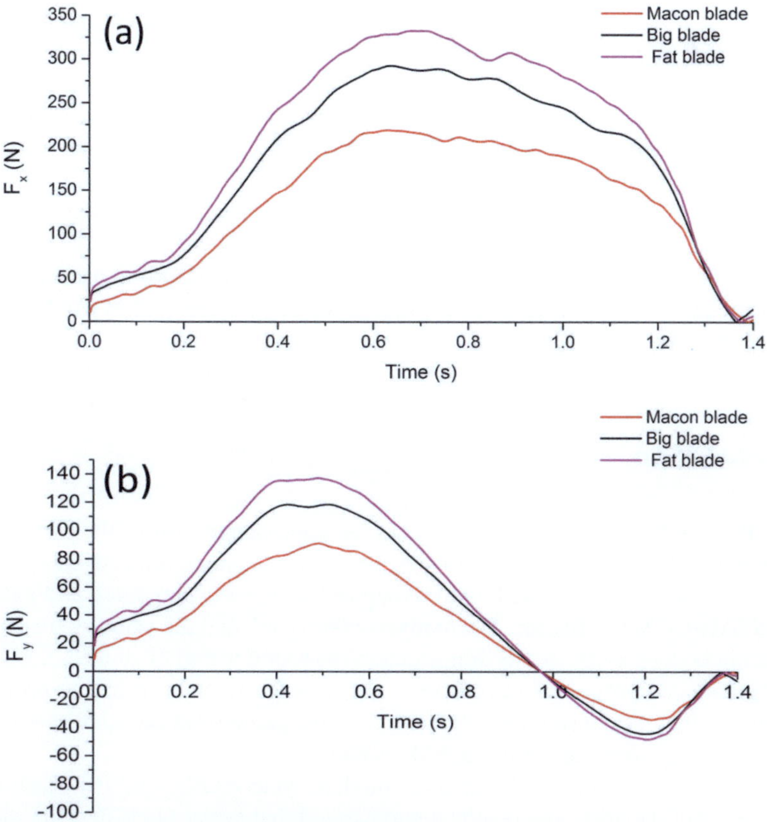

Fig. 7 **a** Force x and **b** Force y acting on the blade

According to Caplan et al. (2007b), the lift force generated by the blade contributes to boat performance, functioning similarly to a wing's aerodynamics (Caplan and Gardner 2007b). However, while this concept helps explain the flow over the blade, it does not fully describe how it generates thrust for the boat. The pressure on the blade surface produces a higher drag force than lift force, as the lift acts parallel to the surface of the blade, and the drag acts perpendicular to it (see Fig. 2). This study identifies rowing propulsion mechanics as drag-induced, consistent with the findings of Chung (2008) and Cabrera et al. (2006).

Chung (2008) explained that thrust force responds to the generated drag force acting on the blade before accelerating the boat (Chung 2008). That blade movement is a combination of linear and rotational velocity. The drag on the blade resists its movement through the water, ensuring continuous thrust is applied to the boat. This situation creates a "virtual fulcrum," or pivot point at the submerged blade, transferring force to the boat through the lock. However, the magnitude force acting on the oarlock was less than the force at the blade due to the leverage created by the oar.

The current findings related to LEV can extend to other domains and sports that deal with hydrodynamics, such as kayaking, canoeing, or pedalling. In these activities, LEV plays the same role as rowing. Creating a region of low pressure and a pressure boundary layer on one side of the blade or paddle improves propulsion efficiency. In kayaking, for instance, the paddle blades rely on blade and water flow interaction to ensure maximum driving thrust. Understanding the LEV dynamics makes it possible to optimize the blade shape and the attack angle of kayak paddles. This will result in greater power transfer, less energy loss, and more efficient propulsion performance for kayakers. Similarly, in canoeing and stand-up paddling, using LEV principles to obtain greater propulsion efficiency with less effort would not only increase the speed of progress but also extend travel time.

4 Conclusion

This study employed a dynamic computational approach to explore the performance of commercially available rowing blades under unsteady, incompressible, turbulent flow conditions. The generated hydrodynamic forces were simulated and analyzed using STAR-CCM+ software. The analysis confirmed that the primary mechanism facilitating boat propulsion was drag-induced, as found in earlier studies. Separating the drag-induced propulsion mechanism suggests the LEV is critical to propel the body using drag. A considerable amount of thrust came from the LEV that formed at the blade's tip as it passed through the water.

The simulation compared three blade models: Macon, Big, and Fat. Each varies in size and surface area. The results demonstrated that larger blade designs, such as the Fat blade, exhibited stronger fluid circulation and higher propulsive forces due to the enhanced LEV effect. The Fat blade, for instance, showed a 51% increase in maximum force compared to the Macon blade. These findings underscore the importance of blade design in optimizing hydrodynamic performance. By optimizing blade design, it is possible to improve rowing efficiency, providing valuable insights for athletes and equipment manufacturers aiming to maximize performance.

References

Ahmed MR, Faizal M, Lee Y-H (2013) Optimization of blade curvature and inter-rotor spacing of Savonius rotors for maximum wave energy extraction. Ocean Eng 65:32–38

Binning S, Fulton C (2011) Non-lethal measurement of pectoral fin aspect ratio in coral-reef fishes. J Fish Biol 79(3):812–818

Cabrera D, Ruina A (2006) Propulsive efficiency of rowing oars. J Appl Biomech

Caplan N, Gardner TN (2007) Optimization of oar blade design for improved performance in rowing. J Sports Sci 25(13):1471–1478. https://doi.org/10.1080/02640410701203468

Caplan N, Gardner TN (2007a) A fluid dynamic investigation of the big blade and macon oar blade designs in rowing propulsion. J Sports Sci 25(6):643–650

Chung M-H (2008) Numerical study of rowing hydrofoil performance at low Reynolds numbers. J Fluids Struct 24(3):313–335

Coppel A, Gardner TN, Caplan N, Hargreaves D (2010) Simulating the fluid dynamic behaviour of oar blades in competition rowing. Proc Inst Mech Eng Part P J Sport Eng Technol 224(1):25–35

DeVoria AC, Ringuette MJ (2012) Vortex formation and saturation for low-aspect-ratio rotating flat-plate fins. Exp Fluids J Artic 52(2):441–462. https://doi.org/10.1007/s00348-011-1230-z

Fuss FK, Fundel S, Weizman Y, Smith RM (2016) Smart oar blade for hydrodynamic analysis of rowing. Procedia Eng 147:735–740. https://doi.org/10.1016/j.proeng.2016.06.268

Hord K, Lian Y (2016) Leading edge vortex circulation development on finite aspect ratio pitch-up wings. AIAA J 2755–2767

Houzeaux G, Eguzkitza B, Aubry R, Owen H, Vázquez M (2014) A Chimera method for the incompressible Navier-Stokes equations. Int J Numer Meth Fluids 75(3):155–183

Kim D, Gharib M (2011) Characteristics of vortex formation and thrust performance in drag-based paddling propulsion. J Exp Biol 214(13):2283–2291

Leroyer A, Barré S, Kobus J-M, Visonneau M (2008) Experimental and numerical investigations of the flow around an oar blade. J Mar Sci Technol 13(1):1–15. https://doi.org/10.1007/s00773-007-0256-7

Nolte V (2009) Shorter oars are more effective. J Appl Biomech 25(1)

Nolte V (2023) Rowing science. Human kinetics

Novais ML et al (2012) The effect of depth on drag during the streamlined glide: a three-dimensional CFD analysis. J Hum Kinet 33:55–62. https://doi.org/10.2478/v10078-012-0044-2

Phillips N, Knowles K, Bomphrey RJ (2015) The effect of aspect ratio on the leading-edge vortex over an insect-like flapping wing. Bioinspir Biomim 10(5):056020

Sliasas A, Tullis S (2009) Numerical modelling of rowing blade hydrodynamics. Sports Eng 12(1):31–40. https://doi.org/10.1007/s12283-009-0026-3

Sliasas A, Tullis S (2010) The dynamic flow behaviour of an oar blade in motion using a hydrodynamics-based shell-velocity-coupled model of a rowing stroke. Proc Inst Mech Eng Part P J Sport Eng Technol 224(1):9–24

van Nieuwburg W et al (2023) Improving rowing performance by adjusting oar blade size and angle. Front Sport Act Living 5:1109494

Yusof A, Omar AH, Syahrom A, Harun M (2018) Analysis of oar blade hydrodynamics for rowing propulsive mechanism: experiment and computational fluid dynamics simulation. Int Rev Mech Eng 12:920–927

A Systematic Review of Position-Specific for Anthropometric in Basketball Elite Athlete

Gao Yili, Hadafi Fitri Mohd Latip, and Mohd Shahrizal Sunar

Abstract Basketball is a physically demanding sport characterized by body contact. Due to differences in physical attributes and tactics among male basketball players, they are assigned to various playing positions, making anthropometric measurements the strongest discriminative factor among them. This systematic review aims to identify the differences in body composition among elite male basketball players across different playing positions. Abstracts of all articles found through electronic searches were reviewed, and articles that did not meet the inclusion criteria were filtered out. Out of 4050 papers, 4034 studies were excluded. The remaining 16 papers were included in this systematic review to demonstrate the differences in body composition among elite male basketball players in different playing positions. The review of these 16 studies confirms that elite male basketball players exhibit varying body compositions based on their playing positions. In conclusion, the three most common positions in basketball are associated with distinct body compositions.

Keywords Basketball · Elite athletes · Playing positions · Anthropometry · Body composition

Gao Yili
College of Physical Education, Zheng Zhou University of Technology, Zhengzhou, Henan, China

Gao Yili · H. F. M. Latip (✉)
Sport Innovation and Technology Centre, Institute of Human Centered Engineering Universiti Teknologi Malaysia, Johor Bahru, Johor, Malaysia
e-mail: hadafifitri@utm.my

H. F. M. Latip
School of Education, Faculty Social Science and Humanity, Universiti Teknologi Malaysia, Johor Bahru, Johor, Malaysia

M. S. Sunar
Department of Emergent Computing, Faculty of Computing, Universiti Teknologi Malaysia, Johor Bahru, Johor, Malaysia

M. H. A. Hassan et al. (eds.), *Proceedings of the 3rd International Conference on Innovation and Technology in Sports*, Lecture Notes in Bioengineering,
https://doi.org/10.1007/978-981-95-0804-4_29

1 Introduction

Basketball originated in 1892 and has been one of the oldest events included in the Olympic Games since 1939. It is also one of the most popular team sports globally. An athlete's success in basketball depends on physical fitness, psychological abilities, tactical skills, and morphological characteristics such as height and weight (Cui et al. 2019). When the differences in physical fitness and technical skills among athletes are relatively small, variations in anthropometric measurements may have greater significance and have been proven to be key determinants in the selection process (Hoare 2000). Basketball player success is largely related to their anthropometric characteristics (Ziv and Lidor 2009), because some morphological characteristics are greatly influenced by genetic factors and can't be significantly altered through training. However, other characteristics can be modified through training.

Due to tactical reasons, players have their designated positions, and playing positions significantly affect basketball players' body composition, anthropometric measurements, and physical fitness scores (Čović et al. 2023). However, Considering the specific height of the basket in reality and the number of shots that can win a basketball game, body composition takes the highest priority in determining playing positions (Garcia-Gil et al. 2018). Therefore, Basketball players with higher muscle content, longer arm spans and longer legs will play closer to the basketball hoop, while players with shorter height and weight will generally play farther from the hoop because they are more flexible. (Gryko et al. 2019). The three most common playing positions are guards, forwards, and centers. The purpose of this systematic review is to determine the body composition of male basketball players in these three playing positions.

2 Methodology

2.1 Study Selection

A search was conducted from August 1 to 2, 2024, across three electronic databases (Web of Science, Google Scholar, ScienceDirect) for publications from 2008 to 2024. The keywords used were basketball, elite athletes, playing positions, anthropometry, body composition, and other synonyms of these terms. A total of 18 articles were identified, including 3 quasi-experimental studies, 3 cross-sectional comparative studies, 11 comparative studies, and 1 study combining cross-sectional and longitudinal comparisons, all meeting the criteria and evidence for the research question. Among the 18 primary studies, 10 were focused on Europe, 2 on Asia, 1 on North America, 2 on Africa, 1 on South America, 1 on global data, and 1 on both North America and Europe. The timeframe for the searched databases was from 2005 to 2020. Data were extracted in a standardized and pre-specified manner (Table 1).

Table 1 The result of the studies

Studydata	Participants	Location	Measurement content	Design of study	Play position	Result
Te Wierike et al. (2015)	43 males, age 14.66 ± 1.09	Holland	Height, weight, leg length, sitting height, lean body mass, body fat percentage	An experimental study	Guard, forward, center	Comparisons between guards, forwards and centers show that guards are shorter and mature later, while centers are the tallest and mature earliest
Gryko et al. (2019)	109 men, ages U-14, U-15, U-16, U-18 and U-20	Poland	Height, weight, body fat percentage, BMI, arm span, sitting height, torso length, limb length, etc.	An experimental study	Guard, center forward	Height ($p < 0.001$) and wingspan ($p < 0.001$) showed significant differences; the center has the largest measurements, the forward has the second largest measurements, and the guard has the smallest measurements
Hoffman (2020) (Abdelkrim et al. 2010)	1000 male and female youth	USA and Europe	Height, weight, arm span, hand length, body fat percentage, etc.	Cross-sectional comparative study	Power forward and center, point guard, small forward, point guard	Centers are significantly taller, heavier, and have a greater wingspan compared to guards and forwards
Charni et al. (2022)	30 females, age 21.4 ± 2.3	Tunis	BMI, weight, height, skinfold thickness, etc.	An experimental study	Guard, forward, center	Weight (forwards: $p < 0.001$; guards: $p < 0.01$), height, and lower limb length (guards: $p < 0.001$; forwards: $p < 0.001$) were significantly higher in centers. Centers also had significantly higher BMI, leg and thigh muscle volume, and maximum thigh cross-sectional area compared to forwards
Gál-Pottyondy et al. (2021)	39, male and female, juvenile	multinational	Physical parameters such as height and weight	Comparative study	Guard, forward, center	Forwards and centers are taller, guards are the shortest, and centers are the heaviest

(continued)

Table 1 (continued)

Studydata	Participants	Location	Measurement content	Design of study	Play position	Result
Cormery et al. (2008)	68 males	France	Height, weight, skinfold thickness	Comparative study	Guard, forward, center	Centers are taller and heavier, while guards are smaller and lighter, and there is no significant difference in their average body fat percentage
Anđelić et al. (2021)	773 males, age 24–27	Five European countries (Montenegro, Russia, Italy, Spain, France)	Height, weight, body mass index (BMI)	Comparative study	Point guard, small forward, point guard, power forward and center	The height and weight of centers are significantly higher than those of point guards and forwards. Point guards have the lowest BMI, while centers have the highest BMI
Pojskić et al. (2014)	55, age 19.09 ± 3.13	Bosnia	Height, weight, percent body fat, percent defatted weight, hand length, foot length, leg length, arm length, shoulder width	Comparative study	Guard, forward, center	The height, weight, bone width, bone length and body circumference of the center are the largest, followed by the height, weight and limb length of the forward, and the smallest of the guard. There are no significant differences in body fat percentage, muscle mass percentage or body mass index between different positions
Cui et al. (2019)	1160 males, aged >19 years	Top global players	Height, arm length	Comparative study	Point guard, power forward, small forward and center	Height and wingspan are key determining factors for selecting guards
Čović et al. (2023)	47 males, age 14–15	Bosnia and Herzegovina	Height, weight, skinfold thickness, arm spread, hand spread and standing reach	Cross-sectional comparative study	Guard, forward, center	Centers are significantly taller, heavier, and fatter than forwards, and also heavier than guards. Centers have a higher body fat content

(continued)

Table 1 (continued)

Studydata	Participants	Location	Measurement content	Design of study	Play position	Result
Vaquera et al. (2015)	110 men's professional leagues (ACB, LEB, EBA) and national youth teams (U18 and U20)	Spanish	Weight, height, body mass index (BMI), skinfold thickness	Cross-sectional comparative study	Guard, forward, center	Guards were lighter than centers and forwards ($p < 0.05$), and forwards were 12.9% lighter than centers ($p < 0.05$). In terms of player height, guards were shorter than centers and forwards, and centers were higher than forwards ($p \leq 0.03$). The center had a higher BMI than the forwards and guards. Body fat content was no significant difference
Köklü et al. (2011)	45 males, age 24.0 ± 3.8	Turkish	Height, weight, body fat percentage	Comparative study	Guard, forward, Center	Guards are significantly lower than forwards and centers. Centers are significantly heavier than forwards and guards
Fields et al. (2018), Te Wierike et al. (2015)	323 females, age 18–24	USA	Height, weight, body composition	Cross-sectional and longitudinal study	Guards and forwards	Defenders are significantly smaller and leaner than forwards
Boone and Bourgois (2013)	144, age 26 ± 5, gender not mentioned	Belgium	Height, weight, body fat percentage	Comparative study	point guard, point guard, power forward and center, small forward	Centers are usually heavier than forwards and guards

(continued)

Table 1 (continued)

Studydata	Participants	Location	Measurement content	Design of study	Play position	Result
Nikolaos (2015)	22 males, age 15.33	Athens, Greece	Height, weight, body fat, BMI	Comparative study	Guard, forward, center	Centers were the heaviest, tallest, and had the highest levels of body fat and BMI, and forward players were those with the least body fat and lowest BMI. Defenders differed significantly in height, weight, and BMI centers
Erol et al. (2014)	30 male, U15	Turkish	Height, weight, leg length, body mass index, hand length, arm length, upper body length, hand width, arm span, body fat removal mass	Comparative study	Guard, forward, center	The defender has the smallest size and weight
Abdelkrim et al. (2010)	Under 18, under 20 and adult teams	Tunisia, Africa, Arab	Height, weight and body fat percentage	Comparative study	Point guard, point guard, power forward and center, small forward	Power forwards and centers are the tallest and heaviest, followed by guards and small forwards.
Sallet et al. (2005)	58 players, adult	French	Height, weight and body fat percentage	Comparative study	Guard, forward, center	The height (203.9 $\pm$ 5.3 cm) and weight (103.9 $\pm$ 12.4 kg) of centers were the highest, while forwards had the highest height (195.8 $\pm$ 4.8 cm) and weight (89.4 $\pm$ 7.1 kg), and guards had the highest height (185.7 $\pm$ 6.9 cm) and weight (82.0 $\pm$ 8.8 kg). Centers also had the highest body fat percentage, followed by forwards, and guards had the lowest

2.2 Inclusion and Exclusion Criteria

This systematic review included studies that met the criteria of investigating body composition differences among elite basketball players in different playing positions.

The exclusion criteria were: 1. Not in English; 2. Participants are basketball players with limb disabilities or college basketball players; 3. Studies that focus mainly on testing basketball skills and tactics, or psychological and physical tests; 4. Studies whose main evaluation is based on subjective measurements; 5. Studies that compare basketball with other sports; 6. University theses and dissertations.

2.3 Data Collection and Analysis

The information extracted from the selected literature includes 1. Study/data; 2. Participants; 3. Location; 4. Measurement content; 5. Design of study; 6. Play position; 7. Result. The extracted data can be found in Supplementary Material 1. We identified data that could support distinguishing playing positions and body characteristics. Each topic is described as a separate section, and we evaluate different positions and body compositions of basketball players in the article.

2.4 Research Quality

To ensure high-quality literature, articles were downloaded and assessed for quality using EasyScholar, a tool installed via Google Chrome. Articles were selected based on their quality evaluations provided by EasyScholar.

3 Result

The search strategy is illustrated in Fig. 1. Out of 4050 papers, there were 3790 excluded articles. Finally, only 18 articles were included in this systematic review. These 16 studies provided evidence of the body composition of elite basketball players in different playing positions. The participants were randomly selected, and the majority were elite athletes from Europe, Most of them are elite athletes from Europe, but there is relatively little research on athletes from other continents. Including both male and female athletes, as well as adults and minors.

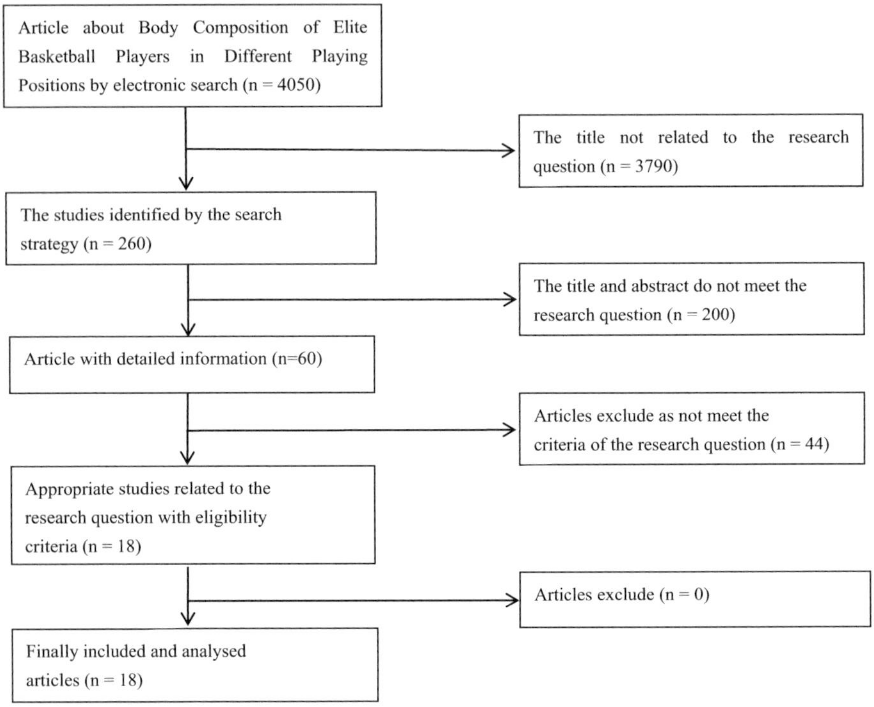

Fig. 1 Research methodology flowchart

4 Discussion

This review aimed to find evidence of differences in body morphology across playing positions. Our findings from 18 studies demonstrate that elite basketball players, whether in Europe, Asia, North America, Africa or globally, play different positions and body morphology, with several researchers finding that elite athletes are taller and heavier at the center position than forwards and guards, followed by guard basketball players being the shortest and lightest (Čović et al. 2023; Gryko et al. 2019; Gál-Pottyondy et al. 2021; Cormery et al. 2008; Anđelić et al. 2021; Pojskić et al. 2014; Vaquera et al. 2015; Köklü et al. 2011; Abdelkrim et al. 2010; Sallet et al. 2005). The position of centers requires them to have a significant advantage in height and weight to dominate inside confrontations. Centers usually have the largest absolute size and their role in rebounding, capping, and scoring inside is crucial.

Karol Gryko's research in 2019 mentions that elite basketball centers players are taller and have longer arms than forwards and guards, and Athletes who are taller usually have longer arms, mainly because their arms can reach farther (Gryko et al. 2019). These characteristics are very important in the game of basketball, including defending spacing and jumping, as well as dunking or shooting (by shorter players). From this, it can be seen that the height of a basketball player has a very big advantage in the game, in all aspects of basketball (Zarić et al. 2020). Milan Anđelić in 2021 found that point guards had the lowest BMI and centers had the highest (Anđelić et al. 2021); At the same time, another study conducted in two phases found consistent results, also finding that centers had the highest BMI, while point guards had the lowest BMI.(Vaquera et al. 2015; Nikolaos 2015; Charni et al. 2022). It was found that all European leagues had similar BMI values and that BMI proved to be very important (Cormery et al. 2008), So the BMI is consistent among elite basketball players around the world.

Cormery et al. (2008(found it is no significant difference in body fat ratios (Cormery et al. 2008), however, four other studies found that centers had higher body fat than guards and forwards, and guards had the lowest body fat (Čović et al. 2023; Vaquera et al. 2015; Nikolaos 2015; Boone and Bourgois 2013; Sallet et al. 2005), it has been explained that smaller players weigh less and therefore accelerate more easily and exert less force.

Charni (2022) found that centers not only have higher height and weight BMI than forwards and guards but also have significantly higher leg and thigh muscle volume as well as maximum thigh cross-sectional area than forwards (Charni et al. 2022). Wierike et al. (2014) found that guards are the shortest to mature later, and centers are the tallest to mature earlier. This phenomenon may explain the need to take maturity into account in the selection process of youth players. There is a dearth of literature on the body morphology of elite basketball players at their playing positions, with studies focusing on European players, Relatively little research has been done in other continents. In addition, some of the studies on elite basketball players have a relatively small sample size.

This systematic review aimed to identify the playing positions of basketball players with different body morphology and studies that could support the differentiation between playing levels and playing positions. We also sought to find out and answer whether the results of field tests such as basketball are correlated with game statistics. In 18 studies, it can be demonstrated that height, weight, BMI, arm span and body fat are key indicators for basketball players.

In addition, maturity is an important influencing factor for U18 teams. Although premature selection of playing positions is not recommended, We attempted to summarize the physical morphology of basketball players' playing positions to support the differentiation of basketball playing positions (Table 1). The physical morphology of basketball players' playing positions may help coaches find the most suitable position for players in talent identification. The results showed that the physical morphology of basketball players is particularly important, especially it is the most powerful factor in differentiating basketball players in different positions, In addition, research in other continents besides Europe also needs to pay attention to the body morphology of high-level athletes in different positions.

5 Conclusion

These 18 studies have shown that body shape has been shown to be related to different playing positions and has a high number of influencing factors. Height, weight, BMI, body fat and arm span were all affected. In this systematic evaluation, we aimed to identify the specific body form needed for specific positions; however, for players under 18 years of age, their maturity level should also be considered.

References

Abdelkrim NB et al (2010) Positional role and competitive-level differences in elite-level men's basketball players. J Strength Cond Res 24(5):1346–1355

Anđelić M, Joksimović M, Kukrić A, Nikšić E, D'Angelo S, Zlojutro N, Skrypchenko I, Ćeremidžić D (2021) Body height, body mass, body mass index of elite basketball players in relation to the playing position and their importance for success in the game. Acta kinesiologica 15(2):74–79

Boone J, Bourgois J (2013) Morphological and physiological profile of elite basketball players in Belgium. Int J Sport Physiol Perform 8(6):630–638

Charni Y, Hammami M, Gaamouri N, Aouadi R, Gaith A, Oranchuk DJ, Souhaiel Chelly M, Bragazzi N (2022) Postural profile of female basketball players and differences among playing positions. Eur Rev Med Pharmacol Sci 26(17)

Cormery B, Marcil M, Bouvard M (2008) Rule change incidence on physiological characteristics of elite basketball players: a 10-year-period investigation. Br J Sport Med 42(1):25–30

Čović N, Čaušević D, Alexe CI, Rani B, Dulceanu CR, Abazović E, Lupu GS, Alexe DI (2023) Relations between specific athleticism and morphology in young basketball players. Frontiers in sports and active. Living 5:1276953

Cui Y, Liu F, Bao D, Liu H, Zhang S, Gómez MÁ (2019) Key anthropometric and physical determinants for different playing positions during National Basketball Association draft combine test. Front Psychol 10:2359

Erol E, Özen G, Koç H (2014) General anthropometric and selected motor skills of elite young male basketball players according to position on the court of players. J Athl Perform Nutr 1(1):1–9

Gál-Pottyondy A, Petró B, Czétényi A, Négyesi J, Nagatomi R, Kiss RM (2021) Field testing protocols for talent identification and development in basketball—a systematic review. Appl Sci 11(10):4340

Garcia-Gil M, Torres-Unda J, Esain I, Duñabeitia I, Gil SM, Gil J, Irazusta J (2018) Anthropometric parameters, age, and agility as performance predictors in elite female basketball players. J Strength Cond Res 32(6):1723–1730

Gryko K, Stastny P, Kopiczko A, Mikołajec K, Pecha O, Perkowski K (2019) Can anthropometric variables and maturation predict the playing position in youth basketball players? J Hum Kinet 69(1):109–123

Hoare DG (2000) Predicting success in junior elite basketball players—the contribution of anthropometic and physiological attributes. J Sci Med Sport 3(4):391–405

Köklü Y, Alemdaroğlu U, Koçak F, Erol A, Fındıkoğlu G (2011) Comparison of chosen physical fitness characteristics of Turkish professional basketball players by division and playing position. J Hum Kinet 30(2011):99–106

Nikolaos K (2015) Anthropometric and fitness profiles of young basketball players according to their playing position and time. J Phys Educ Sport 15(1):82

Pojskić H, Separovic V, Muratovic M, Uzicanin E (2014) Morphological differences of elite Bosnian basketball players according to team position. Int J Morphol 32(2):690–694

Sallet P et al (2005) Physiological differences in professional basketball players as a function of playing position and level of play. J Sport Med Phys Fit 45(3):291

Te Wierike SCM, Elferink-Gemser MT, Tromp EJY, Vaeyens R, Visscher C (2015) Role of maturity timing in selection procedures and in the specialisation of playing positions in youth basketball. J Sport Sci 33(4):337–345

Vaquera A, Santos S, Villa JG, Morante JC, García-Tormo V (2015) Anthropometric characteristics of Spanish professional basketball players. J Hum Kinet 46(1):99–106

Zarić I, Kukić F, Jovićević N, Zarić M, Marković M, Toskić L, Dopsaj M (2020) Body height of elite basketball players: do taller basketball teams rank better at the FIBA world cup? Int J Environ Res Public Health 17(9):3141

Ziv G, Lidor R (2009) Physical attributes, physiological characteristics, on-court performances and nutritional strategies of female and male basketball players. Sport Med 39:547–568

Comparative Study of Anthropometry and Body Composition of University Basketball Players in China and Malaysia

Gao Yili, Hadafi Fitri Mohd Latip, and Mohd Shahrizal Sunar

Abstract The aim of this study was describe and compare anthropometric and body composition of university basketball players in China and Malaysia. The study involved 60 male participants, divided into two groups: 30 basketball players from the School of Physical Education at Zhengzhou University of Technology (ZUT) China, and 30 players from the university-level basketball team at Universiti Teknologi Malaysia (UTM). All participants were between the ages of 19–24. Morphological characteristics were assessed using four variables: height, weight, leg length, and foot size. Body composition was evaluated using two variables: body mass index (BMI) and body fat percentage. The central parameters and dispersion parameters of all variables were calculated. Non-parametric tests and correlation analyses were conducted to determine the differences and correlations in anthropometric measurements and body composition between ZUT and UTM basketball players. The results showed that ZUT athletes have significantly greater height, weight, and leg length compared to UTM athletes ($p < 0.001$ and $p = 0.009$). There is no significant difference in BMI and foot size between the two groups ($p = 0.684$, $p = 0.790$). The body fat percentage in the ZUT group is slightly higher than that of the UTM group. Both groups show a significant positive correlation between height and weight, foot size, and leg length, and the ZUT group shows higher correlations between BMI, body fat percentage, height and weight, and height and foot size. Therefore, these findings

G. Yili
College of Physical Education, Zheng Zhou University of Technology, Zhengzhou, Henan, China

G. Yili · H. F. M. Latip (✉)
Sport Innovation and Technology Centre, Institute of Human Centered Engineering Universiti Teknologi Malaysia, Johor Bahru, Johor, Malaysia
e-mail: hadafifitri@utm.my

H. F. M. Latip
School of Education, Faculty Social Science and Humanity, Universiti Teknologi Malaysia, Johor Bahru, Johor, Malaysia

M. S. Sunar
Department of Emergent Computing, Faculty of Computing, Universiti Teknologi Malaysia, Johor Bahru, Johor, Malaysia

M. H. A. Hassan et al. (eds.), *Proceedings of the 3rd International Conference on Innovation and Technology in Sports*, Lecture Notes in Bioengineering,
https://doi.org/10.1007/978-981-95-0804-4_30

may provide coaches with insights into the significant anthropometric differences between the two countries.

Keywords Basketball · Morphological characteristics · Anthropometry · Body composition · University students

1 Introduction

College basketball plays a significant role in physical education programs (Li and Nasiruddin 2023). Today, it stands as one of the most popular sports globally (Li et al. 2017), boasting a history that spans 128 years (Sansone et al. 2022). Additionally, basketball enjoys one of the highest participation rates among youth, widely applied in university campus physical activities. The widespread popularity of basketball is primarily due to its ease of learning, especially the accessibility of technical facilities and equipment (Li et al. 2017). Not only does it have high participation, but it is also an entertaining sport (Masanovic et al. 2019).

Anthropometry is defined as 'the scientific procedure and process of measuring the anatomical dimensions of the human body surface using specialized equipment, such as length, width, circumference, and skinfold thickness' (Slice 2005). In certain sports, specific anthropometric and body composition parameters required for success are crucial. Success in basketball is closely related to specific anthropometric characteristics, body composition, and somatotype (Guimarães et al. 2019; Gutnik et al. 2015). Moreover, anthropometry is also a key determining factor in the selection process of basketball players (Vaquera et al. 2015).The success of basketball players is inseparable from their excellent anthropometric indicators (Hoare 2000). However, body composition is also an important factor influencing athletes' performance and health, including body fat and body mass index (Keating et al. 2017). Mathur and Salokun (1985) proposed that physical characteristics and body composition form the foundation for achieving superior athletic performance (Mathur and Salokun 1985).

For a long time, the anthropometric characteristics and body composition of athletes have been the subject of numerous studies, as many researchers believe that with long-term systematic training, the body will adapt morphologically and functionally to the influences it undergoes (Masanovic 2018).It is noteworthy that many of the measurements reported by Carter and Eiben et al. in 1981 were conducted during the 1970s and 1980s (Carter 1981; Eiben 1981).However, in many places, more time has been spent enhancing athletes' physical fitness without considering the assessment of their body composition and nutritional status (Triki et al. 2012).When differences in fitness and skills among athletes are small, disparities in anthropometric indices may become more pronounced, and these have been shown to be key determining factors in the selection process (Hoare 2000). Basketball players' success is largely correlated with their anthropometric characteristics (Ziv and Lidor 2009).

Accurate assessment of body composition in sports is crucial, as incorrect evaluation may lead to errors in training prescriptions and dietary formulations, thereby affecting athletic performance (Popovic et al. 2014).

However, despite extensive research on the anthropometry and body composition of athletes from various countries and regions, there is relatively little research on Asian collegiate basketball players; furthermore, Asian countries are unevenly ranked in the world, and from the Asian University Basketball Championship (AUBC), the rankings show that China ranks No.1 (men) while Malaysia ranks No. 25 (men) (AUBC 2024), with Chinese basketball players placed at the top and Malaysian players at a moderate level, so it seems extremely important to determine the reasons behind this gap in their performance levels.

This study will select Chinese collegiate male basketball players and Malaysian collegiate male basketball players, and by comparing their anthropometric and body composition data, it aims to identify the differences in these measurements between China (ZUT) and Malaysian (UTM) university basketball players, thereby filling this research gap and providing new data and insights with high academic value and innovation. Therefore, this study aims to compare the anthropometric and body composition characteristics of Chinese and Malaysian basketball players in order to reveal significant differences and provide a better understanding.

2 Materials and Methods

2.1 Participants

This article uses purposive sampling, mainly screening from three factors: age, gender, and basketball player status. Then, according to the rule of thumb, 30 subjects is sufficient for statistical analysis. A total of 60 participants were selected, comprising 30 male basketball players from the School of Physical Education at Zhengzhou University of Technology (ZUT) and 30 male basketball players from the basketball team of Universiti Teknologi Malaysia (UTM).

Participants were between 19 and 24 years old. The tests included anthropometric measurements and body composition, conducted over the course of one week. The analysis used the average of three measurements. At the time of testing, all participants were in good clinical health, with no recent infectious diseases, asthma, or cardiopulmonary disease history, nor any injuries preventing them from participating in intense physical activity. Prior to the testing, all basketball players expressed their willingness to participate and provided verbal consent. They were instructed to refrain from engaging in strenuous exercise and to maintain a normal diet on the day preceding the test. At 13 min warm-up was conducted prior to the measurements.

The measurement time of the two groups of samples will be carried out in the second half of 2023. 30 ZUT basketball players will be measured in September, and 30 UTM basketball players will be measured in December. The advantage of this is that in order to minimize the impact of the temperature difference between the two countries, the measurement temperature of the two groups of athletes is kept as consistent as possible to reduce errors. Both groups of tests are around 30 degrees Celsius, and the test period lasts for one week.

2.2 Research Procedure

Anthropometric measurements were performed according to the standards proposed by the International Society for the Advancement of Kinanthropometry (International Society for the Advancement of Kinanthropometry ISAK 2001). This test covered seven variables (height, weight, body mass index, leg length, foot length, and skinfold measurements). The instruments were calibrated prior to testing. Participants were measured on an outdoor basketball court while wearing light clothing.

Height was measured, with a fixed stadiometer to the nearest 0.1 cm. Weight was measured, with a standard scale or portable balance to the nearest 0.1 kg. The body mass index (BMI) was calculated by dividing body weight (kg) by the square of body height (m), with the following ranges: BMI < 18.5 kg/m^2 (underweight), 18.5–24.9 kg/m^2 (normal), 25.0–29.9 kg/m^2 (overweight), and BMI $\geq$ 30.0 kg/m^2 (obese) (WHO 2000). Leg length was then measured to the nearest 0.1 cm with a tape measure and foot length was measured to the nearest 0.1 cm with a caliper. Skin fold thickness was measured to the nearest 0.1 cm at two locations (triceps brachii and subscapularis) with a caliper.

To minimize measurement errors, each athlete was tested three times (sports anthropometric measurements) and the average was taken as the final measurement. When measuring skinfold thickness, the thumb and index finger are used to pinch the skinfold first, and then a caliper is applied to clamp the subcutaneous fat. Skinfold thickness is measured on the right side of the body, with each site measured three times to obtain the thickness in those three areas. All skinfold measurements are conducted on the basketball court, and to reduce measurement error, the same researcher measures all participants.

Body fat percentage was calculated using the equation developed by (Slaughter et al. 1988).

$$\text{Body fat\% (Male)} = 6.93 + 0.428(\text{triceps} + \text{subscapular angle})$$

2.3 Statistical Analysis

In this study, all paper-based measurement results were entered into an Excel spreadsheet and then processed using IBM SPSS Statistics 26.0 software on my computer. Descriptive statistics were first used to calculate the central parameter and discrete parameter of each anthropometric variable to understand the basic distribution of the data. To determine whether the data conformed to a normal distribution ($P > 0.05$), the Mann-Whitney U test was used to compare the two groups of variables, with significance set at $P < 0.05$. Finally, Spearman correlation analysis was performed to explore the relationship between different anthropometric measurements and body composition indicators, with $P < 0.05$ considered statistically significant.

3 Results

As shown in Table 1, using central parameters, dispersion parameters, and the Shapiro-Wilk test revealed that the height and foot length of both UTM and ZUT basketball players followed a normal distribution ($P > 0.05$). The tallest ZUT player was 1.98 m, whereas the tallest UTM player was 1.88 m. The average height of the UTM group was 1.76 m, compared to 1.82 m for the ZUT group, indicating a slightly taller average height among the ZUT group. Moreover, the tallest ZUT player is 10 cm taller than the tallest UTM player.

In addition, the weight of the UTM group was normally distributed: skewness = 0.540, kurtosis = −0.601, P = 0.057. The average weight of the UTM group was 66.5 kg, while that of the ZUT group was 73.3 kg. This shows that Chinese basketball players (ZUT) weigh more than Malaysian basketball players (UTM), which aligns with the fact that Chinese players are generally taller and thus slightly heavier.

Table 1 Descriptive statistics and normality tests of physical measurement and body composition variables[a]

Group	Variables	Min	Max	Median	Skewness	Kurtosis	Shairo-Wilk sig
UTM	Height (m)	1.63	1.88	1.76	−0.192	−0.402	0.670
ZUT		1.72	1.98	1.82	0.240	0.072	0.760
UTM	Weight (kg)	52.0	87.0	66.5	0.540	−0.601	0.057
ZUT		63.0	96.2	73.3	0.924	0.532	0.032
UTM	Foot size (cm)	23.6	29.5	26.3	0.045	−0.173	0.829
ZUT		23.8	28.0	26.5	−0.878	0.654	0.113
UTM	Leg length (cm)	86.0	103.0	92.0	0.912	1.092	0.036
ZUT		88.2	110.4	97.3	0.605	0.163	0.271
UTM	BMI	17.8	29.8	21.3	0.955	0.792	0.020
ZUT		18.0	29.0	21.7	1.209	2.126	0.014
UTM	Body fat%	11.6	25.7	14.2	0.720	−0.903	0.002
ZUT		11.1	23.1	16.2	0.429	0.237	0.692

Note [a]Grouping variable: ZUT and UTM (N = 30); P > 0.05, normal distribution; BMI—Body Mass Index

Although the two countries' athletes differed in average weight, Table 1 indicates that ZUT players are heavier than UTM players; however, their BMIs are quite close: the UTM group's BMI was 21.3, and the ZUT group's was 21.7. Thus, the higher average weight in the ZUT group compared to the UTM group can be attributed to their relatively greater height. Both groups had BMI values within the normal range, and their maximum and minimum BMI values were similar, indicating that the overall BMI levels of the two groups were nearly the same.

Leg length in the ZUT group was normally distributed, with a maximum value of 110.4, a skewness of 0.605 (slightly positive skew), a kurtosis of −0.163 (indicating a flatter distribution than normal), and $P = 0.271$ ($P > 0.05$). Body fat percentage in the ZUT group was also normally distributed ($P = 0.692$). These results suggest that leg length and body fat percentage in the ZUT group both followed normal distributions, and the ZUT group's leg length exceeded that of the UTM group, possibly due to differences in overall height.

However, the body mass index of both UTM and ZUT basketball players did not conform to a normal distribution ($P < 0.05$). The ZUT group's weight slightly deviated from normality ($P = 0.032$), falling just below 0.05. Moreover, leg length and body fat percentage in the UTM group did not follow a normal distribution ($P < 0.05$).

From Table 1, it can be observed that height and foot length in the UTM and ZUT groups generally follow a normal distribution, whereas weight, body mass index (BMI), and body fat percentage show clear deviations from normality in certain cases. This indicates that nonparametric statistical methods should be used for subsequent analyses of these variables.

As shown in Table 2, nonparametric tests reveal significant differences between ZUT and UTM. Height ($p < 0.001$), leg length ($p < 0.001$), and weight ($p < 0.009$) all exhibit significant differences, with height and leg length being highly significant. This finding suggests that ZUT basketball players are taller than UTM players, which also leads to a leg-length advantage. However, there are no significant differences in foot length, BMI, or body fat between ZUT and UTM. In summary, the variables with significant differences are height, weight, and leg length (all p values <0.05). The variables without significant differences are BMI, foot length, and body fat (all p values >0.05). These results suggest that, in the comparison of anthropometric measures, Chinese and Malaysian students differ significantly in height, weight, and leg length, whereas their BMI, foot length, and body fat do not show significant differences.

As shown in Table 3, there are notable positive correlations in both groups. For instance, there is a strong positive correlation between weight and BMI (UTM: 0.85, ZUT: 0.71), indicating that changes in weight are directly reflected in BMI. There is also a clear positive correlation between height and both foot length and leg length, suggesting that taller athletes generally have longer legs and larger feet. Weight and body fat percentage demonstrate a moderate positive correlation (UTM: 0.47, ZUT: 0.50). Furthermore, height shows a significant positive correlation with both foot length and leg length, reinforcing that taller athletes tend to have longer legs and

Table 2 The non-parametric test results for anthropometric and body composition variables[a]

	Height	Weight	BMI	Foot size	Leg length	Body fat
Mann-Whitney U	176.0	274.0	422.5	432.0	134.5	428.5
Wilcoxon W	641.0	739.0	887.5	897.0	599.5	893.5
Z	−4.06	−2.60	−0.41	−0.27	−4.67	−0.32
Asymp. sig. (2-tailed)	<0.001	0.009	0.684	0.790	<0.001	0.750

Note [a]Grouping variable: ZUT and UTM, $p < 0.05$ indicates statistical significance

bigger feet, and foot length is also significantly positively correlated with leg length.

In both groups, the correlation between BMI and height is low: there is no apparent correlation between BMI and height in either group (UTM: 0.15, ZUT: −0.28). Body fat percentage does not show any notable correlation with height, foot length, or other measured variables.

Table 3 Correlation between anthropometric and body composition of UTM and ZUT students

Correlation coefficient		Height	Weight	BMI	Foot size	Leg length	Body fat
Height	UTM	1.00	0.57**	0.15	0.63**	0.66**	−0.10
	ZUT	1.00	0.39*	−0.28	0.55**	0.50**	0.05
Weight	UTM	0.57**	1.00	0.85**	0.55**	0.32	0.47**
	ZUT	0.39*	1.00	0.71**	0.45*	0.62**	0.50**
BMI	UTM	0.15	0.85**	1.00	0.22	−0.04	0.57**
	ZUT	−0.28	0.71**	1.00	−0.04	0.16	0.62**
Foot size	UTM	0.63**	0.55**	0.22	1.00	0.36*	0.30
	ZUT	0.55**	0.45*	−0.04	1.00	0.62**	−0.06
Leg length	UTM	0.66**	0.32	−0.04	0.36*	1.00	−0.17
	ZUT	0.50**	0.62**	0.16	0.62**	1.00	0.06
Body fat	UTM	−0.10	0.47**	0.57**	0.30	−0.17	1.00
	ZUT	0.05	0.50**	0.62**	−0.06	0.06	1.00

Note Correlation is significant at the 0.01 level (2-tailed) for **, and at the 0.05 level (2-tailed) for *

4 Discussion

4.1 Comparison of Anthropometry and Body Composition

Based on the results of descriptive statistics and normality tests, the two groups of athletes showed some differences in height, weight, BMI, and body fat percentage. For example, ZUT athletes had significantly greater height ($P < 0.001$), weight ($P = 0.009$), and leg length ($P < 0.001$) compared to UTM athletes. This may be related to geographical and genetic factors, as well as differences in nutrition and living conditions between the regions. This result aligns with previous studies, indicating that basketball players from different regions indeed have significant differences in height (Salmón-Gómez et al. 2023).

The results showed that the weight of ZUT students was significantly higher than that of UTM students ($p < 0.009$). On the other hand, this may have been caused by dietary practices and the level of sports coaching. Baseball players who are taller may include heavier weights, which may be related to height differences, as well as environmental factors and training intensity. Numerous components, like exercise and nutrition, are thought to have an impact on weight, due to research (Zhang et al. 2024; Bojarczuk et al. 2024). Additionally, the ZUT group's baseball players were heavier, perhaps as a result of their greater muscle mass, which is very useful in sports.

There was no discernible difference between the ZUT and UTM groups in terms of foot length, according to the foot length statistics ($p = 0.790$). There were no significant differences in foot length between the ZUT and UTM parties, which suggests that both parties had different heights and weights. In general, foot length is usually very correlated with height, but this connection is not always absolute. Gryko et cetera (2018) pointed out that there was no significant difference in foot length among basketball players in different geographical regions (Gryko et al. 2018). In sports or games, the freedom of foot length plays an important role in performance, but in this review, foot length did not show significant differences, which may indicate that the distinction in foot length among basketball players in various regions may be tiny, that is, the environment has many effects on foot length.

The data on the leg length of basketball players showed that the leg length of basketball players in the ZUT group was significantly longer than that of basketball players in the UTM group ($P < 0.001$). In sports events, the leg length of players is a key factor influencing explosive power and is also related to activity performance (Papla et al. 2022). For hockey people, leg length is just as crucial. Baseball players in the ZUT group have a dynamic edge on the court because they have long legs, particularly when fighting and recovering, which necessitate fast responses and long-distance going. The analysis of baseball players 'leg length additionally supports the hypothesis that height variations affect their athletic performance (Taylor et al. 2020).

The basketball players in the ZUT and UTM groups had a small BMI difference, according to the BMI data analysis ($p = 0.684$). When examining the weight-to-height amount, the body mass index (BMI) is an important sign and is often used to decide whether a woman's weight is at a good amount. Only a select few players in this study had regular BMI values above the normal range, whereas the regular BMI values of the individuals in both parties were within the ordinary range. Despite this, the two parties 'BMI did not significantly differ between the students.

This demonstrates that the entire weight ratio remained almost constant despite the significant weight and height differences. In contrast, it even shows that the BMI indicators of both teams of players remained secure and within the normal range. But, each group's diet and training regimen varied. High-intensity education, which helped both student groups gain muscle while reducing body fat and maintaining a good BMI, may have contributed to this persistence. Ramos-Campo and others agree. Increasing muscle large helps improve strength and power, which are necessary for sports performance, according to (Ramos-Campo et al. 2014), but too much weight can affect sports results.

According to the study's findings, ZUT athletes and UTM athletes had no statistically significant differences in body fat percentage ($P = 0.750$). The two teams of players may have had similar body weight percentages because of their diverse diet and exercise regimens. Vigo et cetera. Athletes 'endurance, speed, and agility, which are indicators of high-level performance, are affected by higher fat mass (FM) (Vigo and Viviani 2020). Despite significant differences in the athletes 'height and weight, the body large ratios between the two teams of baseball players were nearly the same during intense training.

4.2 Correlation Analysis of Anthropometry and Body Composition

In this study, a correlation analysis was conducted on height, weight, BMI, foot length, leg length, and body fat percentage in both UTM and ZUT groups.

First, as seen in Table 3, there are significant differences in correlations between different variables in the UTM and ZUT groups. For the UTM group, height showed a strong positive correlation with weight ($r = 0.569$, $p < 0.01$), consistent with previous research (Gryko et al. 2018). Pointed out that under normal BMI conditions, the height and weight of high-level athletes are usually positively correlated because taller athletes usually need heavier weight to achieve the best athletic performance (Gryko et al. 2018). In contrast, the ZUT group's height and weight correlation was weaker ($r = 0.385$; $p = 0.01$), which could be attributed to the significant height and weight differences between these athletes, which could be brought on by stricter athlete management or negligence.

BMI showed a significant positive correlation with body fat percentage in the UTM group ($r = 0.570$, $p < 0.01$), indicating that athletes with higher BMI also tend to have higher body fat percentage. This finding is in line with Ranasinghe et al.'s conclusions. Ranasinghe et al. (2013), who also found a significant positive correlation between BMI and body fat percentage, especially in adolescent basketball players. In contrast to the present study, a stronger correlation was found between BMI and body fat percentage in the ZUT group of basketball players ($r = 0.618$, $p < 0.01$), a finding also supported by (Mazic et al. 2009).

Additionally, sports players 'height and foot length were significantly related to their weight and leg length in the UTM type. Large people so have extended feet. Longer arms make athletes heavier. Ran et al. (2020) study the relationship between body type and entire figure type was confirmed by 2020. According to the research that polled Chinese people, there was a significant positive correlation between height, foot length, and weight. According to the research, feet and leg length may be explained in part by height and weight.

Eventually, the association between body fat percentage weight and BMI was stronger in the ZUT group than in the UTM group. The ZUT group's statistics, on the other hand, demonstrated a weaker relationship between body fat percentage and different variables. Based on this difference, hockey players in the ZUT group do pay more attention to overall physical fitness and skill performance, while runners in the UTM group may pay more attention to body fat power. Who emphasized that different training methods have different effects on athletes 'bodies (Xiao et al. 2021).

5 Conclusion

The mean values of height, weight, and leg length of Chinese ZUT basketball players were statistically significantly higher than those of Malaysian UTM basketball players ($P < 0.001$ and $P = 0.009$, both), showing that geographic location and ethnic background may be critical factors influencing these indicators. Between the two groups of basketball players, there were no significant differences in BMI and foot length ($p = 0.684$, $p = 0.790$, respectively). Yet, the body fat percentage of ZUT baseball players was slightly higher than that of UTM baseball people, which may be due to the different diet and training methods of the two parties.

The correlation analysis of the research showed that although there were significant differences between BMI and body fat percentage between the two teams of sports people, there were major positive correlations between height and weight, foot length and leg length. There were major connections between body fat percentage, height and weight, and height and foot length and body fat percentage in the ZUT group, showing that these indicators do have a greater effect on their sports performance.

References

AUBC (2024) Asian University Basketball Championship https://www.timeshighereducation.com/world-university-rankings/2024/regional-ranking?page=2#

Bojarczuk A, Egorova ES, Dzitkowska-Zabielska M, Ahmetov II (2024) Genetics of exercise and diet-induced fat loss efficiency: a systematic review. J Sport Sci Med 23(1):236
Carter JL (1981) Somatotypes of female athletes. In: The female athlete, vol 15. Karger Publishers, pp 85–116
Eiben OG (1981) Physique of female athletes—anthropological and proportional analysis. In: The female athlete, vol 15. Karger Publishers, pp 127–141
Gryko K, Kopiczko A, Mikołajec K, Stasny P, Musalek M (2018) Anthropometric variables and somatotype of young and professional male basketball players. Sports 6(1):9
Guimarães E, Baxter-Jones A, Maia J, Fonseca P, Santos A, Santos E, Janeira MA (2019) The roles of growth, maturation, physical fitness, and technical skills on selection for a Portuguese under-14 years basketball team. Sports 7(3):61
Gutnik B, Zuoza A, Zuozienė I, Alekrinskis A, Nash D, Scherbina S (2015) Body physique and dominant somatotype in elite and low-profile athletes with different specializations. Medicina 51(4):247–252
Hoare DG (2000) Predicting success in junior elite basketball players—the contribution of anthropometric and physiological attributes. J Sci Med Sport 3(4):391–405
International Society for the Advancement of Kinanthropometry (ISAK) (2001) International standards for anthropometric assessment. Underdale, SA, Australia
Keating SE, Johnson NA, Mielke GI, Coombes JS (2017) A systematic review and meta-analysis of interval training versus moderate-intensity continuous training on body adiposity. Obes Rev 18(8):943–964
Li L, Simiyu WW, Liao T, Feng Y (2017) Selected demographic characteristics of male basketball players: the case of China and the USA. J Phys Educ Sport 17(4):2678–2684
Li M, Nasiruddin NJBM (2023) The effect of blended learning in basketball teaching in Chinese universities. Revista Ibérica de Sistemas e Tecnologias de Informação E62(5):84–594
Masanovic B (2018) Comparative study of anthropometric measurement and body composition between junior basketball and volleyball players from Serbian national league. Sport Mont 16(3)
Masanovic B, Popovic S, Bjelica D (2019) Comparative study of anthropometric measurement and body composition between basketball players from different competitive levels: elite and sub-elite. Pedagog, Psychol, Med-Biol Probl Phys Train Sport 4:176–181
Mathur DN, Salokun SO (1985) Body composition of successful Nigerian female athletes. J Sport Med Phys Fit 25(1–2):27–31
Mazic S, Djelic M, Suzic J, Suzic S, Dekleva M, Radovanovic D, Starcevic V (2009) Overweight in trained subjects—are we looking at wrong numbers? (body mass index compared with body fat percentage in estimating overweight in athletes). Gen Physiol Biophys 28:200–204
Papla M, Perenc D, Zając A, Maszczyk A, Krzysztofik M (2022) Contribution of strength, speed and power characteristics to change of direction performance in male basketball players. Appl Sci 12(17):8484
Popovic S, Bjelica D, Jaksic D, Hadzic R (2014) Estudio Comparativo de Medidas Antropométricas y Composición Corporal entre Jugadores Élite de Fútbol y Voleibol. Int J Morphol 32(1):267–274
Ramos-Campo DJ, Martínez Sánchez F, Esteban García P, Rubio Arias JÁ, Bores Cerezal A, Clemente-Suarez VJ, Jiménez Díaz JF (2014) Características antropométricas en función del puesto en jugadores profesionales de equipo: Baloncesto, balonmano y fútbol sala. Int J Morphol 32(4):1316–1324
Ran L, Luo H, Zhao C, Zhang X, Hu H, Wang Z (2020) A study on the correlation of foot data with body height and weight of Chinese adults. In: Advances in physical ergonomics and human factors: proceedings of the AHFE 2019 international conference on physical ergonomics and human factors, vol 10. Springer International Publishing, Washington DC, USA, pp 197–203
Ranasinghe C, Gamage P, Katulanda P, Andraweera N, Thilakarathne S, Tharanga P (2013) Relationship between body mass index (BMI) and body fat percentage, estimated by bioelectrical impedance, in a group of Sri Lankan adults: a cross-sectional study. BMC Public Health 13:1–8

Salmón-Gómez L, Catalán V, Frühbeck G, Gómez-Ambrosi J (2023) Relevance of body composition in phenotyping the obesities. Rev Endocr Metablc Disord 24(5):809–823

Sansone P, Makivic B, Csapo R, Hume P, Martínez-Rodríguez A, Bauer P (2022) Body fat of basketball players: a systematic review and meta-analysis. Sport Med-Open 8(1):26

Slaughter MH et al (1988) Skinfold equations for estimation of body fatness in children and youth. Hum Biol:709–723

Slice DE (2005) Modern morphometrics. In: Modern morphometrics in physical anthropology. Springer, US, Boston, MA, pp 1–45

Taylor JB, Hegedus EJ, Ford KR (2020) Biomechanics of lower extremity movements and injury in basketball. BasketbL sport. Med Sci 37–51

Triki M, Rebai H, Abroug T, Masmoudi K, Fellmann N, Zouari N, Tabka Z (2012) Comparative study of body composition and anaerobic performance between football and judo groups. Sci Sport 27(5):293–299

Vaquera A, Santos S, Villa JG, Morante JC, García-Tormo V (2015) Anthropometric characteristics of Spanish professional basketball players. J Hum Kinet 46(1):99–106. https://doi.org/10.1515/hukin-2015-0038

Vigo A, Viviani F (2020) The adolescent basketball player: the importance of some anthropometric characteristics for speed, resistance, power and agility. Antrocom: Online J Anthropol 16(2)

WHO (2000) Obesity: preventing and managing the global epidemic. Report of a WHO consultation. In: WHO consultation on obesity. World Health Organization, Geneva, Switzerland

Xiao W, Soh KG, Wazir MRWN, Talib O, Bai X, Bu T, Gardasevic J (2021) Effect of functional training on physical fitness among athletes: a systematic review. Front Physiol 12:738878

Zhang M, Ward J, Strawbridge RJ, Celis-Morales C, Pell JP, Lyall DM, Ho FK (2024) How do lifestyle factors modify the association between genetic predisposition and obesity-related phenotypes? A 4-way decomposition analysis using UK biobank. BMC Med 22(1):230

Ziv G, Lidor R (2009) Physical attributes, physiological characteristics, on-court performances and nutritional strategies of female and male basketball players. Sport Med 39:547–568

A Low-Cost Goal Line Technology (GLT) Solution for Grassroots Football: Performance Analysis of Ball Detection across Variable Speeds

Ilmi Mohd Ariffin, Zulkifli Mohamed, Mohd Hanif Mohd Ramli, Mohd Rozaiman Aziz, and Azlan Aziz

Abstract This study introduces a portable and affordable Goal Line Technology (GLT) system tailored for grassroots and lower-tier football leagues. The system is built using open-source tools, with OpenCV handling image processing and YOLOv8 enabling real-time ball detection. Designed for a compact Raspberry Pi setup paired with a Pi Camera, the system's performance was rigorously evaluated at various ball speeds to determine its detection accuracy as a key factor for reliable goal confirmation during fast-paced football matches. The results show the system achieves 95% accuracy for ball speeds up to 15 m/s. However, at higher speeds, such as 25 m/s, accuracy declines to 65%, highlighting hardware limitations in handling rapid movements. These findings highlight the potential of developing low-cost and accessible GLT solutions and suggest further hardware and algorithmic optimizations to improve detection robustness at high speeds. This study contributes valuable insights into the development of affordable officiating technology, offering practical implications for its adoption in football at various levels.

Keywords Goal line technology · YOLO object detection · OpenCV

I. M. Ariffin
Department of Electrical Engineering, Politeknik Sultan Salahuddin Abdul Aziz Shah, Shah Alam, Selangor, Malaysia
e-mail: Ilmi.ariffin@psa.edu.my

Z. Mohamed (✉) · M. H. M. Ramli
Faculty of Mechanical Engineering, Universiti Teknologi Mara, Shah Alam, Selangor, Malaysia
e-mail: zulkifli127@uitm.edu.my

M. R. Aziz
Faculty of Mechanical Engineering, Universiti Teknologi MARA Cawangan Pulau Pinang, Permatang Pauh, Pulau Pinang, Malaysia

A. Aziz
Faculty of Engineering, Universiti Teknologi Brunei, Bandar Seri Begawan, Brunei Darussalam

M. H. A. Hassan et al. (eds.), *Proceedings of the 3rd International Conference on Innovation and Technology in Sports*, Lecture Notes in Bioengineering,
https://doi.org/10.1007/978-981-95-0804-4_31

1 Introduction

Goal Line Technology (GLT) has transformed football officiating by offering a dependable way to determine whether the ball has completely crossed the goal line. By reducing human error, GLT assists referees in making quick, accurate decisions that can profoundly impact match outcomes (Ungruhe and Schmidt 2020). Well-known systems like Hawk-Eye (Sport and Hawk-Eye—technology in sport n.d.) and GoalControl (Mani 2022), rely on sophisticated technologies, including high-speed cameras and magnetic sensors, to monitor ball movement with exceptional precision (Wilson 2022). While highly effective, these systems are often prohibitively expensive and require complex installations, limiting their use to professional leagues with substantial financial resources (Sport and Hawk-Eye—technology in sport n.d.). This creates a significant gap for grassroots and lower-tier leagues, which remain dependent on human judgment and are more susceptible to officiating errors.

To address these accessibility challenges, this study introduces a portable and affordable GLT system built with open-source software and low-cost hardware. The proposed system leverages OpenCV (Bradski and Kaehler 2008) for video preprocessing and YOLO (You Only Look Once) (Talaat and ZainEldin 2023) for real-time ball detection, running on a compact Raspberry Pi setup with a Pi Camera. This design aims to democratize goal-line technology, making it feasible for lower-budget leagues to adopt fair and accurate officiating standards. However, a key challenge lies in ensuring the system maintains detection accuracy across a wide range of ball speeds, particularly in dynamic match scenarios where ball velocities can vary dramatically during rapid plays or direct kicks.

This paper evaluates the performance of the portable GLT system by analyzing its detection accuracy at different ball speeds. The goal is to assess its operational reliability and identify limitations, with a focus on areas where algorithmic refinements or hardware upgrades could improve real-world performance. The findings demonstrate the potential of low-cost, open-source GLT solutions to bring advanced officiating technology to a broader audience. Moreover, the study highlights future opportunities to enhance affordable sports technologies, paving the way for wider adoption in grassroots and lower-tier football leagues.

2 Methodology

2.1 System Setup and Hardware Configuration

The Portable GLT system developed in this study comprises a Raspberry Pi 4 Model B, which serves as the processing unit, and a Pi Camera v2 for video capture as shown in Fig. 1. The Raspberry Pi was selected for its balance of cost-effectiveness and computational power, capable of handling real-time image processing with OpenCV and YOLO. The Pi Camera is positioned to provide a clear, direct view of the goal

Fig. 1 Raspberry Pi 4 (left), Pi camera (right)

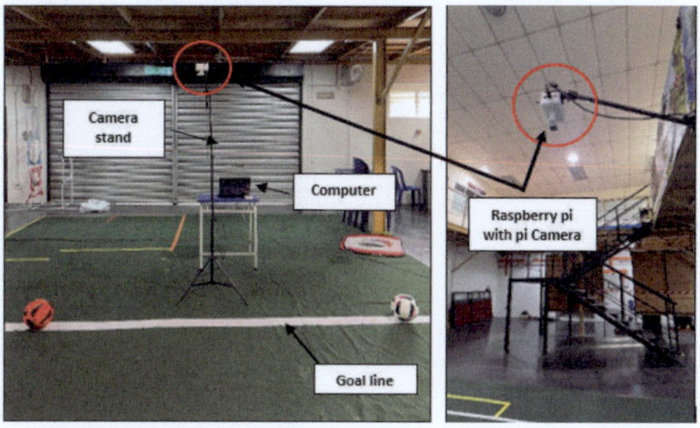

Fig. 2 Experimental setup

line, maximizing visibility of the ball as it approaches and crosses the line. The setup was mounted on a stable stand at a fixed height to minimize camera movement and ensure consistent detection performance. Figures 2 and 3 shows the experimental setup for the study camera view angle respectively.

2.2 *Software Framework and Object Detection*

The GLT system's software integrates OpenCV for video processing and YOLOv8 for real-time object detection. OpenCV manages the video feed, preprocessing the data before directing it from the Pi Camera to the Raspberry Pi. The Raspberry Pi then relays the video for additional processing on a computer equipped with a GPU. YOLOv8, recognized for its speed and efficiency, was trained on a specialized dataset designed to accurately identify a standard football under various lighting conditions and camera perspectives. Figure 4 shows the Portable GLT System detection and goal confirmation visualization.

Fig. 3 Camera view angle

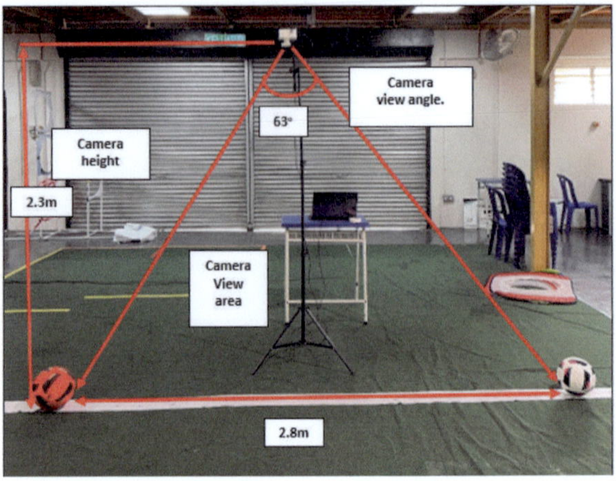

Fig. 4 Ball detection and goal confirmation from portable GLT system

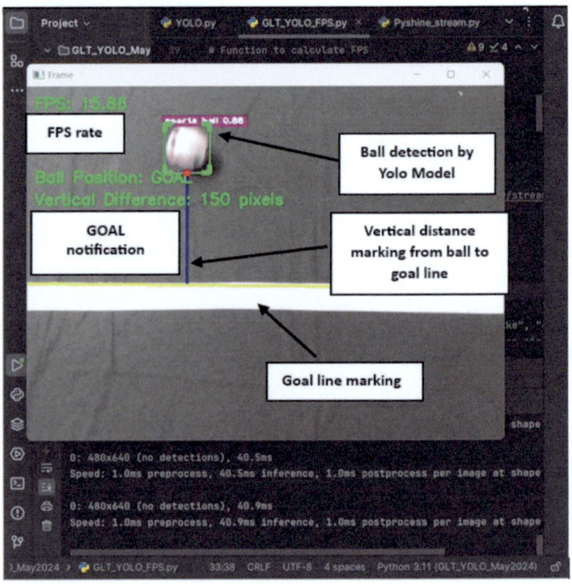

To optimize the detection pipeline, the video frames are resized for efficient processing while retaining sufficient resolution to maintain detection accuracy. YOLO detects the ball within each frame and tracks its movement relative to the goal line. When the ball crosses the predefined goal line position, the system registers a "goal" event. The detection pipeline maintains a stable frame rate, carefully optimized to ensure both real-time responsiveness and efficient processing. Figure 5 shows the flow chart for video capture and image processing for the Portable GLT System.

Fig. 5 Flowchart for video streaming using Raspberry Pi and Pi Camera (left) and ball–detection and goal line monitoring (right)

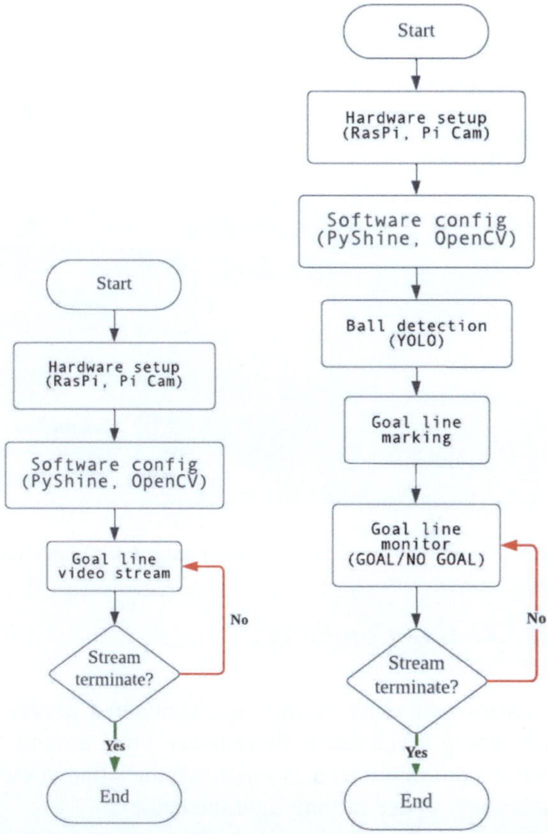

2.3 Ball Speed Measurement and Control

To evaluate detection accuracy across a range of ball speeds, the system used laser sensors to measure ball velocity. Two sensors were placed on either side of the goal line, measuring the time taken for the ball to pass between them and calculating speed based on the distance covered. This setup enables accurate measurement across a spectrum of speeds, from slow rolls to high-speed kicks.

The variable ball speed test was designed to assess the Portable Goal Line Technology (GLT) system's ability to detect the ball accurately at different speeds. This evaluation is vital, as football matches involve kicks at a wide range of velocities, requiring the system to maintain consistent detection accuracy despite these variations. The test is conducted based on FIFA GLT TEST 05 section on FIFA GLT Testing Manual (Fédération Internationale de Football Association 2019). Figure 6 shows the setup for the test. The ball is kicked in the direction of red brown and purple arrow for respective goal area target.

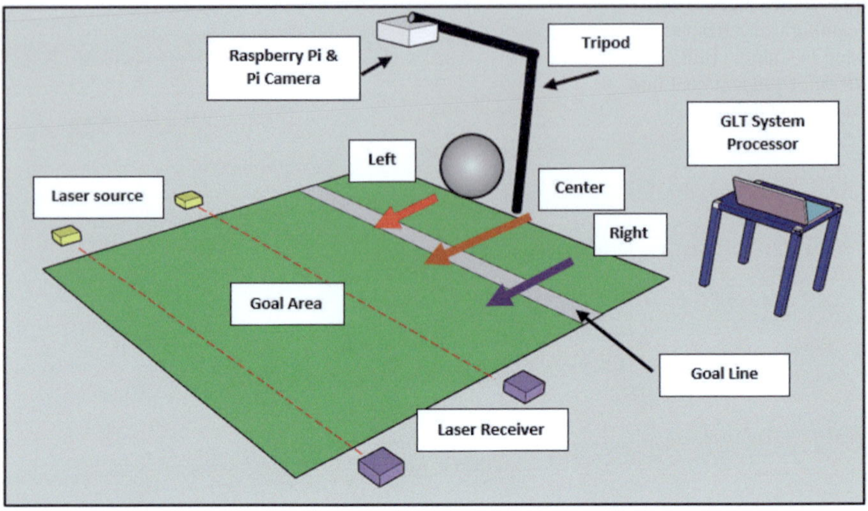

Fig. 6 Variable ball speed experimental setup using unrestricted kick method

2.4 Testing Conditions

To ensure reliability, testing was conducted under controlled lighting conditions, maintaining a consistent illuminance level around the goal area. This controlled environment minimized external variables that could affect the detection accuracy, particularly under varying light intensities.

Series of direct unrestricted kicks are performed towards the goal at different speeds. The ball's speed is measured using two sets of laser sensors positioned on the field, which capture the speed as the ball moves towards the goal. This setup allows for precise measurement of the ball's velocity during each trial. Figure 7 shows the ball speed recoded for goal area target of left, right and center.

2.5 Performance Metrics

To assess the performance of the Portable Goal Line Technology (GLT) system, we used the detection rate as a standardized metric to evaluate its reliability, consistency, and adaptability under real-world conditions. The detection rate reflects how effectively the system identifies both the ball and the goal line in video frames—an essential requirement for accurate goal confirmation. It is calculated as the percentage of frames where the YOLO model successfully detects both objects, serving as a direct indicator of the system's accuracy. Several factors, including lighting conditions, ball speed, and camera angles, influence this metric. For instance, detecting fast-moving

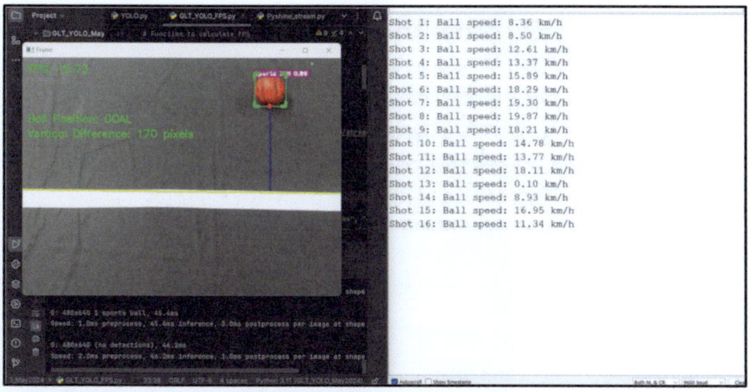

(a)

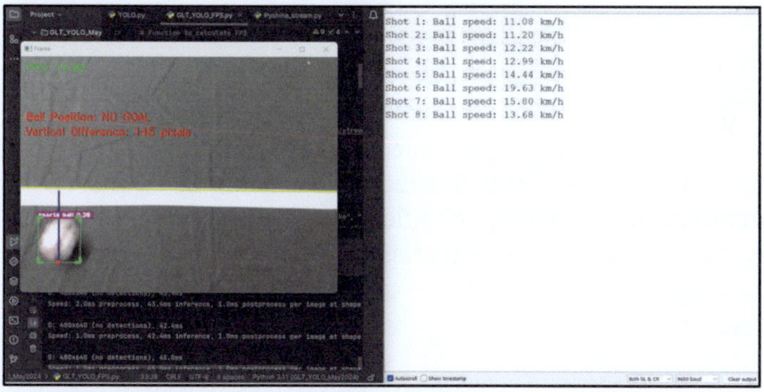

(b)

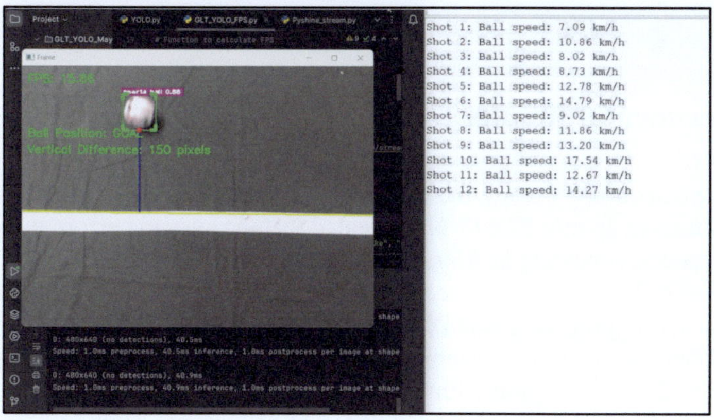

(c)

Fig. 7 Ball speed measurement for **a** right, **b** left and **c** center position of goal area

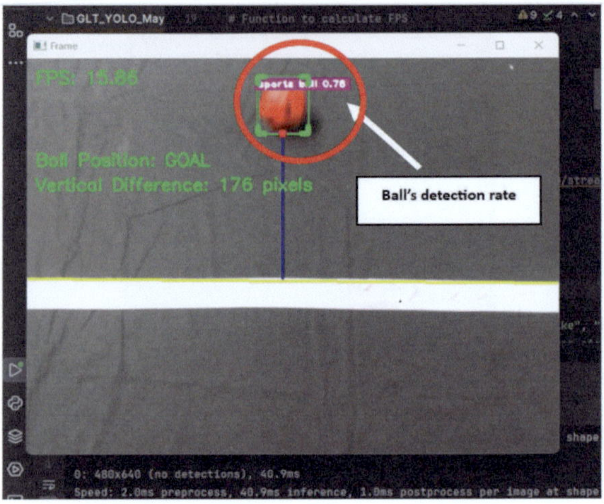

Fig. 8 Ball detection rate using YOLOv8 image recognition model

balls or those approaching the goal line at sharp angles can be particularly challenging. Figure 8 illustrates the ball detection rate achieved by the YOLOv8 image recognition model under varying conditions. By testing the system across different ball speeds, camera positions, and lighting environments, we can better under-stand its ability to adapt to the dynamic nature of live play. The formula for calculating the detection rate is:

$$Detection\ Rate(\%) = \frac{Number\ of\ Correct\ Detection}{Total\ Number\ of\ Frames} \times 100 \tag{1}$$

3 Experimental Results

To evaluate detection accuracy, the Portable GLT system was tested with ball speeds ranging from 5 to 25 m/s. The results revealed that the system performed reliably at moderate speeds, achieving an average accuracy of 95% for ball speeds up to 15 m/s. However, as the ball speed increased beyond this range, the system's accuracy gradually declined, dropping to 81% at 20 m/s and further to 65% at 25 m/s. This decrease in performance reflects the system's challenges in handling fast-moving objects, likely due to the computational limitations of the Raspberry Pi. Figure 9 illustrates the detection rate of the system across different ball speeds, including unrestricted shots, providing insight into the system's strengths and areas for improvement.

The results show that the system's ability to detect the ball is significantly influenced by its speed and position. Detection rates decrease consistently as the ball's speed increases, with the steepest drop in accuracy observed at speeds exceeding 18 m/s. This indicates that the GLT system may need calibration or algorithmic

Fig. 9 Detection rate of variable ball speed for right, left and center goal area

improvements to maintain reliable performance at higher speeds. Techniques such as reducing motion blur or increasing frame rates in image processing could help address these challenges.

The testing revealed that while the Portable GLT system performs reliably at moderate speeds, its accuracy and consistency decline as the ball's speed approaches the upper limit of the tested range. These findings emphasize the need for hardware enhancements or algorithmic optimizations to improve detection in high-speed scenarios. However, despite these limitations, the system shows strong potential as a low-cost GLT solution, particularly for grassroots football leagues, where ball speeds typically remain within its reliable detection range.

4 Future Research Directions

Future research could focus on integrating multi-camera setups to capture additional angles, minimizing the likelihood of missed detections caused by perspective limitations. Testing the system in real-world scenarios, such as outdoor environments with changing lighting conditions and variable field surfaces, would also provide valuable insights into its robustness and adaptability. Furthermore, advancements in AI accelerators and sophisticated tracking algorithms could significantly enhance the system's detection capabilities, pushing it closer to becoming a more comprehensive and reliable solution for goal-line technology.

References

Bradski G, Kaehler A (2008) Learning OpenCV. In: O'Riley Media, Inc. 1005 Gravenstein highway north, Sebastopol, CA

Fédération Internationale de Football Association (2019) FIFA quality programme for goal-line technology testing manual 2014

Mani A (2022) Goal-line technology in soccer: discussion and evaluation of systems. Academia

H Sport Hawk-Eye—technology in sport. https://www.harrodsport.com/advice-and-guides/hawkeye-technology-in-sport. Last accessed 05 Feb 2024

Talaat FM, ZainEldin H (2023) An improved fire detection approach based on YOLO-v8 for smart cities. Neural Comput Appl 35(28):20939–20954

Ungruhe C, Schmidt MB (2020) Why are east african players absent in european football? Localizing african football migration along structural constraints, colonial legacies and voluntary immobility. J Sport Soc Issues

Wilson M (2022) Sony's AI Hawk-Eye tech could finally solve soccer's VAR controversies. https://www.techradar.com/news/sonys-skeletrack-tech-could-finally-solve-soccers-var-offside-controversies. Last accessed 21 Dec 2023

The Relationship between Dynamic Balance and Soccer Skill Performance among Female Soccer Players

Nurul Nabilah Azizul, Nur Hafizatul Alya Yahya, Nur Syazliana Suhaimi, Aimie Shyzleena Che Suhaimi, and Siti Azilah Atan

Abstract The present study investigates the relationship between balance and soccer skills performances among female soccer players. Twenty-three female soccer players (n = 23, weight: 49.7 ± 6.9 kg, height: 1.60 ± 0.4 m, age: 18.4 ± 1.9 years) from sports schools and academies volunteered to participate in this study. The Johnson Modification Balance Test (JMBT) and the Star Excursion Balance Test (SEBT) were used to assess dynamic balance ability and Loughborough Soccer Passing Test (LSPT) and Loughborough Soccer Shooting Test (LSST) were used to assess soccer skills performances. A Pearson Correlation was used to measure the correlation between dynamic balance and soccer skills performance. No significant was found between dynamic balance and soccer skills performances: JMBT and LSPT ($r = 0.10$, $P = 0.963$), JMBT and LSST ($r = 0.10$, $P = 0.963$) as well as the SEBT and LSPT ($r = -0.32$, $P = 0.129$) and SEBT and LSST ($r = 0.78$, $P = 0.724$). At this point, within the study context, equilibrium has no significant impact on the ability to execute passes and shoot accurately among female soccer players. More research is required to determine the influence of balance on motor skills, balance ability may be related to competition level for some sports with the more proficient athletes displaying greater balance ability.

Keywords Football · Female · Soccer skills · Youth · Balance

1 Introduction

Soccer is a sport that demands a combination of physical fitness, technical skills, and tactical understanding (Salter et al. 2021). Owing to its popularity and financial interest, it is also one of the most extensive research in team sports (Atan et al. 2023). Even so, this is largely based on gathering match analysis data, physiological demands, equipment, and training methodology (González-Rodenas et al.

N. N. Azizul · N. H. A. Yahya · N. S. Suhaimi · A. S. C. Suhaimi · S. A. Atan (✉)
Defence Fitness Academy, National Defence University of Malaysia, Kuala Lumpur, Malaysia
e-mail: sitiazilah@upnm.edu.my

M. H. A. Hassan et al. (eds.), *Proceedings of the 3rd International Conference on Innovation and Technology in Sports*, Lecture Notes in Bioengineering,
https://doi.org/10.1007/978-981-95-0804-4_32

2022; Jakiwa et al. 2024; Hodgson et al. 2014), the information on balance and soccer performance is still scarce. While some research has explored the dynamics of balance in other sports and has traditionally focused on male players (Brachman et al. 2017), there is an increasing need to examine these factors in female athletes. There remains a gap in the literature specifically exploring the relationship between dynamic balance and soccer skills performance among female soccer players. Soccer skills refer to the technical abilities required to control the ball effectively, playing a crucial role in a team's success (Atan et al. 2023). Fundamental skills such as dribbling, passing, and shooting involve frequent turns and rapid changes in direction. Maintaining balance is essential for executing these movements efficiently without losing stability.

Balance is often regarded as motor ability, which refers to an ability that is specifically related to the performance of a motor skill (Zemková 2013). Balance is generally categorized into two types: static and dynamic. Static balance is the ability to maintain a stable position while the body is stationary and a fundamental component of sports like gymnastics, shooting, and archery (Zemková 2013; Hrysomallis 2011). Solovey et al. (2020) Study showed volleyball athletes with superior static balance demonstrate better performance indicators, including tactical strikes and defensive maneuvers. Besides, static balance is also essential for maintaining body position during skill execution, particularly in sports requiring precision, such as gymnastics and archery (Zemková 2013; Omorczyk et al. 2019). Athletes with better static balance can adapt to varying conditions, enhancing their overall performance (Zemková 2013; Omorczyk et al. 2019; Sanghavi et al. 2021). Meanwhile, dynamic balance is the ability to maintain stability and control of the body while in motion or transitioning between movements (Chtara et al. 2018). The relationship between dynamic balance and athletic performance is significant across various sports, influencing athletes' ability to execute complex movements and maintain stability during competition (Chtara et al. 2018; Ozmen 2016). In martial arts performance such as karate, judo, taekwondo, and Brazilian jiu-jitsu requires a high level of dynamic balance to perform actions like kicks, strikes, grappling, and evasive maneuvers (Cherepov et al. 2021). Recent studies on volleyball players found that dynamic balance is a predictor of successful performance in several key areas. For instance, research has shown that players with better dynamic balance often achieve higher jump heights, which are essential for blocking and spiking (Solovey et al. 2020). In general, dynamic balance is crucial for athletic performance, as it enables athletes to execute sport-specific movements effectively.

Dynamic balance plays a crucial role in various soccer-specific skills, such as dribbling, passing, shooting, and sudden directional changes (Jadczak et al. 2018). For instance, in shooting and passing, players need to be able to shift their weight and adjust their body position quickly in order to make accurate and powerful shots and passes (Ateş 2019). In addition, dynamic balance is also important for defensive players who need to be able to quickly adjust their body position to defend against opponents and win possession of the ball (Ozmen 2016). Moreover, an athlete's ability to maintain dynamic balance not only supports skill execution but also contributes to injury prevention, as improper balance control is linked to an increased risk of common soccer injuries, such as knee or ankle sprains, particularly among female players who tend to have higher injury rates than their male counterparts (Han et al. 2015). Given that these skills frequently demand rapid movement, shifting balance, and adjustment to opponents, dynamic balance is likely to have a strong correlation with overall skill execution.

In soccer, information on female players is still lacking compared to their male counterparts. Much of the existing research on soccer skills performance has been centered on male players (Gomez-Gonzalez et al. 2023). It is important to note that research on female players plays an equally critical role for enhancing this population performance. To date, female and male soccer players differ in various physical, physiological, biomechanical aspects and strength development patterns that can influence their performance, training needs, and injury risk (Bradley et al. 2013; Schache et al. 1999; Hewett et al. 2006). Male soccer players typically cover between 10 and 13 km per game while female soccer players generally cover between 8 and 11 km in a 90 min match duration. Lesser distance of running showed a different physical attributes due to physiological capacity such as maximum oxygen uptake (V02max) (Bradley et al. 2013). Males has greater muscle mass and strength which allows for more high-intensity sprints and rapid directional changes (Bradley et al. 2013). This contributes to covering greater distances and having higher speeds throughout the game. In terms of balance ability, male and female soccer players exhibit notable differences largely due to anatomical and biomechanical factors (Schache et al. 1999). Even though female athletes demonstrated better static balance performance than males, particularly in visual input interpretation (Akinoğlu et al. 2023) males generally outperform females in dynamic balance tasks, particularly in reach distances during the Y-balance test (Ewais et al. 2024). Furthermore, female soccer players are 2–8 times more likely to suffer Anterior Cruciate Ligament (ACL) injuries than their male counterparts as they display different landing mechanics, often with greater knee valgus (inward knee movement) during landings or sudden stops (Schache et al. 1999; Hewett et al. 2006). Understanding these differences could inform the development of tailored training programs that account for gender-specific physiological and biomechanical factors, enhancing performance outcomes and addressing injury prevention more effectively for female players.

Hence, this study aims to investigate the relationship between dynamic balance and soccer skills performance among female soccer players. To assess this relationship, dynamic balance will be measured using specific test, Star Excursion Balance Test (SEBT) and Johnson Modification Balance Test (JMBT) while soccer skill performance will be evaluated through specific drills in dribbling and passing, the Loughborough Soccer Passing Test (LSPT) and shooting accuracy test Loughborough Soccer Shooting Test (LSST). By identifying key areas where balance may enhance or limit skill execution, the study seeks to provide evidence-based insights that could benefit athletes, coaches, and sports scientists. A clearer understanding of this relationship could support the implementation of balance-specific training interventions in soccer training programs for female athletes, fostering a more holistic approach to skill development and performance enhancement. This research contributes to the growing effort to address the unique needs of female soccer players, promoting a more inclusive approach to athletic training and sports science.

2 Methodology

2.1 Participants

A total of twenty-three female soccer players from sports schools and academies volunteered to participate in this study. (n = 23, height: 1.60 ± 0.4 m and body weight 49.7 ± 6.9 kg; age 18.4 ± 1.9). All participants gave their consent prior to the study. The inclusion criteria were outfield players (players that are positioned outside the goal area), currently active in soccer training and competitive in local and international tournaments, free from injury, and healthy.

2.2 Procedure

All participants attended three sessions for assessment of dynamic balance and soccer skill performances. The first session was to familiarise themselves with the dynamic balance test: Star Excursion Balance Test (SEBT) (Plisky et al. 2009), Johnson Modification Balance Test (JMBT) (Atan 2009) and soccer skills tests: the Loughborough Soccer Passing Test (LSPT) and Loughborough Soccer Shooting Test (LSST) (Ali et al. 2007). In the second session participants performed the dynamic balance test in full and in the third session the participants performed the soccer skills test in full.

2.3 Star Excursion Balance Test (SEBT)

The SEBT is a reliable tool to measure dynamic balance and stability (Plisky et al. 2009). The SEBT requires participants to perform a variety of reaching exercises with one limb while maintaining balance with the other. Participants were required to stand in the center of the line (see Fig. 1), hands firmly placed on their hips. On the go signal, participants need to maintain balance on one leg while reaching as far as possible with the other leg in multiple directions.

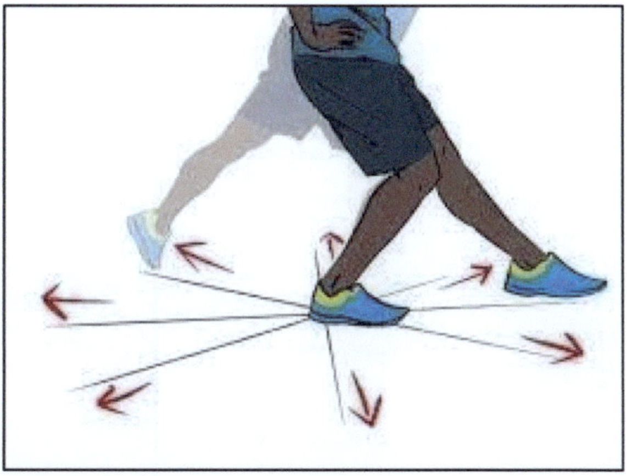

Fig. 1 Star execution balance test

2.4 *Johnson Modification Balance Test (JMBT)*

The JMBT is an assessment of dynamic balance, in which the participants alternate foot hopping and holding a static position for five seconds at each point (see Fig. 2). The individual stands on one leg at the starting mark. They then hop from one mark to the next on the same leg, attempting to land on each mark as accurately as possible. After landing on each mark, the individual must maintain balance for a set period, often about 5 s, before hopping to the next mark. This continues until the course is completed. At each point, the sole of the foot must completely cover each tape mark so that it cannot be seen. 5 points are awarded for successfully landing on a tape mark during the test, and 1 point is awarded for each second that the participant maintains a solid position on the tape mark. One can get a maximum of 10 points for each tape mark and 100 points for the test. Landing mistakes 5 points, such as failing to halt, contacting the ground, or not fully covering the tape mark (Atan 2009).

2.5 *Loughborough Soccer Passing Test (LSPT)*

The LSPT is a valid and reliable test developed to assess soccer skill performance such as passing, dribbling, control, and decision-making (Ali et al. 2007). The LSPT requires players to make 16 passes (8 long passes and 8 short passes) into four standard gym benches with a different-coloured marker: green, blue, red, and yellow. The player starts in the center of the grid and has 43 s to complete the test. The goal is to pass the ball accurately into each target zone in a specified sequence. Players use one touch to receive and one touch to pass, adhering to specific rules regarding passing accuracy and control. Each successful pass is recorded, and the time taken to complete each pass is recorded. Incorrect passes, poor control, or passes that do

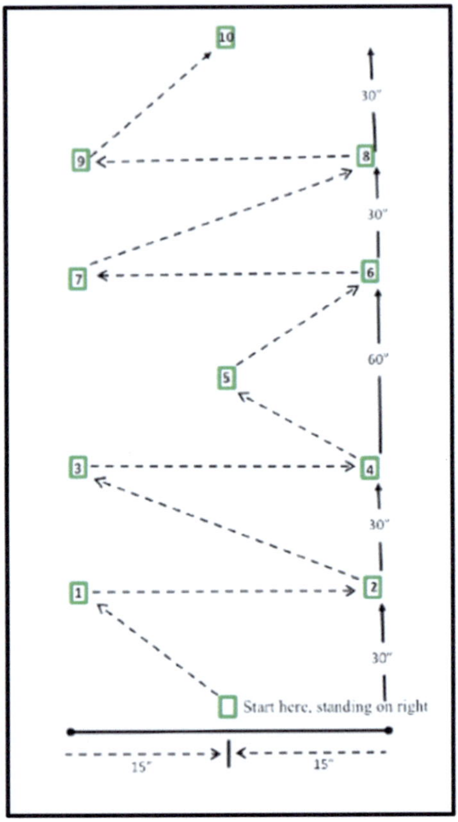

Fig. 2 Johnson Modification Balance Test (JMBT)

not reach the target zones result in penalty points, which are deducted from the total score. The penalty included:

- Add 5 s for missing the bench fully or passing the wrong bench.
- Add 3 s to compensate for missing the target regions (0.6 × 0.3 m).
- Allow 2 s for ball handling and an additional 2 s for passing the ball outside of the allocated area.
- If the ball touch any cone, 1 s will be added.

If the ball strikes the 10 cm strip in the center of the target, a second is deducted from the overall time for every second spent beyond the allotted 43 s to complete the test (Fig. 3).

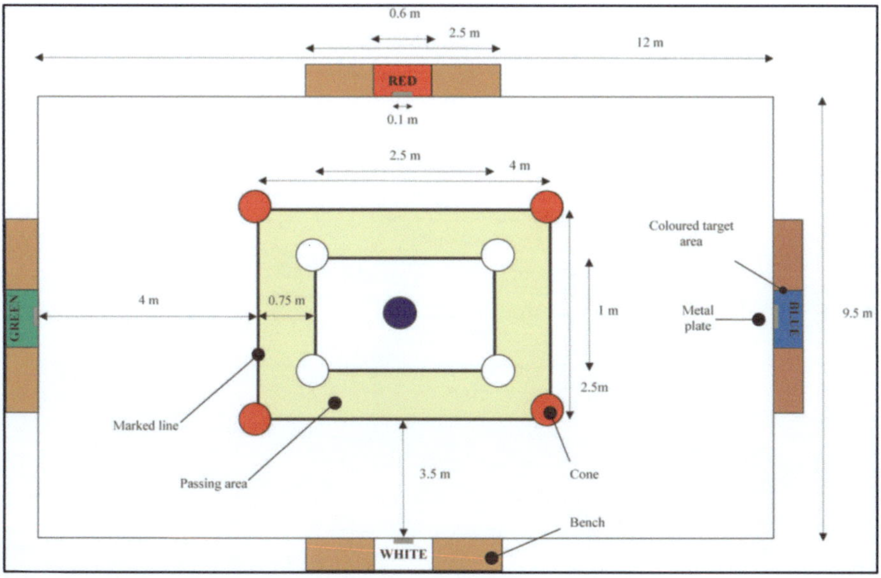

Fig. 3 Schematic representation of the LSPT

2.6 Loughborough Soccer Shooting Test (LSST)

The LSPT was developed as a test to assess shooting ability, reaction time, decision-making, and running speed (Ali et al. 2007). Figure 4 illustrates the layout of the LSST. The participant stood (facing away from the goal) in the middle of the shooting zone and a ball was placed in front of the participant. When the investigator called, the participant sprinted to the appropriate cone, touched the top of it, returned to the middle of the shooting area, kicked the ball to the rebound board, controlled the ball if necessary, and made the shot to the goal within the shooting area. Then, the participant followed up her shot by sprinting past the goalkeeper. Each participant made up to 10 shots with 1 min rest periods between each shot sequence.

3 Statistical Analysis

All results were reported as mean ± standard deviation. The Pearson Correlation was used to determine the relationship between dynamic balance and soccer skill performances. All statistical analyses were performed using SPSS software (version 21.0, SPSS inc, Chicago, IL) with the level of significance set at $P \leq 0.05$.

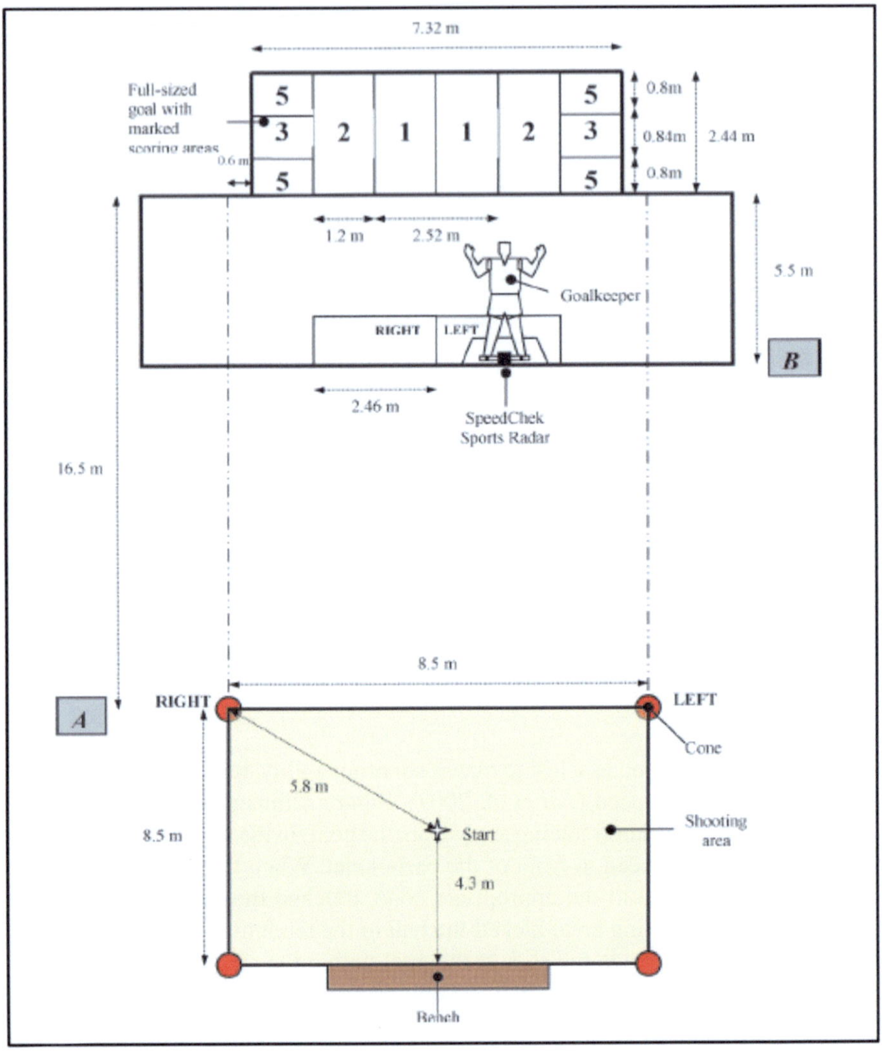

Fig. 4 Schematic representation of the LSST

4 Results

Table 1 represented the mean ± SD of the SEBT and JMBT. The JMBT scores indicate very high performance values (mean = 91.45) close to the ceiling of 100 points. The SEBT scores show lower values indicate lower dynamic balance in the right and left leg in each direction (Table 2).

This table summarized the mean ± SD of the LSPT and LSST. Participants took longer than the allocated time, exceeding 15 s to complete 16 passes. Notably, the

Table 1 Mean and standard deviation in balance test

	N	Mean ± SD
JMBT	23	91.7 ± 7.9
SEBT	23	115.2 ± 2.6

Table 2 Mean and standard deviation in soccer skills test

LSPT	Mean ± SD	LSST	Mean ± SD
LSPT 16 passes	58.8 ± 4.3 s	LSST completion time	13.4 ± 0.9 s
LSPT with penalty time (total time)	99.5 ± 8.9 s	LSST speed of the ball	33.7 ± 3.2 km h
		LSST point	2.1 ± 0.4

time taken to complete the LSPT significantly increased when including penalties, adding an average of 8.9 s, indicating that penalties can substantially impact overall performance time.

In contrast, the LSST demonstrated a much quicker average completion time for the 10 shooting attempts compared to the LSPT, with low variability of only 0.9 s, reflecting consistent performance among the players. The average speed at which the ball travelled when shot was measured at 33.7 ± 3.2 km/h. However, despite this speed, the accuracy of the shooting was notably low, making it easier for goalkeepers to save the shots, which was a critical point in the assessment.

These results highlight the differences in performance metrics between passing and shooting, suggesting that while players may exhibit consistent shooting speed, challenges in accuracy can hinder overall effectiveness during competitive scenarios.

4.1 Correlation Between Dynamic Balance and Soccer Skills Performance

Table 3 summarizes the Pearson correlation results between different balance tests (SEBT and JMBT) and various performance metrics related to the LSPT and LSST. It was found that no significant relationship with SEBT, indicating that performance on the SEBT does not appear to influence the time taken for the LSPT however the LSPT total time has a positive weak relationship with SEBT, indicating a weak link between the SEBT performance and overall time taken, including penalties. The correlation between the LSPT time, penalty time, and total time showed positive weak relationships with the JMBT scores, suggesting light, but not significant, associations with the LSPT performance.

In the LSST, the total time showed a positive moderate relationship, implying that a better balance on the SEBT could moderately impact the LSST completion time. However, no relationship, meaning the SEBT balance ability does not correlate with

Table 3 Summary of pearson correlation result between dynamic balance and soccer skills performances

Balance test	LSPT time (s)	LSPT penalty time (s)	LSPT total time (s)
SEBT	No relationship ($r = -0.32$, $P < 0.129$)	Positive moderate relationship ($r = 0.42$, $P < 0.043$)	Positive weak relationship ($r = 0.11$, $P < 0.595$)
JMBT	Positive weak relationship ($r = 0.10$, $P < 0.963$)	Positive weak relationship ($r = 0.57$, $P < 0.04$)	Positive weak relationship ($r = 0.28$, $P < 0.194$)
Balance test	LSST time (s)	LSST speed (km h)	LSST point
SEBT	Positive moderate relationship ($r = 0.78$, $P < 0.724$)	No relationship ($r = -0.06$, $P < 0.781$)	Positive moderate relationship ($r = 0.35$, $P < 0.094$).
JMBT	No relationship ($r = -0.10$, $P < 0.963$)	Positive weak relationship ($r = 0.18$, $P < 0.408$)	No relationship ($r = -0.18$, $P < 0.407$)

the speed of the ball during the LSST. In the LSPT point, there was a positive moderate relationship, suggesting that higher SEBT scores may lead to slightly better LSST performance. Meanwhile, the relationship between the JMBT and LSST showed no relationship with the LSST time, indicating that JMBT performance does not relate to the time taken in the LSST. Positive weak relationship, implying a minor link between the JMBT balance and LSST speed and no relationship between the JMBT and LSPT point scoring. Overall, SEBT appears to have a stronger connection to soccer-specific performance metrics compared to JMBT, although these relationships are moderate or weak.

5 Discussion

This study examined the relationship between dynamic balance and soccer skill performance among female soccer players. The findings indicate no statistically significant correlation between these variables within the studied cohort. Thus, the results do not provide strong evidence that dynamic balance directly influences soccer skills in this sample. These findings align with previous research (Kartal 2020), which also reported inconsistent correlations between dynamic balance, as measured by the SEBT, and performance metrics such as sprint speed and jumping ability in junior soccer players. Collectively, these results suggest that while dynamic balance plays a role in sports performance, its influence may vary depending on factors such as age, skill level, and specific performance demands.

The absence of a significant relationship between balance abilities and soccer skill performance observed in this study may be attributed to several factors. First, unstructured training could be a contributing factor. Athletes who do not engage in structured balance training may exhibit reduced postural control and functional performance, leading to less effective outcomes in competitive environments. Previous research (Schedler et al. 2020) highlights the importance of structured balance training in enhancing postural stability and movement efficiency, both of which are critical for optimal soccer performance. Consequently, the lack of dedicated balance training may explain the limited correlation with skill performance observed in this study.

Another factor influencing the results may be the competitive level of the participants. Research by (Brayne 2021) suggests that balance abilities vary considerably by skill level, with elite players typically exhibiting superior balance compared to recreational players. Studies further support this, showing a stronger relationship between dynamic balance and performance in young elite athletes, where balance is essential for precise movement execution and rapid directional changes. Assessments such as the Y-Balance Test, which measures balance in both dominant and non-dominant legs, could offer a more comprehensive perspective on how balance contributes to skill performance (Chtara et al. 2018). These findings imply that the link between balance and soccer skills may be more pronounced at higher competitive levels, underscoring the importance of dynamic balance in elite play.

The participants' skill level, experience, and competitive background likely influenced both their dynamic balance and soccer-specific skills. Playing at a school or amateur level may result in less developed balance and skill proficiency due to limited exposure to structured training programs. Additionally, the small sample size in this study may have limited statistical power, making it challenging to detect subtle but potentially meaningful effects (Bland and Altman 1995). This limitation suggests that any observed relationship between dynamic balance and soccer skill performance may not be strong enough to establish definitive conclusions. To enhance generalizability and yield more meaningful results, future studies should aim to include a larger and more diverse sample of female soccer players, allowing for a more accurate assessment of these relationships.

Although this study did not find a significant relationship between dynamic balance and soccer skill performance among female players, existing research generally supports the role of balance in enhancing sports performance. For example, studies by (Falces-Prieto et al. 2022; Rouissi et al. 2018) demonstrated that dynamic balance is positively correlated with agility, particularly in tasks requiring rapid directional changes. These findings suggest that improved balance supports agility, which is crucial in soccer. Similarly, Pana's study (Pană et al. 2023) on U13 players from Otopeni Sports Club found positive associations between balance abilities and technical soccer skills such as the instep kick, outside kick, and long passes. These findings highlight that while balance may not universally impact all performance measures, it often plays a significant role in supporting agility and certain technical skills, particularly in younger or developing athletes.

Balance training could be a valuable component of athletic development programs aimed at enhancing overall sports performance, though its impact may vary based on specific skills and demographics. Atan (2009) demonstrated that a structured balance training program significantly improved both dynamic balance and soccer skill performance among young soccer players. Participants were divided into an experimental group receiving balance training and a control group, with findings showing notable improvements in the experimental group, as assessed through the JMBT and LSPT. This suggests that balance training not only enhances stability but also positively impacts soccer-specific skills.

Furthermore, additional research found that specialized balance training programs significantly improved both static and dynamic balance in adolescent soccer players. The experimental group showed a remarkable 37.82% improvement in static balance and a 24.98% improvement in dynamic balance after an eight-week program (Mitrousis et al. 2023). These findings underscore the critical role of balance in executing technical skills in soccer, particularly for younger players who are still developing their abilities. The research suggests that incorporating balance training into regular soccer training could be beneficial for skill development, ultimately contributing to improved performance on the field.

6 Conclusion

It can be concluded that this present study revealed no significant correlation, potentially due to factors such as the participants' amateur playing level, limited exposure to structured balance training, and a small sample size. These limitations may have affected both balance and skill development, reducing the study's ability to detect meaningful effects. To build on these results, future research should aim to incorporate a larger, more diverse sample and consider the influence of competitive level and structured training on balance and soccer performance. This approach may provide more definitive insights into the role of dynamic balance in soccer skills proficiency among female athletes.

References

Akinoğlu B, Acar HY, Paköz B, Kocahan T (2023) Comparison of static and dynamic balance ability according to gender in athletes- a cross sectional study. Turk J Kinesiol 9(2):91–98. https://doi.org/10.31459/turkjkin.1277503

Ali A, Williams C, Hulse M, Strudwick A, Reddin J, Howarth L et al (2007) Reliability and validity of two tests of soccer skill. J Sport Sci 25(13):1461–1470. https://doi.org/10.1080/02640410601150470

Atan SA (2009) The effects of a four-week balance training programme on dynamic balance and scoccer skill performances

Atan SA, Azli MS, Jakiwa J, Rustam S (2023) Relationship between match running performance and physical capacity in Malaysia young soccer player. Sport Mont 21(1):55–59. https://doi.org/10.26773/smj.230209

Ateş BA (2019) Dynamic balance performance of professional Turkish soccer players by playing position. Phys Educ Stud 23(5):223–238. https://doi.org/10.15561/20755279.2019.0502

Bland JM, Altman DG (1995) Statistics notes: multiple significance tests: the Bonferroni method. BMJ 310(6973):170. https://doi.org/10.1136/bmj.310.6973.170

Brachman A, Kamieniarz A, Michalska J, Pawłowski M, Słomka KJ, Juras G (2017) Balance training programs in athletes—a systematic review. J Hum Kinet 58(1):45–64. https://doi.org/10.1515/hukin-2017-0088

Bradley PS, Dellal A, Mohr M, Castellano J, Wilkie A (2013) Gender differences in match performance characteristics of soccer players competing in the UEFA champions league. Hum Mov Sci 33:159–171. https://doi.org/10.1016/j.humov.2013.07.024

Brayne LC (2021) An investigation into football-specific dynamic balance measures. https://doi.org/10.7190/shu-thesis-00463

Cherepov E, Eganov A, Bakushin A, Ya N, Dmitry Y (2021) Maintaining postural balance in martial arts athletes depending on coordination abilities. J Phys Educ Sport 21(06). https://doi.org/10.7752/jpes.2021.06464

Chtara M, Rouissi M, Bragazzi NL, Owen AL, Haddad M, Chamari K (2018) Dynamic balance ability in young elite soccer players: implication of isometric strength. J Sport Med Phys Fit 58(4). https://doi.org/10.23736/s0022-4707.16.06724-4

Ewais NF, Abdel-Aziem AA, Hassan SM (2024) Effect of gender, leg dominancy and body mass index on Y-balance performance among young healthy adults. SPORT TK-Revista EuroAmericana De Ciencias Del Deporte 13:8. https://doi.org/10.6018/sportk.523751

Falces-Prieto M, González-Fernández FT, García-Delgado G, Silva R, Nobari H, Clemente FM (2022) Relationship between sprint, jump, dynamic balance with the change of direction on young soccer players' performance. Sci Rep 12(1). https://doi.org/10.1038/s41598-022-16558-9

Gomez-Gonzalez C, Dietl H, Berri D, Nesseler C (2023) Gender information and perceived quality: an experiment with professional soccer performance. Sport Manag Rev 27(1):45–66. https://doi.org/10.1080/14413523.2023.2233341

González-Rodenas J, Pedrera G, Dorado V, Aranda-Malavés R, Tudela DA, De Matías-Cid P (2022) Effects of age and playing tactics on the individual tactical behavior in U10 and U12 elite Spanish soccer players. Sports 10(11):185. https://doi.org/10.3390/sports10110185

Han J, Anson J, Waddington G, Adams R, Liu Y (2015) The role of ankle proprioception for balance control in relation to sports performance and injury. Biomed Res Int 2015:1–8. https://doi.org/10.1155/2015/842804

Hewett TE, Myer GD, Ford KR (2006) Anterior cruciate ligament injuries in female athletes. Am J Sport Med 34(2):299–311. https://doi.org/10.1177/0363546505284183

Hodgson C, Akenhead R, Thomas K (2014) Time-motion analysis of acceleration demands of 4v4 small-sided soccer games played on different pitch sizes. Hum Mov Sci 33:25–32. https://doi.org/10.1016/j.humov.2013.12.002

Hrysomallis C (2011) Balance ability and athletic performance. Sport Med [Internet] 41(3):221–232. https://doi.org/10.2165/11538560-000000000-00000

Jadczak Ł, Grygorowicz M, Dzudziński W, Śliwowski R (2018) Comparison of static and dynamic balance at different levels of sport competition in professional and junior elite soccer players. J Strength Cond Res 33(12):3384–3391. https://doi.org/10.1519/jsc.0000000000002476

Jakiwa J, Din MFM, Rustam S, Azli MS, Atan SA, Maliki ABHM et al (2024) The development of mini timing gate for sprint performance assessment. In: Lecture notes in bioengineering, pp 443–458. https://doi.org/10.1007/978-981-97-3741-3_39

Kartal A (2020) The relationships between dynamic balance and sprint, flexibility, strength, jump in junior soccer players. Pedagog Phys Cult Sport 24(6):285–289. https://doi.org/10.15561/26649837.2020.0602

Mitrousis I, Bourdas DI, Kounalakis S, Bekris E, Mitrotasios M, Kostopoulos N et al (2023) The effect of a balance training program on the balance and technical skills of adolescent soccer players. J Sport Sci Med:645–657. https://doi.org/10.52082/jssm.2023.645

Omorczyk J, Zajac A, Puszczalowska-Lizis E (2019) Relationships between postural stability in standing and handstand and psychological factors in athletes practicing artistic gymnastics. Balt J Health Phys Act:96–106. https://doi.org/10.29359/bjhpa.11.4.11

Ozmen T (2016) Relationship between core stability, dynamic balance and jumping performance in soccer players. Turk J Sport Exerc 18(1):110. https://doi.org/10.15314/tjse.93545

Pană BA, Potop V, Mihaila I, Trandafirescu G, Mihai I, Mihailescu LE et al (2023) Determining the influence of dynamic balance in the technical training in football game at U13 level. Pedagog Phys Cult Sport 27(5):429–438. https://doi.org/10.15561/26649837.2023.0510

Plisky PJ, Gorman PP, Butler RJ, Kiesel KB, Underwood FB, Elkins B (2009) The reliability of an instrumented device for measuring components of the star excursion balance test. PubMed. https://pubmed.ncbi.nlm.nih.gov/21509114

Rouissi M, Haddad M, Bragazzi NL, Owen AL, Moalla W, Chtara M et al (2018) Implication of dynamic balance in change of direction performance in young elite soccer players is angle dependent? J Sport Med Phys Fit 58(4). https://doi.org/10.23736/s0022-4707.17.06752-4

Salter J, De Ste Croix MBA, Hughes JD, Weston M, Towlson C (2021) Monitoring practices of training load and biological maturity in UK soccer academies. Int J Sport Physiol Perform 16(3):395–406. https://doi.org/10.1123/ijspp.2019-0624

Sanghavi AD, Chotai KT, Patil SC, Rayjade AS, Sawant JM (2021) Comparison of static balance among cricket, badminton, football and track and field athletes. J Evol Med Dent Sci 10(34):2915–2919. https://doi.org/10.14260/jemds/2021/594

Schache AG, Bennell KL, Blanch PD, Wrigley TV (1999) The coordinated movement of the lumbo–pelvic–hip complex during running: a literature review. Gait Posture 10(1):30–47. https://doi.org/10.1016/s0966-6362(99)00025-9

Schedler S, Tenelsen F, Wich L, Muehlbauer T (2020) Effects of balance training on balance performance in youth: role of training difficulty. BMC Sport Sci Med Rehabil 12(1). https://doi.org/10.1186/s13102-020-00218-4

Solovey O, Hunchenko V, Solovey D, Wnorowski K (2020) Influence of static balances level on competitive performance indicators of athletes 17-21 years old in beach volleyball. Phys Educ Stud 24(6):332–339. https://doi.org/10.15561/20755279.2020.0605

Zemková E (2013) Sport-specific balance. Sport Med [Internet] 44(5):579–590. https://doi.org/10.1007/s40279-013-0130-1

Investigation on Golfer's Stability Towards the Swing Profile

Atiqullah Huzaifah, Muhammad Firdaus Mohd Nor, Mohd Nadzeri Omar, Nasrul Hadi Johari, Mohd Hasnun Arif Hassan, Mahadzir Ishak, and Zuazizi Aziz

Abstract This study aims to investigate the relationship between a golfer's perceived stability and their swing profile, using Inertial Measurement Unit (IMU) sensors to gather quantitative data. The IMU sensors were attached to the wrist and waist of a male golfer, who performed 5 swings. After each swing, the golfer rated their perceived stability on a scale of 1–5, with 5 indicating the highest stability. The analysis revealed interesting patterns when comparing the most stable and least stable swings. It appears that an earlier rotation of the hips, before ball impact, contributes to a smoother and more stable swing. The findings suggest a clear correlation between the golfer's stability rating and certain kinematic features of the swing profile, which could provide valuable insights for optimizing performance and preventing injury. This study highlights the potential of using wearable IMU sensors in sports biomechanics for real-time feedback and performance improvement.

Keywords Golf swing profile · Stability · IMU'S · Motion tracking device · Kinematic analysis · Biomechanics

1 Introduction

Golf is a popular sport defined by its intense rotational swings. Players use clubs to hit a small ball into a series of holes on a course, aiming to complete the course in the fewest strokes. The standard format features 18 holes, though 9-hole courses also exist. Golf courses challenge players' skills with various elements, such as teeing grounds, fairways, rough areas, hazards, and greens.

Similar to precision sports like archery and shooting, golf demands fine motor control and accuracy, which rely on good postural stability for optimal performance

A. Huzaifah · M. F. M. Nor · M. N. Omar (✉) · N. H. Johari · M. H. A. Hassan · M. Ishak
Mechanical and Automotive Engineering Technology, Universiti Malaysia Pahang Al-Sultan Abdullah, Pekan, Pahang, Malaysia

Z. Aziz
ALJ-ULF Golf Club Fitter, Kuantan, Pahang, Malaysia

M. H. A. Hassan et al. (eds.), *Proceedings of the 3rd International Conference on Innovation and Technology in Sports*, Lecture Notes in Bioengineering,
https://doi.org/10.1007/978-981-95-0804-4_33

(Parikh et al. 2023). According to Professional Golfer's Associate Canada Class "A" instructor, Jonathan Kim Moss, a proper golf swing follows a specific sequence: starting with a weight shift, followed by hip rotation, then the chest and arms,and finally, the club (Moss 2024). This sequence helps provide the golfer with better stability, contributing to more consistent and accurate shots.

Biomechanics, the study of how mechanical principles apply to human movement, plays a crucial role in improving athletic performance in golf (Hume et al. 2005). By examining golfers' anatomy and physiology, biomechanics helps refine technique and optimize swing mechanics. Qualitative methods in golf biomechanics in–volve observing and analyzing specific skills, allowing coaches and golfers to break down key swing phases, such as the address, backswing, downswing, and follow-through. This detailed analysis helps identify actions that significantly impact performance.

In contrast, quantitative kinematic analysis in golf biomechanics provides numerical data on body angles and movements throughout the swing, offering insights into golfer stability. Advanced motion tracking devices, like swing motion trainers with gyroscopes and accelerometer, measure critical movements, enabling an in-depth understanding of swing mechanics and stability (Hume et al. 2005).

Acceleration is a crucial factor influencing golfer stability during the swing, as it affects the kinetic chain—a series of interconnected body segments that work together. In golf, high acceleration can generate significant force and torque leading to greater distance. However, high acceleration can also potentially lead to instability if not managed effectively (Lynchjo 2008; You et al. 2023). Conversely, acceleration and deceleration phase requires robust muscle control to maintain stability throughout the swing.

In this study, motion tracking devices which is an inertial sensor (IMU's) were attached to the wrist and waist of golfer to measure acceleration data during the golf swing, elucidating the relationship between swing mechanics and stability (Kim and Park 2024).

2 Methodology

This research utilized a multifaceted approach, incorporating both quantitative and qualitative techniques, to examine the linkage between a golfer's stability and the outcomes of their swing performance.

2.1 Devices

This investigation utilized two Inertial Measurement Unit (IMU) sensors, specifically an ESP32 board and an MPU6050, to gather comprehensive kinematic data on the golfer's swing profile (Huzaifah et al. 2024).

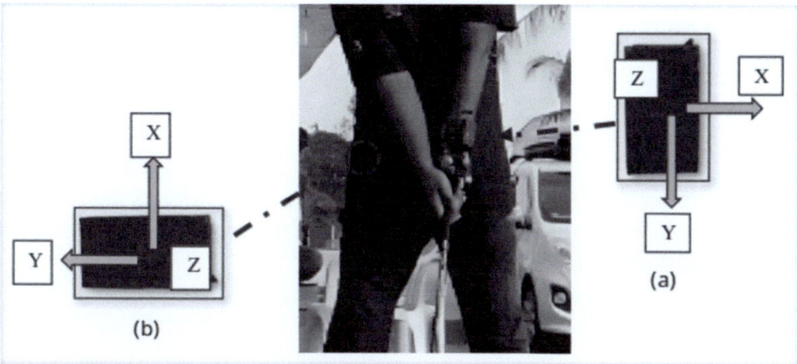

Fig. 1 Devices placement on **a** wrist and **b** waist

These sensors were attached to the participant's wrist and waist as shown in Fig. 1, facilitating the acquisition of various performance metrics, such as acceleration, rotational velocity, and gyroscopic measurements. This setup enabled a thorough examination of the golfer's movements, allowing for the identification of patterns associated with swing mechanics and stability.

2.2 Golfer Profile and Performance Metrics

The study recruited a 36-year-old male golfer, standing 176 cm tall and weighing 85 kg, who is right-handed and has a handicap level of 12, suggesting a moderate level of skill in the sport.

2.3 Swing Movement and Stability Data Collection

The experiment was conducted in a controlled setting, the golfer needs to perform 5 shots with driver club only and maintain consistent environmental conditions for each swing performed. During each swing, the IMU sensors collected real-time kinematic data, capturing the golfer's movements from the initiation to the completion of the swing.

Following each swing, the participant provided a subjective rating of his perceived stability, and this score was documented alongside the corresponding IMU data. The combination of these quantitative and qualitative data points provided a comprehensive perspective on the golfer's performance and the underlying mechanical factors that influenced the swing dynamics.

Table 1 Stability score of each swing trials

Swing trial	T1	T2	T3	T4	T5
Stability score	3	3	4	4	3

To evaluate the golfer's stability for each swing, the golfer is asked to rate his performance on a scale of 1–5 which 5 representing the highest level of stability and 1 indicating significant instability.

3 Result

The study recorded golfer's perceived stability for each of the five full swing trials, as presented in Table 1. The stability scores spanned a range from 3 to 4, suggesting a moderate to high level of stability throughout the swings. Notably, trials T3 and T4 were rated as the most stable, receiving a stability score of 4, while trials T1, T2, and T5 were perceived slightly less stable, earning a score of 3.

Figure 2 illustrates the primary planes relative to the human body, which are used to describe the movements of the golfer's wrist and waist. For the wrist's profile, the x-axis acceleration represents lateral (left-right) movement within the frontal plane, the y-axis captures vertical (up-down) movement in the trans- verse plane, and the z-axis reflects anterior-posterior (front-back) movement in the sagittal plane. The waist profile's acceleration data offers information about the golfer's torso motion. The x-axis acceleration indicates vertical (up-down) movement in the transverse plane, the y-axis captures lateral (left-right) movement in the frontal plane, and the z-axis measures anterior-posterior (front-back) movement along the sagittal plane.

To enhance the clarity of the data, the quantitative findings from the IMU sensors were visualized through charts. As the literature review highlighted, acceleration is a critical factor influencing golfer stability (Lynchjo 2008; You et al. 2023). Therefore, the acceleration data were employed to analyze the swing profiles and these acceleration graphs displayed in Figs. 3 and 4.

4 Discussion

The analysis of the acceleration data aligns with previous studies, such as those by Myeongsub Kim and Sukyung Park, who used acceleration charts to identify distinct swing phases (Kim and Park 2024). Shown in Fig. 5, we also applied similar methods to interpret our acceleration charts, adding an impact line to denote the moment of ball impact during each swing phases. The swing phases consist of address (ADD), back swing top (BST), impact (IMP) and lastly finish (FIN).

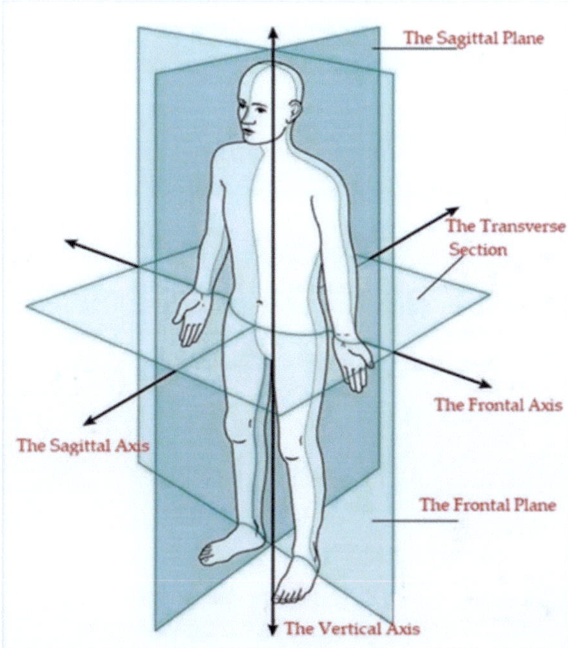

Fig. 2 Basic axes of the human body (Duan et al. 2020)

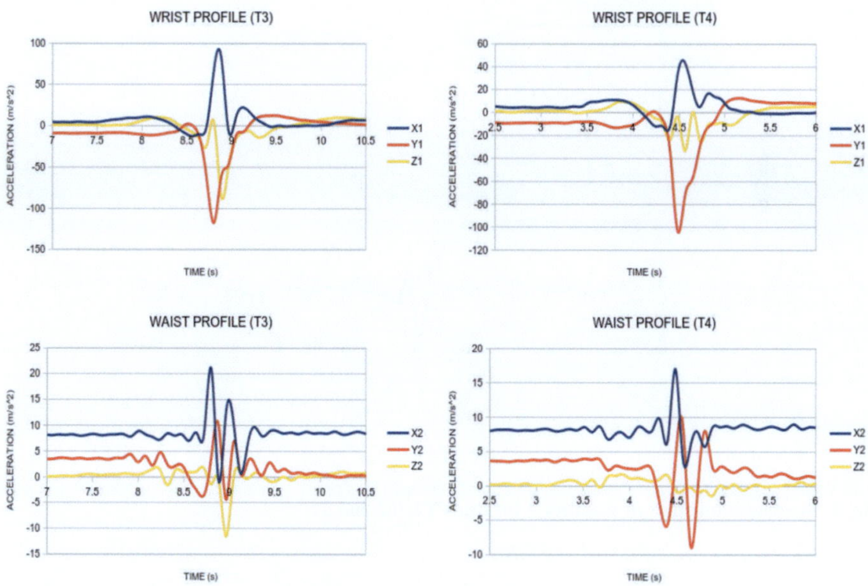

(a) T3 (b) T4

Fig. 3 Acceleration chart of stability 4 for trials T3 and T4

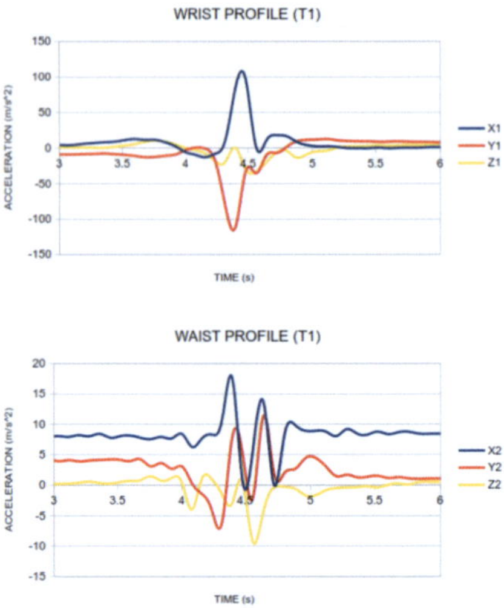

(a) T1

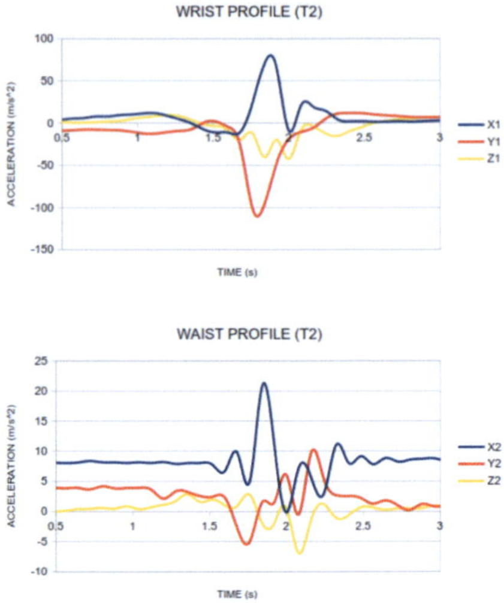

(b) T2

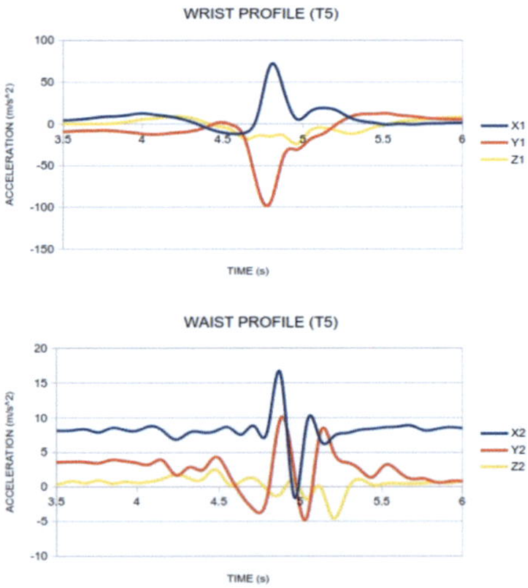

(c) T5

Fig. 4 Acceleration chart of stability 3 for trials T1, T2 and T5

Figure 6 compares the wrist and waist movement patterns of the lowest stability trial (T2, rated 3) with the highest stability trial (T3, rated 4) by adding im- pact line approached. A significant difference is observed in the waist movement align the x-direction (waist-x), which corresponds to the up-and-down motion. The first

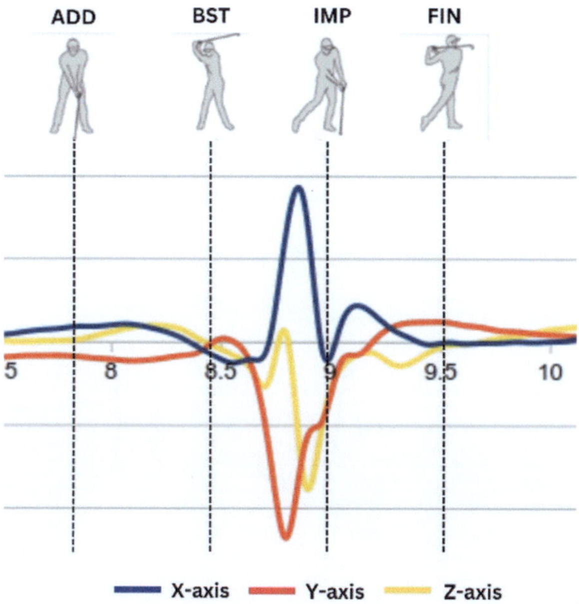

Fig. 5 Wrist acceleration signal of IMU during the entire swing phase

positive peak in waist-x indicates that the waist is lifted during the backswing, while the second positive peak shows that the waist rotates and then lifts again.

It can be seen in Fig. 7, the impact point in T3 occurred after the second peak of the waist-x, indicating that the golfer's hips rotated ahead of the moment of impact. This early hip rotation likely contributed to a smoother, transfer of energy which is a critical component that influences the shot (Gryc et al. 2015; Lynn et al. 2013).

In contrast, as shown in Fig. 8, the T2 chart showed that the impact point occurred before the second peak of the waist-x axis, signifying delayed hip rotation. This misalignment likely disrupted the flow of energy through the golfer's body, leading to instability during the swing. These findings reinforce the observations of coach Jonathan Kim Moss, who emphasized that a proper golf swing follows a specific sequence: starting with a weight shift, followed by hip rotation, then the chest, arms, and finally the club (Moss 2024).

Further justification for this pattern can be also seen by comparing other trials. Refer in Fig. 9, T4, which also had a high stability rating (4), followed a similar pattern to T3, where the hips rotated before impact, contributing to better overall stability. On the other hand, T5, with a lower stability rating (3), exhibited a delayed hip rotation like T2.

This comparative analysis suggests that the timing and coordination of hip and wrist movements play a crucial role in determining the stability of a golf swing. Early hip rotation appears to contribute to better stability and smoother acceleration patterns, as observed in the swings with higher stability ratings.

The findings align with previous research by Frederick Mun and colleagues, which demonstrates that hip rotation is a crucial factor in determining golf swing stability

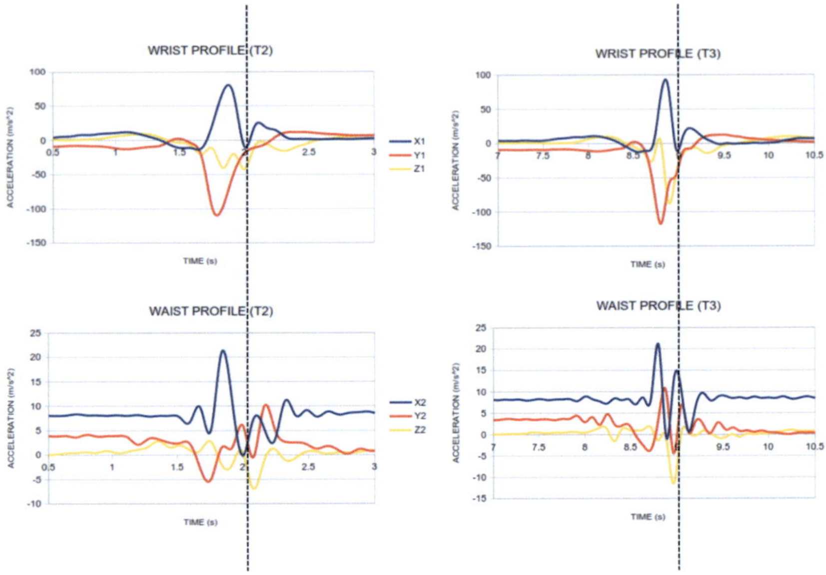

(a) T2 with stability score 3.

(b) T3 with stability score 4.

Fig. 6 Comparing T2 with T3 impact line chart

Fig. 7 Waist-x peak points in T3

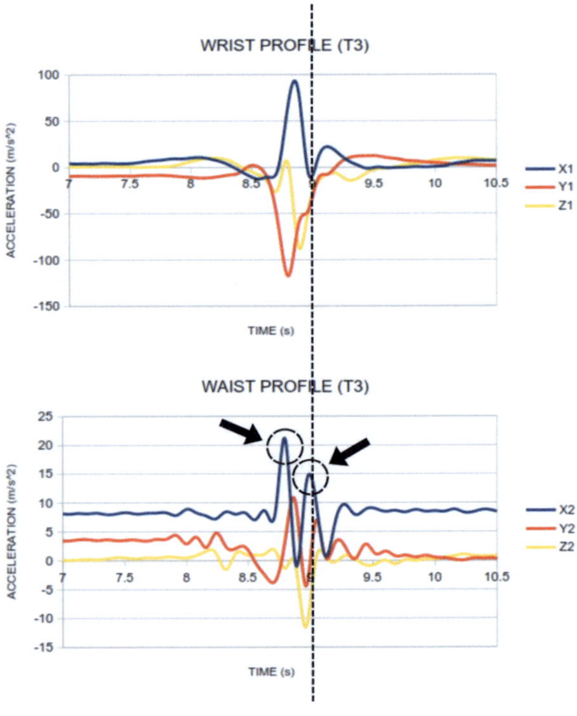

Fig. 8 Waist-x peak points in T2

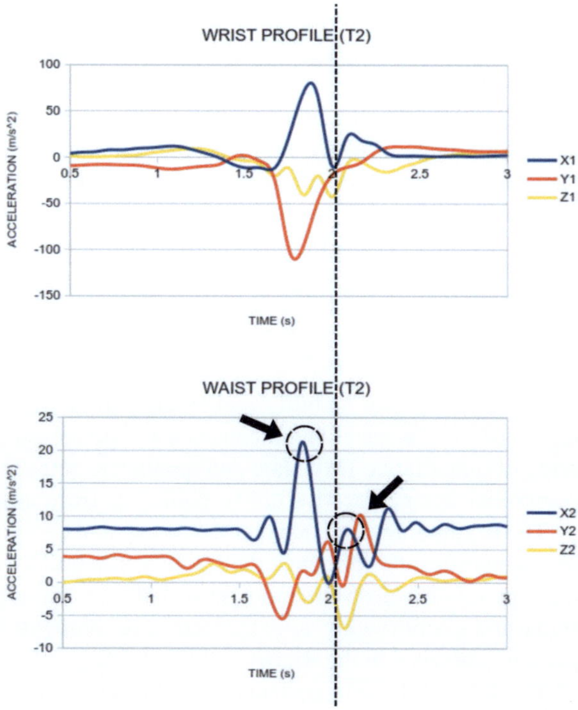

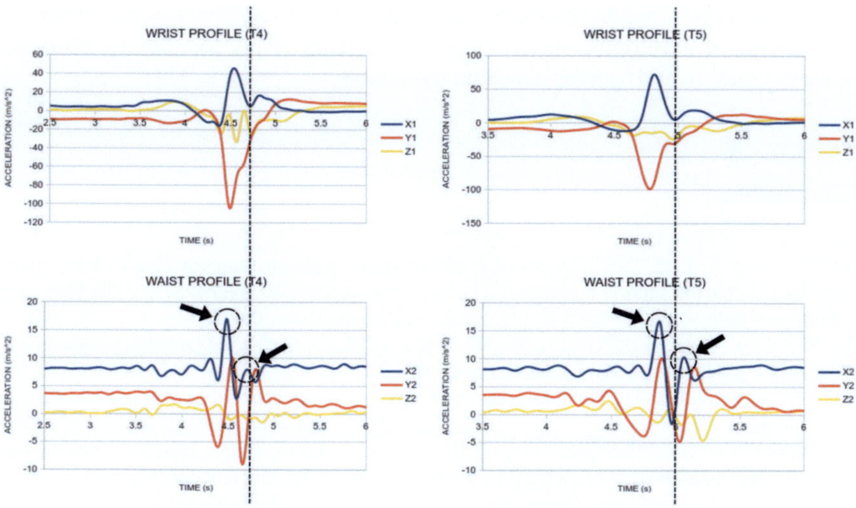

(a) T4 Peak points. (b) T5 Peak points.

Fig. 9 T4 and T5 Waist-x peak points

(Mun et al. 2015). Additionally, by examining the coordination of multiple body parts, identified as critical by Kusan and colleagues (Reveendran et al. 2023), this study provides valuable insights into the timing between hip rotation and the moment of ball impact, offering guidance for golfers to enhance their swing mechanics.

Despite these positive findings, the qualitative evaluation of stability remains subjective and may be inconsistent. Further research using a quantitative tool, such as a force plate, could help provide an objective measure of stability.

5 Conclusion

The study on golfer stability and swing profile yielded valuable insights into the relationship between swing mechanics and perceived stability. The systematic gathering of quantitative IMU data and qualitative self-assessment feedback provided a comprehensive perspective on the golfer's performance.

The key findings indicate that the timing and coordination of hip and wrist movements significantly influence the stability of a golf swing. Early hip rotation, where the hips rotate before the moment of ball impact, appears to contribute to a more stable and smooth swing, as evidenced by the higher stability ratings and acceleration patterns observed in the trials.

These results have important implications for golf instruction and training, emphasizing the need for a focus on proper sequencing and coordination within the kinetic chain. By incorporating these insights into their coaching approaches, instructors can assist golfers in developing more stable and efficient swing mechanics, ultimately enhancing their overall performance on the course.

References

Duan Y, Zhang X, Li Z (2020) A new quaternion-based Kalman filter for human body motion tracking using the second estimator of the optimal quaternion algorithm and the joint angle constraint method with inertial and magnetic sensors. Sensors 20:6018

Gryc T, Zahálka F, Maly T, Malá L, Hrasky P (2015) Movement's analysis and weight transfer during the golf swing. J Phys Educ Sport 15:781–787

Hume PA, Keogh J, Reid D (2005) The role of biomechanics in maximizing distance and accuracy of golf shots

Huzaifah A, Omar MN, Johari NH, Hassan MHA, Ishak M, Aziz Z (2024) Development of motion tracking device to determine swing profile of a golf player. In: Mohamed Z, Ngali MZ, Sudin S, Ibrahim MF, Casson A (eds) Proceedings of the 2nd international conference on innovation and technology in sports, ICITS 2023, 27–28. Springer Nature Singapore, Malaysia, Singapore, pp 167–183

Kim M, Park S (2024) Enhancing accuracy and convenience of golf swing tracking with a wrist-worn single inertial sensor enhanced reader

Lynchjo (2008) Physics of the golf swing–instruction and playing tips–the sand trap.com

Lynn SK, Frazier BS, New KN, Wu WFW, Cheetham PJ, Noffal GJ (2013) Rotational kinematics of the pelvis during the golf swing: skill level differences and relationship to club and ball impact conditions 2(2):116–125

Moss JK (2024) Sequence of the golf swing. https://youtu.be/lXEmb8TRNgg?feature=shared

Mun F, Suh SW, Park HJ, Choi A (2015) Kinematic relationship between rotation of lumbar spine and hip joints during golf swing in professional golfers. Biomed Eng Online 14(1):41–41

Parikh H, Phillips A, Stone M, Yoshida R, McCullough M, Herman M, Shin S (2023) Keeping the swing on par: golfers and upper extremity injuries they face. Open Orthop J 17(1)

Reveendran K, Omar MN, Johari NH, Hassan MHA, Aziz A (2023) Investigation on the acceleration of wrist and waist during a golf swing towards the ball trajectory. In: Omar SFS, Hassan MHA, Casson A, Godfrey A, Anwar P, Abdul Majeed P (eds) Innovation and technology in sports. Springer Nature Singapore, Singapore, pp 223–232

You X, Xu Y, Liang M, Baker JS, Gu Y (2023) The relationship between ground reaction forces, foot positions and type of clubs used in golf: a systematic review and meta-analysis. Appl Sci (Switzerland) 13(12):7209

Development of Low-cost LED-Driven Boxing Machine

Muhammad Muhaimim Mahmud, Muhammad Haniff Gusrial, Nur Aqilah Othman, and Mohd Hasnun Arif Hassan

Abstract This project aims to examine the usage of different sensors in the development of a cost-friendly boxing workout machine with the incorporation of light-emitting diodes (LEDs) sequence for indication purposes. This development is an alternative to the usage of a traditional monotonous punching bag by boxing athletes for training purposes where unattractive and not interactive traditional style makes the athletes felt less motivation. Lack of real-time feedback also affects the athlete's performances. Suitable sensors to be incorporated are selected based on several criteria such as sensitivity and precision, durability, practical–ity and costing. LED strip and LED ring using LED sequence approach will be included in the design which function as a real-time feedback that controlled by push buttons to indicates different reaction time by varying the LED strip speed as well as its ON and OFF states. The system's testing results indicated that the proposed sensor is suitable and practical for the boxing training purposes in real applications.

Keywords Boxing · LED sequence · ESP32

1 Introduction

Boxing training is a physically demanding and highly disciplined form of exercise that focuses on developing strength, speed, agility, endurance, and overall fitness (Wu 2022; Zhang and Cheng 2023). It combines several methods, including conditioning workouts, punches, footwork, and defensive maneuvers. Boxing training has historically been done with a variety of tools, such as focus mitts (Nasser and Ahmed

M. M. Mahmud · M. H. Gusrial · N. A. Othman
Faculty of Electrical and Electronics Engineering Technology, Universiti Malaysia Pahang Al-Sultan Abdullah (UMPSA), Pekan, Pahang, Malaysia

N. A. Othman (✉) · M. H. A. Hassan
Centre for Sports Technology Innovation (STIC), Universiti Malaysia Pahang Al-Sultan Abdullah (UMPSA), Pekan, Pahang, Malaysia
e-mail: nuraqilah@umpsa.edu.my

M. H. A. Hassan et al. (eds.), *Proceedings of the 3rd International Conference on Innovation and Technology in Sports*, Lecture Notes in Bioengineering,
https://doi.org/10.1007/978-981-95-0804-4_34

2022), speed bags (Watts et al. 2023), punching bags (Munir et al. 2023), and skipping ropes. The punching bag is a common tool used in traditional boxing training. Boxers can practice punching and improve their power, accuracy, and technique by using a punching bag. It enhances general body mobility, footwork, and hand-eye coordination. When practicing with a punching bag, boxers frequently mix and match punches, defensive techniques, and footwork drills.

A monotonous action when performing the boxing training tend to make the user felt less motivation due to its unattractive style of training. There is no output that can be visually seen when doing the traditional boxing practices where this can lower the user's motivation especially for the beginners. The speed of the punches would also be inconsistent due to the lack of instant real-time feedback.

In order to increase the user's performance, an interactive, engaging and cost-friendly boxing machine is developed by incorporating suitable sensor that able to withstand an impact form the punches, customizable speed timing to enhance the user's agility and visual indication using LED-based technology using LED sequence approach to provide instant instruction and feedback. The development is suitable for the purpose of boxing training which can be done anywhere due to its simple design.

This paper is structured as follows. Section 2 shows the sensor selection for the system, Sect. 3 defines the overall procedures in LED sequence approaches, Sect. 4 provides the prototype result and Sect. 5 concludes the contribution of this paper.

2 Sensor Selection

The sensor to be incorporated in the development of boxing workout machine should have the ability to detect and withstand the impact from the punch. Several criteria are considered before choosing the most suitable sensor for boxing application.

2.1 Selection Criteria

Criteria that are considered are in terms of sensitivity and precision, durability, practicality and costing. All of the criteria are explained in Table 1.

2.2 Sensor Testing

Several sensors are tested in order to fulfill the criteria listed in Table 1. Sensors that involved in the testing are as follows:

a. Accelerometer (Tollenaar 2023)

Table 1 Sensors selection criteria

Criteria	Remarks
Sensitivity and precision	Sensors able to be operated under the covered punch plate and precisely detect changes occurs between each state
Durability	Sensors able to withstand the continuous punch forces while maintaining a constant input detection
Practicality	Sensors that consist of simple connections and installation procedures
Costing	Sensors that are cost-friendly

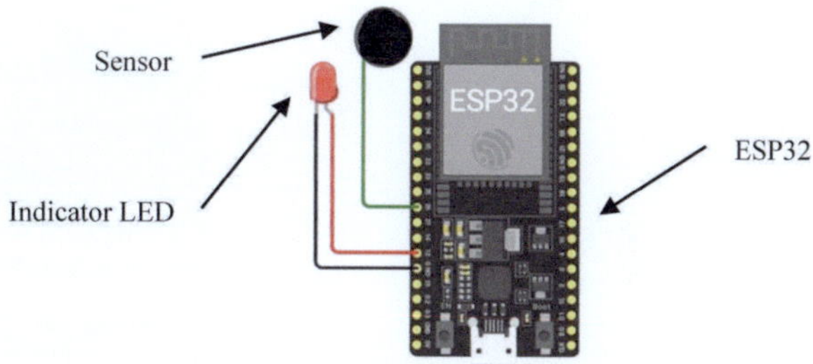

Fig. 1 Connection for sensor testing

b. Touch sensor (Noh and Jeong 2021)
c. Piezoelectric sensor (Mukhopadhyay 2021)
d. ESP32 built-in touch sensor (Andreas et al. 2019).

The sensors are hands-on tested with the used of ESP32 as the microcontroller and programmed in Arduino IDE software. A single LED is connected to ESP32 as an indicator for detecting precise timing when the sensor received an input as well as indicating the system is functioning well. Sensor, LED and ESP32 are connected as shown in Fig. 1. For testing, all sensors are place under a plate cover as in Fig. 2. A punch is given to the sensors and the sensor reading displayed on the serial monitor is evaluated for testing its sensitivity and precision as shown in Fig. 3. In order to observed its durability, multiple moderate punches are given to the sensors with an LED as an indicator for indicating the system is still functioning well. However, practicality is tested in terms of their simplicity in installation while costing is not tested.

3 LED Sequence Development

In order to develop the LED-driven boxing workout machine, the LED sequence is introduced as an interactive feature to the ordinary boxing routine. Aside from interactive feature, the LED provide a visual feedback, instruction for the user, timing and

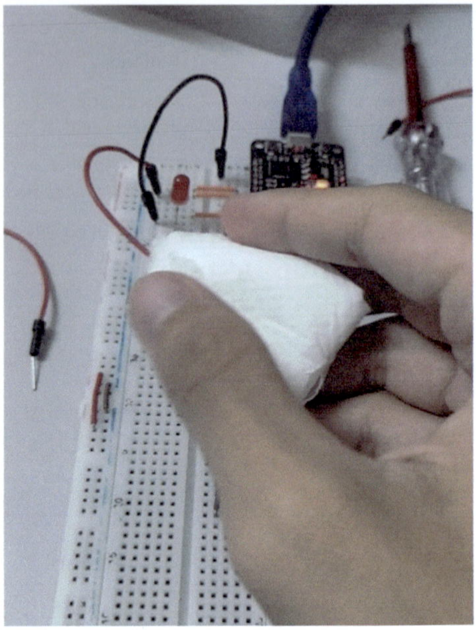

Fig. 2 Sensor is placed under the plate for testing

Fig. 3 Sensor reading displayed on serial monitor

punches intensity indicator. The LED sequence is developed by integrating ARGB LED strip and ARGB LED ring with ESP32.

3.1 System Design

ESP32 is chosen as the microcontroller due to the fact that it can work in same environment as Arduino but with better functionality (Othman et al. 2024). ESP32 consist of nine (9) built-in touch pins that is customizable into punch sensor and 30 pinouts for digital and analog input-output that compatible with Arduino IDE software. ESP32 also works with a higher processing speed which is around 240MHz in clock speed (Hercog et al. 2023). Pinout for ESP32 Do-It development board is as shown in Fig. 4.

As for LED strip, there are various types and choices available but WS2812B addressable RGB (ARGB) LED strip is chosen due to several advantages that it offers (Zainuddin et al. 2024). This LED strip comes with an integrated driver where each addressable LED in WS2812B strip can be assigned independently in terms of its brightness and colour. This function makes it possible to create interactive and interesting visual effects. The LED strip is compatible with fastLED.h library in Arduino IDE software. The strip also available in a variety of versions with varying sizes, sealants, and LED densities which gives sufficient options for various applications to suits the user's need.

Other than LED strip, this development also included 24 ARGB LED ring which makes the looks of the final prototype to be neater as well as safely securing all connections and installations for every component. Both LED strip and LED ring used is as shown in Fig. 5. All of the components are working together to perform the boxing system as shown in Table 2. As presented in Table 2, ESP32 will act as the microcontroller, the touch sensor and push button as the input to receive punch and control the mode for the system. ARGB LED strip and ARGB LED ring will be the output of the system for the punching detection and its timing indication.

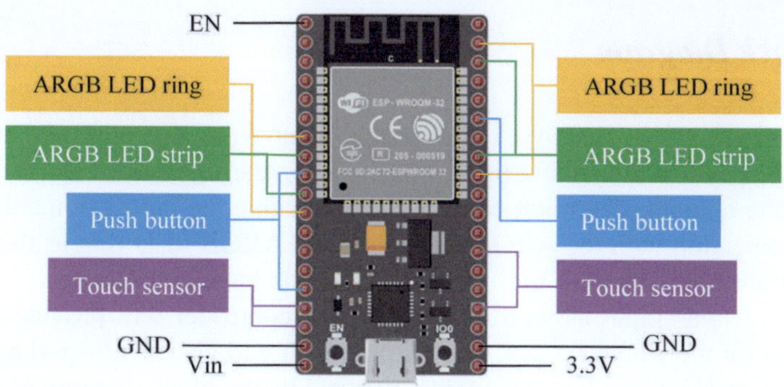

Fig. 4 ESP32 pinouts

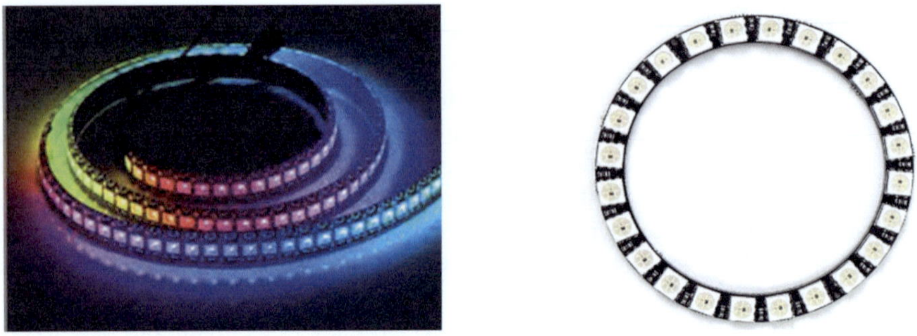

(a) (b)

Fig. 5 **a** WS2812 ARGB LED strip and **b** SK6812 ARGB LED ring

Table 2 Component listing for the system

	Components	Function
Processor	ESP32 microcontroller	Coordinate all components
Input	ESP32 built-in touch sensor	Punch receiver
	Blue push button	System power switching (START/STOP)
	Red push button (no. 1)	ARGB LED strip mode changer
	Red push button (no. 2)	ARGB LED strip power switching
Output	ARGB LED strip	Punch timing and speed indication
	ARGB LED ring	Punch detection indication
Power	Power module	Power step down (from 9 to 3.3 V)
	9 V Battery	Power supply

3.2 Block Diagram

The system begins by receiving the punch input from the user and necessary processing occurs within the ESP32 and it ready the input data for the integration with the outputs. The microcontroller ensures that the input compatibility and its synchronization with the ARGB LED strip and ARGB LED ring as the outputs. These outputs will generate the result as what has been programmed. The touch sensor that acts as input as well as the system's controller will provide feedback mechanism for the whole system. The output result will also be affected by the role of feedback mechanism. The system is illustrated by the block diagram as shown in Fig. 6.

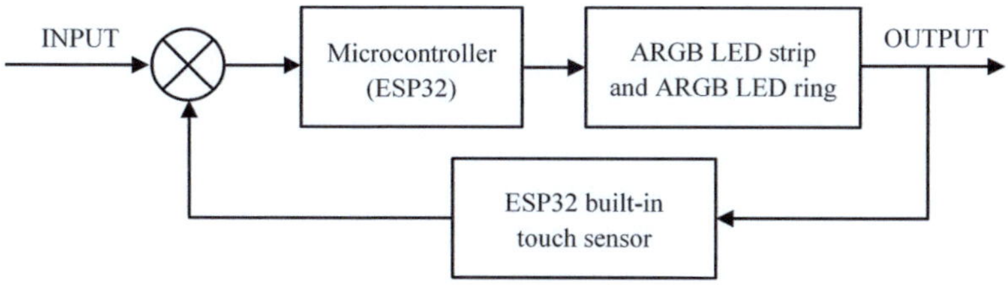

Fig. 6 Block diagram for the system

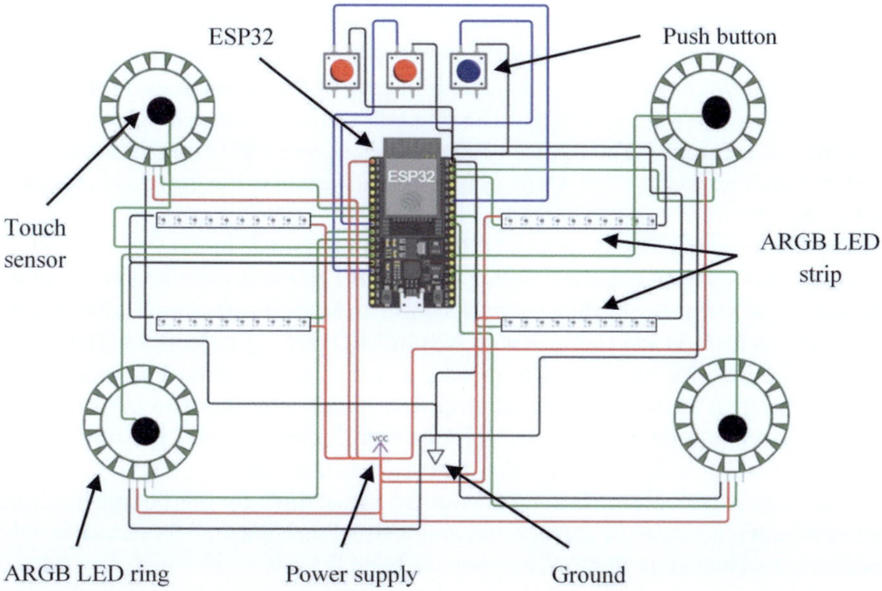

Fig. 7 Schematic diagram for the system

3.3 Circuit Diagram

Connections between components is presented in the schematic diagram as shown in Fig. 7 where includes ESP32, ARGB LED strip (4 nos.), ARGB LED ring (4 nos.), touch sensors (4 nos.), push buttons (3 nos.), and the power supply.

3.4 Programming Design

The development involved the use of Arduino IDE software for the ESP32 programming building with C++ as the programming language. The system starts by switching ON the START/STOP push button (blue push button). The procedure followed by choosing the preferred mode which is Mode 1 and Mode 2 using the second push button (red push button). Mode switching allowed user to choose their preferred visual reaction either with the help of ARGB LED strip (Mode 1) or not (Mode 2). Mode 1 allowed users to choose the desired running speed (Speed 1 and Speed 2) for the ARGB LED strip (blue light for Speed 1 and red light for Speed 2) before triggered the ARGB LED ring to give response with green light while Mode 2 is directly triggered the ARGB LED ring without allowing the speed selection and involvement of ARGB LED strip.

The system evaluates the input received (punch) from the touch sensor as the feedback. An internal timer will start to count the ARGB LED ring reaction time which is 0.5s (as soon as the timing from the ARGB LED strip is finish as for Mode 1). Once the sensor is triggered by the punch within 0.5 s, the system is stop and indicating the whole process is completed. In case of the sensor does not triggered by any punch within 0.5 s, ARGB LED ring will response with red light indicating punching time is finish. The ARGB LED strip can be turn OFF using the second red push button if necessary. The whole system is represented in the flowchart as shown in Fig. 8.

The total reaction time is set to be different for each mode in terms of the ARGB LED strip timing but a constants timing for ARGB LED ring according to the mode chosen by the user using the first red push button. In Mode 1 with Speed 1, the ARGB strip will run for 0.5s and continue with ARGB LED ring to run for 0.5 s. Therefore, the reaction time for Mode 1 with Speed 1 is 1.0 s. As for Mode 1 with Speed 2, slightly different at the ARGB LED strip time which is changed to 0.3 s. Hence, the total reaction time will be 0.8 s. As for Mode 2, only ARGB LED ring time is considered which is 0.5 s. All reaction time as shown in Table 3.

The purpose of choosing 0.3 and 0.5 s as the visual timer for this boxing machine development is because the average maximum visual reaction time for human is only around 0.2 s (Jain et al. 2015). Therefore, 0.3 and 0.5 s is a suitable timing for the purpose of this prototype.

3.5 *Hardware Design*

Interactive hardware design is incorporated in the development where each of the punch location or called target needed to withstand impact and provide resistance when it being punched. It is to mimic the traditional punching bag properties but with a low-cost material which suites for training purposes. The proposed design is illustrated in Fig. 9.

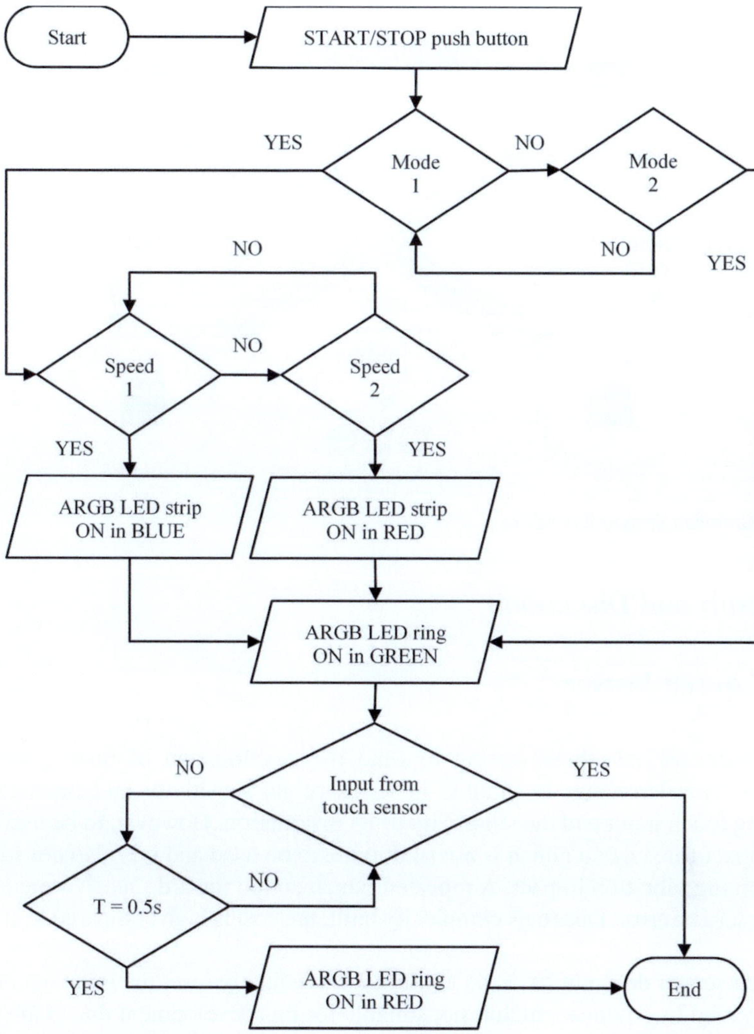

Fig. 8 Flowchart for the system

Table 3 Speed mode reaction time

	Mode 1		Mode 2 (s)
	Speed 1 (s)	Speed 2 (s)	
ARGB LED strip timer	0.5	0.3	–
ARGB LED ring timer	0.5	0.5	0.5
Total reaction time	1.0	0.8	0.5

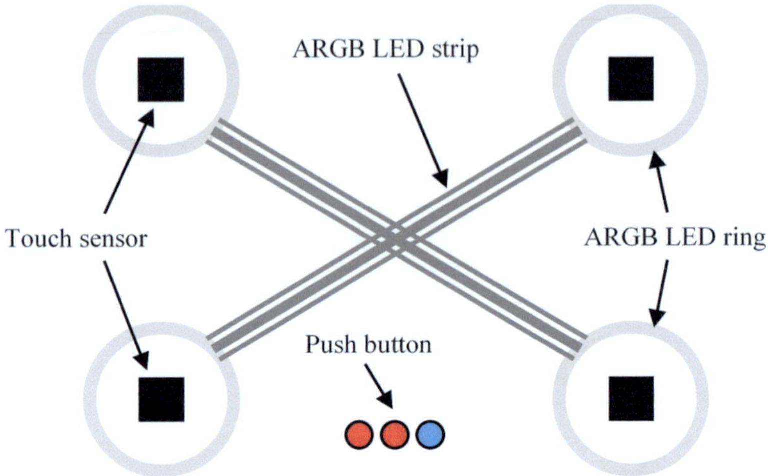

Fig. 9 Hardware design illustration

4 Result and Discussion

4.1 Chosen Sensor

Several sensors have been testing to suite the development of boxing workout machine. Accelerometer is suitable in terms of its sensitivity and precision for detecting touch input and the simplicity of its installation. However, to be used for a high impact force like a punch is not so durable to be used and it is also not suitable for receiving vibration impact. A repeated punch action towards accelerometer also causing it to be error. Due to its complexity built, the cost is high compared to another sensor.

Touch sensor does not have the ability to immediately reset the input again after it is triggered by a punch which is not suitable for this development due to the needs of fast and dynamic reaction. In order to solve that, an additional part is needed to be installed which causing complexity in terms of practicality. This sensor also not so precise as it can sometimes triggered earlier than it should be. In terms of cost is low due to its availability in the market.

Piezoelectric sensor is also suitable in terms of its sensitivity and precision for detecting touch input where the analog input value is easily managed and a function threshold is used in coding development which makes it produce good feedback. The price for this sensor is also affordable. However, this sensor required many wiring during its installation and powered by different voltage from ESP32. As for its durability, the detection plate is easily crack and bend after receiving several punch inputs.

Table 4 Sensors testing summary

Criteria	Sensitivity and precision	Durability	Practicality	Costing
Accelerometer	✓	✗	✓	✗
Touch sensor	✗	✗	✗	✓
Piezoelectric sensor	✓	✗	✗	✓
ESP32 built-in touch sensor	✓	✓	✓	✓

Finally, the ESP32 built-in touch sensor is acceptable for all criteria which makes it the most suitable sensor to be used. It is good in inputs coordination, simple installation, durable sensor plate for receiving punch impact and cost-friendly compared to other sensors that has been tested. The summary for the sensor testing is shown in Table 4 where ✓ means suitable and ✗ means not suitable.

4.2 LED Sequence Result

The LED sequence using ARGB LED strip and ARGB LED ring is functioning well with the desired sequenced. The sequence started by pressing the blue push button that will turn ON the system. As soon as the system is ON, one (1) of the four (4) ARGB LED strip will randomly light up (in running LED sequence) with the default mode which is Mode 1 with Speed 1 (blue colour ARGB LED strip). According to its LED reaction time, the ARGB LED ring will be triggered next and followed as what has been explained in Sect. 3.4. The sequence result is as shown in Fig. 10.

4.3 Boxing Machine Prototype

The final look of the LED-driven boxing machine prototype is as shown in Figs. 11, 12 and 13 where LED sequence is the based of this development.

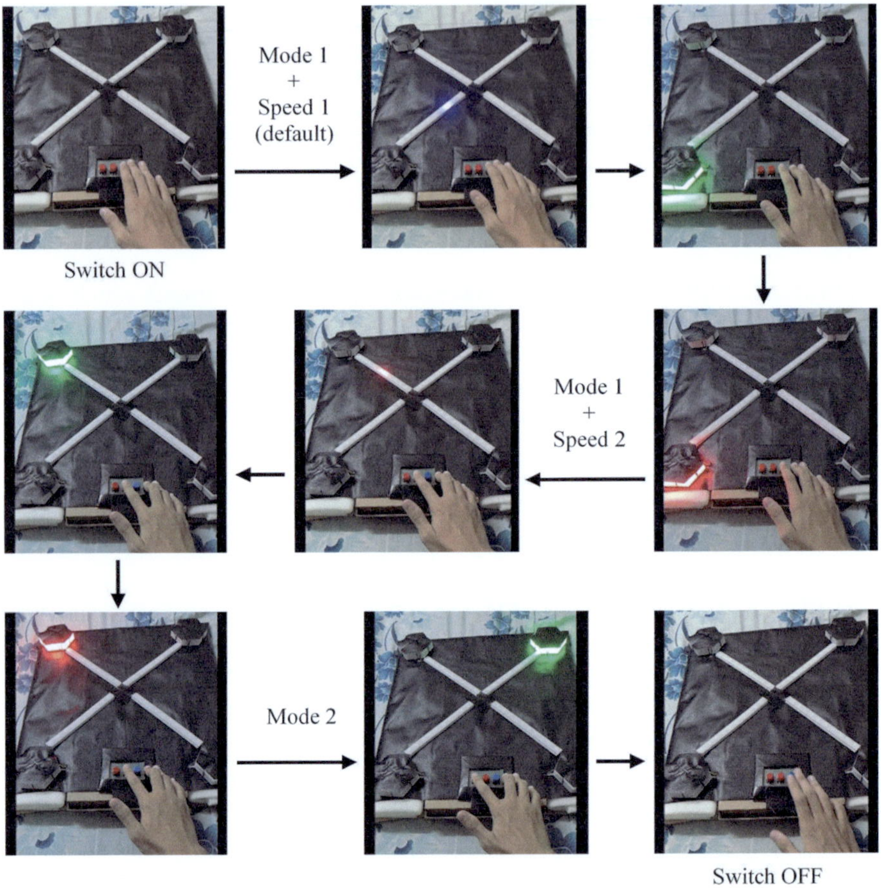

Fig. 10 LED sequence result

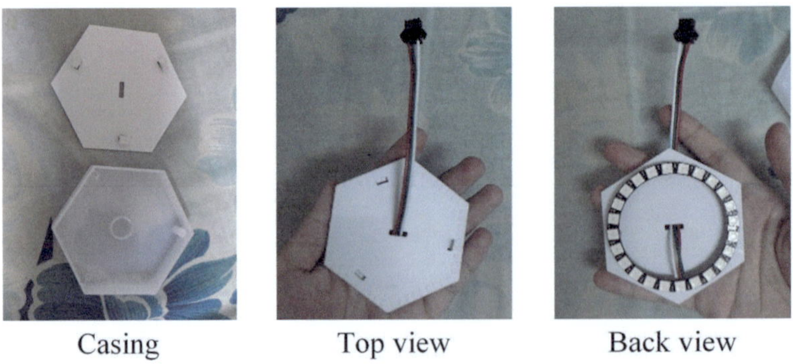

Fig. 11 ARGB LED ring casing

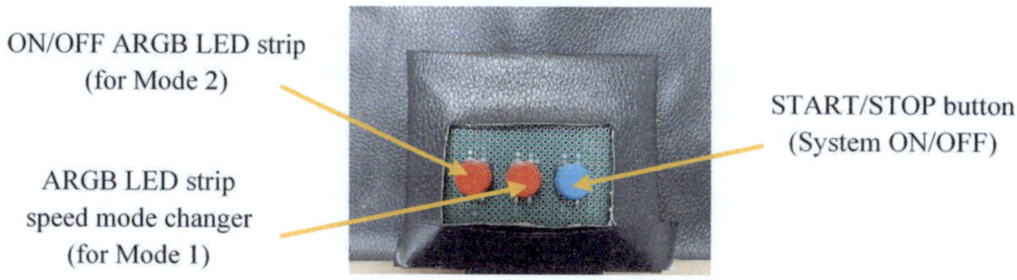

Fig. 12 Control push button

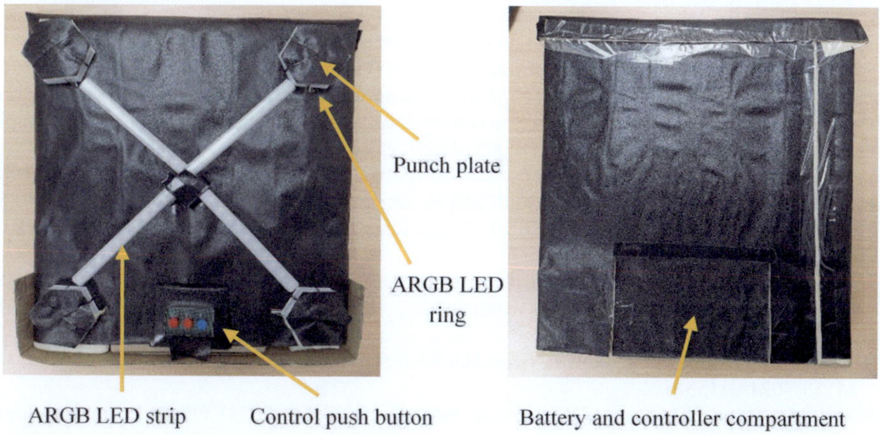

Fig. 13 Prototype front (left) and back (right) overview

5 Conclusion

In conclusion, the development of LED-driven boxing machine helps to improve the user's performances compared to the used of the traditional punching bag. This is due to its advantages in terms of visual engagement where colour indication from the used of ARGB LED strip and ARGB LED ring makes it more interesting as well as entertaining. The different reaction time between the colour makes user improved their timing and coordination due to the synchronization of their punches and motions with the desired timing. In terms of feedback, the visual interpretation from the LEDs can provide instant punch speed for the user to monitor their agility performance. A side from the system, this development is flexible where it can be placed anywhere that is needed due to its simple design.

Some improvements can be done for future development such as adding more punch plate in order to increase the user's performance and agility level. In terms of power consumption, a reliable and more environment friendly approach can be instilled such as using solar energy. Other than that, integration of IoT can be a good approach as it can be controlled and monitored by the user via wireless devices where more accurate reading can be analyzed. Some additional feature such as sound can also be added as the reaction for the system. However, this development is still considered a significant contribution to the society where it improvised the traditional monotonous boxing experience into a modern and entertaining new system.

References

Andreas CR, Aldawira HW, Putra N, Hanafiah SS, Wibisurya A (2019) Door security system for home monitoring based on ESP32. Procedia Comput Sci 157:673–682. https://doi.org/10.1016/j.procs.2019.08.218

Hercog D, Lerher T, Truntič M, Težak O (2023) Design and implementation of ESP32-based IoT devices. Sensors 23(15):6739. https://doi.org/10.3390/s23156739

Jain A, Bansal R, Kumar A, Singh K (2015) A comparative study of visual and auditory reaction times on the basis of gender and physical activity levels of medical first year students. Int J Appl Basic Med Res 5(2):124. https://doi.org/10.4103/2229-516X.157168

Mukhopadhyay K (2021) Application of sensors in sports. Int J Multidiscip Res Publ 4(1):103–108

Munir HMW, Latif A, Mehmood N, Patafi MAM, Ahmed N, Malika (2023) Effect of punching bag exercises on Body Mass Index (BMI). J Heal Rehabil Res 3(2):965–969. https://doi.org/10.61919/jhrr.v3i2.215

Nasser AJ, Ahmed DM (2022) The effect of battle rope exercises on increasing the number of single, double and multiple punches in elite boxing. Rev Iberoam Psicol Del Ejerc Y El Deport 17(5):260–263

Noh SM, Jeong C (2021) Wearable devices. Nat Biotechnol 39(12):1511. https://doi.org/10.1038/s41587-021-01153-6

Othman MA, Amri LH, Azmi MNF, Othman NA, Hassan MHA (2024) Wireless timing system using esp-to-esp communication protocol for precision timing in sports. In: Lecture notes in bioengineering, pp 577–586. https://doi.org/10.1007/978-981-97-3741-3_50

Tollenaar D (2023) Boxing punch classification with accelerometer data. Tilburg University. http://arno.uvt.nl/show.cgi?fid=169829

Watts CR, Thijs Z, King A, Carr JC, Porter R (2023) A pilot study of the effect of a non-contact boxing exercise intervention on respiratory pressure and phonation aerodynamics in people with parkinson's disease. J Clin Med 12(14):4806. https://doi.org/10.3390/jcm12144806

Wu Q (2022) Boxing speed and muscle scientific training. Rev Bras Med Do Esporte 28(2):82–84. https://doi.org/10.1590/1517-8692202228022021_0428

Zainuddin MAH, Gusrial MH, Othman NA, Ahmad H, Yatim NM (2024) Development of LED-based pacing running system. In: Lecture notes in bioengineering, pp 287–296. https://doi.org/10.1007/978-981-97-3741-3_27

Zhang X, Cheng N (2023) Special training for boxer's athletic ability and body control. Rev Bras Med Do Esporte 29:3–5. https://doi.org/10.1590/1517-8692202329012022_0331

A 3-Dimensional Finite Element Modelling of Insole for Flatfoot Children

Mohd Firdaus Mahamad Ali and Nadzifah Yaakub

Abstract The Finite Element Method (FEM), a digital modeling technique widely used in foot biomechanics research, is utilized to model the structure of the foot insole. FEM allows detailed analysis of mechanical behaviors and stress distributions, providing insights into the effectiveness of insole de-signs for alleviating flatfoot issues in children. Technological innovations that have revolutionized evaluation methods include 3D foot kinematic analysis. The objective of this study is to investigate the impact of various factors such as weight, height, and BMI on the finite element analysis (FEA) model of insoles designed for children with flatfoot and to provide insights into the mechanical behaviors and stress distributions of foot-insoles, and to evaluate the effectiveness of different insole designs in alleviating flat-foot issues. 3 years old, and 6 years old, children with different weights that have flatfoot problems have been involved in this study. The three-dimensional configurations of children's insoles were created utilizing SolidWorks software, and the finite element analysis (FEA) was performed with ANSYS software. The material considered for manufacturing the insole is Thermoplastic polyurethane (TPU) 95A. The model for the insole is specifically analyzed according to the foot size of the children. This study is conducted to investigate the significant correlation with variables, such as weight, height, and BMI, that affect the FEA model analysis while standing. The simulation results indicate that the Von Mises stress distribution varies significantly based on the child's weight, with 6-year-old children experiencing higher stress levels than 3-year-old children due to greater weight loading. The peak plantar pressure was found to be 3.9526×10^{-5} Pa for 3-year-old children and 3.762×10^{-5} Pa for 6-year-old children, highlighting the correlation between weight and foot pressure. It was also demonstrated that thicker insoles provide better pressure distribution, reducing the

M. F. M. Ali (✉)
Faculty of Creative Technology and Heritage, University Malaysia Kelantan, Kelantan, Malaysia
e-mail: mfirdaus.ma@umk.edu.my

N. Yaakub
Faculty of Bioresources and Food Industry, University Sultan Zainal Abidin, Terengganu, Malaysia
e-mail: nadzifah@unisza.edu.my

M. H. A. Hassan et al. (eds.), *Proceedings of the 3rd International Conference on Innovation and Technology in Sports*, Lecture Notes in Bioengineering,
https://doi.org/10.1007/978-981-95-0804-4_35

strain on the medial longitudinal arch. These findings confirm that FEA is an effective tool for optimizing insole designs based on individual weight and foot structure, which can significantly enhance comfort and support for children with flatfoot.

Keywords Flat-foot · Children · Insole · Prototype · Sole materials

1 Introduction

A lower medial arch of the foot is a characteristic of flatfoot, which is usually asymptomatic (Oerlemans et al. 2023). Individuals with flat feet have the full sole of their foot in contact with the ground. This circumstance arises when the arch fails to form or collapses. In asymptomatic cases, a child with flat feet experiences no symptoms aside from the physical appearance of the foot. The condition is often recognizable by the foot lying flat on the ground with little to no visible arch. When viewed from behind, the outer toes are usually visible, but not the big toe. There are two varieties of flatfoot: rigid and flexible (Pfeiffer et al. 2006). Less than 1% of people have inflexible flatfoot, which is defined by a depressed arch when both weight-bearing and not (Michaudet et al. 2018). In contrast, flexible flatfoot is more prevalent. Foot orthoses have limited effectiveness for rigid flatfoot due to the restricted range of motion. Some children with flexible flatfoot may report lower limb pain and demonstrate reduced lower limb function. Children's flatter foot structure is due to the immaturity of complex bone, soft tissue, and neurological function, which typically improves over the first decade of life (Uden et al. 2017).

Flatfoot can be influenced by a variety of congenital and acquired factors (Michaudet et al. 2018), with age being a significant risk factor. The normal development of the foot in children is associated with physiological flatfoot (Evans and Rome 2011). Most infants are born with flat foot, and the arch typically elevates spontaneously within the first decade of life (Staheli 1987). Studies have shown that flatfoot is present in 54% of 3-year-olds and 26% of 6-year-olds (Pfeiffer et al. 2006). This condition is common in infants and toddlers up to the age of 3–4 as their arch develops. According to Mickle et al. (2006), flatfoot is prevalent among children aged 2–5 years as the arch of the foot forms (Mickle et al. 2006). If flat feet persist beyond the age of 5–6, it may warrant further evaluation. Research by Martin Pfeiffer found that flatfoot was detected in 54% of children in the 3-year-old group, compared to only 24% in the 6-year-old group. The highest prevalence of flatfoot is observed in children aged 2–6 years, ranging from 21 to 57%, and this prevalence decreases to between 13.4 and 27.6% in primary school-aged children. The incidence continues to decline with increasing age, reaching a prevalence of 5–14% in adulthood (Pfeiffer et al. 2006). A study in Nigeria found that approximately one in every five children aged 6–10 years old had flatfoot, with obesity further increasing the risk (Ezema et al. 2014).

In the field of footwear, the insole is a component situated between the foot and the shoe sole (Aracil et al. 2018). Being in direct contact with the foot (Buldt et al.

2015), the insole supports body weight and directly affects the biomechanics of the foot and body. Its design strategy aims to minimize potentially harmful tissue stress on the foot's plantar region. Insoles provide a critical interface between the foot and the shoe and, when combined with outsole modifications, offer a direct approach to reducing potentially damaging tissue stresses on the plantar aspect of the foot (Collings et al. 2020). The insoles also play a keyrole for the researcher to optimising foot insole in design and comfort. Concretely, research has developed in three directions: to increase comfort when wearing different type of shoes, for daily activities, to increase athletic performance, to correct some foot malformation escpecially on children (Alcantara 2019). Traditionally, orthotic insoles have been produced using footprints in a foam box to obtain the foot's shape (Vicenzino 2004). However, this method often lacks precision and accuracy when matching the insole to the patient's foot geometry, frequently resulting in orthotic footwear with low comfort (Anggoro et al. 2021).

Additionally, the traditional method often incurs higher costs and longer production times. Currently, advancements in technology, such as 3D scanners, allow for the acquisition of the virtual geometry of the foot in both static and dynamic modes. These scanners, in conjunction with pressure platforms that provide barefoot plantar pressure maps, enable the precise modeling of insole properties, enhancing both accuracy and comfort (Li et al. 2017).

The finite element method (FEM) has been widely utilized in foot-related biomechanics studies and has proven to be a promising tool for predicting stress distributions at the foot–insole interface, provided that the complex geometry and material properties of both the foot and insole are accurately considered (Chen et al. 2003). Most of the previous studies on footwear midsole design via FE simulation are based on adult foot diseases, and few are designed for obese children to distribute the bone stress distribution and reduce plantar pressure (Zhou et al. 2023). Simulating the foot and footwear on computers is essential not only for generating realistic animations but also as a tool for economically evaluating the design of functional footwear. The most common method for this simulation is the Finite Element Method (FEM). Finite element analysis enables the safe simulation of potentially hazardous or destructive load conditions and failure modes, allowing engineers to understand the physical response of a system at any point (Song et al. n.d.).

This paper develops and validates a 3D finite element model of an insole specifically designed for children with flatfoot. This model aims to address the limitations of traditional insole fabrication methods by utilizing advanced technologies such as 3D scanning and finite element analysis (FEA) to achieve a precise fit and optimal support. The objectives of this study are to analyze the effects of weight, height, and BMI on the stress distribution and mechanical behavior of the insole, evaluate the plantar pressure variations under different loading conditions, and determine the effectiveness of different insole designs in alleviating flatfoot conditions in children. By understanding these factors, this study seeks to improve the biomechanical function and comfort of insoles for pediatric flatfoot, ultimately contributing to better clinical outcomes and enhancing the quality of life for affected children.

2 Methodology

2.1 Participants

A total of two male flat foot child involved in this study are 3 and 6 years of age with the weight of 3 years children is 12 kg with 24.5 cm shoe size and for 6 year is 27.4 kg with 24 cm shoe size. Figure 1 shows that back position of children having flatfoot. The heel of the children not straight as orange line and curved as white line. Figure 2 show the soles of the feet is totally flat, no arch and unstable medial column. The foot print in Fig. 3 is identified the degree of flatfoot. The 3 years old experience flat foot first degree and 6 years old children is 3rd degree.

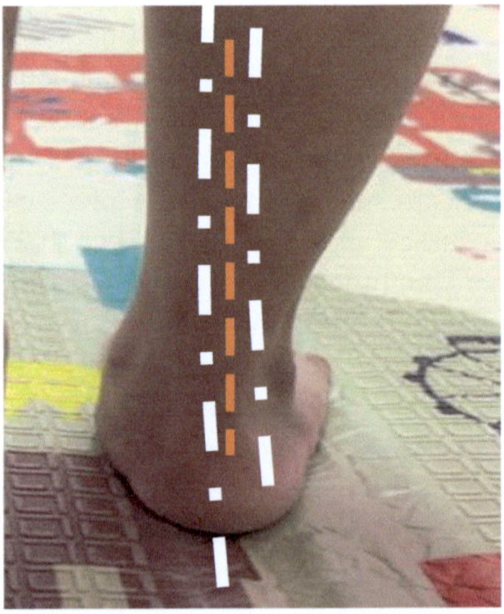

Fig. 1 Back foot position of flatfoot children

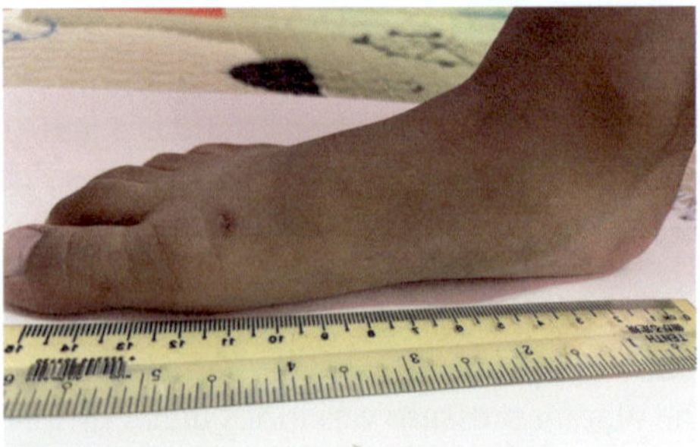

Fig. 2 Side foot position of flatfoot

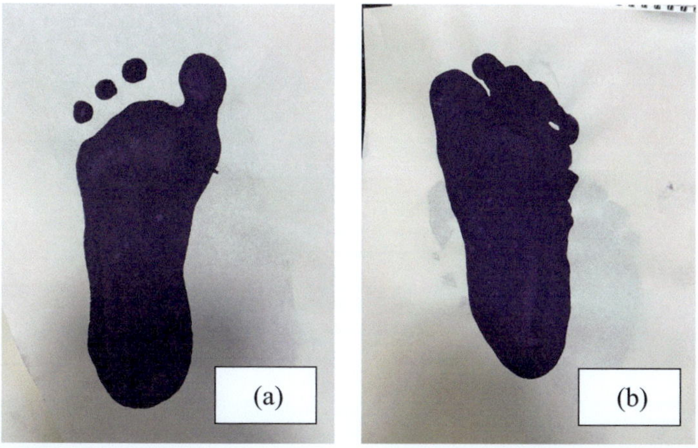

Fig. 3 Foot print **a** 3 years old and **b** 6 years old

2.2 *Study Design*

Figure 4 show the research flow chart for this study. The flowchart effectively delineates the structured approach of a research project aimed at addressing flatfoot in children through the FEA analysis and the development of a custom 3D-printed insole. It highlights the critical steps from obtaining ethical approval, collecting and analyzing data, to the design and production of the final product. Each stage involves specific activities and reviews to ensure the reliability and validity of the research outcomes.

2.3 *Data Analysis*

The 3D shapes of insole have been built using Solid works software. These models must be exported and recognized by the interface of a software program type FEA (Finite Element Analysis). In addition to preparing these models and transforming it for analysis method, FEA, information has been added, for the measurement's units, the physical and mechanical properties. This analysis has been done with ANSYS software.

2.4 *Insole Modelling*

The insole is designed according to the size and shape of the 3 years old, and 6 years old, with flatfoot. It was designed with a 3 mm thickness, as thicker insoles

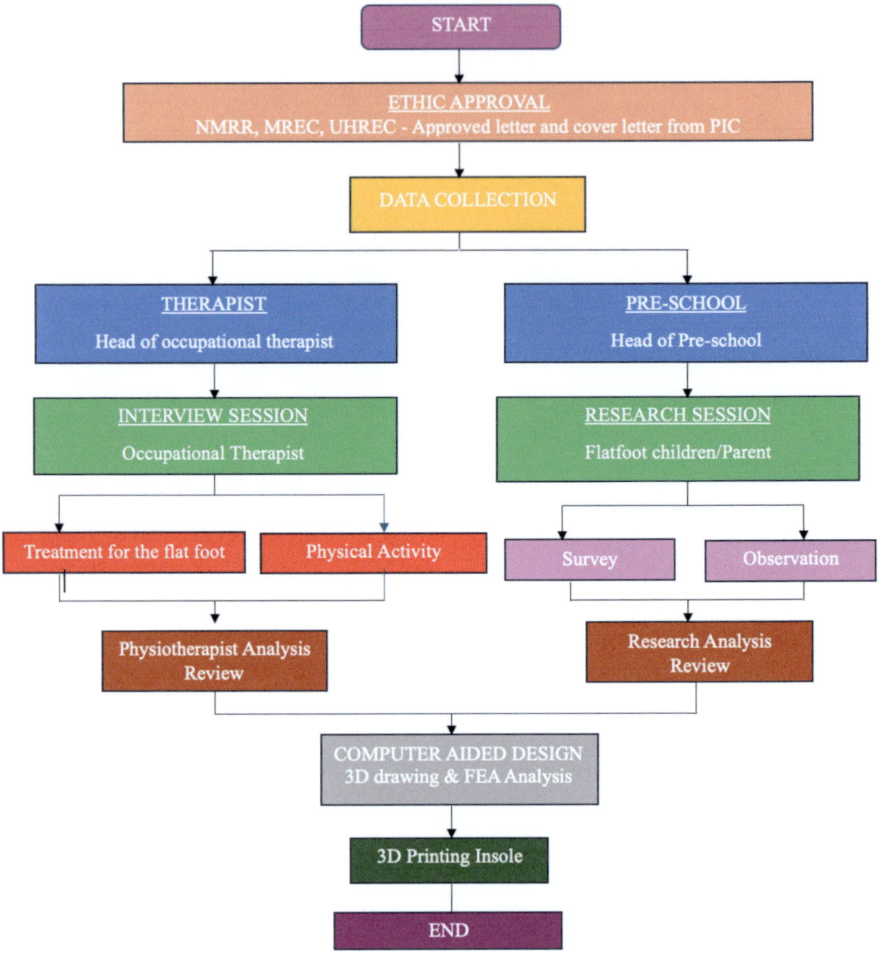

Fig. 4 Research flow charts

provide better pressure distribution and support (Goskea et al. 2006). Figure 5 shows the traditional custom-made insole, primarily made from soft-foam material, which is commonly used in casual, sports, and orthopedic footwear. The raised section is designed to correct or support the foot arch, essential for managing flatfoot conditions. Figure 6 illustrates the 3D insole model prototype, which incorporates advanced design technologies to create customized solutions for flatfoot. The material used for the insole is Thermoplastic Polyurethane (TPU) 95A, known for its excellent mechanical properties (Thermoplastic et al. 2017), high resistance to abrasion, wear, and tear, making it highly suitable for 3D printing applications in engineering plastics (Faculty 2021).

Insole is designed according to the size and shape of the 3 years old, and 6 years old, flat foot children. It was designed of 3 mm thickness. the thicker the

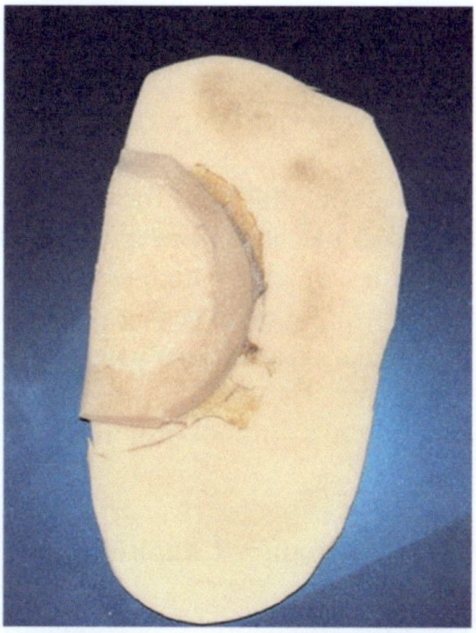

Fig. 5 Tradisional insole

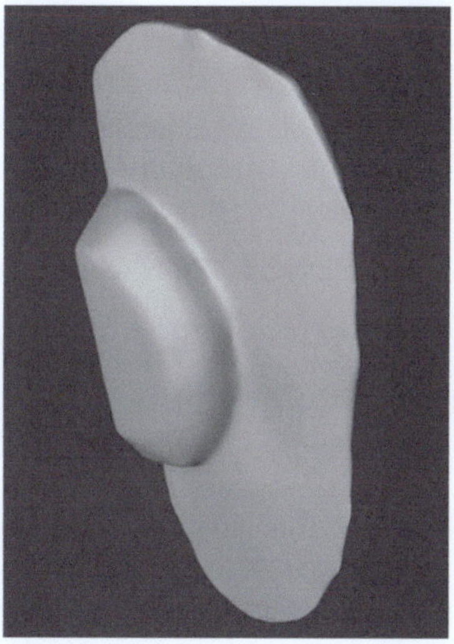

Fig. 6 3D insole

insole, the more efficiently is (Goskea et al. 2006). Figure 5 show the tradisional custom made insole for use in footware. The insole is primarily made from a soft-foam material. It can be used in casual shoes, sports shoes, or even specialized orthopedic footwear. The raised section is likely intended to correct or support the foot arch, which is particularly important for conditions such as flatfoot. This helps in maintaining proper foot alignment and reducing strain on the foot muscles and ligaments. Figure 6 illustrates the 3D insole model prototype, which exemplifies the use of advanced design technologies to create customized solutions for foot conditions like flatfoot.

2.5 *Finite Element Analysis (FEA)*

2.5.1 Material Properties

The material that used for the insole is thermoplastic polyurethane (TPU) since it has good mechanical properties (Thermoplastic et al. 2017). The combination of properties which are high resistance to abrasion, wear and tear make it suitable used in 3D printing engineering plastic (Faculty 2021). The material properties used in the simulation are young's modulus (E) 25 MPa, Poisson's ratio (ν) 0.49 and Density 1.22 g/cm^3. These values were chosen based on experimental data from existing TPU materials used in foot orthotics.

2.5.2 Boundary Conditions and Load Application

The FEA simulation was conducted under a static standing condition to replicate real-life weight-bearing scenarios. The boundary conditions applied were:

Fixed support: The bottom surface of the insole was constrained in all directions to simulate contact with the shoe sole.

Load application: A distributed load was applied on the insole's top surface to represent the pressure exerted by the child's foot. The loads were calculated based on body weight and plantar pressure distribution:

3-year-old child: 120 N (corresponding to 12 kg body weight)

6-year-old child: 270 N (corresponding to 27.4 kg body weight)

2.5.3 Analysis Outputs

The FEA simulation provided key outputs, including Von Mises stress distribution to determine critical stress regions, total deformation to assess insole flexibility and plantar pressure distribution to evaluate weight distribution across the foot. The results from the FEA simulations were analyzed to determine the impact of weight, height, and BMI on insole performance and to optimize the design for improved comfort and support.

3 Results and Discussion

Figure 7 presents the Von Mises stress distribution for the insoles of flatfoot children while standing. Figure 7a illustrates the stress distribution for the 3-year-old child, with a maximum stress value of 19,706 Pa, while Fig. 7b represents the 6-year-old child, with a maximum stress of 22,412 Pa. To ensure a direct and accurate comparison, the color contour scale has been standardized, allowing for a clearer visualization of stress variations between the two cases.

Figure 8 displays the plantar pressure distribution for both subjects Fig. 8a corresponds to the 3-year-old child, while Fig. 8b represents the 6-year-old child. The position and peak value of planar pressure can be inferred from the location of maximum deformation, which is indicated by the red region. The peak planar pressure value is 3.9526×10^{-5} Pa for 3-year-old children and 3.762×10^{-5} for 6-year-old children. The results indicate that the shape and density of plantar pressure distribution are influenced by the child's weight, with heavier children exerting higher pressure on the insole. This observation aligns with findings from Vergara-Amador et al. (2021), which established a correlation between flatfoot prevalence and factors such as age, gender, and BMI (Amador et al. 2012). Key contributors to flatfoot development include age, overweight nutritional status, gender, body composition, ligament weakness, footwear type, and family history. These refinements in the analysis provide a more precise evaluation of stress and pressure distribution, ensuring that insole designs are optimized for different weight conditions in children with flatfoot.

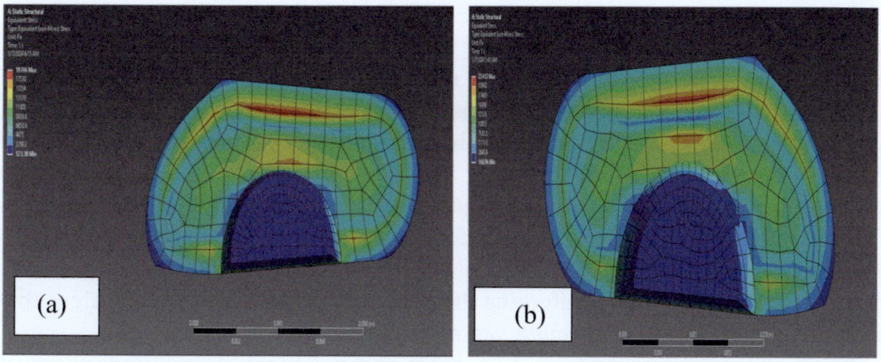

Fig. 7 Equivalent stress standing simulation insole of flatfoot children **a** 3 years old and **b** 6 years old

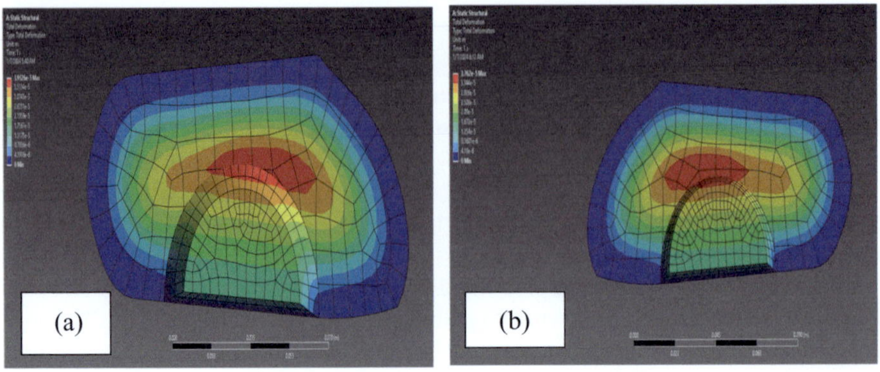

Fig. 8 Von Misses stress standing simulation insole of flatfoot children **a** 3 years old and **b** 6 years old

4 Conclusions

This study examined the effects of weight and height on plantar pressure distribution and stress variation in flatfooted children using Finite Element Analysis (FEA). The results highlight that increasing body weight not only amplifies the overall plantar pressure but also alters the stress distribution pattern across the foot. While it was expected that heavier children would exert greater force, the study found that pressure distribution is not solely dependent on weight but also influenced by foot morphology and surface contact area. The 6-year-old child, despite being heavier, exhibited only a slight increase in peak plantar pressure compared to the 3-year-old, suggesting that a larger foot surface area plays a role in redistributing forces more effectively.

Furthermore, the Von Mises stress distribution analysis revealed that the highest stress concentrations occurred in the medial arch and heel regions, with the 6-year-old experiencing greater internal stress due to increased loading. This indicates that as weight increases, the foot structure is subjected to higher mechanical strain, potentially accelerating arch collapse in flatfooted individuals. This study also emphasizes that height influences foot pressure distribution by modifying the load-bearing characteristics of the foot, reinforcing the need for personalized insole designs that account for both weight and height variations.

From a biomechanical perspective, these findings highlight the importance of insole optimization for children with flatfoot, where thicker and more supportive insoles may be necessary for heavier children to enhance pressure distribution and reduce internal stress. The use of Thermoplastic Polyurethane (TPU) 95A in the insole design was found to be beneficial in maintaining structural integrity and durability under varying load conditions. Future studies should focus on dynamic simulations during walking or running, as real-world foot loading conditions may further refine insole design for improved comfort and clinical outcomes.

In summary, this study demonstrates that plantar pressure distribution and stress variation in flatfooted children are influenced by both weight and height, not just body mass alone. These findings provide valuable insights for the development of biomechanically optimized insoles, ensuring that children with flatfoot receive effective support to mitigate stress-related discomfort and long-term complications.

Acknowledgments The author is grateful to thank UMK for their support through the UMK Prototype (UMK PRO) research grant under project no. R/PRO/A0200/01207A/002/2022/01132.

References

Alcantara M (2019) Influence of shoe characteristics on the development of Valgus Foot in children. J Clin Med 8(1):85. https://doi.org/10.3390/jcm8010085

Amador VE, Sánchez RFS, Posada JRC, Molano AC Guevara OA (2012) Prevalence of flatfoot in school between 3 and 10 years. Study of two different populations geographically and socially. Colomb Med (Cali) 43(2):141–146. PMID: 24893055; PMCID: PMC4001940

Anggoro PW, Bawono B, Jamari J, Tauviqirrahman M, Bayuseno AP (2021) Advanced design and manufacturing of custom orthotics insoles based on hybrid Taguchi-response surface method. Heliyon 73. ISSN 2405-8440

Aracil MD, Pérez JJH, Morenilla AJ, Mora HM (2018) 3D printing of functional anatomical insoles. Comput Ind 95 38–53. ISSN 0166-3615. https://doi.org/10.1016/j.compind.2017.12.001

Buldt AK, Levinger P, Murley GS, Menz HB, Nester CJ, Landorf KB (2015) Foot posture is associated with kinematics of the foot during gait: a comparison of normal, planus and cavus feet. Gait Posture 421:42–48. ISSN 0966-6362. https://doi.org/10.1016/j.gaitpost.2015.03.004

Chen WP, Ju CW Tang FT (2003) Effects of total contact insoles on the plantar stress redistribution: a finite element analysis. Clin Biomech 18(6):S17–S24. ISSN 0268-0033. https://doi.org/10.1016/S0268-0033(03)00080-9

Collings R, Freeman J, Latour JM, Paton J (2020) Footwear and insole design features for offloading the diabetic at risk foot—a systematic review and meta-analyses. Endocrinol Diabetes Metab 4(1):e00132. https://doi.org/10.1002/edm2.132. PMID: 33532602; PMCID: PMC7831212

Evans AM, Rome K (2011) A Cochrane review of the evidence for non-surgical interventions for flexible pediatric flat feet. Eur J Phys Rehabil Med 47:69–89

Ezema C, Abaraogu U, Okafor G (2014) Flat foot and associated factors among primary school children: a cross-sectional study. Hong Kong Physiother J 32:13–20

Faculty E (2021) Design of 3D printed custom-made orthopedic insoles I

Goskea S, Erdemirb A, Petreb M, Budhabhattib S, Cavanagh PR (2006) Reduction of plantar heel pressures: insole design using finite element analysis. J Biomech 39:2363–2370

Li Y, Falk B, Linke BS, Voet H, Schmitt R, Lam M (2017) Cost, sustainability and surface roughness quality—a comprehensive analysis of products made with personal 3D printers. 16:1–11

Michaudet C, Edenfield KM, Nicolette GW, Carek PJ (2018) Foot and ankle conditions: pes planus. FP Essent 465:18–23

Mickle KJA, Steele JR, Munro BJ (2006) The feet of overweight and obese young children: are they flat or fat? obesity. Silver Spring 14(11):1949–1953

Oerlemans LNT, Peeters CMM, Munnik-Hagewoud R, Nijholt IM, Witlox A, Verheyen CCPM (2023) Foot orthoses for flexible flatfeet in children and adults: a systematic review and meta-analysis of patient-reported outcomes. BMC Musculoskelet Disord 24(1):16. https://doi.org/10.1186/s12891-022-06044-8. PMID: 36611153; PMCID: PMC9825043

Pfeiffer M, Kotz R, Ledl T, Hauser G, Sluga M (2006) Prevalence of flat foot in preschool-aged children. Pediatrics 118:634–639. https://doi.org/10.1542/peds.2005-2126

Song Y, Shao E, Bíró I, Baker JS, Gu Y. Finite element modelling for footwear design and evaluation: a systematic scoping review

Staheli LT (1987) Evaluation of planovalgus foot deformities with special reference to the natural history. J Am Podiatr Med Assoc 77:2–6. https://doi.org/10.7547/87507315-77-1-2

Thermoplastic D, Non-suitable A, Max D, Tpu W, Tpu B, Tpu R (2017) Technical data sheet TPU 95A Test method 1–3

Uden H, Scharfbillig R, Causby R (2017) The typically developing paediatric foot: how flat should it be? a systematic review. J Foot Ankle Res 10:37. https://doi.org/10.1186/s13047-017-0218-1

Vicenzino B (2004) Foot orthotics in the treatment of lower limb conditions: a musculoskeletal physiotherapy perspective. Man Ther 9:185–192

Zhou Q, Niu W, Yick KL, Gu B, Sun Y (2023) Numerical simulation of the effect of different footwear midsole structures on plantar pressure distribution and bone stress in obese and healthy children. Bioengineering 10(11):1306. https://doi.org/10.3390/bioengineering10111306

Data Analytics in Sport Science—Current Applications and Opportunities in an Unexplored Frontiers

Muhammad Nur Farhan Saniman, M. H. Muhammad Sidik, Ahmad Shahir Jamaludin, Mohd Amir Shahlan Mohd Aspar, Devin Babu, Abdul Nasir, and Norain Binti Abdullah

Abstract Data analytics has significantly impacted the sports industry, revolutionizing athlete performance, team strategy, and fan engagement. This review explores the current applications of data analytics in sports and identifies areas that remain underutilized. Ten established areas of application are discussed, including injury prevention, player tracking, performance optimization, and game outcome prediction. These applications demonstrate how data-driven insights have enhanced decision-making, improved player safety, and optimized training regimens. Despite these advances, several aspects of sports remain underexplored by data analytics. This review highlights five key areas with potential for future development: emotional and psychological monitoring of athletes, ethical integrity and fairness oversight, environmental sustainability of sports events, personalized nutrition based on genetic data, and the enhancement of spectator experiences in real-time. Addressing these gaps could unlock new opportunities for innovation and elevate the sports industry. This review provides a comprehensive overview of the current state of data analytics

---

M. N. F. Saniman
Mechanical Engineering Section, Universiti Kuala Lumpur Malaysia France Institute, Bandar Baru Bangi, Malaysia

M. H. M. Sidik
Faculty of Mechanical and Automotive Engineering Technology, Universiti Malaysia Pahang Al-Sultan Abdullah, Pekan, Pahang, Malaysia

A. S. Jamaludin
Faculty of Manufacturing and Mechatronic Engineering Technology, Universiti Malaysia Pahang Al-Sultan Abdullah, Pekan, Pahang, Malaysia

M. A. S. M. Aspar · D. Babu · A. Nasir (✉)
Faculty of Electrical and Electronic Engineering Technology, Universiti Malaysia Pahang Al-Sultan Abdullah, Pekan, Pahang, Malaysia
e-mail: abdnasir@umpsa.edu.my

N. B. Abdullah
Bio-Coke Research Institute, Kindai University, Osaka, Japan

M. H. A. Hassan et al. (eds.), *Proceedings of the 3rd International Conference on Innovation and Technology in Sports*, Lecture Notes in Bioengineering,
https://doi.org/10.1007/978-981-95-0804-4_36

in sports and suggests directions for future research and application, emphasizing the potential for data-driven solutions to shape the future of sports.

Keywords Sports analytics · Performance optimization · Data-driven decision making · Injury prevention · Sport science

1 Introduction

The integration of data analytics in sports has revolutionized the industry, transforming how teams, athletes, and organizations operate. Initially, sports analytics emerged in the form of basic statistical analysis, primarily to assess player performance and team outcomes. Early pioneers, such as Bill James in baseball with his sabermetrics approach, laid the groundwork for a more data-driven perspective in sports (Lewis 2004). However, the last two decades have seen a remarkable evolution with the advent of big data, artificial intelligence (AI), and machine learning (ML), enabling deeper insights into areas like injury prevention, performance optimization, and team strategy (Morgulev et al. 2018).

Today, data analytics is used in virtually every aspect of sports science, from real-time performance tracking to fan engagement (Babu et al. 2024). This review highlights two key areas of focus: the categories where data analytics has already been widely implemented and the underexplored areas where its potential remains untapped. The first aspect will delve into the ten primary applications, including injury prevention, player tracking, and game outcome prediction. These are areas where data analytics have already proven to offer significant benefits, such as improving player safety and refining coaching strategies (Ma 2024; Ogugua Chimezie Obi et al. 2024). The second aspect will address five categories that have seen limited application of data analytics, such as emotional and psychological monitoring and environmental impact assessments.

Data analytics has not only transformed the internal mechanics of sports but has also had a profound impact on the broader sports industry. With tools like machine learning and wearable technology (Abdulmalek et al. 2022), it is now possible to predict injuries, optimize training regimens, and even forecast game outcomes with increasing accuracy (Jin and Zhan 2024). As this review explores both the established and unexplored areas, it will provide a comprehensive view of how data analytics is reshaping the sports world and highlight the future possibilities for its integration.

2 Current Applications of Data Analytics in Sports

The application of data analytics in sports has significantly evolved from simple statistical approaches to sophisticated techniques that now influence various aspects of sports performance, management, and fan engagement. Table 1 summarizes where data analytics have already made a profound impact.

Building on the detailed categorization in Table 1, the current landscape of data analytics in sports is characterized by its ability to drive real-time decision-making and enhance overall performance strategies, as shown in Fig. 1. By leveraging data from wearables and machine learning models, teams are no longer relying solely

Table 1 Categorization of Data Analytics Applications in Sports

Category	Key Insight
Injury prevention	Machine learning models and wearable technology track physiological and biomechanical variables to predict and reduce injury risk by monitoring workload and movement patterns (Sarlis et al. 2024; Seshadri et al. 2021; Zadeh et al. 2021)
Talent identification	Predictive models analyze performance data over time to assess player potential and support recruitment decisions (D'Urso et al. 2023; Kaur et al. 2021; Morgulev et al. 2018)
Performance optimization	Real-time data from wearables and AI provides insights into athletes' performance, allowing for the adjustment of training regimens and tactics (Liu et al. 2023; Ma 2024)
Team strategy and decision making	Advanced data analytics assist in tactical decisions during matches, using real-time player performance and opposition data to adjust strategies (Goes et al. 2021; Legg et al. 2012; Michelsen et al. 2015)
Fan engagement	Analyzing fan behavior through data collected from digital platforms helps teams enhance marketing strategies and tailor content for better fan engagement (Ogugua Chimezie Obi et al. 2024; Yoshida et al. 2024)
Game outcome prediction	Models analyze performance metrics and external conditions to predict game outcomes and inform team strategies (Ma et al. 2023; Mendes-Neves et al. 2024; Rahman 2020)
Data-driven coaching	Coaches use analytics to customize training plans and monitor health in real time, optimizing game preparation (Changqing Liu and Xie 2024; Fujii 2021; Ma 2024; Zattoni et al. 2022)
Sports medicine and rehabilitation	Data analytics in rehabilitation allows monitoring of recovery progress, optimizing rehabilitation plans, and ensuring athletes' long-term health (Kovoor et al. 2024; Olsen et al. 2024; Wu 2024)
Player tracking and motion analysis	Player tracking and motion analysis systems offer real-time data for tactical adjustments based on player movements and workload (Dawson et al. 2024; Midoglu et al. 2024; Modena et al. 2023)
Data visualization tools	Visualization tools translate complex datasets into actionable insights, helping teams make faster, data-driven decisions during matches (Liu et al. 2023; Liu and Liu 2017; Prabhu 2024)

on traditional methods; they now incorporate precise, data-driven insights to tailor everything from athlete performance optimization to fan engagement (D'Urso et al. 2023; Kaur et al. 2021; Ma et al. 2023). The advent of real-time data processing enables teams to make quick adjustments during games, further proving the importance of integrating analytics into both tactical and strategic domains (Michelsen et al. 2015).

One of the most transformative applications of data analytics lies in its ability to unify multiple facets of sports operations—from injury prevention to fan behaviour analysis—into a seamless decision-making framework (Fujii 2021; Zattoni et al. 2022). As teams continue to adopt more advanced analytics for game performance, they are also finding novel ways to translate insights into improved recovery strategies and personalized player training regimens. These data-driven innovations extend beyond physical performance, positioning data analytics as a central pillar for sports organizations aiming to not only enhance player outcomes but also optimize commercial strategies for long-term success (Dawson et al. 2024; Olsen et al. 2024).

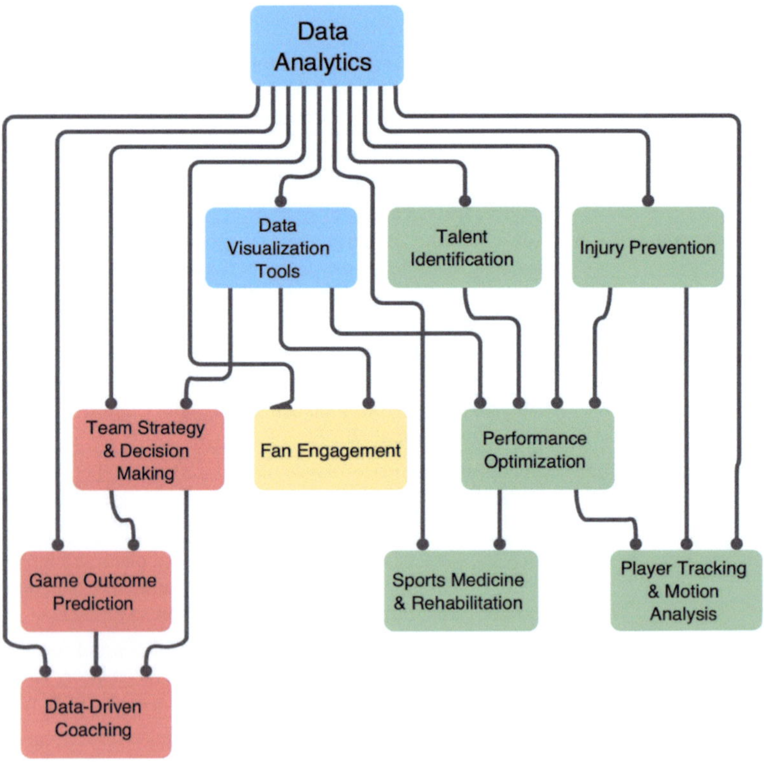

Fig. 1 Interconnected applications of data analytics in sports, highlighting key areas like performance (green), strategy (red), and fan engagement (yellow)

3 Unexplored Areas of Data Analytics in Sports

Despite the significant strides made in sports analytics, there remain several areas where the full potential of data analytics has yet to be realized. This section highlights five key unexplored or underutilized areas where data analytics can offer substantial benefits to the sports industry.

3.1 Emotional and Psychological Real-Time Monitoring

While data analytics has been widely applied to physical performance metrics, there is relatively little research focused on using it to monitor athletes' emotional and psychological states. Emotional and mental well-being play critical roles in an athlete's overall performance, yet real-time tools for tracking mood, stress levels, and mental fatigue remain underdeveloped in sports (Nässi et al. 2017). Integrating data from psychological surveys, heart rate variability, and other biofeedback devices could provide real-time insights (Abdulmalek et al. 2024; Nasir et al. 2024) into an athlete's emotional state. As an example, EmotionWatch (Kempter et al. 2014) uses visual summaries of public emotions during the tennis final between Roger Federer and Andy Murray (see Fig. 2). A similar concept could inspire graphics for monitoring athletes' emotions in real-time. This data could be used to prevent burnout, improve mental toughness, and provide better emotional support for athletes.

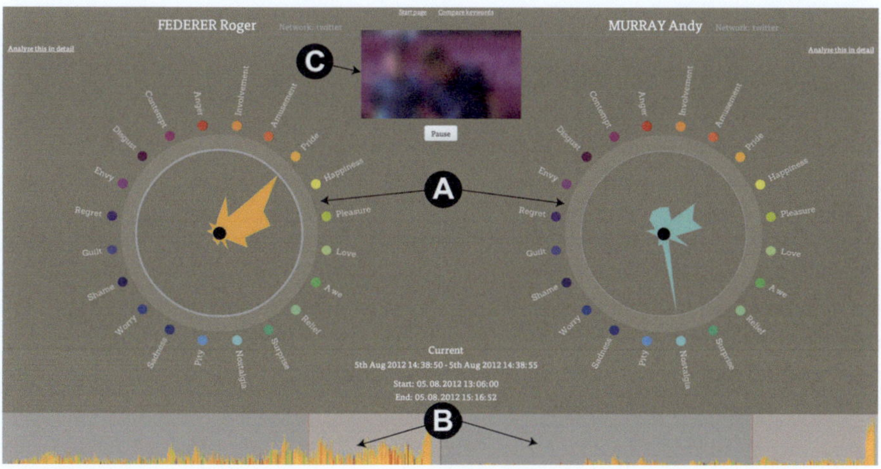

Fig. 2 Real-time emotion wheels that visualize the public emotions by EmotionWatch (Kempter et al. 2014)

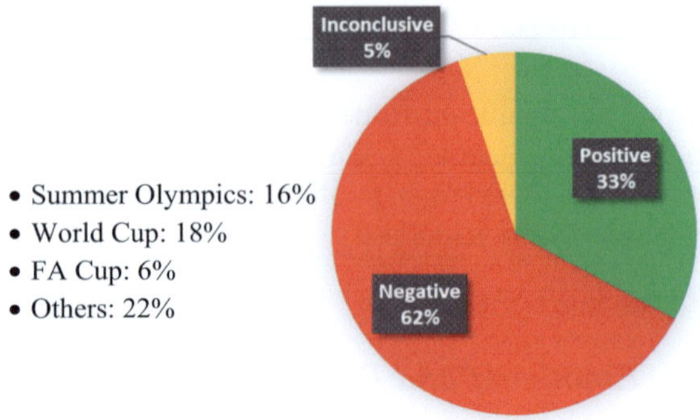

Fig. 3 Environmental impact of world major events from 2000 to 2021 (Cerezo-Esteve et al. 2022)

3.2 Ethical and Integrity Monitoring

Another area that remains underexplored is the use of data analytics for monitoring ethical practices and integrity within sports. Although sports organizations have begun using technology to monitor on-field performance, there has been little focus on how analytics could help detect and prevent issues like match-fixing, doping, and unfair practices in athlete recruitment (Forrest and McHale 2019). Advanced analytics could be used to track anomalies in player or team performance that may suggest unethical behavior, thereby safeguarding the integrity of sports. Furthermore, data-driven audits of financial transactions and player transfers could enhance transparency in sports organizations.

3.3 Environmental Impact of Sports Events

The environmental impact of large sporting events such as Olympics and World Cup has become a growing concern (see Fig. 3) (Cerezo-Esteve et al. 2022), but data analytics has not yet been fully utilized to minimize this impact. While some major events like the Olympics have introduced sustainability initiatives, there is little research focused on leveraging data analytics to optimize the logistics, energy consumption, and waste management of sports events (Zhang and Yang 2023). By using predictive models to track the carbon footprint of events, analyze transportation logistics, and monitor waste, event organizers could make more informed decisions to reduce environmental impact.

3.4 Nutritional Analytics Based on Genetic Data

The intersection of genetic data and sports performance is another area where data analytics remains underutilized. While some research has explored personalized

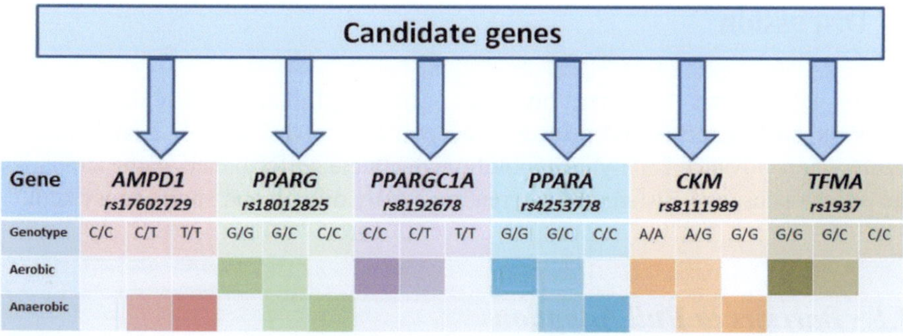

Fig. 4 Matrix chart showing the influence of various genetic markers (AMPD1, PPARG, PPARGC1A, PPARA, CKM, and TFMA) on aerobic and anaerobic performance in athletes. The chart illustrates how different genotypes of each gene contribute to either aerobic or anaerobic capabilities, with darker colors indicating a stronger influence on performance (Balberova et al. 2021)

training programs based on an athlete's physical performance, the application of genetic data to tailor nutrition plans, so-called nutrigenomics, is still in its infancy (Guest et al. 2019). By analyzing genetic markers that influence metabolism, muscle fiber composition, and injury susceptibility (see Fig. 4), sports organizations could develop personalized nutrition and training regimens to maximize performance and recovery. Nutrigenomics could become a crucial part of sports science, allowing for more individualized approaches to athlete care.

3.5 *Data-Driven Spectator Experience Enhancement*

Although fan engagement is increasingly data-driven, the live spectator experience has not yet fully benefited from the potential of data analytics. Many sports organizations use analytics to track online fan behavior, but real-time personalization of in-stadium experiences, such as tailored seating arrangements, customized food and beverage offers, and real-time updates on favorite players, remains largely untapped (Morrison et al. 2009). Implementing analytics to enhance the live experience could transform how spectators engage with sports events, creating more interactive and immersive environments. Real-time data collection from fans could be used to optimize stadium logistics, improve crowd control, and enhance overall satisfaction.

4 Discussion

Section 4.1 explores the current obstacles preventing full adoption of data analytics in sports, such as costs and privacy concerns. If resolved, Sect. 4.2 envisions the future where advanced analytics revolutionize athlete performance, game strategies, fan experiences, and sustainability, creating a fully data-driven sports ecosystem.

4.1 Barriers to Full Adoption

Despite success in these areas, Table 2 outlines the major barriers to adopting data analytics in sports. These challenges primarily revolve around cost, data privacy, and IT resources, limiting smaller organizations' ability to fully implement advanced systems.

Cost remains one of the primary challenges. High-quality data analytics systems, including the hardware and software required for data collection and analysis, can be prohibitively expensive for smaller teams or leagues. While major professional sports organizations can afford these investments, lower-tier teams often lack the resources to implement such advanced technologies (Grow 2020).

Data privacy concerns are another significant barrier. As data analytics becomes more prevalent, especially in the context of athlete performance and health monitoring, issues surrounding the privacy and security of sensitive information have emerged. There is a growing need for clear regulations and guidelines on how athlete data should be collected, stored, and used (Rush and Osborne 2022). Additionally, the misuse or misinterpretation of data can lead to negative consequences, such as biased decisions regarding athlete selection or player performance assessments (Devriendt et al. 2018). Figure 5 illustrates the data flow from collection through analysis and usage, highlighting key points where potential misuse may occur, such as unauthorized data collection, insecure transmission, and biased algorithms. These issues emphasize the need for improved data policies and protocols.

Table 2 Key barriers to adopting data analytics in sports

Barrier	Description	Examples/Impact
Cost	High cost of implementing data analytics systems, including hardware and software	Smaller teams and lower-tier leagues struggle to afford these systems
Data privacy concerns	Issues regarding the collection, storage, and use of athlete data	Lack of clear regulations may lead to misuse or bias in decision-making
IT resources	Inadequate infrastructure for real-time data processing and analysis	Teams without IT expertise cannot fully leverage data analytics

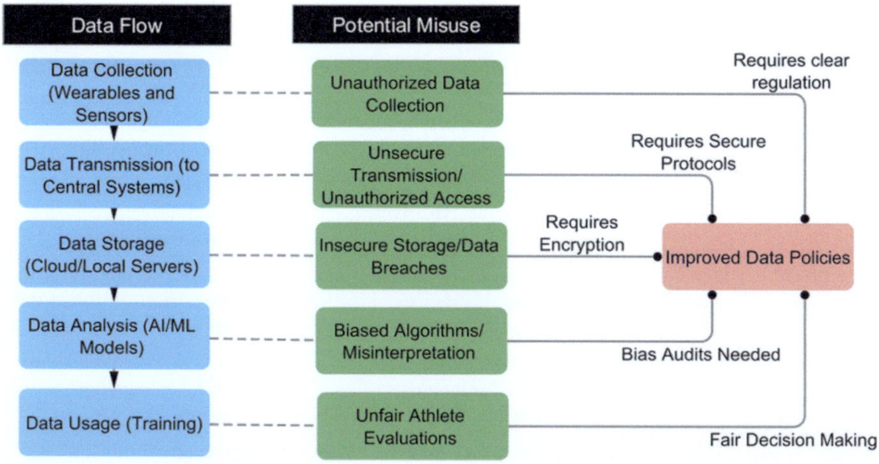

Fig. 5 Visualization of data flow in sports analytics and potential misuse points, emphasizing the need for improved data policies

Technological limitations also hinder the widespread use of data analytics in sports. While advanced tracking and analysis tools are available, not all sports organizations have the infrastructure to support these technologies. For example, real-time data processing requires robust IT systems and specialized personnel to manage and interpret the data (Legg et al. 2012; Michelsen et al. 2015). Without these resources, teams may struggle to fully leverage the potential of data analytics in their operations.

4.2 The Future of Sports with Data Analytics

As data analytics continues to evolve, the sports industry is on the verge of several major shifts that will redefine athlete training, in-game strategy, and fan engagement. The transition from current methods to future applications will involve moving from data-enhanced decision-making to fully data-driven ecosystems (Amin et al. 2024), where every element of sports experience is optimized through advanced analytics, as summarized in Table 3.

From Reactive to Proactive Personalized Athlete Optimization. Today, personalized athlete training programs are based largely on real-time physiological data from wearables, such as heart rate, speed, and recovery metrics. This allows coaches to adjust training loads and recovery plans based on how an athlete is performing at that moment. While effective, this model remains reactive adjustments are made after an athlete shows signs of fatigue or strain.

The future will see a shift towards proactive, predictive personalization. Rather than waiting for signs of overtraining or injury, advanced AI models will predict

Table 3 Comparison of current and future trends in sports data analytics, highlighting key paradigm shifts

Current trend	Future trend	Impact/Insight
Reactive athlete training adjustments	Proactive, predictive athlete optimization	AI-driven models to foresee dips in performance and prevent injury
Real-time coach decisions	Instantaneous AI-driven strategy adaptations	Faster, data-driven decisions with fewer human errors
Generic fan engagement	Fully personalized, immersive spectator experiences	Tailored AR/VR engagement, real-time customization of the experience
Post-event ethical monitoring	Continuous, automated oversight for integrity monitoring	AI detects unethical behavior in real-time, enhancing transparency
Reactive sustainability efforts at events	Data-driven, proactive sustainability management	Real-time environmental impact monitoring for events
Physical metrics dominate performance evaluation	Psychological metrics integrated into performance profiles	Mental resilience, cognitive sharpness considered in recruitment and strategy

performance dips, injury risks, and mental fatigue before they occur by analyzing vast datasets including not only real-time physiological metrics but also long-term trends, genetic predispositions, and even environmental factors. These AI models will constantly learn and evolve, creating training programs that adjust daily based on each athlete's unique trajectory. In this paradigm, athletes won't just be reacting to their bodies—they'll be training with precise foresight into how their bodies will perform over time. This future will focus on peak readiness rather than merely preventing injury.

Real-Time Strategy and Game Evolution. Currently, in-game strategy adjustments occur in real-time but are still reliant on human interpretation of the data. Coaches use tools like GPS trackers and wearables to gather information about player positioning, fatigue, and opposition tactics. However, the final decision still rests with the coach's intuition and analysis, which takes valuable time and leaves room for human error or delay.

The future will see real-time strategy move beyond human capabilities, powered by instantaneous AI-driven decisions. Instead of relying solely on coaches to interpret and act on the data, AI systems will process thousands of data points per second—everything from player stamina to weather conditions—and offer real-time tactical recommendations to coaches or even directly to the players on the field. These systems will be able to adapt strategy in milliseconds, allowing teams to adjust to changing conditions and opposition tactics faster than human analysis ever could. Dynamic play-calling, where plays or strategies change based on minute-to-minute developments in the game, will become the norm. In this future, teams that adopt

AI-powered strategy systems will play with an unprecedented level of precision and fluidity.

Personalized Fan Engagement and Immersive Spectator Experiences. Fan engagement today is already influenced by data analytics, with personalized marketing and targeted content based on fans' digital interactions. However, this primarily focuses on digital touchpoints like social media, apps, and websites, offering generalized experiences across large groups of fans.

The future will bring a much deeper level of personalization, where every moment of a fan's journey—whether digital or in-person—will be tailored to their preferences. Imagine arriving at a stadium and being greeted by a personalized AR overlay showing real-time stats of your favorite player, guiding you to your seat, and recommending merchandise or food options based on your past preferences. Your entire experience, from purchasing tickets to watching the game, will be data-driven and tailored in real-time. In the digital sphere, fans will be able to engage with games through interactive VR simulations that mirror live events, allowing them to virtually step into the game from a player's perspective.

This shift goes beyond traditional fan engagement and into the realm of immersive, interactive fan experiences, where data is continuously collected and acted upon to create a seamless, individualized journey.

Automated Oversight of Ethical and Integrity Monitoring. In the present, ethical monitoring in sports—whether for doping, match-fixing, or corruption—relies heavily on post-event investigations or whistleblowing. While some real-time monitoring systems are being explored, they are limited in scope and effectiveness.

The future will see automated, continuous oversight through data analytics. Every play, pass, and point will be tracked not just for performance but for ethical consistency. AI will be capable of identifying suspicious behavior patterns in real-time, flagging possible anomalies in player performance or financial transactions that suggest unethical activity. This means that instances of match-fixing, doping, or unfair practices will be caught as they occur, making sports fairer and more transparent. Teams and governing bodies will be able to run predictive models to evaluate the likelihood of corruption before it becomes a major issue.

Data-Driven Sustainability in Sport. Large-scale sports events today face increasing scrutiny over their environmental impacts, with many organizations adopting basic sustainability measures like reducing waste or offsetting carbon emissions. However, these efforts are often reactive and not fully optimized for maximum impact.

In the future, data analytics will enable the proactive management of sustainability efforts, using predictive models to optimize everything from transportation logistics to energy usage at events. Imagine a system that tracks not only energy consumption in real-time but also predicts the environmental impact of various logistics decisions, allowing organizers to adjust in advance. Stadiums could be designed based on historical data to use renewable energy more efficiently, while AI-driven logistics systems could direct fans to use the most eco-friendly transportation options. Sports organizations will no longer simply mitigate their environmental impact—they will actively optimize it in real time.

Psychological Well-being as New Metrics of Performance. At present, success in sports is measured largely through physical metrics like speed, strength, agility, and wins. While psychological well-being and emotional resilience are recognized as important, they remain secondary considerations that are difficult to quantify.

In the future, new metrics will emerge that integrate mental and emotional well-being as core components of an athlete's performance profile. Advanced data analytics will measure resilience, focus, emotional regulation, and even cognitive sharpness during competition. Teams will no longer focus solely on physical attributes; instead, they will seek holistic athletes who excel both physically and mentally. These new metrics will be factored into recruitment, player development, and even in-game strategy. For example, a player with high emotional resilience might be favored during high-pressure moments, while another with exceptional cognitive sharpness could be relied upon for strategic plays. This shift will redefine what it means to be an elite athlete and elevate mental health as a key element of success in sports.

5 Conclusion

Data analytics has fundamentally reshaped how sports teams, athletes, and organizations approach performance, strategy, and fan engagement. Current applications in areas such as injury prevention, performance optimization, and game strategy have demonstrated the power of data-driven insights, allowing for more precise decision-making. These advancements have offered significant benefits, from enhancing player safety to improving in-game tactics and deepening fan interaction.

However, the full potential of data analytics in sports remains untapped. Unexplored areas like emotional and psychological monitoring, ethical oversight, and environmental sustainability offer new opportunities for innovation. The future of sports will likely see a shift towards proactive, predictive analytics that will redefine athlete management, game strategy, and even the fan experience. As data becomes more integrated into every facet of the sports industry, the need for robust ethical frameworks and responsible data usage will grow.

To fully capitalize on the potential of data analytics, sports organizations must invest in more advanced AI models, explore new areas of application, and establish ethical standards that protect athlete privacy while fostering transparency. The future of sports is not just about refining physical performance but also embracing a holistic, data-driven approach that optimizes all aspects of the game, from strategy to sustainability.

References

Abdulmalek S, Nasir A, Jabbar WA (2024) LoRaWAN-based hybrid internet of wearable things system implementation for smart healthcare. Internet of Things 25:101124. https://doi.org/10.1016/J.IOT.2024.101124

Abdulmalek S, Nasir A, Jabbar WA, Almuhaya MAM, Babu D (2022) An internet of wearable things (IOWT) based system for smart healthcare monitoring. IET Conf Proc 2022(22):255–261. https://doi.org/10.1049/ICP.2022.2622

Amin MRRM, Nasir A, Muhammad Sidik MH, Jamaludin ASB, Rosli AM (2024) Applying convolutional neural networks (CNNs): a machine learning method in modern sports. Lecture Notes Bioeng, 405–415. https://doi.org/10.1007/978-981-97-3741-3_36

Babu D, Hakim S, Rosli AM, Yasin MRM, Rosle MH (2024) Harnessing solar energy: a green technology approach in sports facility management. Lecture Notes Bioeng 363–376. https://doi.org/10.1007/978-981-97-3741-3_33

Balberova OV, Bykov EV, Medvedev GV, Zhogina MA, Petrov KV, Petrova MM, Al-Zamil M, Trefilova VV, Goncharova PS, Shnayder NA (2021) Candidate genes of regulation of skeletal muscle energy metabolism in athletes. Genes 12(11):1682. https://doi.org/10.3390/genes12111682

Cerezo-Esteve S, Inglés E, Segui-Urbaneja J, Solanellas F (2022) The environmental impact of major sport events (Giga, Mega and Major): a systematic review from 2000 to 2021. Sustainability 14(20):13581. https://doi.org/10.3390/su142013581

Dawson L, Beato M, Devereux G, McErlain-Naylor SA (2024) A review of the validity and reliability of accelerometer-based metrics from upper back-mounted GNSS player tracking systems for athlete training load monitoring. J Strength Cond Res 38(8):e459–e474. https://doi.org/10.1519/JSC.0000000000004835

Devriendt T, Chokoshvili D, Favaretto M, Borry P (2018) Do athletes have a right to access data in their Athlete Biological Passport? Drug Test Anal 10(5):802–806. https://doi.org/10.1002/dta.2380

D'Urso P, De Giovanni L, Swartz T (2023) Editorial: big data and data science in sport. Ann Oper Res 325(1):1–7. https://doi.org/10.1007/s10479-023-05344-z

Forrest D, McHale IG (2019) Using statistics to detect match fixing in sport. IMA J Manag Math 30(4):431–449. https://doi.org/10.1093/imaman/dpz008

Fujii K (2021) Data-driven analysis for understanding team sports behaviors. J Robot Mechatron 33(3):505–514. https://doi.org/10.20965/jrm.2021.p0505

Goes FR, Kempe M, van Norel J, Lemmink KAPM (2021) Modelling team performance in soccer using tactical features derived from position tracking data. IMA J Manag Math 32(4):519–533. https://doi.org/10.1093/imaman/dpab006

Grow N (2020) Free agency for the front office: how data analytics and non-compete agreements threaten to disrupt competitive balance in U.S. Professional Sports Leagues. SSRN Electronic Journal. https://doi.org/10.2139/ssrn.3649711

Guest NS, Horne J, Vanderhout SM, El-Sohemy A (2019) Sport nutrigenomics: personalized nutrition for athletic performance. Front Nutr 6. https://doi.org/10.3389/fnut.2019.00008

Jin N, Zhan X (2024) Big data analytics for image processing and computer vision technologies in sports health management. Technol Health Care 32(5):3167–3187. https://doi.org/10.3233/THC-231875

Kaur A, Kaur R, Jagdev G (2021) Analyzing and exploring the impact of big data analytics in sports sector. SN Computer Science 2(3):184. https://doi.org/10.1007/s42979-021-00575-y

Kempter R, Sintsova V, Musat CC, Pu P (2014) EmotionWatch: visualizing fine-grained emotions in event-related tweets. Proceedings of the international AAAI conference on web and social media. https://api.semanticscholar.org/CorpusID:1921011

Kovoor M, Durairaj M, Karyakarte MS, Zair Hussain M, Ashraf M, Maguluri LP (2024) Sensor-enhanced wearables and automated analytics for injury prevention in sports. Measur Sensors 32:101054. https://doi.org/10.1016/j.measen.2024.101054

Legg PA, Chung DHS, Parry ML, Jones MW, Long R, Griffiths IW, Chen M (2012) MatchPad : interactive glyph-based visualization for real-time sports performance analysis. Comput Graphics Forum 31(3pt4):1255–1264. https://doi.org/10.1111/j.1467-8659.2012.03118.x

Lewis M (2004) Moneyball: the art of winning an unfair game. W. W. Norton & Company

Liu A, Mahapatra RP, Mayuri AVR (2023) Hybrid design for sports data visualization using AI and big data analytics. Complex Intell Syst 9(3):2969–2980. https://doi.org/10.1007/s40747-021-00557-w

Liu S, Liu Y (2017) Application analysis of the data visualization technology in sports equipment. Proceedings of the 2017 international conference on education, culture and social development (ICECSD 2017). https://doi.org/10.2991/icecsd-17.2017.2

Liu C, Xie Y (2024) Innovative application of computer vision and motion tracking technology in sports training. EAI Endorsed Trans Pervasive Health Technol 10. https://doi.org/10.4108/eetpht.10.5763

Ma S (2024) Enhancing sports education through big data analytics: leveraging models for improved teaching strategies. Appl Comput Eng 57(1):184–189. https://doi.org/10.54254/2755-2721/57/20241331

Ma Y, Guo H, Sun Y, Liu F (2023) Real-time prediction algorithm and simulation of sports results based on internet of things and machine learning. Int J Inf Technol Manage 22(3/4):386–406. https://doi.org/10.1504/IJITM.2023.131845

Mendes-Neves T, Meireles L, Mendes-Moreira J (2024) Forecasting events in soccer matches through language

Michelsen T, Brand M, Cordes C, Appelrath H-J (2015) Herakles: real-time sport analysis using a distributed data stream management system. Proceedings of the 9th ACM international conference on distributed event-based systems, 356–359. https://doi.org/10.1145/2675743.2776775

Midoglu C, Kjæreng Winther A, Boeker M, Dahl Pettersen S, Pedersen S, Ragab N, Kupka T, Hicks SA, Bredsgaard Randers M, Jain R, Dagenborg HJ, Pettersen SA, Johansen D, Riegler MA, Halvorsen P (2024) A large-scale multivariate soccer athlete health, performance, and position monitoring dataset. Scientific Data 11(1):553. https://doi.org/10.1038/s41597-024-03386-x

Modena R, Salaorni F, Schena F, Pellegrini B, Bortolan L (2023) GPS devices for match analysis in soccer. 2023 IEEE International workshop on sport, technology and research (STAR), 5–7. https://doi.org/10.1109/STAR58331.2023.10302650

Morgulev E, Azar OH, Lidor R (2018) Sports analytics and the big-data era. Int J Data Sci Analyt 5(4):213–222. https://doi.org/10.1007/s41060-017-0093-7

Morrison A, Bell M, Chalmers M (2009) Visualisation of spectator activity at stadium events. 2009 13th International conference information visualisation, 219–226. https://doi.org/10.1109/IV.2009.64

Nasir A, Babu D, Muhammad Sidik MH, Abdullah NB, Jabbar WA (2024) Advancement of IoT-based smart systems for real-time sports performance analysis: potential and challenges. Lecture Notes Bioeng 377–391. https://doi.org/10.1007/978-981-97-3741-3_34

Nässi A, Ferrauti A, Meyer T, Pfeiffer M, Kellmann M (2017) Psychological tools used for monitoring training responses of athletes. Perform Enhancement Health 5(4):125–133. https://doi.org/10.1016/j.peh.2017.05.001

Obi OC, Dawodu SO, Onwusinkwue S, Osasona F, Atadoga A, Ifesinachi Daraojimba A (2024) Data science in sports analytics: a review of performance optimization and fan engagemen. World J Adv Res Rev 21(1):2663–2670. https://doi.org/10.30574/wjarr.2024.21.1.0370

Olsen RJ, Hasan SS, Woo JJ, Nawabi DH, Ramkumar PN (2024) The fundamentals and applications of wearable sensor devices in sports medicine: a scoping review. Arthroscopy: J Arthroscopic Related Surgery. https://doi.org/10.1016/j.arthro.2024.01.042

Rahman MdA (2020) A deep learning framework for football match prediction. SN Appl Sci 2(2):165. https://doi.org/10.1007/s42452-019-1821-5

Rush C, Osborne B (2022) Benefits and concerns abound, regulations lack in collegiate athlete biometric data collection. J Legal Aspects Sport 32(1):62–94. https://doi.org/10.18060/25479

Sarlis V, Papageorgiou G, Tjortjis C (2024) Injury patterns and impact on performance in the NBA league using sports analytics. Computation 12(2):36. https://doi.org/10.3390/computation12020036

Seshadri DR, Thom ML, Harlow ER, Gabbett TJ, Geletka BJ, Hsu JJ, Drummond CK, Phelan DM, Voos JE (2021) Wearable technology and analytics as a complementary toolkit to optimize workload and to reduce injury burden. Front Sports Active Living 2. https://doi.org/10.3389/fspor.2020.630576

S Prabhu S (2024) Inaccurate and misinformative data visualisations in sports-field. Int J Sci Res (IJSR) 13(7):1434–1437. https://doi.org/10.21275/SR24718153653

Wu B (2024) Real time monitoring research on rehabilitation effect of artificial intelligence wearable equipment on track and field athletes. EAI Endorsed Trans Pervasive Health Technol 10. https://doi.org/10.4108/eetpht.10.5150

Yoshida M, Biscaia R, Uhrich S, Gordon BS, Huettermann M, Nakazawa M (2024) Fan engagement behavior: validation of a theory-based scale. J Sport Manag 38(2):133–150. https://doi.org/10.1123/jsm.2023-0075

Zadeh A, Taylor D, Bertsos M, Tillman T, Nosoudi N, Bruce S (2021) Predicting sports injuries with wearable technology and data analysis. Inf Syst Front 23(4):1023–1037. https://doi.org/10.1007/s10796-020-10018-3

Zattoni L, Ercolin L, Mandini S, Piva T, Zerbini V, Raisi A, Zaccagni L (2022) Video-analysis of "High-Risk" on-field movements biomechanics in amateur male futsal players: preliminary findings. 2022 IEEE International workshop on sport, technology and research (STAR), 180–184. https://doi.org/10.1109/STAR53492.2022.9860010

Zhang L, Yang Y (2023) Towards sustainable energy systems considering unexpected sports event management: integrating machine learning and optimization algorithms. Sustainability 15(9):7186. https://doi.org/10.3390/su15097186

Current Trend in Preferred Deadlift Stance Among Elite Powerlifters

Fareedzul Hareez Mohd Pudzi, Mohd Hanifa Sariman, Raja Mohammed Firhad Raja Azidin, Raihana Sharir, and Radin Rafeeuddin

Abstract The deadlift is a fundamental exercise in strength training, especially powerlifting, with conventional and sumo variations being the accepted stances in powerlifting competitions. Despite numerous biomechanical studies, recent data on deadlift stance preferences among competitive powerlifters remain limited. This study seeks to provide an updated understanding of these preferences, mainly focusing on trends in stance selection across different weight classes in male and female athletes over the past six years. This study utilized a non-experimental observational design, performing a notational analysis of international powerlifting competitions held by the International Powerlifting Federation (IPF) from 2017 to 2023. The analysis included athletes who completed at least one deadlift attempt. Data on conventional and sumo stances were collected and subjected to linear trend analysis across various weight classes of both genders. The study showed a clear pattern for men's and women's categories: the sumo stance was more prevalent in the lighter-weight classes. In contrast, the conventional stance was more prevalent in the heavier-weight classes. There has also been a significant shift in deadlift stance preferences over time, with an increasing preference for the sumo stance, particularly among lighter-weight classes. Nevertheless, the utilization of the sumo stance has risen steadily in most of the weight classes, indicating a broader trend in powerlifting competition strategies. These results enhance understanding of current powerlifting practices and may guide coaching strategies to optimize performance based on an athlete's weight class and biomechanical considerations.

Keywords Powerlifting · Deadlift · Deadlift stance · Conventional deadlift · Sumo deadlift · Trend · IPF · Strength training · Biomechanics

---

F. H. M. Pudzi · M. H. Sariman · R. M. F. R. Azidin · R. Sharir · R. Rafeeuddin (✉)
Faculty of Sports Science and Recreation, Universiti Teknologi MARA, Shah Alam, Malaysia
e-mail: rrafeeuddin@uitm.edu.my

R. M. F. R. Azidin · R. Sharir · R. Rafeeuddin
Sports and Exercise Biomechanics Special Interest Group, Universiti Teknologi MARA, Shah Alam, Malaysia

R. Sharir
School of Sport and Exercise Sciences, Liverpool John Moores University, Liverpool, UK

M. H. A. Hassan et al. (eds.), *Proceedings of the 3rd International Conference on Innovation and Technology in Sports*, Lecture Notes in Bioengineering,
https://doi.org/10.1007/978-981-95-0804-4_37

1 Introduction

The deadlift is an exercise in which both hands grip a barbell placed on the floor, and the weight is lifted by extending the knees, hips, and back while keeping the arms straight (McGuigan and Wilson 1996). Strength athletes often include the deadlift in their training programs for performance enhancement and rehabilitation (Escamilla et al. 2000). In terms of performance, the deadlift is primarily used to strengthen the hips, thighs, and back (Escamilla et al. 2002). In powerlifting competitions, the deadlift is one of three lifts used to calculate the total weight lifted alongside the squat and bench press. Competitors are permitted to use two variations of the deadlift in these events: the conventional and the sumo deadlift (Escamilla et al. 2001).

Extensive research has been conducted to understand and optimize deadlift performance and its variations, including studies on deadlift kinematics and kinetics (Brown and Abani 1985), 2D and 3D biomechanical analyses comparing the conventional and sumo deadlift (McGuigan and Wilson 1996; Escamilla et al. 2000, 2001), electromyographic analyses of these two variations (Escamilla et al. 2002), investigations into the 'sticking points' during the lift (Kompf and Arandjelović 2017), and the effects of using weightlifting belts and lifting straps on deadlift performance metrics (Fong et al. 2022; Jukic et al. 2021, 2022a, 2022b, 2022c; Omcirk et al. 2020). Additionally, two studies have explored the relationship between handgrip strength and deadlift performance in powerlifting (Schoffstall et al. 2010; Suazo and DebelİSo 2021).

The conventional stance requires a person to lift a loaded barbell from the ground while maintaining a shoulder-width stance, with the hands positioned just outside the knees. In contrast, the sumo stance is characterized by a wider foot placement and more vertical trunk orientation, has been found to reduce mechanical work, predicted energy expenditure, L4/L5 moment, and L4/L5 shear force compared to the conventional stance. Sumo stance reduced 20–25% mechanical work and approximately 18% predicted energy expenditure compared to the conventional stance (Escamilla et al. 2000; Cholewicky et al. 1991). Conversely, knee moment and moment arm generated during the sumo deadlift are significantly greater than those during the conventional deadlift. In contrast, hip moment and arm are not significantly different for both. In terms of lifting phases, the sumo stance generally has a 10% longer acceleration phase than the conventional stance, suggesting it is more challenging to initiate the lift. Further electromyographic analysis suggests that the sumo stance may more effectively engage the quadriceps and ankle dorsiflexor than the conventional stance (Escamilla et al. 2002). This difference is mainly due to the mechanical demands, as described earlier. However, despite these insights, there remains a lack of comprehensive studies documenting current trends in deadlift stance preferences among elite powerlifters, particularly over the past two decades.

To the best of the authors' knowledge, there is limited published data on the preferred deadlift variations among elite and recreational powerlifters (McGuigan and Wilson 1996; Escamilla et al. 2000). It was previously believed that sumo deadlifts were more commonly favored among elite powerlifters (McGuigan and

Wilson 1996). However, Escamilla, Francisco (Escamilla et al. 2000) found that the frequency of use for both deadlift variations was nearly identical among lifters with lighter body weights. Interestingly, unpublished data shared on social media comparing deadlift stances at the IPF Open Worlds 2018 and IPF Classic Worlds 2019 indicated an increasing trend in the use of the sumo deadlift among lighter lifters (thestrengthathlete. 2019).

From a biomechanical perspective, the sumo deadlift's advantages, including reduced mechanical work and shorter bar path, may offer specific benefits depending on an athlete's anthropometric characteristics (Escamilla et al. 2000). Nonetheless, there is a critical need for comprehensive studies that evaluate how these biomechanical advantages translate into long-term performance outcomes, especially in competitive powerlifting among elite athletes. This gap in the literature is particularly evident in the context of elite powerlifting, where updated data on stance preferences could significantly affect training approaches and competition strategies.

Therefore, the study primarily aims to determine an updated trend in elite powerlifters' preferred deadlift stance. The updated trend will then serve as a baseline to gain further understanding of the deadlift and how it can be practically utilized in powerlifting and various other sports that demand various athletic abilities.

2 Methodology

2.1 Research Design

This study will use a non-experimental observational research design to determine the trend for the last five (years) on preferred deadlift stance amongst powerlifters competing in international powerlifting championships. The study will implement a notational analysis adapted from the protocol and procedure in previous studies (Barreira and Chiminazzo 2020; Hossner et al. 2019; Lignell et al. 2020).

2.2 Sampling Technique

The study will use a purposive sampling technique with a selection focusing on the latest tournament within a 6-year frame from 2017–2023 from live recordings of the International Powerlifting Federation (IPF) World Classic Powerlifting Championship and cross-checked with the official results released on the IPF website (). All athletes who participated in 2017–2023 IPF-organized competitions and performed at least one successful deadlift attempt will be included in the data collection process. Athletes who did not perform a deadlift attempt at all due to any reason or failed to get one successful deadlift attempt (thus declared disqualified) will be excluded.

The current study obtained ethical clearance from the Universiti Teknologi MARA's ethical committee.

2.3 Instrumentation

The live recordings of all powerlifting championships organized by the International Powerlifting Federation (IPF) from 2017 to 2023 that are available from IPF's official YouTube channel (IPF) will be observed by two independent powerlifters with competition experience to determine the deadlift stance of the athletes involved. The data will be verified with the official results released on the IPF website (Results). An inter-rater reliability analysis will be conducted prior to the main study. Two experienced powerlifters will observe the same events from the latest IPF World Classic Powerlifting Championship live recordings ($n = 10$) and evaluate the deadlift stance of each athlete. The intraclass correlation coefficient will then be calculated to assess the consistency of their assessments.

2.4 Data Collection Procedure

Two independent experienced powerlifters will observe the official live recordings to collect data on the distribution of athletes utilizing conventional and sumo deadlift stances. Figure 1 summarizes the data collection procedure for RO1. The same operators will always conduct the data collection throughout the process. The following criteria are described by Escamilla, Francisco (Escamilla et al. 2000) will be taken into account when determining whether a deadlift was done in conventional or sumo stance: (1) in the sumo style, the feet are wider apart and turned out, (2) the arms are positioned inside the knees in sumo deadlift stance, or outside of the knees in the conventional deadlift stance. Athletes who, for any reason, did not attempt a deadlift or who did not have at least one successful deadlift attempt will not be included. Microsoft Excel will be used to record the data. The official results published on the IPF website will then be accessed to verify that the lifter, competition weight class, weight attempted, and placings are correctly observed. To confirm the reproducibility of data collection, 10% of the championship events recordings will be randomly chosen and examined by a third evaluator.

2.5 Data Analysis

For data analysis, a linear trend analysis will be used for each deadlift variant (conventional and sumo deadlifts) to observe the trends across the years (2017–2023). An independent t-test will be used to determine the significance of the preferred deadlift

Fig. 1 Conventional deadlift stance

stance and performance. The athlete's deadlift weight-to-bodyweight ratio will be compared to two (2) deadlift stances (conventional and sumo deadlifts).

3 Results and Findings

3.1 Total of Participants Inclusion and Exclusion

From 2017–2019, Sub Junior, Junior, and Masters categories were included in IPF Classic Championships and were organized at the same time and venue. From 2022 onwards, the Sub Junior and Junior Championship and Masters Championship were organized separately from the Classic Championship. Therefore, for 2017–2019, the data was extracted from a single competition, while for 2022 and 2023, the data was extracted from 3 separate competitions (World Classic Championship, World Sub Junior and Junior Championship, and World Masters Championship) but all were held in the same respective year. There is no data for 2020 as there was no international competition in 2020 due to COVID-19 policy restrictions (Tables 1 and 2).

Table 1 Men's category: total number of athletes included and not included in the count

Year	Not included (n)	Not included (%)	Included (n)	Included (%)	N
2017	52	10.99	421	89.01	473
2018	20	5.15	368	94.85	388
2019	19	3.62	506	96.38	525
2021	22	6.08	355	93.92	377
2022	14	2.97	458	97.03	472
2023	28	4.59	582	95.41	610
Grand total	155	5.48	2690	94.52	2845

Table 2 Women's category: total number of athletes included and not included in the count

Year	Not included (n)	Not included (%)	Included (n)	Included (%)	N
2017	57	16.52	288	83.48	345
2018	2	0.65	305	99.35	307
2019	34	8.56	363	91.44	397
2021	10	3.41	283	96.59	293
2022	6	1.47	403	98.53	409
2023	13	2.74	462	97.26	475
Grand total	122	5.48	2104	94.52	2226

3.2 *Men's Category*

Within six years (2017, 2018, 2019, 2021, 2022, and 2023), 2846 participations were recorded in the Men's categories. 2690 athletes are included in the count. However, there were 2055 distinct male athletes, as the same athlete might have participated multiple times over the years. For example, an athlete might have competed in 2017, 2018, and 2019. However, all participants are included in the count because athletes might change their deadlift stance over the years.

Figure 2 shows the graphical representation of the data in Table 3. Based on the graph (Fig. 2), throughout 2017–2023, the sumo stance is visually more prevalent among lighter weight classes, while the conventional stance dominates heavier weight classes. The slightest difference in the number of deadlift stances can be seen within the under 83 kg and 105 kg weight classes.

Table 4 and Fig. 3 show the total percentage of conventional versus sumo stances regardless of weight class from 2017 to 2023. Overall, without looking into the details of each men's weight class, 46.56% of lifters used the conventional stance, and 53.44% used the sumo stance in 2017. In 2018, the stances intersected, with the sumo stance surpassing conventional usage by a small margin. Following 2018, the percentage of sumo users increased consistently, while the use of the conventional

Fig. 2 Sumo deadlift stance

Table 3 Number and percentage of conventional versus sumo stances across Men's weight classes from 2017 to 2023

Year/Weight class	Conventional (n)	Conventional (%)	Sumo (n)	Sumo (%)	N
2017	*196*	*46.56*	*225*	*53.44*	*421*
<53	0	0.00	7	100.00	7
<59	6	19.35	25	80.65	31
<66	18	35.29	33	64.71	51
<74	22	31.88	47	68.12	69
<83	27	39.71	41	60.29	68
<93	35	51.47	33	48.53	68
<105	29	53.70	25	46.30	54
<120	24	88.89	3	11.11	27
120+	35	76.09	11	23.91	46
2018	*185*	*50.27*	*183*	*49.73*	*368*
<53	1	16.67	5	83.33	6
<59	7	21.88	25	78.13	32
<66	14	31.82	30	68.18	44
<74	22	36.07	39	63.93	61
<83	24	43.64	31	56.36	55
<93	22	52.38	20	47.62	42
<105	30	65.22	16	34.78	46
<120	42	76.36	13	23.64	55
120+	23	85.19	4	14.81	27
2019	*265*	*52.37*	*241*	*47.63*	*506*
<53	2	28.57	5	71.43	7
<59	7	18.92	30	81.08	37
<66	13	30.95	29	69.05	42
<74	30	35.71	54	64.29	84
<83	38	47.50	42	52.50	80
<93	62	57.41	46	42.59	108
<105	39	70.91	16	29.09	55
<120	44	77.19	13	22.81	57
120+	30	83.33	6	16.67	36
2021	*169*	*47.61*	*186*	*52.39*	*355*
<53	2	40.00	3	60.00	5
<59	1	6.67	14	93.33	15
<66	6	26.09	17	73.91	23
<74	18	31.58	39	68.42	57

(continued)

Table 3 (continued)

Year/Weight class	Conventional (n)	Conventional (%)	Sumo (n)	Sumo (%)	N
<83	23	42.59	31	57.41	54
<93	19	32.20	40	67.80	59
<105	32	64.00	18	36.00	50
<120	32	69.57	14	30.43	46
120+	36	78.26	10	21.74	46
2022	*209*	*45.63*	*249*	*54.37*	*458*
<53	3	33.33	6	66.67	9
<59	10	26.32	28	73.68	38
<66	7	15.22	39	84.78	46
<74	22	35.48	40	64.52	62
<83	28	37.84	46	62.16	74
<93	31	42.47	42	57.53	73
<105	41	58.57	29	41.43	70
<120	36	75.00	12	25.00	48
120+	31	81.58	7	18.42	38
2023	*254*	*43.64*	*328*	*56.36*	*582*
<53	3	30.00	7	70.00	10
<59	10	25.64	29	74.36	39
<66	10	17.54	47	82.46	57
<74	21	23.86	67	76.14	88
<83	41	41.84	57	58.16	98
<93	44	40.37	65	59.63	109
<105	45	60.81	29	39.19	74
<120	43	67.19	21	32.81	64
120+	37	86.05	6	13.95	43
Total	1278	47.51	1412	52.49	2690

stance steadily declined. By 2023, 56.36% of lifters preferred the sumo stance, while 45% opted for the conventional stance, which is the most significant gap between the stances across the years.

The trend lines demonstrate a positive trend for the sumo stance ($y = 0.0095x + 0.49$), indicating an increasing trend in its utilization. Conversely, the conventional stance exhibits a negative trend ($y = -0.0095x + 0.51$), reflecting a gradual decrease in utilization. The data highlights a clear shift in preference toward the sumo stance in recent years, with sumo users consistently surpassing conventional users after 2018.

Going into the detail of each weight class, the number of lifters utilizing the sumo stance is decreasing in trend as evidenced by the linear trend line with a negative gradient for under 53 kg (a weight category only for Sub Junior and Junior category),

Table 4 Total percentage of conventional versus sumo stance in Men's category from 2017 to 2023

Year	Conventional (%)	Sumo (%)
2017	46.56	53.44
2018	50.27	49.73
2019	52.37	47.63
2021	47.61	52.39
2022	45.63	54.37
2023	43.64	56.36

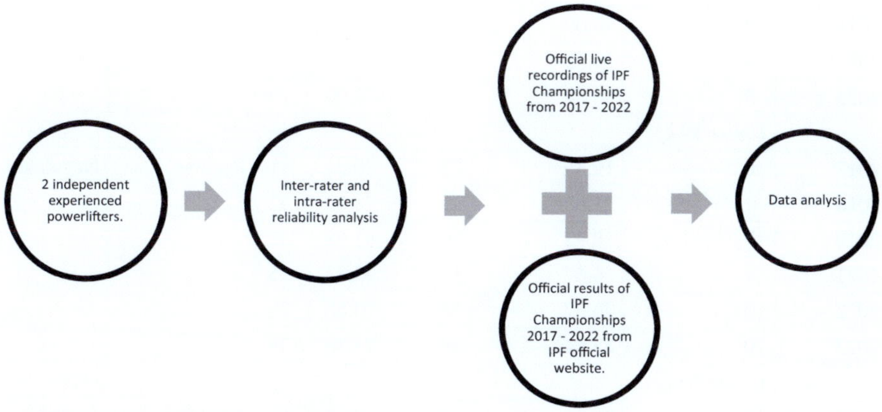

Fig. 3 Data collection procedure diagram

59 kg, 105 kg, and over 120 kg categories. For the rest of the weight categories, the sumo stance has been increasingly utilized over the years based on the positive gradient of the trend line. Table 5 and Figs. 4, 5, 6, 7, 8, 9, 10, 11 and 12 show the conventional versus sumo stance trend for each weight class over 2017–2023.

3.3 Women's Category

Under 69 kg and 76 kg weight categories were added in the Women's category, while the under 72 kg weight category was discontinued from 2021 onwards. 2226 accumulated participations were recorded in 6 years (2017, 2018, 2019, 2021, 2022 and 2023). 2104 female athletes are included in the count. However, there were only 1533 distinct female athletes due to some athletes competing multiple times over the years. Like the men's category, all participations were included in the count, considering that athletes might utilize different stances in competitions.

Table 5 Conventional versus sumo stance trend for each weight class in the men's category from 2017 to 2023

Weight cate-gory/ year	Conven-tional (n)	Conven-tional (%)	Sumo (n)	Sumo (%)	N	Trendline gradient (conventional stance)	Trendline gradient (sumo stance)
	Under 53 kg						
2017	0	0.00	7	100.00	7	Positive	Negative
2018	1	16.67	5	83.33	6		
2019	2	28.57	5	71.43	7		
2021	2	40.00	3	60.00	5		
2022	3	33.33	6	66.67	9		
2023	3	30.00	7	70.00	10		
	Under 59 kg						
2017	6	19.35	25	80.65	31	Positive	Negative
2018	7	21.88	25	78.13	32		
2019	7	18.92	30	81.08	37		
2021	1	6.67	14	93.33	15		
2022	10	26.32	28	73.68	38		
2023	10	25.64	29	74.36	39		
	Under 66 kg						
2017	18	35.29	33	64.71	51	Negative	Positive
2018	14	31.82	30	68.18	44		
2019	13	30.95	29	69.05	42		
2021	6	26.09	17	73.91	23		
2022	7	15.22	39	84.78	46		
2023	10	17.54	47	82.46	57		
	Under 74 kg						
2017	22	31.88	47	68.12	69	Negative	Positive
2018	22	36.07	39	63.93	61		
2019	30	35.71	54	64.29	84		
2021	18	31.58	39	68.42	57		
2022	22	35.48	40	64.52	62		
2023	21	23.86	67	76.14	88		
	Under 83 kg						
2017	27	39.71	41	60.29	68	Negative	Positive
2018	24	43.64	31	56.36	55		
2019	38	47.50	42	52.50	80		
2021	23	42.59	31	57.41	54		
2022	28	37.84	46	62.16	74		

(continued)

Table 5 (continued)

Weight cate-gory/year	Conven-tional (n)	Conven-tional (%)	Sumo (n)	Sumo (%)	N	Trendline gradient (conventional stance)	Trendline gradient (sumo stance)
2023	41	41.84	57	58.16	98		
	Under 93 kg						
2017	35	51.47	33	48.53	68	Negative	Positive
2018	22	52.38	20	47.62	42		
2019	62	57.41	46	42.59	108		
2021	19	32.20	40	67.80	59		
2022	31	42.47	42	57.53	73		
2023	44	40.37	65	59.63	109		
	Under 105 kg						
2017	29	53.70	25	46.30	54	Positive	Negative
2018	30	65.22	16	34.78	46		
2019	39	70.91	16	29.09	55		
2021	32	64.00	18	36.00	50		
2022	41	58.57	29	41.43	70		
2023	45	60.81	29	39.19	74		
	Under 120 kg						
2017	24	88.89	3	11.11	27	Negative	Positive
2018	42	76.36	13	23.64	55		
2019	44	77.19	13	22.81	57		
2021	32	69.57	14	30.43	46		
2022	36	75.00	12	25.00	48		
2023	43	67.19	21	32.81	64		
	120 kg+						
2017	35	76.09	11	23.91	46	Positive	Negative
2018	23	85.19	4	14.81	27		
2019	30	83.33	6	16.67	36		
2021	36	78.26	10	21.74	46		
2022	31	81.58	7	18.42	38		
2023	37	86.05	6	13.95	43		

Table 6 shows the number and percentage of conventional versus sumo stances across Women's weight classes over 2017–2023, while Fig. 13 is the visual representation of Table 6. Compared to the men's category, the women's category shows more contrast in conventional and sumo stance preference, especially in the lighter weight classes. The lighter weight classes (under 43 kg to under 49 kg; under 43 kg is a weight class specific to Junior and Sub-junior categories) in the Women's category

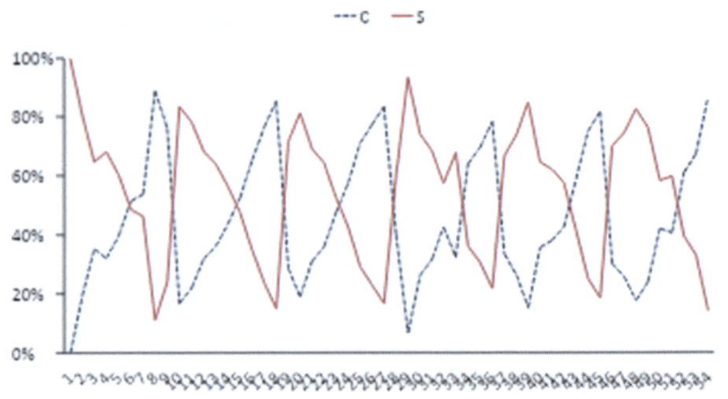

Fig. 4 Percentage of conventional versus sumo stances across Men's weight classes from 2017 to 2023

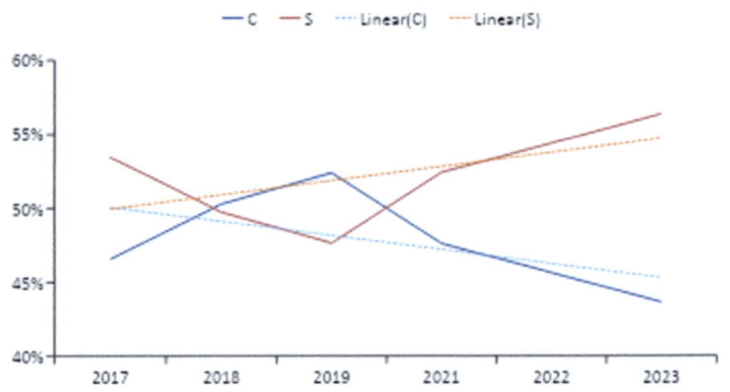

Fig. 5 Total percentage of conventional versus sumo stance from 2017 to 2023

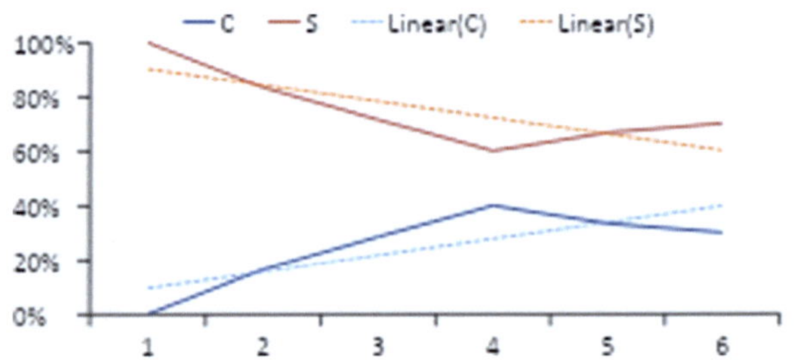

Fig. 6 Men' under 53 kg weight class, sumo versus conventional from 2017 to 2023

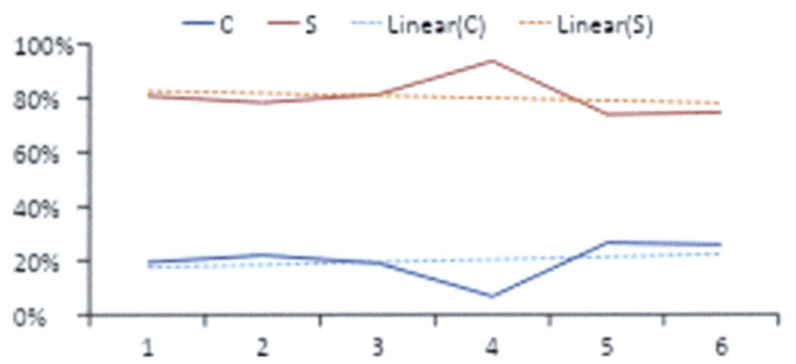

Fig. 7 Men's under 59 kg weight class, sumo versus conventional from 2017 to 2023

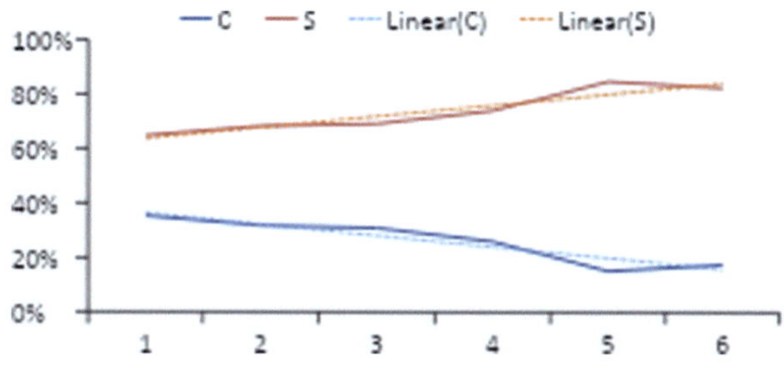

Fig. 8 Men's under 66 kg weight class, sumo versus conventional from 2017 to 2023

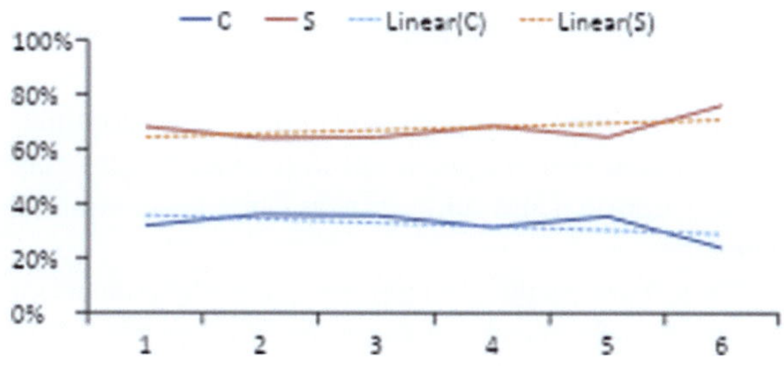

Fig. 9 Men's under 74 kg weight class, sumo versus conventional from 2017 to 2023

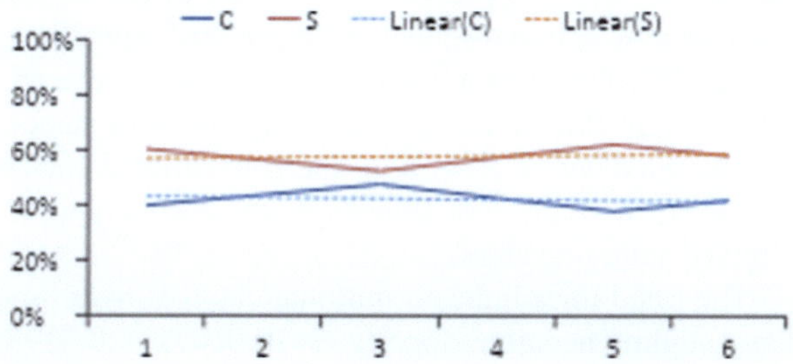

Fig. 10 Men's under 83 kg weight class, sumo versus conventional from 2017 to 2023

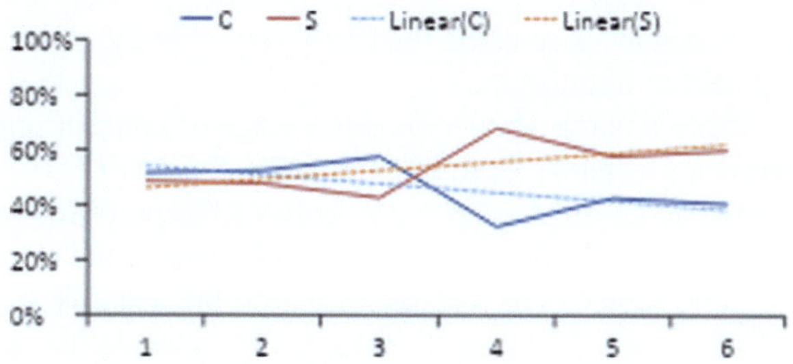

Fig. 11 Men's under 93 kg weight class, sumo versus conventional from 2017 to 2023

recorded sumo stance being utilized exclusively in 4 out of the 6 years. In the under 43 kg weight class, no conventional stance was recorded in 2018, 2019, 2021 and 2022. In 2017 only 1 out of 9 (10%) utilized a conventional stance in the Under 43 weight class, and only 4 out of 33 (10.81%) athletes preferred a conventional stance in the Under 47 weight class. In 2023, no conventional stance was utilized in the

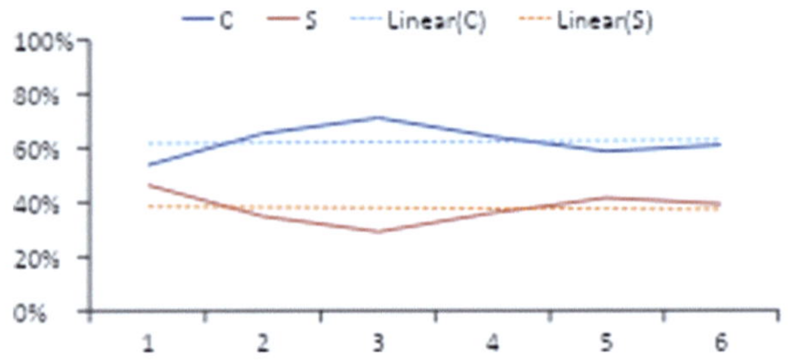

Fig. 12 Men's under 105 kg weight class, sumo versus conventional from 2017 to 2023

under 48 kg and 49 kg weight classes, and 29 out of 34 (71.43%) athletes preferred the sumo stance.

Like the Men's category, the sumo stance dominates lighter weight classes while conventional is more prevalent in heavier weight classes. The discrepancy between the conventional and sumo stances is the least around the under 57 kg to 69 kg weight category.

Table 7 shows the total percentage of the conventional versus sumo stance regardless of weight classes in the women's category. Figure 14 is the graph generated from Table 7. Overall, without considering weight classes, the sumo stance was more prevalent, used by approximately 54.86% of female lifters, compared to 45.14% who used the conventional stance. Between 2017 and 2019, the sumo stance decreased in trend, while the conventional stance slightly increased. During 2020 and 2021, there was a slight fluctuation in popularity for both stances. By 2022, the sumo stance had dropped below 50%, with the conventional stance surpassing the sumo stance. However, in 2023, both stances were nearly equal in usage, with the sumo stance slightly more prevalent.

The trend lines indicate minimal changes over time. The sumo stance trend line has a slightly negative slope ($y = -0.0047x + 0.5293$), reflecting a slight declining trend, while the conventional stance trend line has a positive slope ($y = 0.0047x + 0.4707$), indicating a modest increase. Overall, the difference in usage between the two stances remains relatively tiny over the years, with neither stance showing significant dominance.

Table 8 above shows the percentage of conventional versus sumo stances in the women's category from 2017 to 2023. Figures 15, 16, 17, 18, 19, 20, 21, 22, 23 and 24 below are the visual breakdown of the preferred stances for each weight class. Looking into the details of each weight class in the women's category, the prevalence of sumo stance is decreasing based on the negative trendline gradients in the under 43 kg, 47 kg, 52 kg, and 76 kg weight classes (the under 76 kg weight class was introduced in 2021, replacing the under 72 kg weight class). The trendline gradients are −0.0265, −0.0190, −0.0100, and −0.0560, respectively. In the rest of the weight classes, the sumo stance has become more prevalent over the years, as evidenced by the positive gradient of the trendline. The trendline gradient for the under 57 kg, 63 kg, 69 kg, 72 kg, 84 kg and 84 kg weight classes are 0.0265, 0.0007, 0.0561, 0.0355, 0.0168, 0.0007, respectively.

Table 6 Number and percentage of conventional versus sumo stances across Women's weight classes from 2017 to 2023

Year/Weight class	Conventional (n)	Conventional (%)	Sumo (n)	Sumo (%)	N
2017					
<43	1	10.00	9	90.00	10
<47	4	10.81	33	89.19	37
<52	13	28.89	32	71.11	45
<57	20	46.51	23	53.49	43
<63	28	53.85	24	46.15	52
<72	26	54.17	22	45.83	48
<84	21	75.00	7	25.00	28
84+	17	68.00	8	32.00	25
2018					
<43		0.00	5	100.00	5
<47	3	10.71	25	89.29	28
<52	11	23.91	35	76.09	46
<57	19	41.30	27	58.70	46
<63	27	48.21	29	51.79	56
<72	29	59.18	20	40.82	49
<84	37	80.43	9	19.57	46
84+	26	89.66	3	10.34	29
2019					
<43		0.00	4	100.00	4
<47	7	31.82	15	68.18	22
<52	15	44.12	19	55.88	34
<57	21	34.43	40	65.57	61
<63	37	50.00	37	50.00	74
<72	32	47.06	36	52.94	68
<84	42	67.74	20	32.26	62
84+	30	78.95	8	21.05	38
2021					
<43		0.00	3	100.00	3
<47	2	11.11	16	88.89	18
<52	10	34.48	19	65.52	29
<57	14	28.00	36	72.00	50
<63	22	44.00	28	56.00	50
<72	23	58.97	16	41.03	39
<84	19	50.00	19	50.00	38

(continued)

Table 6 (continued)

Year/Weight class	Conventional (n)	Conventional (%)	Sumo (n)	Sumo (%)	N
<43	19	59.38	13	40.63	32
84+	22	91.67	2	8.33	24
2022					
<43		0.00	5	100.00	5
<47	8	33.33	16	66.67	24
<52	15	35.71	27	64.29	42
<57	22	33.85	43	66.15	65
<63	35	53.03	31	46.97	66
<72	34	54.84	28	45.16	62
<84	42	70.00	18	30.00	60
<43	29	69.05	13	30.95	42
84+	27	72.97	10	27.03	37
2023					
<43	2	28.57	5	71.43	7
<47	5	14.71	29	85.29	34
<52	16	30.77	36	69.23	52
<57	30	33.71	59	66.29	89
<63	31	51.67	29	48.33	60
<72	32	47.76	35	52.24	67
<84	41	61.19	26	38.81	67
<43	33	71.74	13	28.26	46
84+	30	75.00	10	25.00	40
Total	1029	48.91	1075	51.09	2104

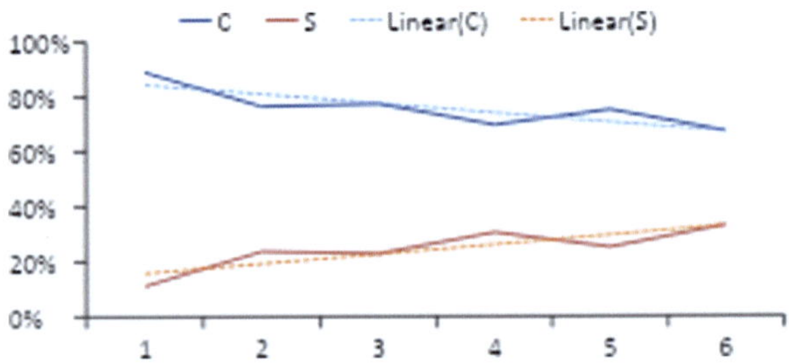

Fig. 13 Men's under 120 kg weight class, sumo versus conventional from 2017 to 2023

Table 7 Total percentage of conventional versus sumo stance in the Women's category from 2017 to 2023

Year	Conventional (%)	Sumo (%)
2017	45.14	54.86
2018	49.84	50.16
2019	50.69	49.31
2021	46.29	53.71
2022	52.61	47.39
2023	47.62	52.38

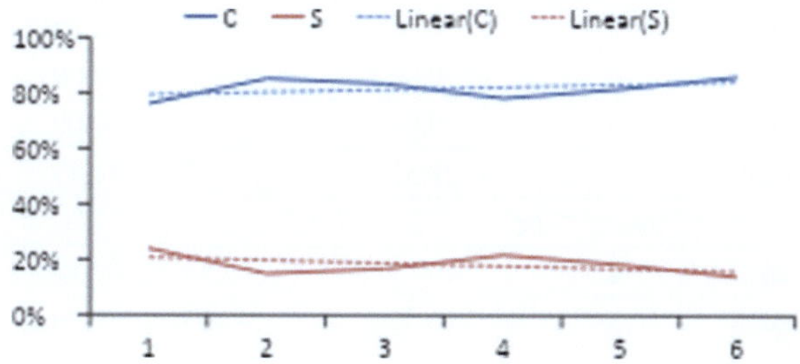

Fig. 14 Men's 120 kg + weight class, sumo versus conventional from 2017 to 2023

4 Discussion

This study's purpose was to describe the prevalence of conventional and sumo deadlift stances utilized by powerlifting athletes within a competition setting, specifically international championships organized by IPF from 2017 to 2023, and to discuss the effectiveness of each. The context underpinning this study was that both deadlift stances are allowed in powerlifting championships despite the differences, particularly in the biomechanical aspect of both deadlift stances. A unique aspect of this current study was the access to a large dataset of athletes, championship detailed results, and the availability of live recordings of the championships.

The relationship between weight class and deadlift stance preference among lifters reveals notable trends, especially when comparing lighter and heavier athletes. This study showed that lighter lifters often prefer the sumo stance, while heavier lifters gravitate toward the conventional stance. This finding is not sporadic, as a consistent pattern can be seen in all weight classes and throughout the six years. This difference between lighter and heavier athletes can be partially explained by biomechanics, as the sumo stance demands less from the lower back musculature (Escamilla et al. 2000, 2002; Cholewicky et al. 1991). This is likely due to the more upright trunk position in the sumo stance in which the sumo deadlift generated a 10% reduction in the L4/L5 moment and an 8% decrease in the L4/L5 shear force (Cholewicky et al. 1991). Smaller and lighter lifters generally have smaller torsos, translating into potentially weaker lower back musculature. The more upright trunk position observed during the sumo deadlift may decrease spinal loads (Cholewicky et al. 1991). The lower back musculature strength may limit lighter athletes' ability for

Table 8 Conventional versus sumo stance trend for each weight class in the Women's category from 2017 to 2023

Weight class/ year	Conven-tional (n)	Conven-tional (%)	Sumo (n)	Sumo (%)	N	Trendline gradient slope (conventional stance)	Trendline gradient slope (sumo stance)
	Under 43 kg						
2017	1	10.00	9	90.00	10	Positive	Negative
2018	0	0.00	5	100.00	5		
2019	0	0.00	4	100.00	4		
2021	0	0.00	3	100.00	3		
2022	0	0.00	5	100.00	5		
2023	2	28.57	5	71.43	7		
	Under 47 kg						
2017	4	10.81	33	89.19	37	Positive	Negative
2018	3	10.71	25	89.29	28		
2019	7	31.82	15	68.18	22		
2021	2	11.11	16	88.89	18		
2022	8	33.33	16	66.67	24		
2023	5	14.71	29	85.29	34		
	Under 52 kg						
2017	13	28.89	32	71.11	45	Positive	Negative
2018	11	23.91	35	76.09	46		
2019	15	44.12	19	55.88	34		
2021	10	34.48	19	65.52	29		
2022	15	35.71	27	64.29	42		
2023	16	30.77	36	69.23	52		
	Under 57 kg						
2017	20	46.51	23	53.49	43	Negative	Positive
2018	19	41.30	27	58.70	46		
2019	21	34.43	40	65.57	61		
2021	14	28.00	36	72.00	50		
2022	22	33.85	43	66.15	65		
2023	30	33.71	59	66.29	89		
	Under 63 kg						
2017	28	53.85	24	46.15	52	Negative	Positive
2018	27	48.21	29	51.79	56		
2019	37	50.00	37	50.00	74		
2021	22	44.00	28	56.00	50		

(continued)

Table 8 (continued)

Weight class/ year	Conven-tional (n)	Conven-tional (%)	Sumo (n)	Sumo (%)	N	Trendline gradient slope (conventional stance)	Trendline gradient slope (sumo stance)
2022	35	53.03	31	46.97	66		
2023	31	51.67	29	48.33	60		
	Under 69 kg						
2021	23	58.97	16	41.03	39	Negative	Positive
2022	34	54.84	28	45.16	62		
2023	32	47.76	35	52.24	67		
	Under 72 kg						
2017	26	54.17	22	45.83	48	Negative	Positive
2018	29	59.18	20	40.82	49		
2019	32	47.06	36	52.94	68		
	Under 76 kg						
2021	19	50.00	19	50.00	38	Positive	Negative
2022	42	70.00	18	30.00	60		
2023	41	61.19	26	38.81	67		
	Under 84 kg						
2017	21	75.00	7	25.00	28	Negative	Positive
2018	37	80.43	9	19.57	46		
2019	42	67.74	20	32.26	62		
2021	19	59.38	13	40.63	32		
2022	29	69.05	13	30.95	42		
2023	33	71.74	13	28.26	46		
	84+						
2017	17	68.00	8	32.00	25	Negative	Positive
2018	26	89.66	3	10.34	29		
2019	30	78.95	8	21.05	38		
2021	22	91.67	2	8.33	24		
2022	27	72.97	10	27.03	37		
2023	30	75.00	10	25.00	40		
Total	1029	48.91	1075	51.09	2104		

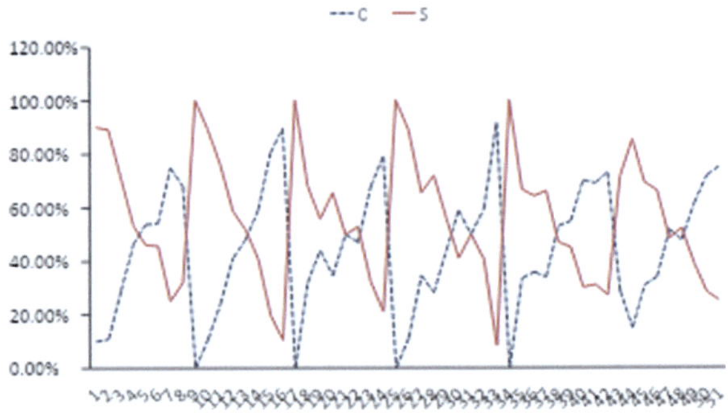

Fig. 15 Percentage of conventional versus sumo stances across Women's weight classes from 2017 to 2023

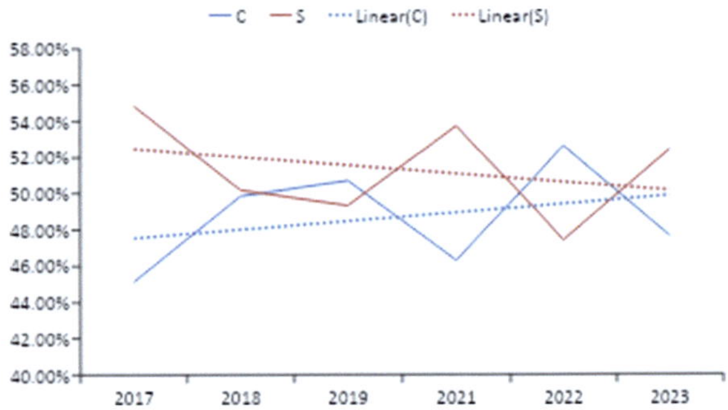

Fig. 16 Total percentage of conventional versus sumo stance in women's category from 2017 to 2023

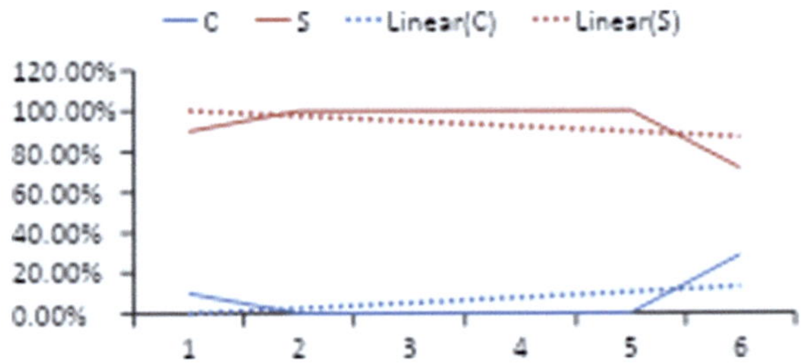

Fig. 17 Women's under 43 kg weight class, sumo versus conventional from 2017 to 2023

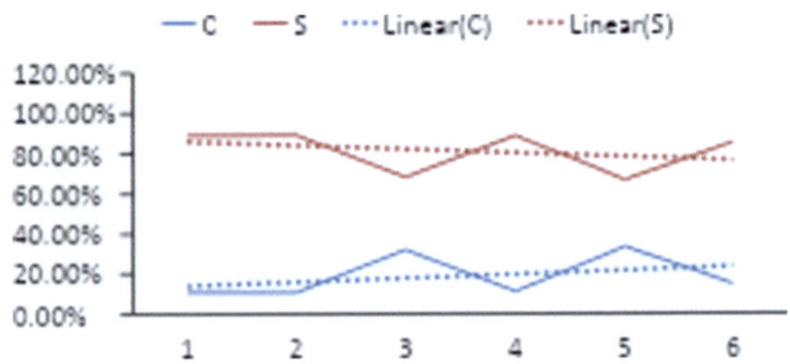

Fig. 18 Women's under 47 kg weight class, sumo versus conventional from 2017 to 2023

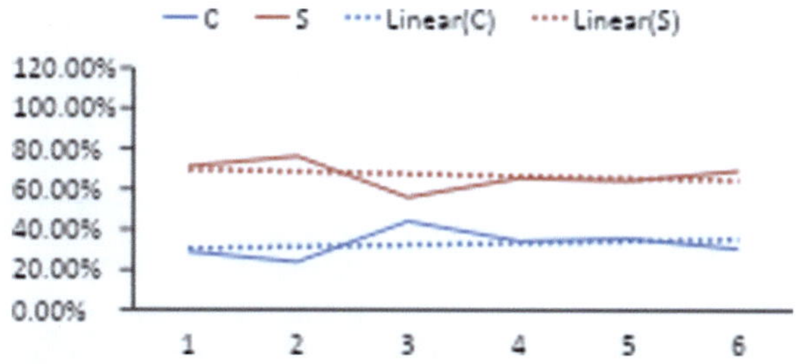

Fig. 19 Women's under 52 kg weight class, sumo versus conventional from 2017 to 2023

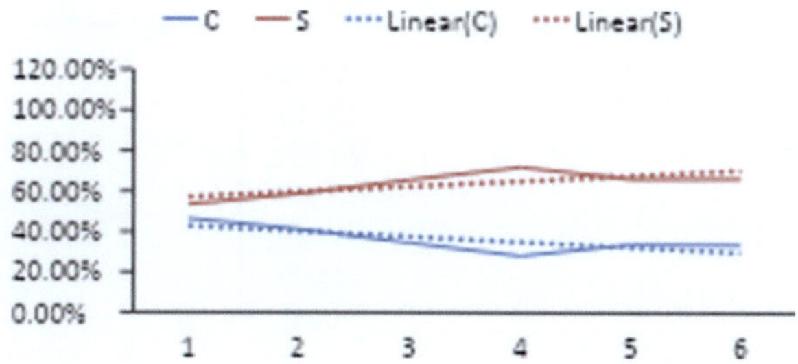

Fig. 20 Women's under 57 kg weight class, sumo versus conventional from 2017 to 2023

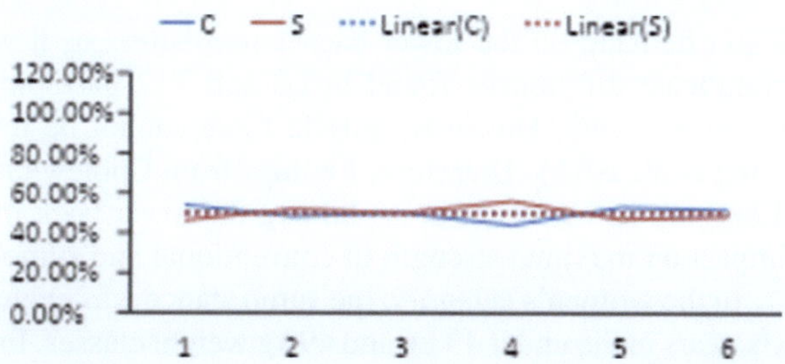

Fig. 21 Women's under 63 kg weight class, sumo versus conventional from 2017 to 2023

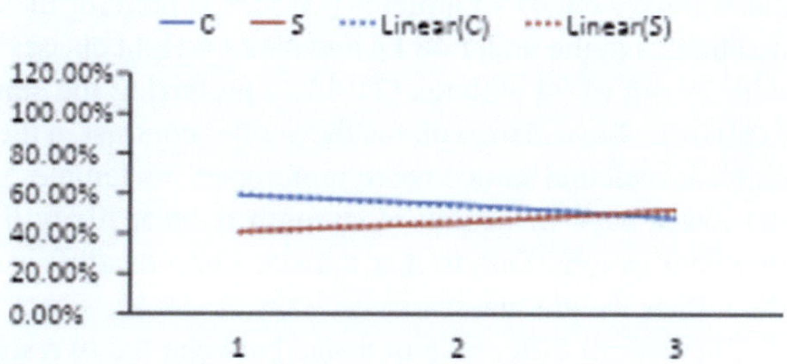

Fig. 22 Women's under 69 kg weight class, sumo versus conventional from 2017 to 2023

maximal deadlifts. Hence, the sumo stance is more favorable for lighter athletes. In contrast, larger-sized athletes with thicker torsos generally have greater lower back musculature strength and, therefore, are more advantageous with a conventional stance. The EMG study by Escamilla, Francisco (Escamilla et al. 2002) did not agree with the findings by Cholewicky, McGill (Cholewicky et al. 1991); sumo put

Fig. 23 Women's under 72 kg weight class, sumo versus conventional from 2017 to 2023

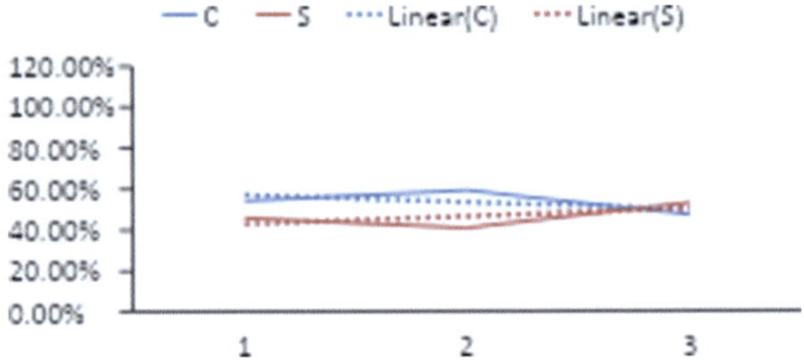

Fig. 24 Women's under 76 kg weight class, sumo versus conventional from 2017 to 2023

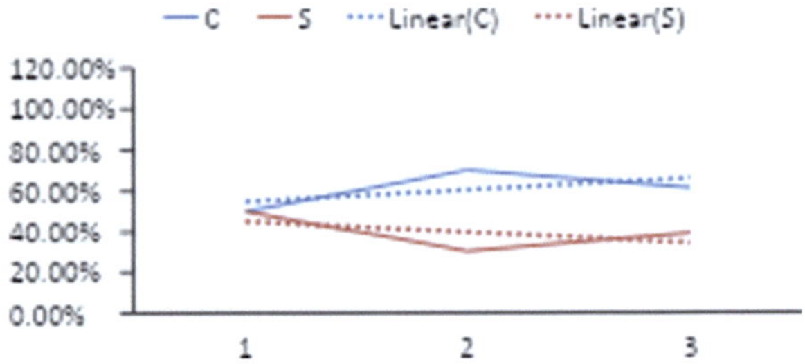

lesser demand on the lower back musculature, as it was found that there were no significant differences found in L3 and T12 paraspinal activity during the ascent based on EMG. However, muscle force cannot be measured directly using EMG (Hug et al. 2015). Therefore, findings from Cholewicky, McGill (Cholewicky et al. 1991) are still relevant in explaining the lower back musculature difference and its impact on maximal strength in conventional and sumo deadlift stances.

In the women's category, the sumo stance was exclusively used in four out of the six years in the under 43 kg and 49 kg weight classes. In 2017, only 1 out of 9 athletes (10%) in the under 43 kg class used the conventional stance, while in the under 47 kg class, only 4 out of 33 athletes (10.81%) opted for the conventional stance. In 2023, no athletes in the under 48 kg and 49 kg weight classes used the conventional stance, with 29 out of 34 athletes (71.43%) preferring the sumo stance. The same concept explained above also explains the greater contrast in the women's category regarding conventional and sumo stance preference. As females generally have smaller torsos, the lower back musculature strength is most likely limiting the ability to deadlift maximal weight. Due to that reason, sumo deadlift is more prevalent, especially in the lighter weight classes.

The overall difference in usage between the two stances has remained relatively small over the years among female athletes, with neither stance demonstrating significant dominance. As females are generally smaller than males, the impact of the lower back musculature strength limitation on maximal deadlift strength is likely more noticeable (less variance in deadlift stances due to limitation in lower back musculature strength). This explains why the difference in usage between the two stances has remained relatively small over the years in the women's category, with neither stance showing significant dominance.

In the sumo stance, the feet are wider apart (hips are abducted more) and turned out (hips are externally rotated more). Therefore, a limited range of motion for hip abduction and external rotation may negatively impact the sumo deadlift performance. This can be a factor determining the preferred deadlift stance among powerlifters. However, to the best of the author's knowledge, studies on the relationship between the range of motion of the hip abduction and external rotation with conventional and sumo deadlift performance are scarce and shall be explored in future studies.

It is worth noting that when comparing conventional and sumo stances, there was no significant difference in hip extension demands at any point in the movement (Escamilla et al. 2000). In another study, EMG readings for the glutes and hamstrings were the same for both deadlift styles, validating this idea (Escamilla et al. 2002). Therefore, differences in hip anatomy, hip flexibility, and anthropometrical measurements may influence deadlift performance regardless of the preferred stance.

The shift from conventional to sumo stance can be seen in most.

Men's weight classes and, to a lesser degree, in most Women's weight classes. Apart from the difference between the stances from a biomechanical standpoint, this shift also suggests that it may be linked to evolving coaching strategies that increasingly tailor deadlift techniques to individual biomechanical advantages and adjustments in competition strategies that prioritize stance efficiency for speed and lockout performance. Escamilla, Francisco (Escamilla et al. 2000) found that lifters using the sumo stance spent significantly longer time in the acceleration phase (slower at the beginning of the lift') (Fig. 25), approximately 10% more time than athletes using the conventional stance (26.1% of total lift time ± 10.0% in sumo stance, 16.1% ± 7.3% of total lift time in conventional stance). On the other hand, ankle and knee moments during the sumo deadlift are relatively large compared to ankle and knee moments generated during the conventional deadlift (Escamilla et al. 2000). This is also supported by the EMG study by Escamilla, Francisco (Escamilla et al. 2002), where it was found that the sumo deadlift had significantly greater EMG activity in the vastus lateralis, vastus medialis, and tibialis anterior. In this case, knee extensors, ankle dorsiflexion, and ankle plantar flexors strength may be a limiting factor for maximal deadlift strength. Due to this reason, coaches may adjust deadlift techniques to align with an individual's biomechanical advantage and modify competition strategies to enhance stance efficiency, speed, and lockout performance (Fig. 26 and 27).

Despite these observations, recent studies investigating competitive powerlifters' stance preferences across different weight classes remain scarce, particularly regarding updated biomechanics research. Future work should delve into these preferences, focusing on the sophisticated interactions between body composition, leverage, and muscle activation patterns to understand why stance selection may vary across lifters of different weights and genders. This gap highlights an unexplored avenue in optimizing deadlifting techniques for competitive advantage.

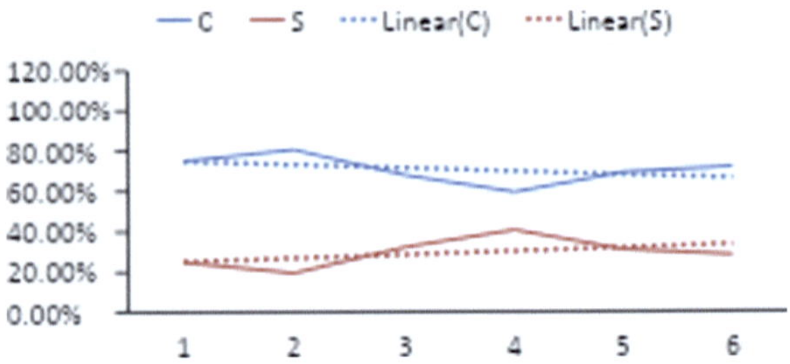

Fig. 25 Women's under 84 kg weight class, sumo versus conventional from 2017 to 2023

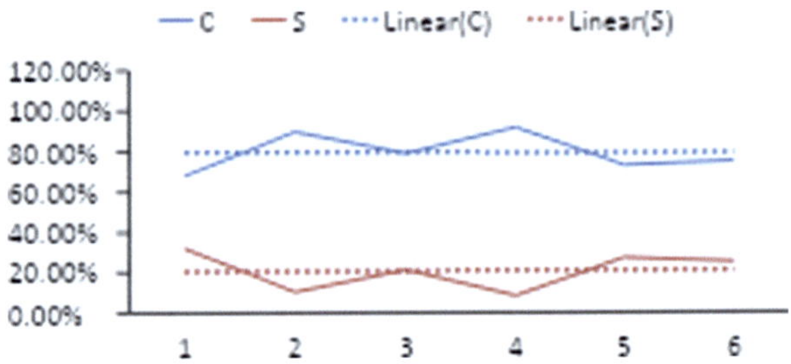

Fig. 26 Women's 84 kg + weight class, sumo versus conventional from 2017 to 2023

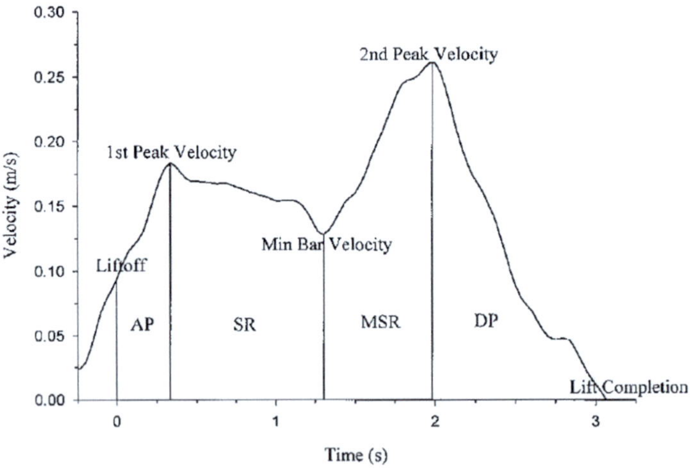

Fig. 27 Representative graph for vertical bar velocity. AP, acceleration phase; SR, sticking region; MSR, maximum strength region; DP, deceleration phase by Escamilla et al. (2000)

5 Conclusion

This study provides insights into the evolving trends in deadlift stance among elite powerlifters over the past six years. A clear pattern emerges across genders: lighter athletes favor the sumo stance, while heavier athletes prefer the conventional stance. These findings suggest that body weight influences stance choice, likely due to biomechanical advantages. The sumo stance's ability to reduce lower back muscle strength demand makes it favorable for athletes with lower body mass, while heavier athletes benefit more from the conventional stance due to stronger lower back muscles (Escamilla et al. 2000; Cholewicky et al. 1991). Despite the rising sumo stance preference in general, particularly among lighter weight classes, the conventional stance remains dominant in heavier athletes. These findings reinforce the importance of individualized training approaches that account for the biomechanical advantages of different body types. While previous studies have outlined the mechanical differences between these stances, the impact of variables like hip flexibility and limb length on stance efficiency across different weights and genders remains an area for further research. The increasing popularity of the sumo stance might also reflect shifts in coaching strategies that emphasize biomechanical efficiency. Tailoring techniques to each lifter's unique strengths allows for more optimized performance, especially in competitive powerlifting. In conclusion, this study highlights the growing trend toward sumo stance usage in lighter-weight classes, while the conventional stance persists among heavier lifters. These findings outline the importance of considering individual biomechanics in training and competition, with further research needed to refine deadlift stance optimization across body types.

References

Barreira J, Chiminazzo JGC (2020) Who, how and when to perform winner points and unforced errors in badminton matches? An analysis of men's single matches in the 2016 Olympic Games. Int J Perform Anal Sport 20(4):610–619

Brown EW, Abani K (1985) Kinematics and kinetics of the dead lift in adolescent power lifters. Med Sci Sports Exerc 17(5):554–566

Cholewicky J, McGill SM, Norman RW (1991) Lumbar spine loads during the lifting of extremely heavy weights. Med Sci Sports Exerc 23(10):1179–1186

Escamilla RF et al (2000) A three-dimensional biomechanical analysis of sumo and conventional style deadlifts. Med Sci Sports Exerc 32(7):1265–1275

Escamilla RF et al (2001) Biomechanical analysis of the deadlift during the 1999 Special Olympics World Games. Med Sci Sports Exerc 33(8):1345–1353

Escamilla RF et al (2002) An electromyographic analysis of sumo and conventional style deadlifts. Med Sci Sports Exerc 34(4):682–688

Fong SSM et al (2022) The influence of weightlifting belts and wrist straps on deadlift kinematics, time to complete a deadlift and rating of perceived exertion in male recreational weightlifters: an observational study. Medicine 101(7):e28918

Hossner EJ et al (2019) The role of viewing distance and viewing angle on referees' decision-making performance during the FIFA World Cup 2014. J Sports Sci 37(13):1481–1489

Hug F, Hodges PW, Tucker K (2015) Muscle force cannot be directly inferred from muscle activation: illustrated by the proposed imbalance of force between the vastus medialis and vastus lateralis in people with patellofemoral pain. J Orthop Sports Phys Ther 45(5):360–365

Jukic I et al (2021) Ergogenic effects of lifting straps on movement velocity, grip strength, perceived exertion and grip security during the deadlift exercise. Physiol Behav 229:113283

Jukic I et al (2022a) Magnitude and reliability of velocity and power variables during deadlifts performed with and without lifting straps. J Strength Cond Res 36(5):1177–1184

Jukic I et al (2022b) Validity of Load-velocity relationship to predict 1 repetition maximum during deadlifts performed with and without lifting straps: the accuracy of six prediction models. J Strength Cond Res 36(4):902–910

Jukic I, García-Ramos A, Tufano JJ (2022c) Velocity-based resistance training monitoring: influence of lifting straps, reference repetitions, and variable selection in resistance-trained men. Sports Health, 19417381221095073

Kompf J, Arandjelović O (2017) The sticking point in the bench press, the squat, and the deadlift: similarities and differences, and their significance for research and practice. Sports Med 47(4):631–640

Lignell E, Rago V, Mohr M (2020) Analysis of goal scoring opportunities in elite male ice hockey in relation to tactical and contextual variables. Int J Perform Anal Sport 20(6):1003–1017

McGuigan MRM, Wilson BD (1996) Biomechanical analysis of the deadlift. J Strength Conditioning Res 10(4):250–255

Omcirk D et al (2020) The effects of lifting straps on the maximal load and mean velocity of deadlift 1RM

Results, IPF international powerlifting federation; Available from: https://www.powerlifting.sport/championships/results

Schoffstall J et al (2010) Grip strength and powerlifting performance

Suazo N, DebelİSo M (2021) The relationship between powerlifting performance and hand grip strength among female athletes. Turkish J Kinesiology

Thestrengthathlete (2019) A few years ago, we thought there was a pretty clear trend of heavier lifters pulling conventional, and lighter lifters favouring sumo. While that still seems to be the case across the male weight classes, that pattern is less noticeable in female athletes. 2019 19 December 2019 [cited 12 October 2022]

Optimizing Stack Height in Running Footwear: Engineering Insights for Enhancing Biomechanical Performance

Abdul Nasir, Norain Binti Abdullah, Devin Babu, Muhammad Haziq, Mohd Amir Shahlan Mohd Aspar, and Muhammad Nur Farhan Saniman

Abstract The role of stack height in running footwear has been critically analyzed with an emphasis on its biomechanical and performance-related impacts. It has been established that stack height influences key factors such as ground reaction forces, foot strike patterns, and running economy. Research has highlighted significant variability in how runners respond to different stack heights, complicating the optimization of footwear design. Various studies have demonstrated both the benefits and drawbacks of high and low stack heights, with ongoing debates about their effects on performance and injury prevention. A major gap in the current literature pertains to the lack of a consensus on how stack height should be adapted to individual biomechanical profiles. Recent developments in footwear engineering, such as carbon plates and advanced foam materials, have further complicated the understanding of stack height optimization. Hypothetical solutions have been proposed, including personalized footwear solutions that tailor stack height to a runner's biomechanics. This method shows promise in bridging the gap by offering individualized footwear designs based on detailed biomechanical profiling. It is suggested that dynamic stack height adjustment technologies and machine learning-based data collection could further improve footwear customization. Moreover, gender-specific responses to stack height require further investigation to develop more precise footwear designs. Future research is recommended to explore the long-term effects of personalized stack height solutions, the integration of smart materials, and the role of psychological factors in footwear design. Such advancements could lead to more effective performance enhancements and injury prevention strategies for runners.

A. Nasir · D. Babu (✉) · M. Haziq · M. A. S. M. Aspar
Faculty of Electrical and Electronic Engineering Technology, Universiti Malaysia Pahang Al-Sultan Abdullah, Pekan, Pahang, Malaysia
e-mail: devinbabu@yahoo.com

M. N. F. Saniman
Mechanical Engineering Section, Universiti Kuala Lumpur Malaysia France Institute, Bandar Baru Bangi, Malaysia

N. B. Abdullah
Bio-Coke Research Institute, Kindai University, Osaka, Japan

M. H. A. Hassan et al. (eds.), *Proceedings of the 3rd International Conference on Innovation and Technology in Sports*, Lecture Notes in Bioengineering,
https://doi.org/10.1007/978-981-95-0804-4_38

Keywords Footwear stack height · Biomechanics · Running performance · Personalized footwear solutions · Injury prevention strategies

1 The Role of Stack Height in Running Footwear

Footwear engineering plays a crucial role in optimizing running performance, particularly through the manipulation of design features such as stack height. Stack height, the distance between the runner's foot and the ground, has gained significant attention due to its direct influence on biomechanical variables such as ground reaction forces, foot strike pattern, and running economy (Lin et al. 2022; Sinclair and Taylor 2023). With rapid advancements in material technology and shoe design, it is increasingly important to understand the interaction between stack height and biomechanical performance. This review focuses on the implications of optimizing stack height in running footwear, with insights drawn from biomechanics, sports engineering, and recent technological innovations. In the past decade, research has revealed how varying the stack height of running shoes can alter both performance outcomes and injury risks. High-stack shoes, often associated with "maximalist" designs, feature thick cushioning, potentially reducing tibial stress fractures and improving overall shock absorption (Sinclair and Taylor 2023). Conversely, minimal stack heights characteristic of "minimalist" shoes—are designed to encourage a more natural foot strike but may increase the risk of injury for some runners due to greater ground reaction forces (Agresta et al. 2022a).

2 Research Gaps and Challenges in Stack Height Optimization

The complexity of optimizing stack height stems from the multitude of factors that interact within the biomechanical system during running. Most notably, the interaction between stack height and performance has been difficult to generalize across various populations and running styles (Bertschy et al. 2023; Ruiz-Alias et al. 2023a, b).

Figure 1 shows the Rocker axis position and rocker radius of current footwear models of different brands with different stack heights. The measurements were made using the free-access software Kinovea, on the sagittal image of each model placed in the same position. The rocker axis was positioned at the center of the rocker radius circle and calculated by its relative position to the total length, i.e., 28 cm for all models. The radius of the rocker designates the amount of upward forefoot curvature of the sole, a larger radius means less upward curvature. A recurring issue in the literature is the lack of consensus on how varying stack heights influence injury risk and performance for different runners (Kiesewetter et al. 2022; Willwacher et al.

2024). Studies have produced contrasting results, with some indicating that higher stack heights enhance energy return and performance, while others argue that these shoes may lead to changes in lower limb mechanics, potentially contributing to overuse injuries (Kiesewetter et al. 2022; Sinclair and Taylor 2023). Another significant challenge in current research is the limited understanding of how individual biomechanics (e.g., foot strike pattern, leg stiffness, and running cadence) affect the optimal stack height for different athletes (Patoz et al. 2022; Wang et al. 2023). While it is well-established that shoe properties such as cushioning and stiffness can alter running mechanics, studies often overlook how specific combinations of runner characteristics interact with stack height (Kiesewetter et al. 2022; Lin et al. 2022). For instance, recent research in the integration of carbon fiber plates inside high-stack shoes has shown mixed results, where some runners reveal improved bio-mechanical efficiency, while others expose higher risks due to changed kinematics (Agresta et al. 2022a).

The stack height of running shoes the distance between the sole material and the ground under the foot as a considerable impact both on economy of running (RE) and injury risk. Maximalist shoes, with a greater stack height, have a larger level of cushioning, and that can make them even more comfortable and, potentially, improve RE through reduced muscle exhaustion over long runs (Esculier et al. 2015). In contrast, a larger level of added cushioning can modify foot mechanics, and, in fact, increase the risk of injury (Pollard et al. 2018). In contrast, shoes with a low stack height, that is, minimalist shoes, have reduced cushioning and a more natural foot strike and increased proprioception, and potentially even greater RE through

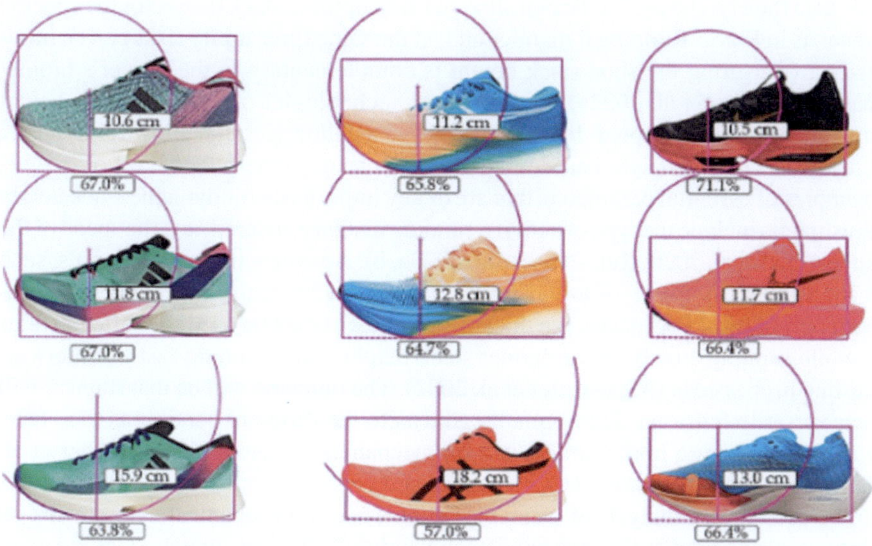

Fig. 1 Rocker axis position (i.e., the starting point of the forefoot upward curvature) and rocker radius of current footwear models of different brands with different stack heights

a reduced-energy gait (Moore et al. 2012) Empirical studies have confirmed that minimalist shoes can maximize RE and strengthening of the foot, but less cushioning can result in increased loading of the ankle and meta-tarsophalangeal joints, and therefore in an increased risk of damaging these structures (Agresta et al. 2022b). In addition, a systematic review confirmed that increased strain in the lower limb could occur with minimalist shoes when transitioning to such a type of shoe, supporting a slow transition (Perkins et al. 2014). Consequently, selection of stack height must be individualized, taking a runner's biomechanics, strength, and medical background into consideration in order to maximize performance and minimize the risk of injury.

2.1 Hypothetical Solutions: Several Possible Best Methods to Close the Gap

Now that the research gaps have been identified, a number of hypothetical solutions can be proposed. The following approaches will aim at the further refinement of the relationship between stack height, biomechanics, and performance outcomes. Firstly, personalized footwear solutions based on biomechanical profiles should be conducted. Development of running footwear that is tailored to the individual biomechanics of the runner is needed, particularly their foot strike pattern, leg stiffness, and gait characteristics (Carranza 2023; Knopp et al. 2023).

This can be achieved by advanced gait analysis techniques, such as motion capture and wearable sensors that could be employed to assess individual running mechanics. This data then can be used to design shoes with an optimal stack height for the specific runner as in Fig. 2. Running dynamics and, at the same time, injury risks can be minimized by ensuring the shoe stack height is complementary to the runner's biomechanics (Agresta et al. 2022a). In this regard, technologies of dynamic stack height adjustment may be adopted. Introduce adaptive technologies that allow for dynamic adjustment of stack height based on the runner's speed, terrain, and level of fatigue. Examples of essential parameters that are of key importance for dynamic stack height adjustments include the type of activity, running distance covered, and the speed of the runner (Gao et al. 2020; Ruiz-Alias et al. 2023a, b). Advances in smart materials, such as shape-memory foams or air chambers, may allow for real-time changes in cushioning properties. For instance, a thicker stack height may be used for softer terrains or while running slowly and a thinner stack height while running fast or otherwise exerting high speeds (Kiesewetter et al. 2022). The outcome will be that runners will wear versatile footwear that optimally adjusts to the different conditions they face, ultimately bettering both comfort and biomechanical efficiency (Willwacher et al. 2024). From then on, more data collection and advanced machine learning models can be used in the analysis of wearable devices and biomechanical assessments on a large scale to predict the optimum stack height of different runners (Casado et al. 2022; Devi et al. 2023). Wearable devices like smart insoles and accelerometers could collect data continuously related to the runners' biomechanics under various

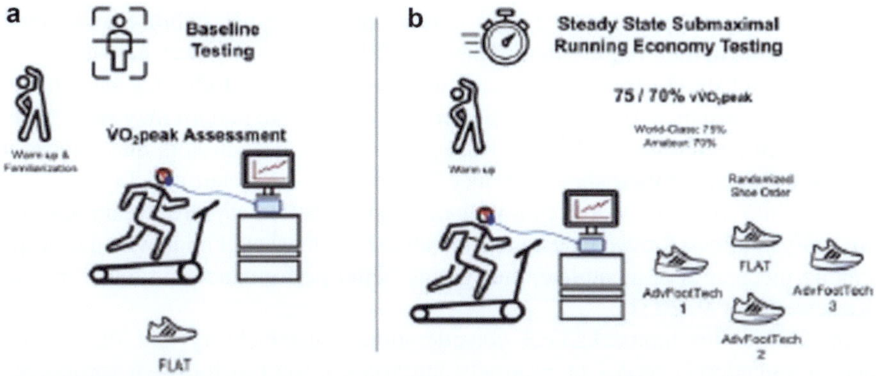

Fig. 2 Methods of baseline data collection from previous studies included maximal oxygen uptake testing during initial visits, and the running economy assessment comparing FLAT to different models of advanced footwear technology

conditions. The datasets would be analyzed by machine learning models deriving personalized recommendations regarding the optimum stack height in consideration of a runner's biomechanical profile and their performance goals (Kiesewetter et al. 2022). The ultimate goal is the creation of a feedback loop where data about individual biomechanics feed into the design of shoes so that the footwear adapts to the dynamic needs of the runner while running (Willwacher et al. 2024).

3 Personalized Footwear Solutions: A Path to Optimized Biomechanics

Of the various hypothetical solutions discussed, personalized footwear solutions based on biomechanical profiles probably offer the best method for stack height optimization in improving biomechanical performance while reducing the risk of injury. It is unique because it caters to the individual variability of running mechanics-the major limitation that has been touted of most prior research studying stack height and running performance (Agresta et al. 2022a; Willwacher et al. 2024).

3.1 Justification of Personalized Footwear Solutions

The main choice for this approach is that biomechanical variability between runners imposes a challenging one-size-fits-all solution. Foot strike pattern, leg stiffness, running cadence, and even gender-specific differences significantly influence the runner's response to stack height variations (Valldecabres 2023). Tailoring shoes to an individual biomechanical profile allow for the incorporation of those variables, ensuring each runner attains the right amount of cushioning, stability, and energy return specific to their requirements (Lin et al. 2022). Research work has shown that runners demonstrate varied kinetic and kinematic alterations following stack height changes. For instance, runners who prefer a forefoot strike pattern may require less stack height to continue with their usual gait, whereas heel strikers may choose to have more stacks in their shoe to reduce impact forces (Sinclair and Taylor 2023). Looking into these individual biomechanics using motion capture, wearables, and gait analysis allows footwear designers to make shoes that better align with the unique running mechanics of an athlete, thus giving better performance and fewer injuries (Kiesewetter et al. 2022).

In addition to biomechanical considerations, sustainability and its environment implications become increasingly important factors in footwear technology. Researcher (Babu et al. 2024) highlight the role of green technology in managing sports infrastructure, an application that could extend to high-stack footwear development. Integration of environmentally friendly materials in running shoes can potentially make them less environmentally taxing and maintain key factors such as comfort and durability. Polyurethane, a polymer often seen in use in sports flooring, has been examined for its engineering behavior and potential use in performance environments (Rosli et al. 2024a, b). In addition, improvements in machining technology for materials can make production in footwear less environmentally taxing through reduced waste and increased efficiency (Rosli et al. 2024a, b). Management of waste is also critical in environmentally friendly sports technology, with examples seen in efforts towards integration of green technology in stadiums and stadiums (Abidin et al. 2024). Similarity, such thinking could extend to footwear production through maximization of processes in production for reduced environment impact. In addition, improvements in sensor technology, such as nano-coating techniques for enhancing durability in piezoelectric sensors, present examples of high-tech materials' ability to maximize performance at the expense of little loss in terms of sustainability (Hakim et al. 2024). Nevertheless, studies investigating impact of environmentally friendly materials in stack height performance have been wanting in terms of availability. Future studies must explore integration of sustainability and biomechanics in a manner that ensures stack height development is in harmony with environmentally friendly practice. The selection of stack height must therefore become personalized, with consideration for a runner's biomechanical make, strength, and medical background for maximization of performance and minimizing chance of injury.

3.2 Implementation: How Personalized Footwear Solutions Work

In real-world application, data collection and processing have to support personalized footwear. In the first place, particular, detailed biomechanical information should be gathered on runners. Such facilities as motion capture systems, force plates, and wearable sensors will help to accomplish this (Hakim et al. 2024). Facilities listed provide measures of ground reaction forces, joint angles, and muscle activation patterns during running. Subsequent wearables, such as smart insoles or accelerometers, would further collect real-time data from training sessions and record dynamic changes in running mechanics across various topographies and velocities (Nasir et al. 2024; Sinclair and Taylor 2023). Biomechanical profiling is made in order to ensure data analytical work for every runner in a constructive way. Profiling variables involve patterns of foot strike, leg stiffness, stride length, and joint kinematics. It also considers personal factors related to a person, like weight, fitness level, and injury history, which all may affect how a runner could respond to particular shoe design features (Willwacher et al. 2024). From there, shoe companies can use that biomechanical profile to determine things like stack height, the characteristics of cushioning, and other design features, all based on the needs of the individual runner. For instance, a high-impact runner might get a higher stack height with softer midsole foam, while an efficient runner would have a lower stack height but firmer cushioning for more energy return (Kiesewetter et al. 2022). Advanced materials, such as carbon plates, can also be integrated into personalization to assist in increasing propulsion without inhibiting natural biomechanics (Lin et al. 2022). One of the major probable advantages is that the design of the shoe can change continuously as the biomechanics of the runner change. As an athlete trains and adapts, his or her biomechanics may change, too, hence changes may be indicated in his or her footwear. Wearable devices and machine-learning algorithms will detect these changes and inform immediate feedback to correct the prototype in real time or in iterations of the prototype design itself (Agresta et al. 2022a). Advanced footwear technology offers an optimized solution to runners through increased performance, improved running economy, and possibly a reduction in overuse injuries. The innovation is novel because it finally allows the gaps in the existing literature to be filled, informing how biomechanical variables interact for different runners' needs and event-specific demands (Willwacher et al. 2024).

4 Summary and Future Directions in Running Footwear Design

In summary, stack height optimization in running shoes is complex and essentially individualistic. The adoption of personalized footwear solutions through biomechanical profiling greatly improves running performance while limiting the risk of injury to a minimum. As pointed out by Bonanno et al. (2023), shoe design may come closer to meet the requirements of the individual runner thanks to state-of-the-art gait analysis technologies, wearables, and machine learning algorithms. This would take into consideration variability in the biomechanical response to stack height, thereby providing a clear line of improvement in both performance and safety. Several areas need further examination. First, longitudinal studies will need to be undertaken to track the effect of such customization on performance and injury rates in the long run. Attention must focus on how runners' bodies adapt to stack height changes over an extended period and whether such adaptations translate into sustained performance improvement or injury risk (Sinclair and Taylor 2023). Enumerate that future research will also explore how smart materials, allowing dynamic variations in stack height, can be integrated during a run. Such materials as shape-memory foams or variable-stiffness carbon plates could enable real-time optimization of cushioning and energy return based on the runner's speed, level of fatigue, and the terrain (Kiesewetter et al. 2022).

While most of the current methodologies of biomechanical profiling are primarily focused on simple variables like pattern of foot strike and leg stiffness, future research overviews should extend this to more specific factors such as pattern of muscle activation, joint loading, and fatigue threshold. Such additional data could further help with optimization of the personalization process and thus result in even more optimized designs of shoes (Agresta et al. 2022a). On the other hand, gender differences in biomechanics-which especially relate to the joints and muscle activation-indicate that there may be needs for gender-specific stack height optimizations. More research needs to be carried out to outline how such differences impact performance and injury risks for both male and female runners, since more specific shoe designs may be necessary (Willwacher et al. 2024). Besides biomechanics, other very important factors involve perceptual factors: a runner's comfort and perceived stability. Future studies may also investigate the interaction of perceptual factors, stack height, and biomechanical performance toward more holistic footwear customization (Lin et al. 2022). These would enable scientists and producers, through research, to work more on designing shoes for runners to satisfy a wide scope of runners. A very important aim in this regard would be to design a shoe that would maximize performance without compromising the long-term health and sustainability of running.

References

Abidin ANSZ, Yasin MRM, Hakim S, Aspar MASM, Jamal NZ (2024) Green technology in sports: the role of waste management in sustainable stadiums. Lecture Notes Bioeng 349–362. https://doi.org/10.1007/978-981-97-3741-3_32

Agresta C, Giacomazzi C, Harrast M, Zendler J (2022a) Running injury paradigms and their influence on footwear design features and runner assessment methods: a focused review to advance evidence-based practice for running medicine clinicians. Frontiers Sports Active Living 4. https://doi.org/10.3389/fspor.2022.815675

Babu D, Hakim S, Rosli AM, Yasin MRM, Rosle MH (2024) Harnessing solar energy: a green technology approach in sports facility management. Lecture Notes Bioeng 363–376. https://doi.org/10.1007/978-981-97-3741-3_33

Bertschy M, Lino H, Healey L, Hoogkamer W (2023) Effects of midsole stack height and foam on the metabolic cost of running. Footwear Sci 15(sup1). https://doi.org/10.1080/19424280.2023.2202942

Bonanno M, De Nunzio AM, Quartarone A, Militi A, Petralito F, Calabrò RS (2023) Gait analysis in neurorehabilitation: from research to clinical practice. Bioengineering (Basel, Switzerland) 10(7). https://doi.org/10.3390/BIOENGINEERING10070785

Carranza VR (2023) Running footwear matters: decoding the influence of running shoe characteristics on physiology, biomechanics and running performance (PhD Academy Award). Br J Sports Med 57(24):1581–1582. https://doi.org/10.1136/BJSPORTS-2023-107452

Casado A, González-Mohíno F, González-Ravé JM, Foster C (2022) Training periodization, methods, intensity distribution, and volume in highly trained and elite distance runners: a systematic review. Int J Sports Physiol Perform 17(6):820–833. https://doi.org/10.1123/IJSPP.2021-0435

Devi DH, Duraisamy K, Armghan A, Alsharari M, Aliqab K, Sorathiya V, Das S, Rashid N (2023) 5G technology in healthcare and wearable devices: a review. Sensors 23(5):2519. https://doi.org/10.3390/S23052519

Esculier JF, Dubois B, Dionne CE, Leblond J, Roy JS (2015) A consensus definition and rating scale for minimalist shoes. J Foot Ankle Res 8(1). https://doi.org/10.1186/S13047-015-0094-5

Gao K, Meng Z, Jiang K, Zhang H, Zeng Q (2020) Shearer height adjustment based on mechanical-electrical-hydraulic cosimulation. IEEE Access 8:222064–222076. https://doi.org/10.1109/ACCESS.2020.3043516

Hakim S, Nasir A, Abidin ANSZ, Yasin MRM, Jabbar WA (2024) Enhancing piezoelectric sensors in sports instrumentation: addressing durability with nano-coating techniques. Lecture Notes Bioeng 471–484. https://doi.org/10.1007/978-981-97-3741-3_41

Kiesewetter P, Bräuer S, Haase R, Nitzsche N, Mitschke C, Milani TL (2022) Do carbon-plated running shoes with different characteristics influence physiological and biomechanical variables during a 10 km treadmill run? Appl Sci 12(15):7949. https://doi.org/10.3390/app12157949

Knopp M, Muñiz-Pardos B, Wackerhage H, Schönfelder M, Guppy F, Pitsiladis Y, Ruiz D (2023) Variability in running economy of kenyan world-class and european amateur male runners with advanced footwear running technology: experimental and meta-analysis results. Sports Med 53(6):1255–1271. https://doi.org/10.1007/S40279-023-01816-1

Lin S, Song Y, Cen X, Bálint K, Fekete G, Sun D (2022) The implications of sports biomechanics studies on the research and development of running shoes: a systematic review. Bioengineering 9(10):497. https://doi.org/10.3390/bioengineering9100497

Moore IS, Jones AM, Dixon SJ (2012) Mechanisms for improved running economy in beginner runners. Med Sci Sports Exerc 44(9):1756–1763. https://doi.org/10.1249/MSS.0B013E318255A727

Nasir A, Babu D, Muhammad Sidik MH, Abdullah NB, Jabbar WA (2024) Advancement of IoT-based smart systems for real-time sports performance analysis: potential and challenges. Lecture Notes Bioeng 377–391. https://doi.org/10.1007/978-981-97-3741-3_34

Patoz A, Lussiana T, Breine B, Gindre C, Malatesta D (2022). Duty factor and foot-strike pattern do not represent similar running pattern at the individual level. Scientific Reports 12(1). https://doi.org/10.1038/S41598-022-17274-0

Perkins KP, Hanney WJ, Rothschild CE (2014) The risks and benefits of running barefoot or in minimalist shoes: a systematic review. Sports Health 6(6):475–480. https://doi.org/10.1177/1941738114546846

Pollard CD, Ter Har JA, Hannigan JJ, Norcross MF (2018) Influence of maximal running shoes on biomechanics before and after a 5K run. Orthopaedic J Sports Med 6(6). https://doi.org/10.1177/2325967118775720

Rosli AM, Azmi AN, Yasin MRM, Nasir A, Abdullah NB (2024a) Exploring the use of polyurethane as an engineering material in sports surfaces. Lecture Notes Bioeng 459–470. https://doi.org/10.1007/978-981-97-3741-3_40

Rosli AM, Jamaludin AS, Razali MNM, Yasin MRM, Hadi MA (2024b) Sustainable manufacturing through materials machining innovations—a review. Lecture Notes Mech Eng 665–674. https://doi.org/10.1007/978-981-97-0169-8_55

Ruiz-Alias SA, Pérez-Castilla A, Soto-Hermoso VM, García-Pinillos F (2023b) Influence of the world athletics stack height regulation on track running performance. J Strength Cond Res 37(11):2260–2266. https://doi.org/10.1519/JSC.0000000000004523

Ruiz-Alias SA, Jaén-Carrillo D, Roche-Seruendo LE, Pérez-Castilla A, Soto-Hermoso VM, García-Pinillos F (2023) A review of the potential effects of the world athletics stack height regulation on the footwear function and running performance. Appl Sci 13(21). https://doi.org/10.3390/APP132111721

Sinclair J, Taylor PJ (2023) Effects of running in minimal, maximal and conventional footwear on tibial stress fracture probability: an examination using finite element and probabilistic analyses. Computation 11(12):248. https://doi.org/10.3390/computation11120248

Valldecabres R (2023) The significance of measuring biomechanical variability in sports performance. Acad J Sports Sci Med 1(3). https://doi.org/10.33552/AJSSM.2023.01.000512

Wang C, Nitschke M, Wannop JW, Stefanyshyn D, Luckfiel T, Schlarb H, Koelewijn A (2023) Prediction of the effect of stack height on running biomechanics using optimal control simulation. Footwear Sci 15(sup1). https://doi.org/10.1080/19424280.2023.2199295

Willwacher S, Mai P, Helwig J, Hipper M, Utku B, Robbin J (2024) Does advanced footwear technology improve track and road racing performance? An explorative analysis based on the 100 best yearly performances in the world between 2010 and 2022. Sports Medicine Open 10(1):1–9. https://doi.org/10.1186/s40798-024-00683-y

Wearable Gait Analysis in Neurological and Musculoskeletal Disorders: From Monitoring to Prevention

Mohd Nizar Mhd Razali, Nurul Hasya Md Kamil, Muhamad Ridzuan Radin Muhamad Amin, Dewan Muhammad Nuruzzaman, and Ahmad Shahir Jamaludin

Abstract Wearable gait analysis technologies have emerged as significant tools in monitoring and preventing gait abnormalities in patients with neurological and musculoskeletal disorders. However, existing gaps related to accuracy, usability, and the transition from monitoring to preventive intervention have been identified. A comprehensive review of the state of wearable devices for gait analysis shows the potential of these technologies, particularly in conditions like Parkinson's disease and osteoarthritis. The need for reliable, real-time feedback and proactive injury prevention remains underexplored, especially in everyday environments. Several hypothetical solutions to improve wearable technology have been proposed, focusing on enhancing accuracy through multi-sensor integration and utilizing machine learning for predictive injury models. These advancements are designed to overcome the limitations of current devices, particularly the inconsistency in data collection and the lack of real-time intervention systems. The integration of multi-sensor wearable systems with machine learning algorithms is proposed as the best method for closing the gap. These systems allow for precise data collection and predictive modeling, enabling early detection of gait abnormalities and timely interventions. Such systems can offer personalized feedback, enhancing both rehabilitation and injury prevention in real-time. Future directions emphasize the importance of improving usability, conducting

M. N. M. Razali (✉) · N. H. M. Kamil · A. S. Jamaludin
Faculty of Manufacturing and Mechatronic Engineering Technology, Universiti Malaysia Pahang Al-Sultan Abdullah, Pekan, Pahang, Malaysia
e-mail: mnizar@umpsa.edu.my

M. R. R. M. Amin
Faculty of Electrical and Electronic Engineering Technology, Universiti Malaysia Pahang Al-Sultan Abdullah, Pekan, Pahang, Malaysia

A. S. Jamaludin
Centre for Advanced Industrial Technology, Universiti Malaysia Pahang Al-Sultan Abdullah, Pekan, Pahang, Malaysia

D. M. Nuruzzaman
Department of Mechanical Engineering, International University of Business Agriculture and Technology, Uttara Model Town, Dhaka, Bangladesh

M. H. A. Hassan et al. (eds.), *Proceedings of the 3rd International Conference on Innovation and Technology in Sports*, Lecture Notes in Bioengineering,
https://doi.org/10.1007/978-981-95-0804-4_39

larger longitudinal studies, and fully integrating wearable technology into healthcare systems. By advancing wearable gait analysis from passive monitoring to an active preventive tool, significant improvements in patient outcomes and quality of life are anticipated.

Keywords Wearable gait analysis · Neurological disorders · Musculoskeletal injury prevention · Multi-sensor systems · Machine learning

1 Wearable Gait Analysis for Neurological and Musculoskeletal Disorders: Current Challenges and Research Gaps

Wearable gait analysis devices have gained significance in the management and observation of neurological and musculoskeletal illnesses. These systems utilize sensors, including accelerometers, gyroscopes, and force sensors, to deliver real-time input on gait parameters, thus enabling early diagnosis, rehabilitation, and injury prevention. Notwithstanding recent breakthroughs, considerable obstacles persist in the translation of these devices into broad clinical practice and in elucidating their long-term effects on disease development.

Wearable technologies for gait analysis have shown significant potential in monitoring and assessing locomotion in persons with diverse illnesses, including Parkinson's disease, stroke, multiple sclerosis, anterior cruciate ligament (ACL) injuries, and osteoarthritis. Current study emphasizes the real-time monitoring capabilities of these devices and the potential benefits of generated feedback for both physicians and patients. For example, sensors like inertial measurement units (IMUs) and pressure-sensitive insoles are frequently employed to collect spatiotemporal aspects of gait, facilitating accurate evaluations in clinical and real-world environments (Mason et al. 2023; Bowman et al. 2021; Spencer et al. 2021).

Despite this, there are a number of ongoing obstacles that prevent the broad implementation of technologies that analyze gait through wearable devices or devices. One of the most important concerns is the precision and dependability of the data that is gathered from the various kinds of various sensors. Inconsistencies in metrics such as stride length, cadence, and gait cycle are frequently brought to light in comparative studies that involve research-grade and commercially available wearables (Liu et al. 2021). Furthermore, the usefulness of these devices in day-to-day living for people who currently have chronic diseases is restricted due to concerns regarding their usability. According to Keogh et al. (2021) and Mancini et al. (2021), accessibility can be hindered by complicated installation procedures and demanding maintenance requirements, particularly for individuals who are elderly or who have motor disabilities.

Rather than putting more of an emphasis on injury prevention, the existing body of research places more of an emphasis on the care of symptoms. This is another significant gap in the research. According to Prasanth et al. (2021), there is a growing demand for long-term research that investigates how wearable devices might be utilized not only to monitor deterioration in gait but also to intervene with the intention of preventing future disease progression or harm. Taking advantage of these gaps presents opportunities to improve the utilization of wearable devices in clinical settings as well as in settings that are situated in the home.

This review focuses on the disparity between gait monitoring and prevention with wearable devices. While wearable devices have demonstrated efficacy in measuring gait characteristics and guiding rehabilitation strategies, their potential as preventive instruments, especially for patients with neurological or musculoskeletal diseases, is still insufficiently investigated. Patients with Parkinson's disease or osteoarthritis may suffer falls or gradual joint degeneration resulting from undiscovered aberrant walking patterns in their early phases (Liu et al. 2021; Keogh et al. 2021). Therefore, it is essential to explore how wearables can transition from diagnostic tools to proactive systems that anticipate and reduce the risk of damage.

2 Hypothetical Solutions: Improving Accuracy and Usability in Wearable Gait Analysis

Various theoretical solutions can elevate wearable gadgets from simple monitoring instruments to proactive systems that aid in the prevention of gait-related injuries. These solutions emphasize the enhancement of precision as well as consistency in gait data, the integration of machine learning for predictive modeling, and the improvement of usability for both physicians and patients. This chapter analyzes these tactics and explores their potential to mitigate the constraints noted in prior research.

2.1 Improving Accuracy and Data Consistency

A primary problem in wearable gait analysis is guaranteeing the accuracy of data collected in nonclinical settings. Future study should implement systematic techniques like to those utilized in clinical research to refine and authenticate findings. An approach that shows promise for increasing data consistency and accuracy is machine learning. Models can be trained to reduce noise from daily activities, hence maintaining the reliability of measurements like as step count, stride length, and joint angles, even under diverse settings (Bowman et al. 2021).

Figure 1, exemplifies a systematic selection method derived from the PRISMA statement, which may facilitate enhancements in data quality within wearable device research. By employing a comparable methodology, researchers may guarantee that the data utilized for predictive modeling and gait analysis adheres to elevated criteria of reliability, hence promoting the development of robust and preventive biofeedback systems.

Another technique to enhance accuracy is to integrate many sensor types within a unified wearable system. Contemporary high-end gadgets incorporate inertial measurement units (IMUs) alongside pressure sensors; yet, opportunities for additional innovation persist. Hybrid systems that integrate wearable sensors with camera-based motion capture can offer a more thorough analysis of gait patterns (Prasanth et al. 2021). This dual strategy may be particularly efficacious for musculoskeletal illnesses, as early identification of minor movement aberrations could avert the emergence of more serious joint issues.

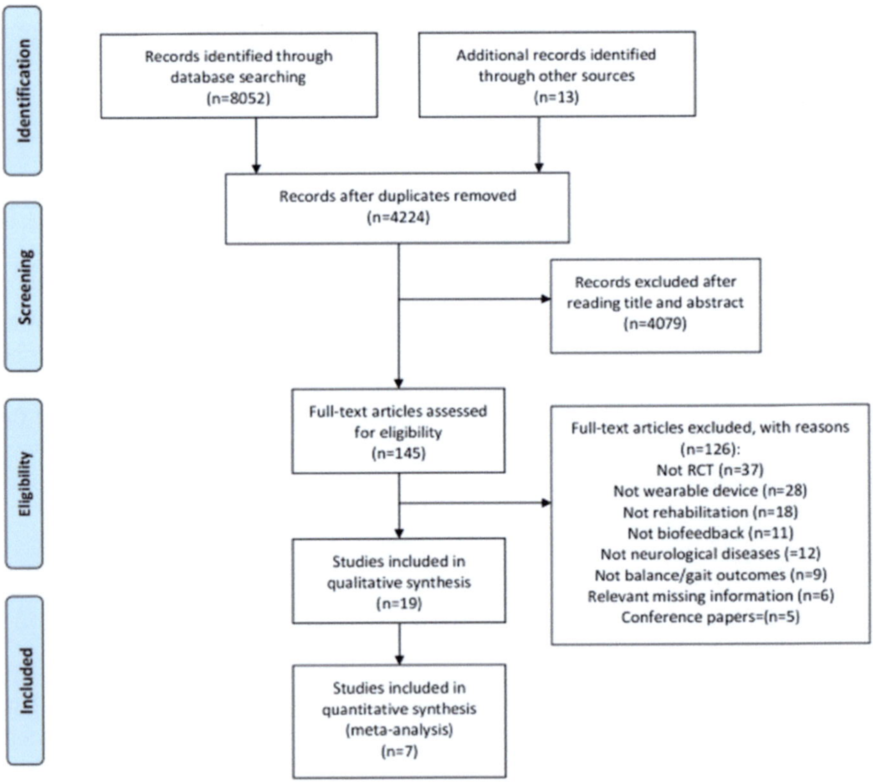

Fig. 1 Flow diagram based on PRISMA statement (Bowman et al. 2021)

2.2 Machine Learning for Predictive Injury Models

Machine learning has the potential to convert wearables into instruments that forecast impending injury or decline in gait. Through the collection of longitudinal data, algorithms can identify patterns or indicators that suggest a patient's gait may be approaching instability or that recurrent joint stress could result in injury (Liu et al. 2021). These prediction models may issue alarms to patients or healthcare practitioners, facilitating prompt action to prevent more severe consequences.

Predictive analytics are especially pertinent in neurological illnesses such as Parkinson's disease, where minor alterations in gait sometimes presage more significant motor complications or recurrent falls. These complications may be prevented or postponed through proactive modifications to therapy, exercise regimens, or medication (Mason et al. 2023; Bendig et al. 2022). Predictive modeling can also assist patients in avoiding specific movements that would otherwise exacerbate their condition, as is the case with musculoskeletal disorders which involve degenerative joints.

2.3 Enhancing Usability and Accessibility

Improving the usability of wearable gait analysis technology is also rather important, particularly for individuals with motor disabilities. Many research-grade technologies are challenging to fit into regular life since they need controlled clinical settings and many times considerable calibration (Keogh et al. 2021). Simplifying these systems and putting sensors into regular clothes or shoes will greatly help uptake in both clinical and domestic settings.

Pressure sensors included into shoe insoles, for instance, can track gait continuously without requiring constant setup over the day. Wireless sensor placements, including those shown in Fig. 2, show how gait and balance can be evaluated without burdening those who already suffer with mobility difficulties (Zampogna et al. 2020).

In addition, wearable devices that can synchronize with smartphone applications provide immediate feedback on gait metrics and propose suggestions for exercises or adjustments to posture (Liu et al. 2021). The ability to combine monitoring with intervention in real time helps to bridge the gap between data collecting and proactive rehabilitation. Wearable gait analysis systems, if designed with the user in mind and considering both technical precision and user-centered design, have the potential to develop into powerful preventative tools for managing neurological and musculoskeletal problems.

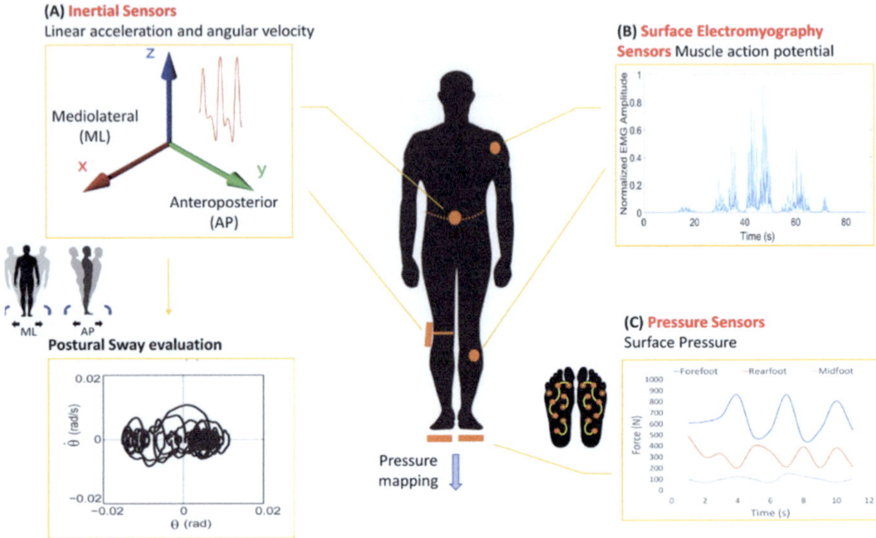

Fig. 2 Wireless sensor arrangement and equilibrium assessment (Zampogna et al. 2020)

3 Multi-Sensor Systems and Machine Learning: The Best Approach for Gait Monitoring and Injury Prevention

After evaluating various options, it has been determined that the best way to monitor gait and prevent accidents is to use a combination of three multi-sensor systems and machine learning. By merging wearable IMUs and predictive modeling techniques, this system ensures a high level of monitoring continuity from walking to prevention. This holistic approach will improve accuracy and give a predictive component that will avoid injuries and improve outcomes for individuals with neurological and musculoskeletal issues.

3.1 Multi-sensor Systems with Machine Learning for Predictive Models

Recent studies have showed the substantial potential that may be reached by mixing wearable IMUs with machine learning models (Bowman et al. 2021; Liu et al. 2021). This technique uses a range of sensors, including as accelerometers, gyroscopes, and pressure sensors, to capture various characteristics of movement. By combining several sensors into a single wearable device, the depth and reliability of the data that is acquired can be considerably improved. This breakthrough makes it easier to monitor a number of biomechanical parameters, such as joint angles, velocities, and ground reaction forces. As a result, the range of information that may be acquired is expanded.

The main advantage of this method is that it can detect small variations in gait that a single sensor system could miss. People who are recovering from musculoskeletal injuries usually notice small changes in the way they walk before they encounter obvious symptoms such as discomfort or weariness. Machine learning algorithms can examine this data to determine the risk of injury, which allows therapists to deliver preventative interventions, such as physical therapy and tailored movement adaptations (Liu et al. 2021).

3.2 Neurological Disorders: An Application of the Concept

This comprehensive technique also seems to be a promising and effective therapy option for neurological disorders, including Parkinson's disease. Gait abnormalities are usually not noticeable in the early stages of Parkinson's disease; nonetheless, they are an indication that motor function is soon to deteriorate. Mason et al. (2023) state that machine learning models that are trained on longitudinal data can detect tiny changes in movement and perform the necessary actions. The experts say that this can reduce the likelihood of falling and other undesirable outcomes. Personalized feedback can significantly improve the odds of recovery and the quality of life given by this model by constantly adjusting to data that is specific to each patient.

It is essential to provide fast feedback to both doctors and patients in order to treat various neurological conditions. Wearable devices that provide immediate support through a smartphone app will allow patients to make intelligent modifications to their movement patterns. This will assist the patient in managing their injuries independently and will lower the chances of future injuries (Keogh et al. 2021). The treatment of gait in individuals with neurological disorders could be greatly improved by a continuous cycle of monitoring, feedback, and action.

3.3 Musculoskeletal Applications and Injury Prevention

Multi-sensor systems that are integrated with machine learning have a lot of potential for treating musculoskeletal issues, including injuries to the anterior cruciate ligament (ACL) and osteoarthritis. Wearable sensors can detect when stress on joints increases by tracking essential biomechanical aspects. They can then provide users with advice on how to change their activity. These gadgets can identify gait asymmetries and compensatory inclinations after surgery. If these problems are not dealt with, it may lead to a longer recovery period or a re-injury (Hinton et al. 2023; Bowman et al. 2021; Skvortsov et al. 2023; Kantan et al. 2023).

Additionally, wearable electronics and environmental sensors can be used to address real-world issues that affect gait. A person's walking style can be affected by altering walking surfaces or by switching from one type of surface to another. As a result, a successful long-term follow-up plan must consider these environmental factors (Di Biase et al. 2020).

3.4 Closing the Gap: Practical Implementation

By integrating multi-sensor systems and predictive machine learning models, wearable devices will evolve from passive monitoring tools to active injury prevention tools. To achieve this transition, following measures can be taken:

Data Collection and Algorithm Development: The first step in gathering data and developing an algorithm is to create a complete database that includes people with a variety of neurological or musculoskeletal problems. This dataset uses gait patterns to improve machine learning algorithms that are designed to predict the risk of injury (Bowman et al. 2021; Liu et al. 2021; Skvortsov et al. 2021).

Real-Time Feedback Systems: After developing this algorithm, the next step is to include it into a wearable device that can provide real-time feedback. If the gait metrics indicate that there may be a problem, the patient will be notified so that appropriate intervention can take place (Hinton et al. 2023; Keogh et al. 2021).

Longitudinal Monitoring and Adaptation: Finally, wearable devices capable of continuous long-term monitoring need to be developed to adapt to changes in gait patterns over time. This approach would be particularly beneficial for people undergoing long-term rehabilitation after musculoskeletal injuries, as well as patients with progressive diseases such as Parkinson's disease (Mason et al. 2023; Di Biase et al. 2020).

This comprehensive technology is specifically designed for the monitoring and prevention of gait-related problems in patients with neurological and musculoskeletal disorders. By focusing on accuracy, adaptation, and proactive intervention effectively combines observational monitoring with proactive injury prevention.

4 Toward the Future of Wearable Gait Analysis: Advancements, Challenges, and Directions

Wearable gait analysis devices have shown great potential to improve the care of neurological and musculoskeletal disorders, particularly in injury rehabilitation and prevention. These devices can go beyond simple monitoring and can provide real-time input to both patients and doctors. This is made possible by integrating machine learning algorithms and multi-sensor systems for predictive modeling. If developed and implemented correctly, this technology can help identify early signs of gait abnormalities and prevent accidents before they happen.

Several critical advancements are essential to incorporate these findings into mainstream therapeutic practice. Improving the accuracy and uniformity of wearable devices by using multiple types of sensors continues to be a focus. The development of machine learning models that provide personalized feedback based on individual gait patterns has great potential to shift gait analysis from a reactive to a preventative approach.

4.1 Future Directions

While the current state of wearable technology for gait analysis is promising, there are several areas where future research should be focused:

Larger, Longitudinal Studies: Future research should encompass larger and more varied cohorts to enhance and confirm the predictive efficacy of machine learning algorithms. Prolonged data collection will be critical for detecting early deviations in gait and ensuring that interventions are timely and effective.

Usability Improvements: Improvements in usability and accessibility are particularly important for individuals with severe motor impairments. Designing systems that integrate seamlessly into everyday items such as footwear or clothing could substantially boost adherence and facilitate long-term use.

Integration with Healthcare Systems: Collaboration among device manufacturers, physicians, and policymakers is essential for the effective incorporation of wearable data into clinical practice. Standardized processes for data exchange, analysis, and interpretation are essential for promoting wider use.

Real-Time Adaptation and Intervention: As machine learning models continue to improve, future wearable devices should adapt in real time to changes in gait, offering personalized interventions. For instance, a device might suggest tailored exercises or modified activities to reduce injury risks.

In conclusion, wearable gait analysis devices possess significant potential for reshaping the management of gait disorders in neurological and musculoskeletal populations. By advancing sensor accuracy, improving usability, and refining predictive modeling, it is possible to develop systems that transition from passive observation to active prevention of gait-related injuries, thereby improving patient outcomes and overall quality of life.

Acknowledgements The authors would like to thank the Ministry of Higher Education (MoHE) Malaysia and Universiti Malaysia Pahang Al-Sultan Abdullah for the financial support provided under the Fundamental Research Grant Scheme with the Grant No: FRGS/1/2024/TK10/UMP/02/2 (UMPSA Internal reference: RDU240103) in completing this work.

References

Bendig J, Wolf A, Mark T, Frank A, Mathiebe J, Scheibe M, Müller G, Stahr M, Schmitt J, Reichmann H, Loewenbrück KF, Falkenburger BH (2022) Feasibility of a multimodal Telemedical intervention for patients with Parkinson's disease—a pilot study. J Clin Med 11(4):1074. https://doi.org/10.3390/jcm11041074

Bowman T, Gervasoni E, Arienti C, Lazzarini S, Negrini S, Crea S, Cattaneo D, Carrozza M (2021) Wearable devices for biofeedback rehabilitation: a systematic review and meta-analysis to design application rules and estimate the effectiveness on balance and GAIT outcomes in neurological diseases. Sensors 21(10):3444. https://doi.org/10.3390/s21103444

Di Biase L, Di Santo A, Caminiti ML, De Liso A, Shah SA, Ricci L, Di Lazzaro V (2020) GAIT analysis in Parkinson's disease: an overview of the most accurate markers for diagnosis and symptoms monitoring. Sensors 20(12):3529. https://doi.org/10.3390/s20123529

Hinton EH, Buffum R, Kingston D, Stergiou N, Kesar T, Bierner S, Knarr BA (2023) Real-time visual kinematic feedback during overground walking improves GAIT biomechanics in individuals post-stroke. Ann Biomed Eng 52(2):355–363. https://doi.org/10.1007/s10439-023-03381-0

Kantan PR, Dahl S, Jørgensen HR, Khadye C, Spaich EG (2023) Designing ecological auditory feedback on lower limb kinematics for hemiparetic GAIT training. Sensors 23(8):3964. https://doi.org/10.3390/s23083964

Keogh A, Argent R, Anderson A, Caulfield B, Johnston W (2021) Assessing the usability of wearable devices to measure gait and physical activity in chronic conditions: a systematic review. J NeuroEngineering Rehabil 18(1). https://doi.org/10.1186/s12984-021-00931-2

Liu X, Zhao C, Zheng B, Guo Q, Duan X, Wulamu A, Zhang D (2021) Wearable devices for GAIT analysis in intelligent healthcare. Front Comput Sci 3. https://doi.org/10.3389/fcomp.2021.661676

Mancini M, Shah VV, Stuart S, Curtze C, Horak FB, Safarpour D, Nutt JG (2021) Measuring freezing of gait during daily-life: an open-source, wearable sensors approach. J NeuroEngineering Rehabil 18(1). https://doi.org/10.1186/s12984-020-00774-3

Mason R, Godfrey A, Barry G, Stuart S (2023) Wearables for running gait analysis: a study protocol. PLoS ONE 18(9):e0291289. https://doi.org/10.1371/journal.pone.0291289

Prasanth H, Caban M, Keller U, Courtine G, Ijspeert A, Vallery H, Von Zitzewitz J (2021) Wearable sensor-based real-time GAIT detection: a systematic review. Sensors 21(8):2727. https://doi.org/10.3390/s21082727

Skvortsov DV, Kaurkin SN, Ivanova GE, Suvorov AY (2023) Targeted training of the function of walking according to the stance and single support phase in patients in the early recovery period of cerebral stroke. J Clin Pract 14(1), 31–43. https://doi.org/10.17816/clinpract112483

Spencer J, Wolf SL, Kesar TM (2021) Biofeedback for Post-stroke GAIT Retraining: A review of current evidence and future research directions in the context of emerging technologies. Front Neurol 12. https://doi.org/10.3389/fneur.2021.637199

Zampogna A, Mileti I, Palermo E, Celletti C, Paoloni M, Manoni A, Mazzetta I, Costa GD, Pérez-López C, Camerota F, Leocani L, Cabestany J, Irrera F, Suppa A (2020) Fifteen years of wireless sensors for balance assessment in neurological disorders. Sensors 20(11):3247. https://doi.org/10.3390/s20113247

Leveraging Virtual Reality to Optimize Mental Imagery and Skill Acquisition for Elite Athletes: A Review

Ainur Munira Rosli, Mohd Nizar Mhd Razali, Muhamad Rusydi Muhamad Yasin, Nurul Hasya Md Kamil, and Ahmad Shahir Jamaludin

Abstract Virtual Reality (VR) technology has emerged as a transformative tool in elite sports training. It offers unprecedented opportunities to enhance mental imagery and skill acquisition. Traditional mental imagery methods, which often rely on static visuals or videos, are limited in their ability to simulate the dynamic pressures of real-world competition. In contrast, VR simulations immerse athletes in high-fidelity environments, replicating emotional stressors such as virtual spectators, time constraints, and crowd noise, which are critical for developing decision-making and mental resilience under pressure. Neurofeedback-enhanced simulations further address gaps in current research by providing real-time cognitive feedback, allowing athletes to refine their focus and emotional regulation. High-fidelity VR environments also improve the transferability of learned skills from virtual scenarios to actual performance settings, bridging a significant gap in traditional methods. While these advancements demonstrate VR's potential to outperform conventional training techniques, challenges persist in standardizing training protocols and assessing long-term impacts on skill retention and psychological resilience. Future research should prioritize longitudinal studies and technological advancements, such as adaptive training systems and neurofeedback integration, to fully realize the potential of VR in revolutionizing elite sports training. These innovations open promising avenues for applications extending beyond skill acquisition to areas like injury rehabilitation and holistic athlete development.

Keywords Virtual reality (VR) · Elite sports training · Mental imagery · Skill acquisition · High-fidelity simulations · Neurofeedback

---

A. M. Rosli · M. N. M. Razali (✉) · M. R. M. Yasin · N. H. M. Kamil · A. S. Jamaludin
Faculty of Manufacturing and Mechatronic Engineering Technology, Universiti Malaysia Pahang Al-Sultan Abdullah, Pekan, Pahang, Malaysia
e-mail: mnizar@umpsa.edu.my

A. S. Jamaludin
Centre for Advanced Industrial Technology, Universiti Malaysia Pahang Al-Sultan Abdullah, Pekan, Pahang, Malaysia

M. H. A. Hassan et al. (eds.), *Proceedings of the 3rd International Conference on Innovation and Technology in Sports*, Lecture Notes in Bioengineering,
https://doi.org/10.1007/978-981-95-0804-4_40

1 The Rise of Virtual Reality in Elite Sports Training: Bridging the Gap Between Mental Imagery and Performance

Virtual Reality (VR) technology has emerged as a transformative tool in elite sports training, revolutionizing how athletes prepare for competition by enhancing mental imagery and skill acquisition. Over the years, mental imagery has been recognized as a critical psychological technique for improving athletic performance. It allows athletes to visualize movements, strategies, and outcomes, helping them in mental rehearsing and preparing for the challenges they will face in real-world competition (Bedir and Erhan 2021; Richlan et al. 2023). However, traditional visual practices, such as using static images, videos, or basic mental rehearsing, have a limitation. These methods lack the ability to replicate the complex environments and emotional pressures that athletes encounter in high-fidelity competitive scenarios. For example, traditional methods cannot simulate the cognitive load, or the emotional stress that is associated with the split-second decision making, crowd noise, or time constraints. What are frequently the determining elements of elite performance (Wu et al. 2023; Szary et al. 2020). The gap between conceptual preparation and practical implementation underscores the necessity for more sophisticated training techniques.

In this context, VR-based training systems are an exceptionally effective tool to improve these gaps. It puts the players into a high-fidelity, dynamic representations of problems that are not only visually realistic but also use important features of competition, a strategic decision under pressure, emotional arousal due to virtual crowd, and required adjustments required by unpredictable scenarios. Such an approach has marked improvement in mental conditioning, decision-making, and concentration; and the transfer of learned skills to the real-world performance contexts (Wu et al. 2023; Huang et al. 2022). This newly advanced neurofeedback VR system has significantly advanced training methodologies by offering real-time cognitive feedback, allowing athletes to track their focus and emotional regulation during training and, thus, improve them. Rectifying significant deficiencies in conventional mental visualization techniques (Farley et al. 2020; Wu et al. 2023). Figure 1 illustrates a user-friendly classification of Sport Performance–Neurofeedback Training schemes, which integrate neurofeedback with VR-based cognitive training. These approaches, including simulated recall, attention-focused tasks, relaxation techniques, and monitoring-guided training, enhance an athlete's ability to regulate focus, adapt to pressure, and optimize performance in unpredictable scenarios.

Moreover, an additional novel characteristic of VR-based training is its adaptivity. Adaptive training systems use machine learning algorithms to set the complexity of scenarios on real-time feedback about the athlete's performance. With such, the training remains to be challenging but not impossible, the approach in avoiding performance plateaus that may hinder the development of skills, and one that enhances continuous skill development (Xie et al. 2021; Strojny et al. 2023). It may also be very successful in creating personalized training sessions. Meeting individual cognitive flexibility and emotional strength, among other athlete demands,

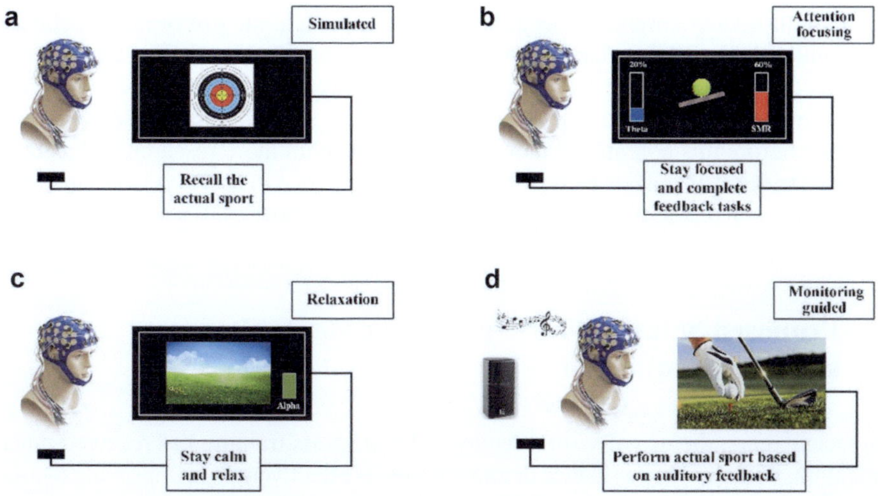

Fig. 1 Classification of Sport Performance–Neurofeedback Training schemes (Gong et al. 2021)

may be effectively provided by such systems. As a consequence, it may enhance the effectiveness of VR training because it addresses the problems related to differences in reactions to traditional forms of training.

Nevertheless, VR-based sports training faces many challenges. One of the major difficulties is the absence of established protocols in conducting VR training and evaluation of such training programs. This hinders the comparison of outcomes between research or the formulation of success criteria (Szary et al. 2020; Strojny et al. 2023). It is also for this reason that long-term effects are still under-researched although short-term studies indicate that VR is highly effective in improving mental imagery and the acquisition of skill. For example, skill retention, psychological resilience, and the transfer of trained skills through VR into actual competition would need further investigation. Longitudinal studies focusing on these aspects are critical to be conducted for a full understanding of VR's potential as a cornerstone of elite sports training (Richlan et al. 2023; van der Meer et al. 2023).

Despite all these problems, the area of VR sports-based training is very fruitful with opportunities for future development. New technologies like AI-based personalization, haptic feedback, and improved neurofeedback systems can be integrated to allow more thorough training experiences. Such breakthroughs will enable athletes to gauge their mental and physical conditions in real time while practicing in the conditions that closely mirror real-world pressures. Furthermore, VR's applications are not limited to skill acquisition; they extend to areas such as injury rehabilitation and recovery training. For example, injured players may use VR to simulate sports movements as activities in an engaging environment to keep up readiness both mentally and physically during inactivity (Huang et al. 2022; Gani et al. 2023).

In summary, VR represents advanced sports-technological-promises in the use of elite sports training. Overcoming the limitations of mental imagery practices by using high-fidelity simulation as an innovative approach, the use of neurofeedback as an enhancement for such practices, and the adoption of adaptive learning systems, is indeed going to change the preparatory processes of an athlete for competition. It, however, has challenges in standardization, individual variation, and long-term efficacy before this potential can ever be achieved. With more research and technological strides, VR may as well revolutionize the practice in sports performance and define new standards of excellence in athletic preparation.

2 Proposed Solutions for Optimizing Mental Imagery and Skill Acquisition

In recent years, the use of virtual reality (VR) in sports training has received much attention, with growing evidence demonstrating its effectiveness in improving mental imagery and skill acquisition. While traditional mental imagery practices based on cognitive psychology have long been used to improve athletic performance by allowing athletes to visualize movements, scenarios, and strategies (Bedir and Erhan 2021 years, Richlan et al. 2023). These methods are limited because they rely on static imagery and mental rehearsals, which cannot simulate the dynamic and complex conditions present during competition. For example, traditional approaches do not include key factors that create stress, such as crowd noise, time pressure, and the need to make quick decisions under pressure. This limitation has led to the search for newer technologies, such as VR, which can provide greater realism and immersion.

2.1 High-Fidelity VR in Sports Training

VR is considered as a tool which would revolutionize the traditional methodologies related to sports training. In VR, athletes or sportspeople are placed within a high-fidelity simulation environment, which can provide them with the optimum level of practicing skills related to competitions. Unlike the mental imagery techniques that were used before, VR ensures a multisensory experience concerning visual and auditory perception and sometimes even haptic feedback to provide a holistic learning environment. This type of immersiveness would not only enhance creativity but help in mental organization as well as decision-making and emotional resilience (Wu et al. 2023; Huang et al. 2022). Studies have emphasized the advantages of VR over traditional training methods, especially in simulating high-pressure competitive scenarios. For instance, stress conditions that can be emulated by VR systems include virtual spectators, unpredictive opponents and a high-stakes time limit. Such simulations have been found to enhance focus of the athletes, cognitive adaptability, and the ability to perform under pressure (Farley et al. 2020; Gani et al. 2023). More than that, the ability of VR to simulate emotional stress fills in a critical gap of traditional training that mostly does not respect the psychological demands of elite competition. Because of this, VR exposes athletes to stress during their training sessions, thus building mental resilience for them, thus preparing them for the high-pressure environment of real events.

2.2 *Adaptive VR Training Programs*

An important advance in VR sports training comes with the innovation of adaptive training systems. These use machine learning algorithms to modify the training difficulty at the very moment the athlete is engaging with it based on his or her level of skill. Adaptive VR training allows athletes to be continuously challenged, promote continuous skill acquisition, and avoid performance plateaus (Xie et al. 2021; Strojny et al. 2023). For example, if an athlete shows improvement in decision-making under time pressure, the system may introduce additional challenges, such as faster pace of play or more unpredictable opponents. This level of personalization not only improves the effectiveness of training but also addresses the needs of each athlete. Adaptive learning has also been widely applied in other fields, such as nursing education and collaborative learning. Showing great success in improving learning outcomes (Liu et al. 2023; van der Meer et al. 2023). The use of similar adaptive VR systems in sports training represents a novel approach to tailoring training experiences for individual athletes while addressing variability in cognitive and emotional responses to stress. This personalization is especially important in elite sports, where even small differences in mental state can have a significant impact on performance outcomes.

2.3 *Neurofeedback-Enhanced VR Systems*

Integrating neurofeedback into VR is an additional novelty in the segment of techniques for the improvement of mental imagery and creativity. Non-invasive brain monitoring technology, such as EEG neurofeedback enhancement systems, gives feedback on real-time information on attention, emotional arousal, and cognitive condition of an athlete. This assumption lets the athlete improve their mental state; mastery over emotional and cognitive responses during stress conditions can be developed. For example, if an athlete shows onset signs of increasing stress during the competition, then the system can immediately alert and, therefore, help the athlete to develop stronger management strategies. Neurofeedback systems have also been applied in other increasingly important fields, like nursing and rehabilitation, showing their use (Liu et al. 2023; Kilcioglu et al. 2023). In sports, these systems are a major advantage in bridging the gap between mental imagery and actual performance under stress. Combining real-time cognitive feedback with high-fidelity VR simulation, neurofeedback enhancement systems provide a powerful tool for enhancing mental strength and physical performance capacity.

2.4 Challenges in Standardizing VR Training

Despite its numerous advantages, VR-based sports training faces several challenges, particularly in the standardization of protocols. The lack of standardized frameworks makes it difficult to compare results across studies, hindering the development of universally accepted best practices (Szary et al. 2020; Strojny et al. 2023). For instance, differences in VR hardware, training scenarios, and evaluation criteria can lead to inconsistent outcomes, making it challenging to assess the true efficacy of VR training programs. Moreover, variability in athlete responses to VR training further complicates standardization efforts. Factors such as prior experience with VR technology, cognitive adaptability, and emotional resilience can significantly influence training outcomes, underscoring the need for tailored solutions that account for these individual differences. Adaptive training systems and neurofeedback mechanisms offer promising approaches to addressing these challenges, providing more personalized and reliable training experience.

2.5 The Need for Longitudinal Studies

While the short-term benefits of VR in sports training are well-documented, the long-term impacts remain underexplored. Questions about the durability of skill retention, the development of psychological resilience, and the overall influence of VR training on an athlete's career trajectory require further investigation. Longitudinal studies focusing on these aspects are essential to fully understand the potential of VR as a transformative training tool (Richlan et al. 2023). Additionally, long-term research can help identify the optimal balance between VR training and traditional methods, ensuring that athletes benefit from the strengths of both approaches. By providing data on the sustained effects of VR-based training, these studies will play a crucial role in shaping the future of elite sports preparation.

3 High-Fidelity VR Environments: The Key to Enhancing Skill Transfer and Decision-Making in Elite Athletes

High-fidelity Virtual Reality (VR) environments have introduced a potentially disruptive innovation in elite sports training by offering unprecedented levels of realism and interactivity. These environments enhance skill acquisition, mental imagery, and decision-making of the athletes. Unlike conventional methods that rely on mainly reliant on passive viewing, repetitive drills, or simple video. VR provides the athlete with realistic environments involving crowd noise, time constraints, and high-pressure decision-making. This helps the athletes to adapt to competitive pressures and enhances their performance (Wu et al. 2023; Farley et al. 2020).

3.1 Realism and Emotional Engagement

A major strength of VR is its ability to induce emotionally meaningful lifelike situations such as the presence of virtual spectators, dynamic behaviors of the opponent, and realistic auditory feedback contribute to raising the levels of immersion, making athletes feel as though they are competing in actual events. This emotional engagement enhances mental arousal, a critical factor for performance under pressure. Findings reveal that VR-trained athletes display better emotional resilience and correctness of decisions in high-pressure situations as compared to their traditionally trained counterparts (Wu et al. 2023; Huang et al. 2022). For instance, VR can simulate high-stakes situations like penalty shootouts or last-second basketball shots, preparing athletes for the psychological demands of competition while improving their ability to focus and perform under stress. These immersive experiences also enhance skill retention by encouraging athletes to internalize behaviors practiced in high-pressure environments (Huang et al. 2022).

3.2 Transferability of Skills

VR allows the athletes to perform an effective knowledge transfer obtained from virtual practice into the real execution. Realism leads to the explicit use of cognitive and motor skills trained under VR in actual competition. Studies prove that VR-trained athletes really enhance their skills related to the reaction spatial abilities, and tactical decision that can be applied to actual situations (Xie et al. 2021; Kilcioglu et al. 2023). This application is very important in the sports sector where quick decisions are required, such as basketball, soccer, or tennis. Athletes trained under VR consistently outperform those trained under conventional methods since VR can stimulate all the conditions of the game and decision-making processes during skill application in a manner analogous to real conditions (Richlan et al. 2023; Liu et al. 2023). Additionally, the heightened engagement and realism in VR enhance long-term skill retention, making it a superior alternative to traditional training.

3.3 Cognitive and Tactical Benefits

Apart from the development of physical skills, the use of VR enhances the development of cognitive and tactical skills amongst athletes. In this context, immersive environment can help athletes develop their decision-making ability as well as adaptability to spot recurrent situations derived from different strategies. For example, VR basketball training may involve passing the ball or shooting within milliseconds, as well as a defensive action keeping the movement of allies and adversaries in consideration. Such training facilitates situational awareness and visual search behavior—important cognitive factors that can enhance anticipation and thus response to dynamic game play (Wu et al. 2023; Farley et al. 2020).

3.4 Addressing Gaps and Limitations

Despite its numerous advantages, the implementation of high-fidelity VR environments comes with their own set of challenges. A major drawback is the varied reactions of athletes towards VR training. Their adaptiveness on cognitive terms, emotional stability, and familiarity with the use of VR will determine how well they will get involved and benefit from the VR simulation. For example, some athletes may not like it because they do not know what is going on and what should be expected. Other athletes may prove very fast learners and increase their performance drastically. It is, therefore, important that VR systems should be adaptive so that the training scenarios can vary according to the needs of each athlete (Xie et al. 2021; Strojny et al. 2023).

Another challenge lies in the absent rule set or standardized procedure for assessing high-fidelity VR training effectiveness. It proves challenging to compare outcomes across studies or to set benchmarks for success as hardware, software, and training methodologies differ. More studies should focus on developing standardized frameworks with regularized performance measures, for instance, considerations of the accuracy of decision-making abilities, reaction time, or skill retention rates (Szary et al. 2020; Richlan et al. 2023). Moreover, while the short-term advantages of VR training are well understood, the long-term impacts are not yet explored. Questions about the durability of skill transfer and the growth of mental strength over time require further investigation through longitudinal studies. Such research would provide good insights into the sustainable effects of VR training, helping coaches and practitioners to optimize its use in athletic development (Richlan et al. 2023).

The future of high-fidelity VR environments in elite sports training lies in the integration of advanced technologies, such as neurofeedback, personalization through AI, and haptic feedback systems. These innovations will ensure that the effectiveness of VR training sessions is maximized as they shall be able to continuously provide feedback on the cognitive and emotional states of athletes, allowing for more accurate

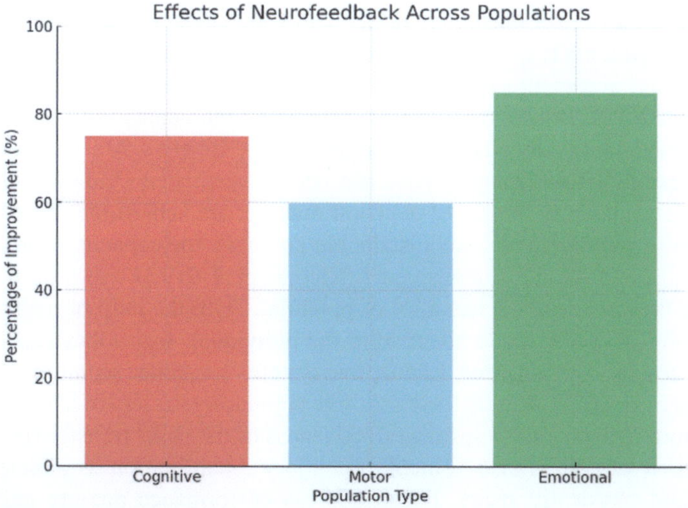

Fig. 2 Neurofeedback-induced improvements in cognitive, motor, and emotional performance in athletes (Tosti et al. 2024)

adjustments of scenarios under training. As an example, during a simulation, neurofeedback could be used to monitor an athlete's stress levels, allowing the system to adapt and provide a specific intervention to improve emotional regulation (Wu et al. 2023). As illustrated in Fig. 2, the highest percentage of improvement was observed in emotional regulation, followed by cognitive and motor performance, highlighting its potential as a tool for optimizing both mental resilience and physical coordination in sports. Also, neurofeedback training was particularly effective in precision-based sports like shooting and golf, while the mixed protocol yielded the highest improvements across multiple sports disciplines (Tosti et al. 2024).

Moreover, VR application in injury rehabilitations can offer exciting opportunities for keeping athletes both mentally and physically prepared during recovery. Real-world sports movements can be simulated in a safe environment, VR helping to keep injured athletes involved in their training, thus guaranteeing a better transition back to full competition (Huang et al. 2022).

4 Summary on Revolutionizing Sports Training with VR

Virtual Reality (VR) has already redefined elite sports training by providing an immersive environment for mental imagery, skill acquisitions, and performance. While it may be clear that it has an advantage over traditional methods, standardization, the variability of individuals, and limited long-term data remains formidable challenges. Overcoming these issues is the key unlock the maximum potential of VR.

A further challenge lies in the standard rules of engagement and assessment benchmarks within which VR sports training should take place. Diverse nature in the hardware, software, and methodology makes it difficult to conduct comparison studies (Szary et al. 2020; Strojny et al. 2023). Moreover, the adaptability and emotional control of the athlete may also influence the result (Xie et al. 2021). The long-term effects of using VR for training have not been investigated. Although immediate positive effects, such as improved decision-making, are achievable, the question of whether these improvements are sustainable and have the capacity to influence an athlete's full career remains unanswered (Richlan et al. 2023).

Despite these issues, VR has a lot of potential. With the help of neurofeedback and AI, training can be made adaptive to the individual, with focus capability and emotions measured in real-time. Its adaptive systems can adjust scenarios to improve results (Wu et al. 2023). Haptic feedback, like simulating the feel of a ball, makes the training more realistic and helps the effectiveness of the skills transfer (Huang et al. 2022). Moreover, VR is also valuable for injury rehabilitation as athletes remain mentally and physically ready. It can address performance anxiety and burnout through therapeutic environments and, thus, proves to be a very versatile tool for athlete development (Huang et al. 2022).

The future of VR lies in developing standardized protocols and conducting longitudinal studies will be essential to ensure consistent evaluations and understand the long-term benefits of VR training (Szary et al. 2020; Richlan et al. 2023). As VR technology evolves, technological development in neurofeedback, AI systems, and haptic feedback systems will deliver more advanced tools for personalization and optimization to further personalize and enhance training experiences. In summary, Further, collaboration between researchers, coaches, and organizations will ensure the fast integration of VR technology within the elite sports training and, hence, establish itself as a cornerstone of modern sports performance.

Acknowledgements The authors would like to thank the Ministry of Higher Education (MoHE) Malaysia and Universiti Malaysia Pahang Al-Sultan Abdullah for the financial support provided under the Fundamental Research Grant Scheme with Grant No: FRGS/1/2024/TK10/UMP/02/2 (UMPSA internal reference: RDU240103) in completing this work.

References

Bedir D, Erhan SE (2021) The effect of virtual reality technology on the imagery skills and performance of target-based sports athletes. Front Psychol 11:2073. https://doi.org/10.3389/fpsyg.2020.02073

Farley ORL, Spencer K, Baudinet L (2020) Virtual reality in sports coaching, skill acquisition and application to surfing: a review. J Hum Sport Exerc 15(3):535–548. https://doi.org/10.14198/jhse.2020.153.06

Gani RA, Setiawan E, Achmad IZ, Aminudin R, Purbangkara T, Hofmeister M (2023) Virtual reality-based Tabata training: a professional method for changing levels of physical fitness and psychological well-being in student-athletes. Pedagog Phys Cult Sport 27(2):91–101. https://doi.org/10.15561/26649837.2023.0204

Gong A, Gu F, Nan W, Qu Y, Jiang C, Fu Y (2021) A review of neurofeedback training for improving sport performance from the perspective of user experience. Front Neurosci 15:638369. https://doi.org/10.3389/fnins.2021.638369

Huang Z, Choi D-H, Lai B, Lu Z, Tian H (2022) Metaverse-based virtual reality experience and endurance performance in sports economy: mediating role of mental health and performance anxiety. Front Public Health 10:789614. https://doi.org/10.3389/fpubh.2022.991489

Kilcioglu S, Schiltz B, Araneda R, Bleyenheuft Y (2023) Short- to long-term effects of virtual reality on motor skill learning in children with cerebral palsy: systematic review and meta-analysis. JMIR Serious Games 11:e42067. https://doi.org/10.2196/42067

Liu K, Zhang W, Li W, Wang T, Zheng Y (2023) Effectiveness of virtual reality in nursing education: a systematic review and meta-analysis. BMC Med Educ 23(1):1–10. https://doi.org/10.1186/s12909-023-04662-x

Richlan F, Weiß M, Kastner P, Braid J (2023) Virtual training, real effects: a narrative review on sports performance enhancement through interventions in virtual reality. Front Psychol 14:1240790. https://doi.org/10.3389/fpsyg.2023.1240790

Szary P, Buchta K, Wrzeciono A, Czech O, & Rutkowski S (2020) What virtual reality can do for sport? A narrative review of the literature. Slovak Journal of Sport Science, 6(1), 108.

Strojny P, Dużmańska-Misiarczyk N (2023) Measuring the effectiveness of virtual training: a systematic review. Computers & Education: X Reality 2:100006. https://doi.org/10.1016/j.cexr.2023.100006

Tosti B, Corrado S, Mancone S, Di Libero T, Carissimo C, Cerro G, Rodio A, da Silva VF, Coimbra DR, Andrade A, Diotaiuti P (2024) Neurofeedback training protocols in sports: a systematic review of recent advances in performance, anxiety, and emotional regulation. Brain Sci 14(10):1036. https://doi.org/10.3390/brainsci14101036

Van der Meer N, van der Werf V, Brinkman W-P, Specht M (2023) Virtual reality and collaborative learning: a systematic literature review. Front Virtual Rity 4:1159905. https://doi.org/10.3389/frvir.2023.1159905

Wu Y, Lukosch S, Lukosch H, Lindeman RW, McKee RD, Fukuden S, Ross C, Collins D (2023) Training mental imagery skills of elite athletes in virtual reality. Front Virtual Rity 4:1189717. https://doi.org/10.3389/frvir.2023.1189717

Xie B, Liu H, Alghofaili R, Zhang Y, Jiang Y, Lobo FD, Li C, Li W, Huang H, Akdere M, Mousas C, Yu L (2021) A review on virtual reality skill training applications. Front Virtual Rity 2:645153. https://doi.org/10.3389/frvir.2021.645153

Data-Driven Strategies for Injury Prevention in Sports: Integrating Biomechanics and Wearable Analytics

Amirul Hakim Sufian, Mohd Nizar Mhd Razali, Nurul Hasya Md Kamil, Ahmad Redza Ahmad Mokhtar, and Ahmad Shahir Jamaludin

Abstract Injury prevention in sports has gained significant attention due to the increasing physical demands placed on athletes. A data-driven approach is seen as essential in mitigating injury risks. The integration of biomechanics, wearable technology, and advanced data analytics presents a promising solution for addressing this issue. Wearable devices equipped with sensor technology can collect real-time physiological and biomechanical data. This information, processed through machine learning models, enables the identification of injury risks and provides personalized insights based on athletes' unique profiles. Research has revealed that wearable technology and machine learning models can predict potential injuries by identifying deviations in biomechanical patterns that traditional methods may overlook. Machine learning algorithms such as unsupervised learning models have proven effective in detecting anomalies and clustering high-risk movements. Despite the advancements in these technologies, a gap remains in translating data into actionable, effective injury prevention strategies. It is suggested that enhancing machine learning models for real-time, personalized injury prediction will bridge this gap. The continuous refinement of predictive algorithms as more data is gathered holds potential for improved accuracy in injury prevention. Real-time feedback systems and the integration of external workload metrics with internal physiological data also contribute to mitigating injury risks. However, challenges such as cost, accessibility, and technical expertise need to be addressed to broaden the applicability of these technologies. Future research should focus on improving the affordability of wearables, enhancing model accuracy, and integrating more sophisticated physiological markers to create even more effective injury prevention strategies.

A. H. Sufian · M. N. M. Razali (✉) · N. H. M. Kamil · A. R. A. Mokhtar · A. S. Jamaludin
Faculty of Manufacturing and Mechatronic Engineering Technology, Universiti Malaysia Pahang Al-Sultan Abdullah, 26600 Pekan, Pahang, Malaysia
e-mail: mnizar@umpsa.edu.my

A. S. Jamaludin
Centre for Advanced Industrial Technology, Universiti Malaysia Pahang Al-Sultan Abdullah, 26600 Pekan, Pahang, Malaysia

M. H. A. Hassan et al. (eds.), *Proceedings of the 3rd International Conference on Innovation and Technology in Sports*, Lecture Notes in Bioengineering,
https://doi.org/10.1007/978-981-95-0804-4_41

Keywords Injury prevention · Biomechanics · Wearable technology · Machine learning · Personalized injury prediction

1 The Role of Data Analytics and Biomechanics in Injury Prevention: Bridging the Research Gap

The risk of injuries to the muscles, bones, and soft tissue increases significantly in competitive sports, as athletes are subjected to intense physical demands. Preventing these injuries has become a major concern. These injuries not only impact athletic performance, but also lead to significant economic and human losses, particularly in professional sports leagues like the NBA, NFL, and FIFA. To address this challenge, AI-driven models have become important tools to understand how injuries happen. It uses data analytics, biomechanics, and wearable technology to provide precise interventions that are specific to each athlete through data visualization (Papageorgiou et al. 2024; Liu et al. 2021).

Biomechanics provides an important perspective on injury prevention by analyzing the mechanical aspects of human movement. Figure 1, shows the biomechanics and wearable data collection framework. By using wearable devices equipped with sensors, it can collect real-time data on athletes' movements, muscle activity, and posture, which provides a detailed picture of how improper form and fatigue can lead to musculoskeletal injuries (Kovoor et al. 2024; Seshadri et al. 2021, by Preotani et al. 2022). This approach integrates subjective coaching with evidence-based interventions by transforming basic biomechanical data into a practical strategy, improving performance and reduces the risk of injury. With the continued development of these technologies, they will provide real-time monitoring and feedback, enabling coaches and athletes to implement immediate adjustments to improve performance and safety of the athletes (Secçkin et al. 2023).

Additionally, wearable technology with sensor-enhanced devices is advancing in its ability to capture a variety of physiological and biomechanical data points, such as heart rate variability, joint movement, and muscle activity. This information can provide insight into injury risk through the integration of machine learning to help risks predictions. For example, sports-specific Inertial Measurement Units (IMUs), a specific device that track shoulder mechanics, have proved that these devices are beneficial in minimizing the risk of injury (Rawashdeh et al. 2016). According to Pretoni et al. (2022), the combination of wearable technology and biomechanics is a step toward more informed preventive strategies in sports science. As shown in Fig. 2, wearable technology operates through a three-layer architecture that integrates sensors, processing units, and network systems to enable real-time monitoring and predictive injury assessment. Moreover, study on basketball injuries, wearable technology can provide early warning signals of soft tissue injuries and poor biomechanics, which allows for fast intervention (Papageorgiou et al. 2024).

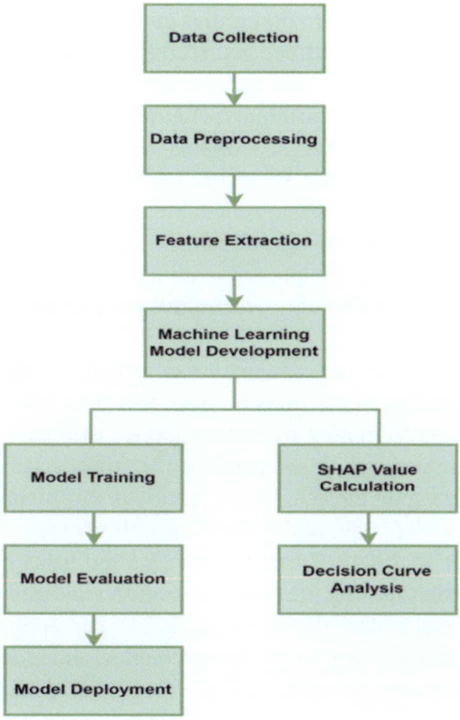

Fig. 1 Biomechanics and wearable data collection framework (Kavoor et al. 2024)

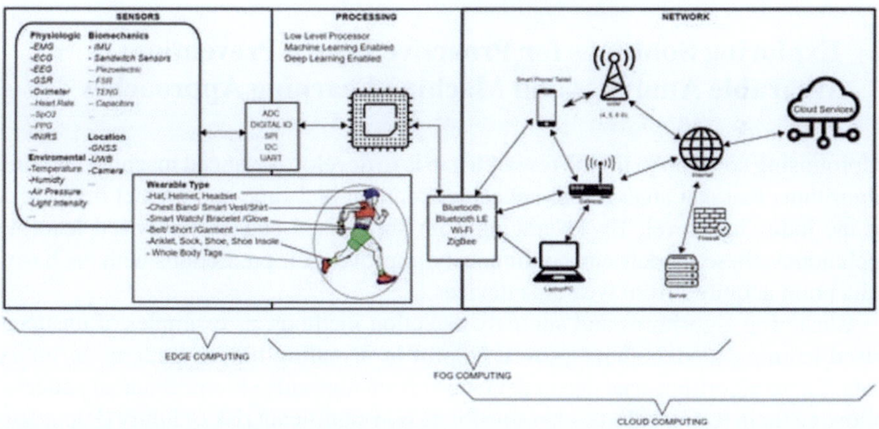

Fig. 2 Three-layer wearable technology, comprising sensor, processing, and network layers for real-time data collection and analysis (Seçkin et al. 2023)

As shown in the studies conducted by Querido et al. (2021), in addition to predictive models, structured recovery protocols play a key role in athletes' recovery. In professional sports such as soccer, a comprehensive approach to post-match recovery incorporates different techniques such as cooling therapy, compression garments and neuromuscular assessments to restore athletic performance and minimize the risk of future injuries. Wearable technologies contribute to this protocol by continuously monitoring physiological indicators and allowing adjustments to recovery strategies based on real-time feedback. In another studies, Hybrid AI and machine learning models enable proactive injury risk assessment by identifying subtle biomechanical patterns, allowing preemptive interventions that minimize overuse injuries (Liu et al. 2021).

Despite advancements in wearable technology and biomechanics, a notable gap persists in the translation of raw data into effective strategies for injury prevention. Contemporary wearables gather extensive biomechanical and physiological data, however, there is a necessity for improved and refined predictive models to accurately interpret these datasets. Although promising trends are emerging, a validated framework for leveraging these insights to reduce injury rates has yet to be developed (Sarlis et al. 2024). Current research often lacks personalized injury prevention strategies. Highlighting the need for an individualized approach tailored to the biomechanics of each athlete (Kovoor et al. 2024). Personalized insights would allow athletes to tailor their training loads, biomechanics, and recovery routines. Representing a significant advance for individual athlete health management (Preatoni et al. 2022).

2 Exploring Solutions for Proactive Injury Prevention: Wearable Analytics and Machine Learning Approaches

A promising solution to fill the research gap is to develop advanced machine learning algorithms that can analyze biomechanical data in real time and predict injury risk at the individual level. By leveraging both supervised and unsupervised learning techniques, these models can continuously improve their predictions with each new data point acquired from wearable devices.

Clustering algorithms and anomaly detection methods, as examples of unsupervised learning models, have proven helpful in revealing hidden patterns in injury data. These algorithms can detect deviations from normative biomechanical patterns, allowing them to issue alerts when an athlete is at significant risk of injury (Papageorgiou et al. 2024). Furthermore, prediction algorithms utilizing neural networks can forecast injuries by analyzing the complex interrelations among many biomechanical factors. An alternative method involves delivering real-time biomechanical feedback during training. Athletes can obtain immediate feedback on their posture, movement dynamics, and physical exertion by combining wearable devices with augmented reality (AR) and a smartphone application (Seshadri et al. 2021). Such devices can alert athletes when they engage in activities that are hazardous or inefficient. Consequently, they can implement modifications prior to the occurrence of any injuries. Athletes can reduce the likelihood of acute injuries, including sprains, strains, and fractures, by incorporating critical data such as joint angles, muscle vibration, and force generation that gathered via wearable sensors with analytics driven by artificial intelligence.

In addition, integrating external workload indicators, for example, mileage, intensity, training frequency with internal physiological response such as heart rate, oxygen consumption, muscle fatigue can also improve injury prevention. Synthesizing both types of metrics in data analysis models can provide a more comprehensive understanding of an athlete's injury risk (Seshadri et al. 2021).

Wearable sensors that track workload and physiological strain can predict fatigue-related injuries by detecting signs of excessive strain without sufficient recovery (Kovoor et al. 2024). Continuous collection of biomechanical and physiological data minimizes the risk of overtraining or undertraining and promotes a balanced training plan. Wearable IMUs are particularly useful in sports with repetitive or intense joint movements. It can collect key information on markers of muscle fatigue and strain (Rawashdeh et al. 2016). This enables personalized adjustments in training intensity, with real-time alarms to avoid overtraining and support balanced motor development and recovery (Preatoni et al. 2022).

Finally, data sharing between sports and leagues may allow better access to more comprehensive data, which in the end might refine strategies for preventing injuries. Acquiring information from the varied disciplines covered could enable the researchers to spot trends across sports and thereby provide an opportunity to develop models that may be applied among different athletic conditions (Seshadri et al. 2021). This technique has already been successfully applied in sports medicine research to provide multi league composite data supporting the biomechanical factors associated with specific types of injuries. Moreover, collaborative data sharing would accelerate the process of validating predictive models in any sports environment.

3 Machine Learning for Personalized Injury Prediction: A Data-Driven Approach to Closing the Gap

Advanced machine learning algorithms which tailored for injury prediction, offer a practical solution to the insufficiency of knowledge concerning the prevention of injuries in sports. These models fuse biomechanical and physiological big data, allowing athletes to build individual profiles reflecting their degree of susceptibility to injuries. By leveraging personal movement patterns, historical injury data, and real-time physiological monitoring (Kovoor et al. 2024; Papageorgiou et al. 2024), these algorithms deliver tailored insights that can predict specific injury risks. As shown in Fig. 3, the big data-based sports visualization model enables structured

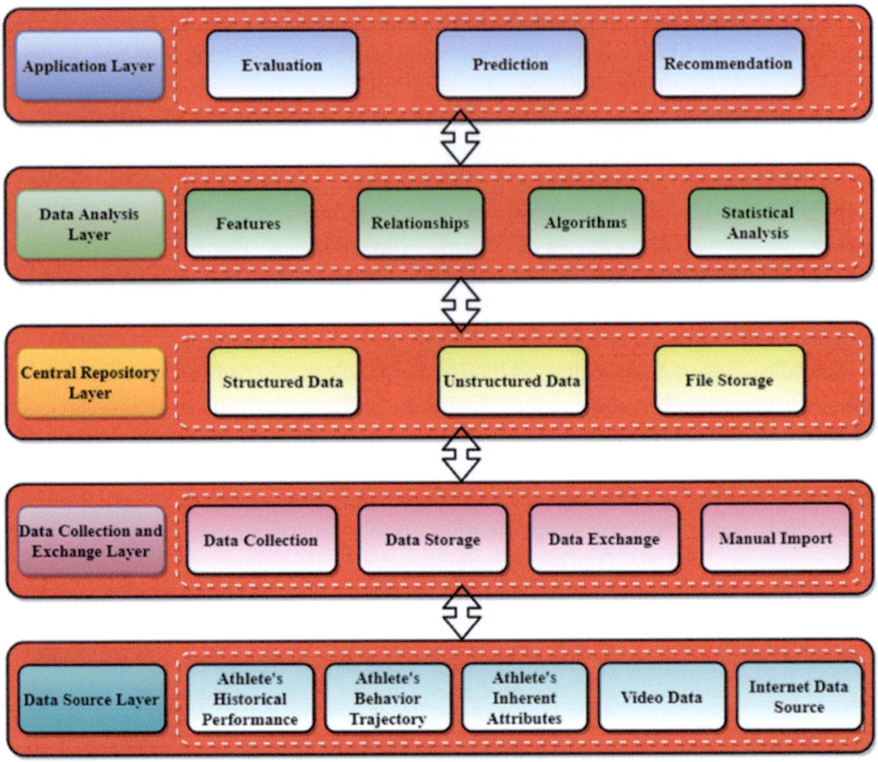

Fig. 3 Big data-based sports visualization model, illustrating data collection, processing, storage, and analysis for injury prevention (Liu et al. 2021)

injury risk analysis by integrating multiple data layers, including athlete history, biomechanical tracking, and AI-driven predictive analytics.

The integration of wearable technology with advanced machine learning is paving the way for highly personalized injury risk assessments. These innovations provide precise injury predictions that can be customized to the unique biomechanical traits of an athlete. This offers pertinent statistics for tailored interventions, hence creating a new baseline for injury prevention measures (Peatoni et al. 2022; Claudino et al. 2019).

Unsupervised machine learning techniques are effective in detecting anomalies in biomechanical data. They frequently exhibit preliminary indicators of potential harm that conventional methods may not reveal. Clustering techniques such as DBSCAN enable the categorization of athletes based on their movement patterns, hence assisting in the identification of anomalies indicative of potential injury risk (Papageorgiou et al. 2024).

Sports-related traumatic injuries may result in repercussions that extend beyond immediate physical discomfort. Moreover, factors such as Mild traumatic brain injuries (mTBIs), prevalent in contact sports, have raised inquiries among sports scientists over their long-term effects on athletes' brains. Research demonstrates that athletes with multiple mild traumatic brain injuries (mTBIs) may require extended recovery periods, persist in experiencing symptoms post-injury, and display cognitive irregularities, according to Brady et al. (2022). Wearable sensors that monitor head impacts and biomechanical stresses effectively identify athletes at risk. The important information obtained is used in both preventive and rehabilitative treatments (Brady et al. 2022).

Advancements in AI and big data analytics have made models vastly capable of offering sizeable insights into the risks of damage. Hybrid artificial intelligence models which integrate biomechanics with data visualization, can assist the management of athletes by detecting and clearly outlining irregularities in the profiles of injury risks (Liu et al. 2021). In addition, recent research also endorses the use of customized recovery procedures aimed at fostering the long-term resilience of athletes (Querido et al. 2021; Van Eetvelde et al. 2021). One emerging need for such research is the development of effective mechanisms for injury prevention that are based on biomechanical data. With historical and real-time data integration, wearables can track accurate biomechanical and physiological measurements and thus continuously improve predictive models by reconciling raw data with practical measures in the process of preventing injuries. This helps to reconcile raw data to pragmatic measures in the process of preventing injuries (Seçkin et al. 2023; Rawashdeh et al. 2016). Integrating internal physiological metrics such as heart rate variability and muscle fatigue with external workload data provides a comprehensive understanding of the mechanisms leading to injury. Machine learning algorithms can provide extensive insights into how athletes adapt to varying training loads through the analysis of both internal and external reactions (Seshadri et al. 2021).

Support from sports organizations and teams is crucial in the investments of wearable technology and big data analytics platforms that capable of handling large volumes of data for the implementation of this method. Equally important, collaboration between medical experts, data scientists, and sports scientists will ensure the required accuracy in the analysis of collected data and how it is used in real-time decisions making process. Such predictive models should be integrated into daily practice, alerting both coaches and athletes to potentially injurious motions or early signs of fatigue. Real-time biomechanical monitoring enhances both athlete performance and injury prevention by offering continuous feedback. He seamless interaction between wearables and athletes enables timely corrective actions, such as technique adjustments or modifications in training intensity. Over time, this proactive approach can significantly reduce the risk of acute injuries, including sprains and strains.

4 Future Directions in Injury Prevention: Integration, Innovation, and Expanded Research

Combining biomechanics, wearable technology and machine learning can provide innovative prevention strategies in preventing sports injuries. While traditional approaches focus only on responding to post injuries, this approach provides a data-driven solution for early detection and prevention. By continuously tracking and monitoring athletes' biomechanical and physiological data, enhanced machine-learning algorithms can predict potential injury risks and provide rapid intervention strategies before injuries occur (Kovoor et al. 2024; Sarlis et al. 2024).

Wearable sensor technology enables this preventative framework. These devices provide a continuous feedback loop that supplies real-time data on injury risks to athletes and coaches (Preatoni et al. 2022). The shift from reactive to proactive injury prevention highlights the significant influence of wearable technology on increasing athlete performance and care (Seçkin et al. 2023).

A notable achievement of this technique is the formulation of personalized injury prevention methods grounded in distinct biomechanical characteristics. This approach facilitates customized risk evaluation across diverse athletic groups, surpassing generalized information. It addresses the constraints of prior models that utilize a "one-size-fits-all" methodology (Seshadri et al. 2021).

Nonetheless, despite these encouraging developments, significant obstacles persist. Two primary challenges are the accessibility and expense of modern wearable devices and data analytics technology. These obstacles hinder the adoption of these technologies, particularly among smaller sports teams and lower-tier clubs. The reliability of a machine learning model is highly dependent on the quality and quantity of the data collected. Data inconsistency or poor quality can lead to inaccurate predictions (Papageorgiou et al. 2024). In addition to the physical outcome, injuries such as ankle sprains have a significant impact on the economics of professional sports activities. This highlights the need for effective preventive measures to mitigate these economic impacts (Hickey et al. 2019).

A major impediment is the requirement for certain technological expertise. The successful implementation of machine learning models requires skilled data scientists capable of managing and analyzing large datasets. Furthermore, the coaches and medical professionals are adept at translating these findings into effective strategies for injury prevention. The full potential of this technology may not be realized due to inadequate collaboration and training (Kovoor et al. 2024).

Future Research Directions

Future research should concentrate on enhancing the price and accessibility of wearable devices to facilitate wider use across all tiers of sports. Enhancing machine learning models to improve forecast accuracy, particularly for intricate injuries such as concussions, would signify significant progress. Furthermore, the expansion of collaborative data-sharing platforms facilitating cross-sport injury analysis could improve model validation and yield comprehensive insights into injury trends across several athletic disciplines. Investigations exploring the correlation between exterior and internal load indicators and injury incidence have demonstrated the possibility for enhanced monitoring methodologies. These developments will facilitate more efficient, preventive strategies in high-risk sports settings, especially for addressing overuse injuries (Jaspers et al. 2018).

Advanced investigation into advanced sensors that can capture intricate physiological data, including biochemical indicators of exhaustion or muscle damage, may enhance model predictions, yielding more precise, real-time insights into athletes' physical states. The integration of genetic data or biomarkers with biomechanical information represents a promising avenue for future research, potentially facilitating hyper-personalized injury prevention techniques (Seshadri et al. 2021).

Future initiatives in injury prevention and recovery must emphasize accessibility and cost to maximize the advantages of these technologies. Wearable technology significantly influences predictive data for models and recovery protocols, facilitating a transition to proactive athlete health care (Liu et al. 2021; Querido et al. 2021). The ongoing enhancement of predictive models, particularly for intricate injuries such as concussions, along with broadened recovery strategies aimed at long-term health and the sustainability of sports careers, will bolster injury preventive methodologies (Brady et al. 2022). Utilizing breakthroughs in data visualization, artificial intelligence, and sensor technologies might create a strong basis for effective athlete health management systems.

Acknowledgements The authors would like to thank the Ministry of Higher Education (MOHE) Malaysia and Universiti Malaysia Pahang Al-Sultan Abdullah for the financial support provided under the Fundamental Research Grant Scheme, Grant No: FRGS/1/2024/TK10/UMP/02/2 (UMPSA internal reference: RDU240103) in completing this work.

References

Brady M, Hume PA, Mahon S, Theadom A (2022) What is the evidence on natural recovery over the year following sports-related and non-sports-related mild traumatic brain injury: a scoping review. Front Neurol 12:756700. https://doi.org/10.3389/fneur.2021.756700

Claudino JG, Capanema D de O, de Souza TV, Serrão JC, Pereira ACM, Nassis GP (2019) Current approaches to the use of artificial intelligence for injury risk assessment and performance prediction in team sports: a systematic review. Sport Med-Open 5(1):28. https://doi.org/10.1186/s40798-019-0202-3

Hickey D, Shield AJ, Williams MD, Opar DA (2019) The financial cost of hamstring strain injuries in the Australian Football League. Br J Sports Med 48(8):729–730. https://doi.org/10.1136/bjsports-2013-092884

Jaspers A, Kuyvenhoven JP, Staes F, Frencken WGP, Helsen WF, Brink MS (2018) Examination of the external and internal load indicators' association with overuse injuries in professional soccer players. J Sci Med Sport 21(6):579–585. https://doi.org/10.1016/j.jsams.2017.10.005

Kovoor M, Durairaj M, Karyakarte MS, Hussain MZ, Ashraf M, Maguluri LP (2024) Sensor-enhanced wearables and automated analytics for injury prevention in sports. Meas: SensS 32:101054. https://doi.org/10.1016/j.measen.2024.101054

Liu A, Mahapatra RP, Mayuri AVR (2021) Hybrid design for sports data visualization using AI and big data analytics. Complex Intell Syst 9(3):2969–2980. https://doi.org/10.1007/s40747-021-00557-w

Papageorgiou G, Sarlis V, Tjortjis C (2024) Unsupervised learning in NBA injury recovery: advanced data mining to decode recovery durations and economic impacts. Information 15(1):61. https://doi.org/10.3390/info15010061

Preatoni E, Bergamini E, Fantozzi S, Giraud LI, Orejel Bustos AS, Vannozzi G, Camomilla V (2022) The use of wearable sensors for preventing, assessing, and informing recovery from sport-related musculoskeletal injuries: a systematic scoping review. Sensors 22(9):3225. https://doi.org/10.3390/s22093225

Querido SM, Brito J, Figueiredo P, Carnide F, Vaz JR, Freitas SR (2021) Post-match recovery practices in professional football: design, validity, and reliability of a new questionnaire. Front Sport Act Living 3:680799. https://doi.org/10.3389/fspor.2021.680799

Rawashdeh SA, Rafeldt DA, Uhl TL (2016) Wearable IMU for shoulder injury prevention in overhead sports. Sensors 16(11):1847. https://doi.org/10.3390/s16111847

Sarlis V, Papageorgiou G, Tjortjis C (2024) Sports analytics and text mining NBA data to assess recovery from injuries and their economic impact. Computers 12(12):261. https://doi.org/10.3390/computers12120261

Seçkin AÇ, Ateş B, Seçkin M (2023) Review on wearable technology in sports: concepts, challenges and opportunities. Appl Sci 13(18):10399. https://doi.org/10.3390/app131810399

Seshadri DR, Thom ML, Harlow ER, Gabbett TJ, Geletka BJ, Hsu JJ, Drummond CK, Phelan DM, Voos JE (2021) Wearable technology and analytics as a complementary toolkit to optimize workload and to reduce injury burden. Front Sport Act Living 2:630576. https://doi.org/10.3389/fspor.2020.630576

Van Eetvelde H, Mendonça LD, Ley C, Seil R, Tischer T (2021) Machine learning methods in sport injury prediction and prevention: a systematic review. J Exp Orthop 8(1):27. https://doi.org/10.1186/s40634-021-00346-x

AI-Driven Performance Metrics: Enhancing Athlete Monitoring and Injury Prevention

Nurul Hasya Md Kamil, Mohd Nizar Mhd Razali, Syukran Hakim Norazman, Mohd Arifpin Mansor, and Ahmad Shahir Jamaludin

Abstract Sports analytics is increasingly incorporating artificial intelligence (AI) to improve athlete monitoring and injury prevention. Creating personalized AI-driven performance assessments with real-time insights into athletes' physiological conditions has been a primary focus. Unlike traditional method, artificial intelligence models may analyze dynamic, complex data to create tailored suggestions that enhance training load while preventing overtraining. Current research indicates that these personalized AI models hold significant promise for solving critical gaps in athlete monitoring, particularly in injury prediction models. However, integrating multimodal data sources—namely, biomechanical, physiological, and environmental data—into coherent AI models remains complex and difficult. Particularly reinforcement learning and neural networks, advances in machine learning methods have showed promise in producing adaptive models that may continuously change depending on the special requirements of individual athletes. Moreover, wearable and sensor data-based prediction models have proven efficacy in seeing early signs of weariness or injury risk, so allowing proactive therapies. From applications of these AI-driven techniques, elite sports have already benefited in training management and injury avoidance. Though progress has been made, further research is needed to enhance data collection methods, mix psychological assessments, and expand

N. H. M. Kamil · M. N. M. Razali · A. S. Jamaludin (✉)
Faculty of Manufacturing and Mechatronic Engineering Technology, Universiti Malaysia Pahang Al-Sultan Abdullah, 26600 Pekan, Pahang, Malaysia
e-mail: shahir@umpsa.edu.my

S. H. Norazman
Faculty of Electrical and Electronic Engineering Technology, Universiti Malaysia Pahang Al-Sultan Abdullah, 26600 Pekan, Pahang, Malaysia

M. A. Mansor
Department of Information and Intelligent Engineering, National Institute of Technology (KOSEN), Sasebo College, 1-1 Okishin, Sasebo-Shi, Nagasaki 857-1193, Japan

A. S. Jamaludin
Centre for Advanced Industrial Technology, Universiti Malaysia Pahang Al-Sultan Abdullah, 26600 Pekan, Pahang, Malaysia

M. H. A. Hassan et al. (eds.), *Proceedings of the 3rd International Conference on Innovation and Technology in Sports*, Lecture Notes in Bioengineering,
https://doi.org/10.1007/978-981-95-0804-4_42

artificial intelligence applications at many levels. Artificial intelligence seems to have promising future prospects in sports analytics in constructing full systems that provide real-time, tailored insights into athlete performance, therefore guaranteeing both optimal performance and low injury risks.

Keywords Artificial intelligence in sports · Athlete monitoring · Injury prevention · Machine learning in sports · Predictive sports analytics

1 Introduction to the Role of AI-Driven Performance Metrics in Modern Sports

Data analytics and artificial intelligence (AI) combined together have transformed sports' athlete performance monitoring and management. AI-driven performance measures have grown recently as it can process massive volumes of data in real-time and offer hitherto unheard-of insights on an athlete's physiological state, tiredness levels, and injury risks (Chmait and Westerbeek 2021; Liu et al. 2021). Unlike conventional approaches depending on subjective assessments and static data, AI systems can measure dynamic performance, incorporating features including movement patterns, physiological reactions, and ambient variables, as depicted in Fig. 1 (Biró et al. 2023). Historically, athletes' performance was formerly assessed by physical testing (VO_2 max), subjective self-reports, or hand-eye gaming viewing (Price et al. 2022). Using these typical methods makes it difficult to sufficiently portray the complexity of physiological and real-time performance data. Wearable sensors and Internet of Things (IoT) devices have greatly expanded the availability and range of data for performance study. Artificial intelligence methods can be applied in measuring heart rate variability, joint acceleration, muscle oxygen levels, and GPS-based movement monitoring, as shown in Fig. 2 (Chen and Dai 2024; Wiecha et al. 2022a, 2023a).

Furthermore, artificial intelligence models that based on machine learning (ML) and deep learning algorithms may now identify underlying trends in performance data

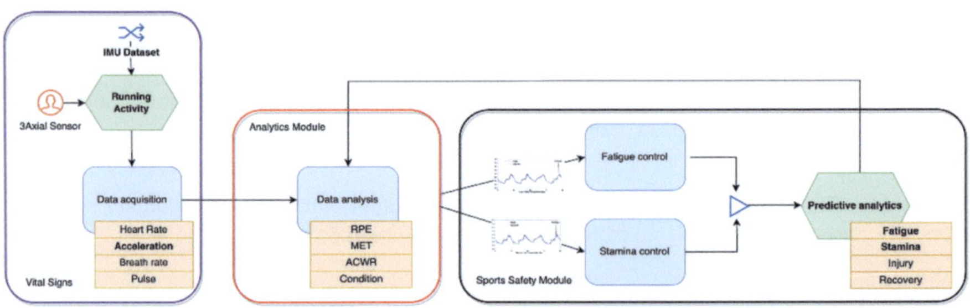

Fig. 1 Sports safety approach on specific use case (Biró et al. 2023)

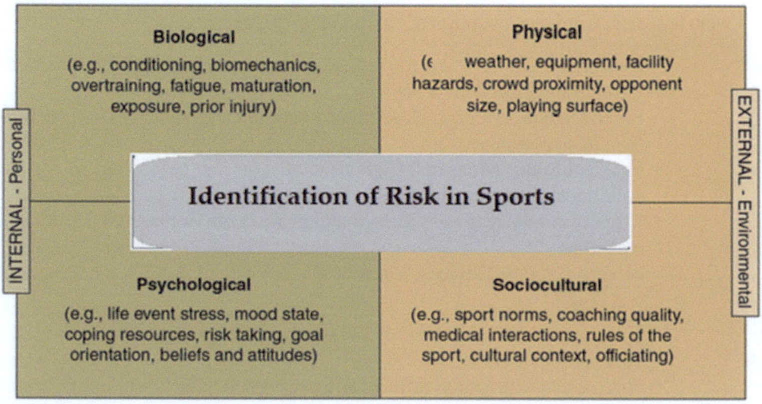

Fig. 2 Variety of data available for performance analysis (Chen and Dai 2024)

that human observers may overlook. These technologies enable users like coaches and medical personnel to make more accurate decisions about an athlete's training regimen, recovery requirements, and competitive readiness (Biró et al. 2023; Sarlis and Tjortjis 2024; Lach et al. 2021). Despite these improvements, the following major research needs still remain:

1. **Personalization of AI Models**: Many current artificial intelligence systems are based on generic models, which may not accurately reflect individual athlete variances in physiology, playing style, or injury history. Creating tailored AI-driven models that may fit personal attributes remains a significant challenge (Biró et al. 2023).
2. **Real-Time Injury Prediction**: Although artificial intelligence has advanced post-injury recovery therapy, only small number of researchs has been done on pre-training/match damage prediction, such as soft tissue injuries etc. Table 1 (Chen and Dai 2024; Wiecha et al. 2022b, 2023b) shows how real-time, AI-driven predictive algorithms might provide preventive treatments before an accident occurs.
3. **Across Data Sources**: Another area that require a lot of development is the integration of multiple data streams—such as biomechanical data, physiological measurements, and environmental conditions—into a coherent model, as shown in Fig. 3. Synthesizing this many inputs into useful insights is difficult for current artificial intelligence systems (Liu et al. 2021, Litchi et al. 2023).

Table 1 Overview of IoT devices in sports (Chen and Dai 2024)

Types of Sensors	• Navigation Units: Track movement patterns and geographical positioning through integrated GPS devices • Force Detection Units: Measure acceleration and velocity changes through dynamic motion sensors • Physical Status Monitors: Capture vital signs and physiological data through continuous heart rate tracking • Climate Analysis Tools: Record and analyze environmental parameters in real-time • Motion Capture Devices: Monitor and analyze joint movement mechanics and range
Installation Methods	• Navigation Units: Seamlessly incorporated into equipment or worn as personal devices • Force Detection Units: Specifically positioned at different body locations for accurate movement analysis • Physical Status Monitors: Adaptable as chest bands, wrist devices, or smart clothing integration • Climate Analysis Tools: Distributed throughout the activity space for comprehensive coverage • Motion Capture Devices: Precisely attached to major joints and movement points
Information Synthesis	Advanced processing systems that combine and analyze data from multiple sensor inputs, creating comprehensive performance insights and actionable assessments
Technical Upkeep	Scheduled system checks, routine calibration procedures, and proactive power management to ensure consistent and reliable data collection
Data Protection	Robust security architecture featuring advanced encryption, controlled user access, and protected storage systems to maintain data confidentiality

2 Advancing Athlete Monitoring: Proposed Solutions Through AI Personalization and Predictive Modeling

Several potential solutions using developments in machine learning algorithms, sensor technology, and data fusion approaches can be suggested to handle the described difficulties. These methods seek to improve the accuracy, personalization, and real-time applicability of performance measures powered by artificial intelligence.

1. Personalization Through Advanced Machine Learning Models

Improving personalizing in artificial intelligence models is one of the most exciting fixes. Recent advances in neural networks and reinforcement learning could be used to build adaptable models learning from the data of individual athletes across time. These systems would not only fit personal physiological traits but also vary depending on performance data acquired with wearable sensors. By continuously

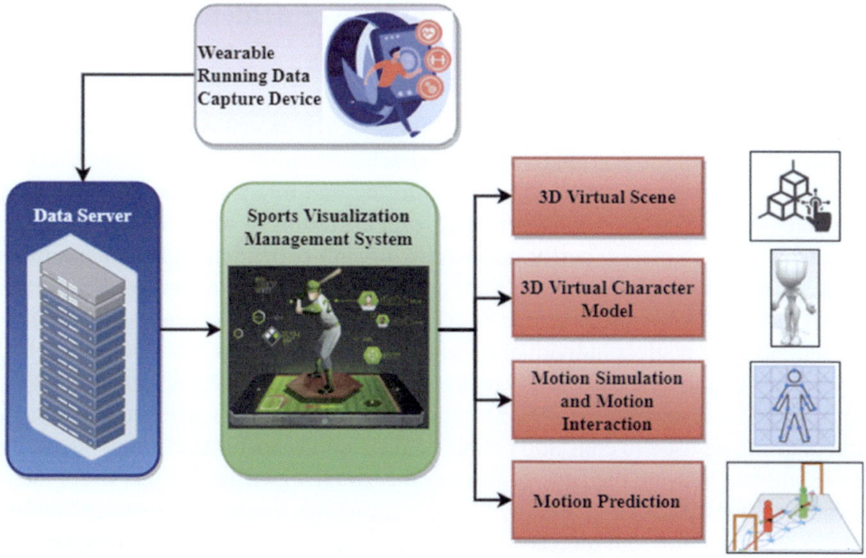

Fig. 3 Sports data visualization management system (Liu et al. 2021)

instructing these models on data including heart rate, muscular tiredness, and environmental conditions (Petek et al. 2021; Biró et al. 2023; Segreti et al. 2023), artificial intelligence might offer tailored training regimens maximizing an athlete's performance and lower injury risk.

2. Predictive Injury Models Using Real-Time Data

Furthermore possible injuries could be projected before their start using artificial intelligence technologies. Integration of real-time data from inertial measuring units (IMUs) and physiological sensors allows artificial intelligence models to track athletes' physical activity and biomechanical deviations before collisions. By using machine learning algorithms, these devices can detect minute changes in gait, joint stress, and muscle weariness, so enabling coaches to act before an injury results (Chen and Dai 2024). Athletes may find that this proactive approach significantly reduces downtime and improves long-term performance sustainability.

3. Multimodal Data Integration for Enhanced Decision-Making

Another crucial step is to include several data sources into a whole artificial intelligence-driven performance model. Combining information from motion sensors, heart rate monitors, GPS trackers, and even environmental sensors—temperature, humidity—would provide a more whole view of an athlete's state. Advanced data fusion methods including convolutional neural networks (CNNs) mixed with recurrent neural networks (RNNs) could be used to integrate these datasets thereby producing more accurate predictions of weariness, preparation, and injury risk (Liu et al. 2021; Sarlis and Tjortjis 2024).

3 Personalized AI Models: The Future of Athlete Monitoring and Injury Prevention

Among the several hypothetical solutions suggested, personalized AI models based on advanced machine learning (ML) approaches stand out as the most effective in bridging the gaps linked to athlete monitoring and injury prevention. Real-time data integration combined with personal level customizing offers an advantage in guaranteeing more precise performance tracking, injury prediction, and intervention actions. Their primary advantage is their ability to adapt and alter in response to each athlete's distinct physiological and biomechanical profile. Generic "one-size-fits-all" models may neglect individual differences in biomechanics, healing rates, and disease susceptibility. However, by using adaptive machine learning methodologies (e.g., reinforcement learning, neural networks), each athlete's historical and real-time data can be used to constantly enhance prediction capability (Biró et al. 2023). For example, choosing the best techniques for balancing effort and rest strengthens learning by boosting training load and recuperation time. Furthermore, understanding how to balance effort and relaxation increases reinforcement learning by increasing training load and recovery. Personalized AI models thus act as virtual performance coaches, constantly evaluating and adjusting plans for performance optimization (Chmait and Westerbeek 2021).

Predictive Power for Injury Prevention

One important area where personalized artificial intelligence systems shine is injury prevention. These systems may track important factors such muscle stress, joint loads, and gait patterns by means of real-time sensor data—from wearable and inertial measurement units. An athlete might show small but noticeable biomechanical alterations. For example, prior to a knee injury. Early signals detected by AI models trained on machine learning classifiers—such as random forests and support vector machines—can notify trainers before the damage results (Chen and Dai 2024). This lets coaches modify the athlete's effort, plan recovery times, or offer successful exercises to lower risk.

Integration of Multimodal Data Streams

Multimodal data integration is another crucial ability of tailored artificial intelligence systems. Combining advanced convolutional neural networks (CNNs) with either long short-term memory (LSTM) architectures or recurrent neural networks (RNNs), can process and combine several data streams, including GPS data, heart rate variability, muscle oxygen levels, and even environmental factors like temperature and humidity. By means of a more accurate and holistic perspective of an athlete's state, this all-encompassing method helps to improve decision-making on training load and recuperation (Liu et al. 2021). Real-time artificial intelligence systems may, for instance, identify an athlete going beyond recommended heart rate variability (HRV) limits in hot settings, implying they run the danger of heat-induced tiredness. The AI model can advise immediate hydration and rest by combining this data with

GPS and muscle oxygenation data, therefore preventing performance reduction or upcoming injuries (Chmait and Westerbeek 2021).

Applications in Elite Sports

Elite sports are already looking at the benefits of AI-powered ideas. Wearable powered by artificial intelligence enable professional teams in endurance sports such as basketball and football to track athlete performance in real time. The NBA uses AI-based data analytics to monitor player workloads and forecast injury risk based on game intensity and frequency of training sessions (Sarlis and Tjortjis 2024). Likewise, artificial intelligence systems applied in cycling and football integrate various data streams to enhance performance monitoring and enable modifications depending on an athlete's real-time physiological and biomechanical status. Long-term athlete development (Chen and Dai 2024) has also been impacted by artificial intelligence-powered insights, which direct data-driven decisions on training cycles, dietary changes, and rest intervals (Kasiak et al. 2023).

4 The Future of AI in Sports: From Predictive Models to Comprehensive Athlete Monitoring

The use of AI-driven performance data has transformative potential in athlete monitoring and injury prevention. These systems use personalized machine learning models to deliver not only real-time insights into athletes' physiological conditions, but also prediction capabilities that can assist prevent injuries and maximize performance.

Summary of Key Points:

- Personalized AI models to individual athletes is the most effective way to improve performance monitoring and prevent injuries. These algorithms may continually learn from real-time data and adapt to each athlete's specific attributes.
- Predictive injury prevention models provide early warning systems that can avoid injuries, particularly those related to fatigue and tiredness.
- Comprehensive AI models that combine multimodal data from many sources (such as environmental monitors, GPS trackers, and physiological sensors) improve decision-making, resulting in more personalized and data-driven training recommendations for athletes.
- These techniques have previously been employed successfully in elite sports, such as basketball and endurance sports, where artificial intelligence is used to assess workload, enhance recovery, and predict injury risks across multiple disciplines.

Future Directions:

While great progress has been done, certain areas of future research and development are critical to fully leveraging AI's promise in sports analytics:

1. **Improving Data Accuracy**: Wearables and IoT devices should capture more accurate and granular data. More precise data inputs, such as more sophisticated movement tracking and physiological monitoring, will result in more accurate AI predictions.
2. **Integrating Psychological Metrics**: Including psychological parameters such as stress, mood, and mental exhaustion in AI models to provide more comprehensive performance analysis and a complete picture of an athlete's total readiness, lessening injury risks caused by mental overload.
3. **Expanding AI's Role in Recovery Management**: Artificial intelligence systems are commonly defined by their performance and ability to prevent injury. Extending these skills to include recovery optimization may improve long-term athlete development even further.
4. **Cross-Sport Applications**: Although artificial intelligence is increasingly being employed in competitive sports, it has only been used sparingly in leisure or lower-level sport. Research should focus on adapting AI technology for broader applications, so that athletes of all types can benefit from advanced monitoring systems.

In conclusion, AI-driven performance metrics have the potential to revolutionize athlete monitoring by providing real-time, personalized insights that can boost performance while decreasing injury risk. To fully achieve this promise, future research should focus on enhancing data collection procedures, including psychological variables, as well as widening the scope of artificial intelligence applications in both elite and amateur sports.

Acknowledgements The authors would like to thank University Malaysia Pahang Al-Sultan Abdullah for the financial support provided under the Postgraduate Research Scheme (PGRS220367) in completing this work.

References

Biró A, Cuesta-Vargas AI, Szilágyi L (2023) AI-assisted fatigue and stamina control for performance sports on IMU-generated multivariate times series datasets. Sensors 24(1):132. https://doi.org/10.3390/s24010132

Chen Z, Dai X (2024) Utilizing AI and IoT technologies for identifying risk factors in sports. Heliyon 10(11):e32477. https://doi.org/10.1016/j.heliyon.2024.e32477

Chmait N, Westerbeek H (2021) Artificial intelligence and machine learning in sport research: an Introduction for Non-data scientists. Front Sport Act Living 3. https://doi.org/10.3389/fspor.2021.682287

Kasiak PS, Wiecha S, Cieśliński I, Takken T, Lach J, Lewandowski M, Barylski M, Mamcarz A, Śliż D (2023) Validity of the maximal heart rate prediction models among runners and cyclists. J Clin Med 12(8):2884. https://doi.org/10.3390/jcm12082884

Lach J, Wiecha S, Śliż D, Price S, Zaborski M, Cieśliński I, Postuła M, Knechtle B, Mamcarz A (2021) HR MaX prediction based on age, body composition, fitness level, testing modality and sex in physically active population. Front Physiol 12. https://doi.org/10.3389/fphys.2021.695950

Lichti J, Maggioni MA, Balcerek B, Becker PN, Labes R, Gunga H, Fähling M, Steinach M (2023) The relevance of body composition assessment for the rating of perceived exertion in trained and untrained women and men. Front Physiol 14. https://doi.org/10.3389/fphys.2023.1188802

Liu A, Mahapatra RP, Mayuri AVR (2021) Hybrid design for sports data visualization using AI and big data analytics. Complex & Intell Syst 9(3):2969–2980. https://doi.org/10.1007/s40747-021-00557-w

Petek BJ, Tso JV, Churchill TW, Guseh JS, Loomer G, DiCarli M, Lewis GD, Weiner RB, Kim JH, Wasfy MM, Baggish AL (2021) Normative cardiopulmonary exercise data for endurance athletes: the cardiopulmonary health and endurance exercise registry (CHEER). Eur J Prev Cardiol 29(3):536–544. https://doi.org/10.1093/eurjpc/zwab150

Price S, Wiecha S, Cieśliński I, Śliż D, Kasiak PS, Lach J, Gruba G, Kowalski T, Mamcarz A (2022) Differences between treadmill and cycle ergometer cardiopulmonary exercise testing results in triathletes and their association with body composition and body mass index. Int J Environ Res Public Health 19(6):3557. https://doi.org/10.3390/ijerph19063557

Sarlis V, Tjortjis C (2024) Sports analytics: data mining to uncover NBA player position, age, and injury impact on performance and economics. Information 15(4):242. https://doi.org/10.3390/info15040242

Segreti A, Picarelli F, Di Gioia G, Coletti F, Crispino SP, Fanale V, Fossati C, Incalzi RA, Pigozzi F, Grigioni F (2023) Athlete's heart or heart disease in the athlete? evaluation by cardiopulmonary exercise testing. J Sport Med Phys Fit 63(7). https://doi.org/10.23736/s0022-4707.23.14536-1

Wiecha S, Kasiak PS, Cieśliński I, Maciejczyk M, Mamcarz A, Śliż D (2022a) Modeling physiological predictors of running velocity for endurance athletes. J Clin Med 11(22):6688. https://doi.org/10.3390/jcm11226688

Wiecha S, Price S, Cieśliński I, Kasiak PS, Tota Ł, Ambroży T, Śliż D (2022b) Transferability of cardiopulmonary parameters between treadmill and cycle ergometer testing in male triathletes—prediction formulae. Int J Environ Res Public Health 19(3):1830. https://doi.org/10.3390/ijerph19031830

Wiecha S, Kasiak PS, Cieśliński I, Takken T, Palka T, Knechtle B, Nikolaidis PT, Małek ŁA, Postuła M, Mamcarz A, Śliż D (2023a) External validation of VO2max prediction models based on recreational and elite endurance athletes. PLoS ONE 18(1):e0280897. https://doi.org/10.1371/journal.pone.0280897

Wiecha S, Kasiak PS, Szwed P, Kowalski T, Cieśliński I, Postuła M, Klusiewicz A (2023) VO2max prediction based on submaximal cardiorespiratory relationships and body composition in male runners and cyclists: a population study. eLife 12. https://doi.org/10.7554/elife.86291

Biomechanical Profiling: Review on Real-Time Data Integration and Predictive Models for Performance Optimization

Noorazliza Sulaiman, Ahmad Shahir Jamaludin, Mohd Nizar Mhd Razali, Muhamad Ridzuan Radin Muhamad Amin, and Abdul Nasir

Abstract Nowadays, optimizing athletic performance in numerous sports mostly depends on biomechanical characterization. Focusing on the cross-sport application of biomechanical concepts in sports including swimming, running, and cycling, this study pays especially attention to injury prevention, performance improvement, and real-time feedback systems. One of the main gaps in current knowledge is the absence of a generalizable framework allowing biomechanical insights from one sport to another. Using universal movement patterns, joint alignments, and muscle activation measures, cross-sport biomechanical profiling fills up this void. In real-time biomechanical data gathering and evaluation, inertial measurement units (IMUs) and artificial intelligence-driven motion tracking systems are very vital tools. By allowing players to maximize characteristics including stride length and cadence across many sports, predictive approaches even more enhance the profiling process. Studies in cycling biomechanics have shown, for example, how saddle height affects knee flexion and power output; similar principles apply to stride mechanics in running and stroke efficiency in swimming. Wearable technology offers real-time technique changes based on constant feedback, hence improving performance. Future directions call for the creation of more generalizable measurements applicable across many sports, the integration of artificial intelligence to offer dynamic, tailored feedback, and longitudinal studies tracking of biomechanical adaptations over time. These developments should make biomechanical optimization more easily available and successful for sportsmen of all levels.

N. Sulaiman · M. R. R. M. Amin · A. Nasir (✉)
Faculty of Electrical and Electronic Engineering Technology, Universiti Malaysia Pahang Al-Sultan Abdullah, Pekan, Pahang, Malaysia
e-mail: abdnasir@umpsa.edu.my

A. S. Jamaludin · M. N. M. Razali
Faculty of Manufacturing and Mechatronic Engineering Technology, Universiti Malaysia Pahang Al-Sultan Abdullah, Pekan, Pahang, Malaysia

A. S. Jamaludin
Centre for Advanced Industrial Technology, Universiti Malaysia Pahang Al-Sultan Abdullah, Pekan, Pahang, Malaysia

M. H. A. Hassan et al. (eds.), *Proceedings of the 3rd International Conference on Innovation and Technology in Sports*, Lecture Notes in Bioengineering,
https://doi.org/10.1007/978-981-95-0804-4_43

Keywords Biomechanical profiling · Multisport optimization · Performance enhancement · Real-time feedback · Predictive modeling

1 Introduction—the Role of Biomechanics in Multisport Performance Optimization

Many sports rely heavily on biomechanics to increase performance. Biomechanics knowledge, which examines athlete movement, can help to optimize methods, reduce injury risks, and increase overall performance. For cycling, for example, differences in saddle height have been closely investigated in terms of knee flexion and force output. These findings correspond to the increasing use of inertial measurement units (IMUs) and accelerometers to track joint angles and body motion in dynamic sports contexts. Tracking knee angles while riding with IMU technology allows athletes to enhance their biomechanics outside of lab settings (Obradović and Stančin 2023; Pérez-Castilla and García-Pinillos 2024). All of which affect performance (Molavian et al. 2023; Burnie et al. 2020), saddle height affects power generation, pedaling technique, and joint kinematics. Universal recommendations are challenging though since individual athletes may need different environments for optimal performance and anthropometry and muscular strength vary. Triathlon and other multisport events significantly compromise biomechanical optimization.

Evans et al. (2021) examined triathlon trunk stability and performance under saddle height and cadence influence. The study indicated the need of a biomechanical approach catered to certain multisport criteria since variations in these elements had a significant effect on trunk acceleration and lower body mechanics. Tracking knee angles while riding with IMU technology can assist athletes improve their biomechanics outside of lab settings (Obradović and Stančin 2023; Pérez-Castilla and García-Pinillos 2024). Recent research show promising results using neural networks to assess lower limb kinematics during stationary cycling with the help of artificial intelligent, AI. Bini et al. (2023) confirmed the accuracy of AI models such as TransPose and MediaPipe, innovate the way of gathering and applying biomechanical data in performance situations.

Particularly cross-sport profiling, biomechanical profiling has the ability to close the distance between sport-specific techniques and generic opti-mizing approaches. Studies on pedaling rates and muscular efficiency in cycling, for example, can help one understand concepts relevant to running and swimming, where joint alignment and force application are equally vital (Yum et al. 2021). By means of this cross-sport strategy, practitioners can create more flexible training programs spanning several sports, therefore enhancing performance holistically. Figure 1 depicts the link between saddle height and knee flexion angle during cycling, as shown by Obradović and Stančin (2023), therefore illustrating the vital part biomechanics play in sports performance. Higher saddle heights apparently lower knee flexion, which results in a more aerodynamic stance and more efficiency.

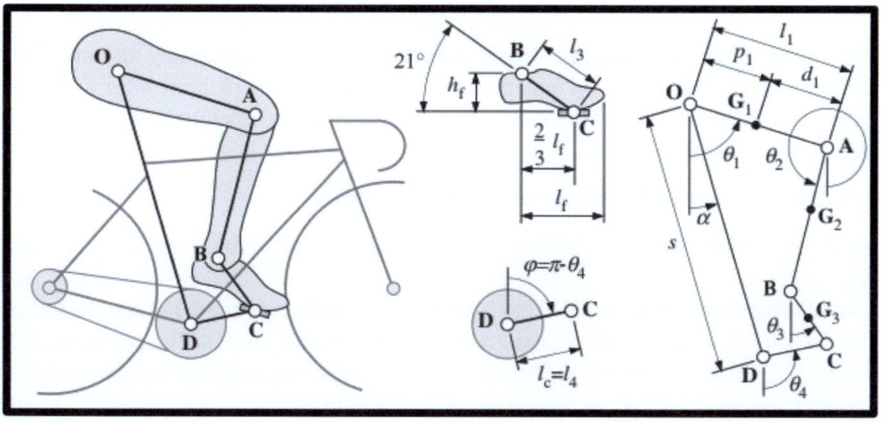

Fig. 1 The skeletal model adopted and gives the basic nomenclature used to describe the kinematics of the system

2 Cross-Sport Biomechanical Profiling

Cross-sport biomechanical profiling offers potential in addressing the challenges and gaps in multisport biomechanics. This method generates detailed biomechanical profiles relevant to a wide range of athletic disciplines by leveraging commonalities in movement patterns, joint mechanics, and force application. Coaches and practitioners can develop more universal, successful training and injury prevention approaches by standardizing the biomechanical ideas involved in many sports, such as cycling, running, and swimming (Walck et al. 2022).

Cross-sport profiling is primarily based on the development of sport-specific metrics that cover many disciplines. For example, in cycling, saddle height and knee flexion angles are crucial for performance optimization. According to Bini (2020) and Palmieri et al. (2020), small changes in saddle height can result in significant alterations in muscle activation and knee joint angles, which have a direct impact on cycling efficiency. Running also follows a similar idea in that running efficiency is improved and impact forces are lowered in great part by stride length, foot strike pattern, and knee alignment. Sport scientists can derive global parameters by cross-application such measurements. Table 1, for example, shows as compared to running and swimming the ideal ranges of joint flexion in cycling.

Table 1 Comparison of joint flexion angles across cycling, running, and swimming (Bini 2020; Yum et al. 2021)

Joint	Cycling (°)	Running (°)	Swimming (°)
Knee flexion	65–75	70–80	60–70
Hip flexion	40–50	45–55	30–40
Ankle flexion	20–30	25–35	20–25

For multisport athletes, especially in triathlon where athletes move from swimming to cycling and running, these measurements are absolutely essential for maximizing performance. By means of smooth transitions in joint alignment and muscle activation patterns, one can lower the risk of injuries including overuse and musculoskeletal stress.

Technology-Assisted Profiling and Data Collection

Improving biomechanical profiling depends critically on technology. For real-time joint angle and movement pattern monitoring, accelerometer-based devices and inertial measurement units (IMUs) have become rather popular (Zong-Hao et al. 2023). Evans et al. (20,211) measured trunk stability and joint mechanics in cycling using accelerometers. Their results showed that variations in saddle height produced notable variations in trunk acceleration, which would have direct bearing on cycling performance. Comparatively to optical motion capture systems (Obradovič and Stančin 2023), IMUs have shown dependability for tracking knee joint angles during cycling with error rates as low as 3.2°, as tabulated in Table 2.

These technologies not only offer the potential for improved data collection in elite sports but also provide accessible, affordable solutions for amateur athletes. The continued development of AI-driven motion analysis, as demonstrated by Bini et al. (2023), further enhances the capacity for biomechanical profiling to deliver precise feedback on joint angles, movement patterns, and muscle activation during performance.

Simulation and Predictive Models

Simulation technologies and prediction models become alternative method to improve cross-sport biomechanical profile. Using kinetic models, Palmieri et al. (2020) simulated how varying pedaling cadences would affect energy consumption during cycling. Their studies showed that optimal cadences change based on the cyclist's body mass index (BMI) and power output; higher cadences are more efficient at higher power levels. Running and swimming are among sports where simulation models can forecast how different stride rates and stroke frequency would affect performance.

Customizing training programs for individual athletes could be achieved with the use of predictive models grounded on such simulations. For example, biomechanical simulations that examine the drag reduction connected with small stroke technique modification would help a swimmer. Similarly, maximizing stride frequency depending on a prediction model including body mass and energy expenditure can

Table 2 Provides a comparison of the reliability of IMU-based systems versus traditional optical systems for joint angle tracking during cycling

Measurement method	Error range (°)	Cost	Real-time monitoring
Inertial measurement units	1–3.2	Low	Yes
Optical motion capture	0.5–2	High	No

improve a runner's performance. Beyond conventional, sport-specific advice, these models enable more customized, data-driven training systems.

Real-Time Feedback Systems and Wearables

Sports biomechanics have witnessed notable increase in the utilization of real-time feedback systems linked with wearable technology. Wearables like LEOMO® and RETul® systems let you instantly modify during training ses-sion by continuously collecting joint angle, force distribution, and muscle activation patterns. The devices are becoming increasingly used in multisport situations to enhance performance across swimming, running, and cycling.

Ho et al. (2023) looked at how torque measurement influenced cycling-assisted electric bikes and showed that adaptive control of motor torque might improve both cycling efficiency and muscular activation patterns under high-load conditions. Such real-time feedback systems offer insights that can be applied in multisport contexts, where players may dynamically modify their biomechanics relying on immediate data.

3 Cross-Sport Profiling: A Method for Multisport Biomechanical Optimization

Cross-sport biomechanical profiling is the best approach for managing the variety of biomechanical demands across sports. This approach combines data-driven insights with practical applications to generate more efficient training and performance results for athletes engaged in swimming, running, and cycling, whereas the joint alignment, muscle activation, and force application can be emphasized. Following are several output can be taken into consideration:

1. Data Consistency Across Sports

Studies show that similar principles apply across various activities when it comes to joint alignment and force production. Assuming as posi-tive an anticlockwise rotation, Fig. 1 shows the absolute orientation from the vertical axis of the four moving segments of the system in the vector $q = [q_1 q_2 q_3 q_4]$. The subscripts i ($i =$ 1:4) refer in the sequence to thigh, shank, foot and crank. l_i marks the length of the ith link. As $p_i + d_i = l_i$, lengths p_i and d_i show the proximal and distal position of the center of mass Gi with regard to the extremities of the ith body segment. Assuming that the pedal spindle is set at two third of the total length of the foot, the length l_3 has been calculated as indicated in Fig. 1; hence, the segment (C − B) forms with a constant angle equal to 21∘. The seat tube angle is chosen at a typical value between road and triathlon, that is, a = 75∘.

2. Biomechanical Insights into Injury Prevention

Cross-sport biomechanical profile offers also insightful analysis of injury prevention. For example, as Table 3 (Evans et al. 2021) shows, poor bike configurations—that is, incorrect saddle height or handlebar position—may cause overuse injuries in cycling. Similar concepts hold true in running, where improper knee and ankle alignment during foot strikes can cause stress fractures or Achilles tendinitis. By tracking and adjusting joint angles and muscular activity using profiling techniques (Obradović and Stančin 2023), athletes can reduce their risk of injury in both sports.

3. Use of Predictive Models for Optimization

Cross-sport profiling is much aided by advanced simulation and predictive models. Palmieri et al. (2020), for instance, simulated the effect of various pedaling cadences on energy efficiency and muscular tiredness during cycles using kinetic models. Predictive models allow one to recreate how different stride rates or foot striking patterns effect energy consumption and exhaustion.

4. Technological Integration: Wearable Devices and AI

Cross-sport profiling depends critically on wearable technologies and real-time feedback systems. Constant data collecting made possible by devices as IMUs, accelerometers, and artificial intelligence-powered motion monitoring systems gives athletes immediate feedback on their biomechanics, as shown in Table 4. IMUs have been effectively utilized for running and cycling, especially for knee angles and muscular activation patterns observation (Obradovič and Stančin 2023). Athlete lower limb kinematics can be quantified utilizing high-accuracy motion analysis systems powered by artificial intelligence, such as the neural networks tested by Bini et al. (2023). These technologies allow one to monitor real-time performance and also imply biomechanical changes depending on training data (Bini et al. 2023). Including the artificial intelligence and real-time feedback systems into training plans would enable athletes to significantly increase their capacity to modify their strategies over numerous sports.

Including real-time biomechanical data into cross-sport profiling reports a revolutionary way to maximize performance. LEOMO® and RETUL® systems are among wearable devices that have showed great potential in providing continuous data on joint angles and muscle activation during training and competition. These technologies provide athletes and coaches with quick feedback, allowing for real-time adjustments. Online data on pedaling force and joint angles can be used during cycling to optimize saddle height and pedal cadence, thereby improving both performance and

Table 3 Presents a comparison of common injuries across different sports and the biomechanical factors contributing to them.

Sport	Common injuries	Biomechanical cause
Cycling	Patellofemoral pain, lower back pain	Incorrect saddle height, poor knee alignment
Running	Stress fractures, Achilles tendinitis	Poor foot strike mechanics, over-pronation
Swimming	Shoulder impingement	Improper stroke technique, excessive internal rotation

Table 4 Compares various wearable technologies used in biomechanical profiling, outlining their key features and benefits.

Device	Function	Application	Accuracy
IMU	Joint angle tracking	Cycling, running, swimming	1–3° error
AI-powered neural networks	Kinematic analysis	Cycling, running	3–12° error
Accelerometers	Trunk and joint stability	Cycling, running, swimming	Moderate

comfort (Ho et al. 2023). Running and swimming also benefit much from such real-time feedback since little technical changes can have major performance improvement effects. While in swimming data on stroke dynamics and body positioning can be used to lower drag and boost speed, in running input on foot strike pat-terns and knee flexion can assist prevent overuse injuries (Yum et al. 2021).

4 Summary and Future Directions

Cross-sport biomechanical assessment provides a transforming approach for maximizing athletic performance in many sports. This method combines real-time feedback technology, predictive modeling, and consistent biomechanical principles to let athletes improve their efficiency while lowering injury risk. Cross-sport profiling offers a flexible framework applicable everywhere for both professional and novice athletes by investigating the similarities among sports including cycling, running, and swimming. Cross-sport profiling turns out to be the best approach for addressing the research gap in multisport performance enhancement. Integration of joint alignment, muscle activation, and force output metrics across various sports has led to important knowledge of performance improvement and injury prevention. By use of inertial measuring units (IMUs), accelerometers, and AI-driven motion analysis systems (Amin et al. 2024), technologies have made biomechanical assessments more accessible and accurate, therefore arming athletes with real-time data on their movement patterns.

Predictive algorithms and advanced simulation models have improved the accuracy of biomechanical profiling still more. By modifying cadence, stride length, and stroke frequency, these models help coaches and athletes to execute high performance scenarios and hence guide optimal performance techniques. An efficient approach for enhancing athletic performance and reducing injury risk in numerous sports is cross-sport biomechanical profiling. This strategy gives a comprehensive approach to maximizing athlete performance by combining universal biomechanical principles, predictive modeling, and real-time feedback systems. Future study should focus on developing more generalizable measurements, including artificial intelligence, and expanding biomechanical profiling to other sports. These advancements will elevate cross-sport profiling to the forefront of sports science, providing coaches and athletes with new techniques for peak performance.

Cross-sport biomechanical profiling requires additional research in four areas: developing standardized metrics that apply across multiple sports rather than sport-specific measurements, extending AI and machine learning integration for real-time analysis and feedback, conducting long-term studies on how athletes adjust to biomechanical changes over time, and understanding how psychological factors influence movement patterns and performance. These enhancements, through individualized feedback, would increase training efficiency and allow professional and amateur athletes to improve their performance using easily available technology.

Acknowledgements The authors would like to thank the Universiti Malaysia Pahang Al-Sultan Abdullah (UMPSA) for providing financial support under Internal Research Grant RDU2303101.

References

Amin MRRM, Nasir A, Muhammad Sidik MH, Jamaludin ASB, Rosli AM (2024) Applying convolutional neural networks (CNNs): a machine learning method in modern sports. In: Lecture notes in bioengineering. pp 405–415. https://doi.org/10.1007/978-981-97-3741-3_36

Bini RR (2020) Acute effects from changes in saddle height in perceived comfort during cycling. Int J Sports Sci Coach 15(3):390–397. https://doi.org/10.1177/1747954120918965

Bini RR, Nascimento VB, Nibali A (2023) Validity of neural networks in determining lower limb kinematics in stationary cycling. Sport Sci Health 20(1):127–136. https://doi.org/10.1007/s11332-023-01075-7

Burnie L, Barratt P, Davids K, Worsfold P, Wheat J (2020) Biomechanical measures of short-term maximal cycling on an ergometer: a test-retest study. Sports Biomech 22(8):997–1015. https://doi.org/10.1080/14763141.2020.1773916

Evans SA, James DA, Rowlands D, Lee JB (2021) Evaluation of accelerometer-derived data in the context of cycling cadence and saddle height changes in triathlon. Sensors 21(3):871. https://doi.org/10.3390/s21030871

Ho P, Yi C, Lin Y, Chung W, Chou P, Yang S (2023) Torque measurement and control for electric-assisted bike considering different external load conditions. Sensors 23(10):4657. https://doi.org/10.3390/s23104657

Molavian R, Fatahi A, Abbasi H, Khezri D (2023) Artificial intelligence approach in biomechanics of GAIT and sport: a systematic literature review. J Biomed Phys Eng 13(5). https://doi.org/10.31661/jbpe.v0i0.2305-1621

Obradović S, Stančin S (2023) Inertial measurement units' reliability for measuring knee joint angle during road cycling. Electronics 12(3):751. https://doi.org/10.3390/electronics12030751

Palmieri G, Tiboni M, Legnani G (2020) Analysis of the upper limitation of the most convenient cadence range in cycling using an equivalent moment based cost function. Mathematics 8(11):1947. https://doi.org/10.3390/math8111947

Pérez-Castilla A, García-Pinillos F (2024) Sports biomechanics applied to performance optimization. Appl Sci 14(9):3590. https://doi.org/10.3390/app14093590

Walck C, Lamb C, Vilches P (2022) Movement optimization through musculoskeletal modeling and multidimensional surface interpolation. In: Lecture notes in computational vision and biomechanics. pp 59–64. https://doi.org/10.1007/978-3-031-10015-4_5

Yum H, Kim H, Lee T, Park MS, Lee SY (2021) Cycling kinematics in healthy adults for musculoskeletal rehabilitation guidance. BMC Musculoskeletal Disord 22(1). https://doi.org/10.1186/s12891-021-04905-2

Ma CZH, Li Z, He C (2023) Advances in biomechanics-based motion analysis. Bioengineering 10(6):677. https://doi.org/10.3390/bioengineering10060677

Flexible Anti-Stabbing Fabrics in Martial Arts Gear: Design, Material Properties, and Performance Evaluation

Ahmad Shahir Jamaludin, Nurul Hasya Md Kamil, Khairul Ikmal Azmi, Mohamad Rusydi Mohamad Yasin, and Mohd Nizar Mhd Razali

Abstract Developing flexible anti-stabbing materials for martial arts equipment becomes even more crucial to raise athlete safety without sacrificing performance. Particularly the usage of multi-layer composites and ultra-high molecular weight polyethylene (UHMWPE)—recent developments in material science have delivered enhanced puncture resistance while preserving flexibility. Still, considering these advances one questions how best to balance safety, airiness, and freedom of movement. Layer design and material composition considerably affect the performance of the fabric in dynamic sports like martial arts, where comfort and flexibility are critical. Including auxetic structures—known for their ability to stretch laterally under pressure—into the fabric will aid to increase its energy absorption and comfort. Furthermore providing safety and extra performance advantages are smart textiles including sensors for real-time athlete physical stress monitoring. Still, the high-impact, dynamic motions unique of martial arts protective gear cannot be effectively duplicated in current testing protocols for it. New, standard tests covering puncture resistance, flexibility, and durability under realistic conditions have to be created. The best approaches to overcome these challenges are improving testing processes and optimizing material stacking with auxetic materials for flexibility and protection. Future study could look into hybridizing materials with nanocomposites to improve the performance of these fabrics even more. Smart material integration with martial arts equipment makes it feasible for the next generation of remarkably high-performance protective sports gear, therefore greatly enhancing athlete performance and safety.

Keywords Smart textiles · Anti-stabbing fabrics · Martial arts gear · Auxetic materials · UHMWPE

A. S. Jamaludin (✉) · N. H. M. Kamil · K. I. Azmi · M. R. M. Yasin · M. N. M. Razali
Faculty of Manufacturing and Mechatronic Engineering Technology, Universiti Malaysia Pahang Al-Sultan Abdullah, Pekan, Pahang, Malaysia
e-mail: shahir@umpsa.edu.my

A. S. Jamaludin
Centre for Advanced Industrial Technology, Universiti Malaysia Pahang Al-Sultan Abdullah, Pekan, Pahang, Malaysia

M. H. A. Hassan et al. (eds.), *Proceedings of the 3rd International Conference on Innovation and Technology in Sports*, Lecture Notes in Bioengineering,
https://doi.org/10.1007/978-981-95-0804-4_44

1 Introduction: The Rise of Smart Protective Fabrics in Martial Arts Gear

Using flexible anti-stabbing materials in martial arts sports equipment marks a major technological innovation in ensuring athlete safety without compromising performance (Wei et al., 2022; Wu et al. 2023a, b). Especially including physical activity like karate, judo, and mixed martial arts (MMA), martial arts generate greatly injury hazards, especially puncture wounds. In sports where agility and flexibility of movement are vital, traditional protective gear is generally heavy and inhibits athletes' mobility. Particularly multi-layer composite fabrics, recent developments in smart materials promise improved protection without compromising flexibility and comfort (Yan and Han-Song 2020).

The development of stab-resistant materials—especially those made from ultra-high molecular weight polyethylene (UHMWPE)—marked a significant progress in this do-main. Good tensile strength and flexibility these materials offer help to balance wearability with protection. Research on these textiles has shown positive results in terms of puncture resistance and durability especially when numerous layers are combined with woven and nonwoven materials, as shown in Fig. 1 (Zhang et al. 2023).

The state of the art reveals a significant research gap even if the ideal con-figuration of these layers—including the number of layers, their arrangement, and the mix of

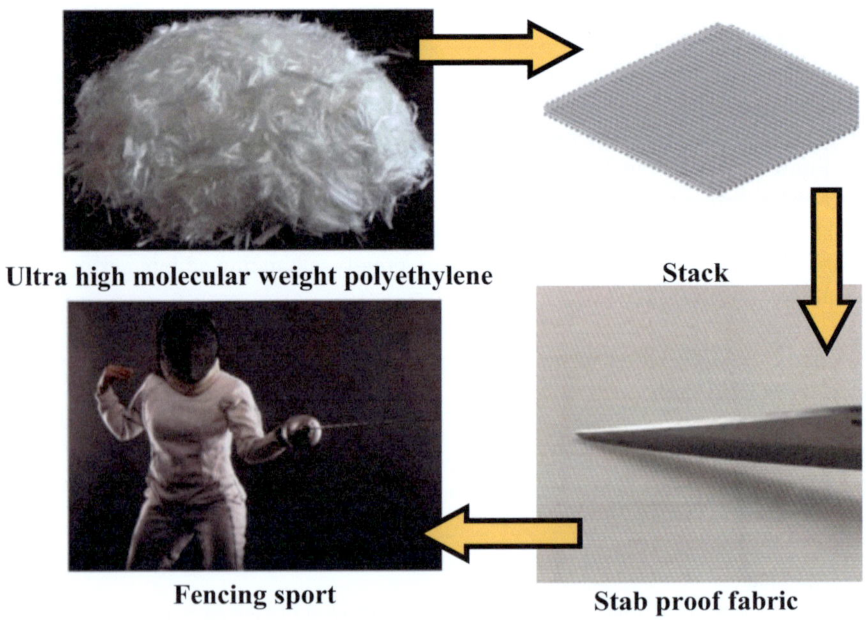

Fig. 1 Numerous layer stack to obtain stab proof fabric

different materials to maximize both comfort and performance—is exposed by their absence. Studies reveal that although more layers usually translate into improved puncture resistance, they also lower ventilation and moisture permeability (Zhang et al. 2023; Dayyoub et al. 2022). This trade-off generates need for more research on how these materials could be best used for martial arts equipment since athletes need not only protection but also maximal mobility of movement and comfort during intense physical exercise. The testing and standardizing of these materials therefore constitute another important challenge. Many of the present guidelines for protective sports equipment ignore the particular needs of martial arts. Particularly underdeveloped are testing methods for puncture resistance and durability during dynamic motions, which has resulted in differences in the stated efficacy of different protective gear (Duncan et al. 2018). Therefore, more thorough testing simulating real-world use situations, such continuous movement and impact, which are typical in martial arts training and competition.

2 State of the Art: Current Innovations and Challenges in Anti-Stabbing Material

Ongoing studies on martial arts protective gear have found several ways in improving performance, comfort, and safety. These methods center on structural design, material optimization, and improved monitoring technology, as shown in Table 1.

3 Optimizing Material Design and Testing Protocols for Improved Performance

Among the various solutions presented, advanced layer optimization combined with auxetic structures and improved testing and standardization protocols seems to offer the most promising approach in the development of flexible anti-stabbing fabrics for martial arts gear.

1. Advanced Layer Optimization:

Studies indicate that the comfort and protection qualities of the textile depend critically on the arrangement of various components. For martial arts clothing, combining woven and nonwoven layers of ultra-high molecular weight polyethylene (UHMWPE) provides outstanding puncture resistance; yet, these materials must be correctly arranged to prevent compromising breathability and flexibility (Zhang et al. 2023). Martial arts practitioners who require maximal mobility throughout training and competition have to maintain flexibility while building tensile strength. Optimizing a four-layer configuration—two woven and two nonwoven—has provided an excellent blend of protection and comfort as well as suitable air and moisture permeability (Zhang et al. 2023; Cheng et al. 2023).

Table 1 Summary of innovative approaches for enhancing the performance and safety of flexible anti-stabbing fabrics in martial arts gear, focusing on material optimization, structural design, and testing protocols

Approach	Description	Key benefits	References
Advanced layer optimization	Investigates the optimal layering of woven, nonwoven, and composite materials to balance protection, flexibility, and breathability	Improved puncture resistance and moisture management; fine-tuned designs for martial arts-specific needs	Zhang et al. (2023)
Incorporating auxetic structures	Utilizes materials that expand laterally when stretched, enhancing energy absorption and adaptability to movement	Increased comfort and flexibility; enhanced impact resistance; better alignment with athlete's dynamic movements	Duncan et al. (2018)
Smart material integration	Embeds sensors into fabrics for real-time monitoring of athlete performance and safety	Dynamic, personalized protection; alerts for potential injury risks; data-driven training enhancements	Barbosa (2018)
Improved testing and standardization	Develops testing protocols that simulate dynamic, high-impact movements typical of martial arts activities	Reliable comparisons between materials; clearer certification guidelines; enhanced safety assurance for real-world use	Duncan et al. (2018), Dayyoub et al. (2022)

Solution Impact: Additionally improving puncture resistance, layer optimization lowers weight and stiffness. This would enable martial arts athletes to keep the necessary agility and mobility while gaining from enhanced safety, hence tackling one of the main difficulties in creating appropriate protective gear (Dayyoub et al. 2022).

2. Incorporation of Auxetic Structures

Including auxetic constructs into martial arts tools helps to raise the protective capability and flexibility even more. Special in their nature, auxetic materials expand laterally when stretched, therefore enhancing energy absorption and adaption to the athlete's activities (Duncan et al. 2018). Form-fitting characteristics of auxetic materials assist sports equipment to be more comfortable and to raise impact resistance. Moreover, their almost perfect fit to the body helps to better disperse pressure in collisions or impacts, hence reducing the likelihood of injuries (Chu et al. 2023).

Solution Impact: Especially in contact-intensive sports like martial arts, auxetic structures would directly aid to reduce injury risks by absorbing more energy during impacts. Mostly this capacity helps to prevent injuries during high-intensity bouts and training sessions (Duncan et al. 2018).

3. Improved Testing and Standardization

Standardized testing procedures simulating real-world use situations have to be created in order to confirm the success of these improved materials. Many times ignoring the dynamic and high-impact character of martial arts activities, current standards produce protective gear inadequate for the demands of the sport (Barbosa 2018). New testing methods for evaluating material performance are critical in contexts like martial arts, which are fast-paced and high-contact. These testing could include dynamic puncture resistance, flexibility under motion, and durability over time, ensuring that the gear meets safety standards while maintaining performance quality (Khuyen et al. 2021).

Solution Impact: Manufacturers should improve testing processes to better develop and certify martial arts-optimized protective gear, giving athlete's confidence in the gear's reliability during intense physical activity (Barbosa 2018). This would also establish industry-wide benchmarks, resulting in higher-quality products throughout the market (Krushna et al. 2023).

By combining sophisticated layer optimization, auxetic materials, and improved testing processes, martial arts protective gear can evolve to provide superior protection, more comfort, and greater agility. These materials will enable the creation of clothing that is lightweight, extremely breathable, and puncture-resistant, representing a major improvement over standard protective fabrics. This integrated approach can solve the existing research gaps in martial arts gear by addressing both the material properties and the lack of comprehensive testing, thus creating a pathway for the development of highly effective and safe protective gear for athletes (Wang et al. 2023).

4 Summary and the Path Forward

Investigating flexible anti-stabbing materials for martial arts equipment has revealed fascinating chances for combining innovative materials with contemporary design ideas. Closing the protection-performance gap calls both layer optimization and auxetic materials as well as better testing techniques.

- Layer optimization strikes a mix between protection and comfort for martial arts athletes who want both freedom of mobility and safety.
- Auxetic materials help to form and absorb energy. In contact-heavy martial arts, appropriate flexibility helps to lower the risk of injury.
- Standardizing and improved testing will help to close the testing gap by ensuring that protective gear is suitable to the specific needs of martial arts and durable.

Following topics can be further as future direction of the study in flexible anti-stabbing materials:

Hybrid UHMWPE: Further research into hybridizing UHMWPE with other materials such nanocomposites or graphene-based materials could provide even thinner, stronger, and more breathable fabrics. For smart gear, these hybrids might also provide fresh features as improved heat control or conductivity.

Integration of Wearable Technology: As smart textiles get more developed, their incorporation into martial arts gear could offer real-time feedback on performance and safety, therefore giving coaches and athlete's vital information on pressure points and stress during battle. By warning sportsmen to perhaps harmful levels of effort, this could also act as a prophylactic against damage.

Long-term Wearability and Durability: Studies should look at how these parts resist environmental factors such sweat, temperature swings, and wear and tear from continuous motion over several years.

By making martial arts equipment safer, more comfortable, and more suited for the physical demands of the practice, flexible anti-stabbing materials could revolutionize it. Still, achieving these goals will need constant material science and product testing study. By using smart design basic concepts, the sports industry may close present gaps and open the road for the future generation of high-performance sports equipment.

Acknowledgements The authors would like to thank the Ministry of Higher Education (MoHE) Malaysia and Universiti Malaysia Pahang Al-Sultan Abdullah for the financial support provided under the Fundamental Research Grant Scheme with Grant No: FRGS/1/2024/TK10/UMP/02/2 (UMPSA internal reference: RDU240103) in completing this work.

References

Barbosa T (2018) Smart sport equipment: reshaping the sports landscape. Motricidade 14(2–3):1–2

Cheng L, Li J, Guo A, Zhang J (2023) Recent advances in flexible noninvasive electrodes for surface electromyography acquisition. npj Flex Electron 7(1):1–26

Chu Y, Zhu B, Li S, Chen C, Huang W, Liu Y, Zhao X, Zhang Z, Fahaduzzaman M, Wang X, Chen C, Chai X, Chen X (2023) A novel flexible and breathable stab-resistance fabric composites with ZnO nanowires growing technology. J Market Res 24:6438–6450

Dayyoub T, Maksimkin AV, Filippova OV, Tcherdyntsev VV, Telyshev DV (2022) Shape memory polymers as smart materials: a review. Polymers 14(17):3511

Duncan O, Shepherd T, Moroney C, Foster L, Venkatraman PD, Winwood K, Allen T, Alderson A (2018) Review of auxetic materials for sports applications: expanding options in comfort and protection. Appl Sci 8(6):941

Khuyen NQ, Han PVD, Nguyen NT, Le QB, Harjo M, Anbarjafari G, Kiefer R, Tamm T (2021) The use of laminates of commercially available fabrics for anti-stab body-armor. Polymers 13(7):1077

Krushna BR, Sharma S, Prasad BD, Francis D, Sridhar C, Misra D, Bose M, Kumar HP, Shetty A, Nagabhushana H (2023) Fabrication of highly flexible luminescent films, hydro gels and anti-counterfeiting applications of La2MoO6:Sm3+ phosphors. J Sci Adv Mater Dev 9(1):100641

Wang S, Chen M, Wei L, Dong J, Sun R (2023) Anti-puncture, frigostable, and flexible hydrogel-based composites for soft armor. J Market Res 21:2915–2925

Wei R, Dong B, Zhai W, Li H (2022) Stab-resistant performance of the well-engineered soft body armor materials using shear thickening fluid. Molecules 27(20):6799

Wu S, Sikdar P, Bhat GS (2023a) Recent progress in developing ballistic and anti-impact materials: nanotechnology and main approaches. Defence Technol 21:33–61

Wu Y, Jia J, Wang Y, Zheng C, Zhao L (2023b) Investigation on hydraulic fracturing and flexible anti-hydrofracturing solution for Xiaowan Arch Dam. Appl Sci 13(16):9302

Yan Z, Han-Song L (2020) Preparation and application of continuous fiber reinforced high-performance thermoplastic composites. DOAJ Directory of Open Access Journals

Zhang J, Ding H, Wang Y (2023) Preparation and sports performance of multilayer flexible anti-stabbing fabric martial arts sports composite conjugate materials. Front Chem 11

Simulating Reality: High-Fidelity Simulators and Its Impact on Cognitive and Physical Performance of an Athlete

Ahmad Shahir Jamaludin, Mohamad Rusydi Mohamad Yasin, Abdul Nasir, Mohd Arifpin Mansor, and Mohd Nizar Mhd Razali

Abstract Proposing a strong approach to enhance both mental and physical performance, high-fidelity simulators are transforming areas including sports, healthcare, and marine training. These simulators create realistic settings that challenge people to enhance their motor abilities and decision-making in real-life circumstances. High-fidelity simulations can be expensive, and it is yet unknown how closely the knowledge gained in these virtual environments applies to real-world scenarios. Models of hybrid training offer a balanced approach to combine the benefits of both approaches. This approach links virtual skills with actual application by combining sophisticated virtual simulations with useful instruction. Studies show that hybrid models improve physical abilities via instantaneous input in addition to cognitive skills including fast decision-making in high-stress events. This all-encompassing strategy provides a more harmonic and effective way for growth and education. Hybrid training provides excellent flexibility. For many sectors, it is adaptable and allows quite cheap scalability. Combining creative VR simulations with more basic physical training techniques allows businesses to build customized, reasonably priced solutions. These hybrid training designs ensure acquired abilities by way of feedback loops allowing customized training to be sustained throughout time. The main focus should be on evaluating over time how well these skills are maintained, improving adaptive simulations, and matching the costs to benefits. The review shows how hybrid training models could be used in numerous fields, therefore generating enhanced and more successful training solutions.

A. S. Jamaludin · M. R. M. Yasin (✉) · M. N. M. Razali
Faculty of Manufacturing and Mechatronic Engineering Technology, Universiti Malaysia Pahang Al-Sultan Abdullah, Pekan, Pahang, Malaysia
e-mail: rusydi@umpsa.edu.my

A. Nasir
Faculty of Electrical and Electronic Engineering Technology, Universiti Malaysia Pahang Al-Sultan Abdullah, Pekan, Pahang, Malaysia

M. A. Mansor
Department of Information and Intelligent Engineering, National Institute of Technology (KOSEN), Sasebo College, 1-1 Okishin, Sasebo-Shi, Nagasaki 857-1193, Japan

M. H. A. Hassan et al. (eds.), *Proceedings of the 3rd International Conference on Innovation and Technology in Sports*, Lecture Notes in Bioengineering,
https://doi.org/10.1007/978-981-95-0804-4_45

Keywords High-fidelity simulators · Hybrid training models · Cognitive performance enhancement · Physical performance optimization · Skill transfer in training environments

1 The Role of High-Fidelity Simulators in Enhancing Cognitive and Physical Performance

High-fidelity simulations are revolutionizing industries such as sports, marine training, aviation, and healthcare (Moliterno et al. 2024). These simulators use powerful virtual reality (VR), augmented reality (AR), and motion capture technology to accurately recreate real-world environments. Their capacity to reproduce complicated cognitive and physical activities provides practical training that improves cognitive skills (such as decision-making under duress) and physical performance, as seen in Fig. 1 (Moliterno et al. 2024). Research on high-fidelity simulation finds major gaps in generalizability, accessibility, and cost-effectiveness, notably in the balance of cognitive and physical gains (Guerrero et al. 2022).

Cognitive training in high-fidelity environments emphasizes mental resilience, problem solving, and decision-making abilities. This is especially important in safety–critical contexts such as aviation, marine, and military training, where split-second decisions may be the difference between success and disaster. Studies on marine simulation fidelity, for example, show that students who train in high-fidelity simulators have considerably higher self-efficacy, motivation, and perceived

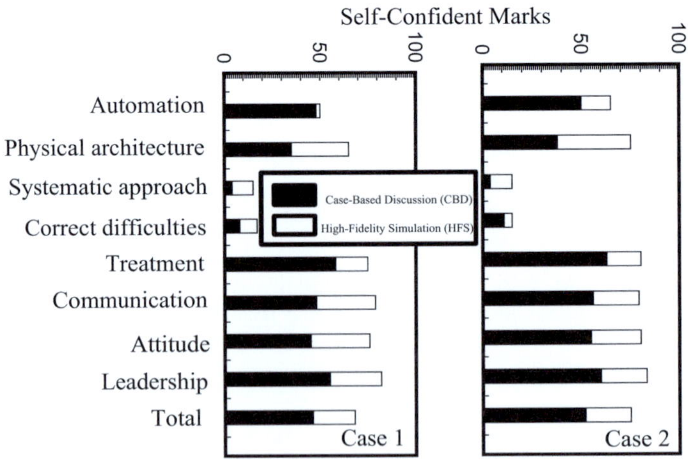

Fig. 1 Crude means of the post-intervention scores of the high-fidelity simulation group (HFS) and the case-based discussion group (CBD) in the simulation checklist of two scenarios (Moliterno et al. 2024)

skill growth (Renganayagalu et al. 2019). These students worked with immersive simulators that recreated real engine-room situations, including complicated decision-making scenarios that needed cognitive agility.

In sports, cognitive performance is critical for enhancing tactical skills. VR-based therapies have demonstrated promise in improving an athlete's cognitive reaction under high-pressure situations. Figure 2 depicts how virtual simulations assist athletes in responding to unexpected events demanding tactical and strategic thinking (Fortes et al. 2021; Richlan et al. 2023). However, whether this cognitive improvement is applicable to real-world situations remains to be seen. According to study, there is uncertainty about the long-term preservation of these cognitive benefits and how well they translate outside of controlled simulations (Choi et al. 2023). From a physical standpoint, high-fidelity simulators allow for comprehensive motion analysis, which can improve biomechanical performance. Motion capture and simulation approaches in sports training are especially useful for improving motor abilities. In basketball training, for example, the combination of Augmented Reality/Virtual Reality (AR/VR) and motion capture technology has enabled athletes to improve their ball shooting, defense strategy, and movement patterns by providing real-time feedback on joint angles and velocities (Liu et al., 2023). This results in a considerable reduction in response times throughout gameplay, indicating physical improvement (Wong and Wong 2023).

Furthermore, motion analysis systems are gaining popularity in combat sports, where they sharpen technical skills and boost physical resilience. These systems combine digital video, optical sensors, and electromechanical transmitters to provide precise feedback on movements and procedures (Polak et al. 2016). The degree of realism in motion simulation helps sportsmen enhance their physical performance as well as injury prevention. Notwithstanding progress, one of the most important research gaps is the practical application of acquired knowledge in simulation systems. For instance, even if athletes can develop in virtual environments, it is usually difficult to translate these improvements to actual competitions. Moreover, although useful, high-fidelity simulations are costly and typically unavailable to smaller-scale training facilities, hence creating an economic disparity in access to sophisticated training technologies (Abbasi et al. 2024; Renganayagalu et al. 2019).

2 Hybrid Training Models: A Synergy Between Simulation and Real-World Practice

Developing hybrid training models that mix conventional teaching techniques with high-fidelity simulators is one way forward. Research comparing virtual training with mannequin-based simulation show the supportive character of both approaches. While virtual approaches are better suited for cognitive components such decision-making (Abbasi et al. 2024), mannequin-based solutions shine in enhancing physical performance in CPR instruction. Applying this mixed approach to sports and other

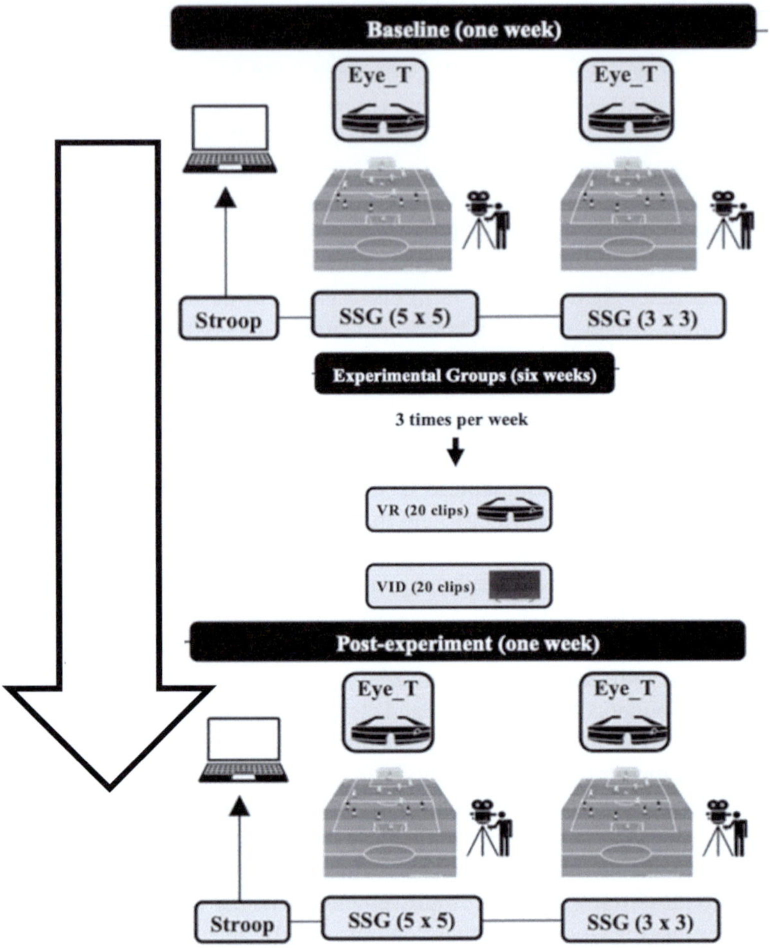

Fig. 2 Experimental design of the study. *Note* SSG = small-sided game; Eye_T = Eye-tracker; VR = virtual reality; VID = video screen (Fortes et al. 2021)

professions ensures that both cognitive and physical skills are sufficiently developed, hence overcoming the skill transfer barrier. In sports, such hybrid models may enable athletes to switch between high-fidelity VR training and on-field practice. This approach can guarantee that physical repetition in real-world settings supports cognitive developments gained in virtual environments. The aim is to facilitate a seamless transfer of abilities from simulation to actual performance, therefore minimizing the difference between training settings and practical uses (Liu et al., 2023).

Another option is to create cost-effective simulators that maintain high fidelity while being less expensive. Integration of hardware and software innovations, such as the use of low-cost sensors and open-source motion capture systems, can result in high-fidelity simulations that are more accessible to a wider range of sectors. For example, reduced versions of motion analysis systems used in elite sports could be developed to satisfy the needs of smaller training programs (Polak et al. 2016). Furthermore, deploying cloud-based simulations may assist lower the cost of physical infrastructure. Training centers can use cloud computing to remotely access high-fidelity simulations, removing the need for expensive on-site equipment. This strategy would increase availability of cutting-edge training techniques in many different fields (Meyers et al., 2020).

Moreover, trainers could incorporate feedback systems that precisely track a person's performance both in real-world environments and during high-fidelity simulations so enhancing the practical relevance of the acquired abilities. Trainers can modify and personalize simulation environments depending on statistics on athletes' or trainees' outside of simulation performance. Wearable sensors would allow one to monitor athlete performance in both training and competition contexts. This data allows VR or AR simulations aimed at specific regions for development to be generated. Such feedback loops would assure a continual process of skill refining and improve the chances of converting simulated skills to real-world performance (Liu et al., 2023).

The cognitive burden placed on trainees is an important aspect in determining the success of high-fidelity simulators. Burnout and a loss of retention of acquired knowledge from training can result from too much cognitive strain. Consequently, it is imperative to control cognitive load by means of adaptable simulation environments catered to user performance (Meyers et al., 2020). Starting with simple scenarios and progressively become more complex, simulations could flattening the learning curve and increase cognitive skill retention by means of flattening of the learning curve and In high-stress fields like athletics or aviation, where trainees must progressively build mental toughness, this approach might be very helpful. Adaptive simulators' tailored training experiences would help users control their stress and improve their under duress judgments (Bhat et al., 2023).

3 Proposing Blending Simulated and Physical Training: Closing the Gap in Cognitive and Physical Skill Transfer

Among the various suggested approaches to close the distance between high-fidelity simulations and real-world performance, both cognitively and physically, hybrid training models stand out as the most promising one. These methods combine the best features of real-world experience and simulation-based training to guarantee that the improvements made in simulated situations are reinforced in actual settings. Following are the possible benefit gained from the approach:

1. **Enhanced skill transfer**—The possibility of divergence between virtual and real-world performance raises the main issues with high-fidelity simulators. Athletes, for example, might shine in a virtual environment where outside variables are under control but their performance might change when erratic real-world events are added. Simulated or real-world hybrid models enable individuals to modify their surroundings, therefore closing this disparity and supporting the cognitive and physical skills developed. Motion capture technology can be used in sports, for example, to track physical performance in real-time during a training session; VR environments can allow athletes improve their tactical skills and decision-making (Liu et al., 2023). After that, the players can use the knowledge acquired in simulated sessions to real-world gameplay or practice. This kind of integration between virtual and real surroundings helps confirm both cognitive and physical improvements, hence lowering the danger of skill atrophy between sessions (Richlan et al. 2023).
2. **Complementary Cognitive and Physical Benefits**—Hybrid training programs combine mannequin-based physical education with VR's ability to simulate cognitive obstacles such as decision-making under duress, providing the benefits of both. This twin approach is especially important in high-stakes industries such as athletics and healthcare, where both physical and cognitive performance are required. Particularly in cardiopulmonary resuscitation (CPR) training, studies comparing mannequin-based simulation with virtual approaches show that the mix of these techniques produces greater overall skill development (Abbasi et al. 2024). Therefore, the physical repetition provided by mannequin-based training enhances the cognitive gains attained by VR (including decision-making and situational awareness).
3. **Adaptability to Various Industries**—The great benefit of hybrid models is their flexibility in many spheres. In the sports sector as much as in the health-care sector, hybrid models provide adaptable, scalable answers fit for particular requirements. In national defense education, for instance, high-fidelity simulators let trainees participate with realistic decision-making activities; hands-on practice strengthens the procedural knowledge acquired in the simulations (Renganayagalu et al. 2019). This adaptability makes hybrid training approaches a good fit for many disciplines, therefore encouraging cognitive resilience as well as physical proficiency (Brown et al. 2020).
4. **Economic Feasibility and Scalability**—Another important benefit of hybrid training is its cost-effectiveness relative to entirely high-fidelity simulations. Although building a whole VR-based environment could need a large initial outlay, combining VR with physical training techniques such as mannequins or real-world practice will maximize available resources. Industries can keep training quality without paying outrageous expenses by investing in fairly high-fidelity simulators for cognitive activities and matching low-tech solutions for physical exercise (Polak et al. 2016).

Moreover, this method can easily be scaled up. New institutions or sports training centers without advanced high-fidelity simulators can benefit from the hybrid model as a cost-effective option. They can use traditional training methods for physical conditioning and utilize VR and motion capture for cognitive training (Armenia et al., 2018; Meyers et al. 2020).

4 Summary—Towards Accessible, Adaptive, and Efficient Training Solutions

Including high-fidelity simulators into training settings particularly in sports sectors marks a major progress in cognitive and physical skill development. To close the shortcomings related to present high-fidelity simulation techniques, the hybrid training model offers the most interesting solution nonetheless. Combining virtual reality system with physical exercise approach ensures that cognitive and motor skills are enhanced. Future study should focus especially on improving hybrid models since longitudinal studies to investigate the long-term retention of taught skills from hybrid training should help us. Then, it is imperative to improve tailored instruction that fits simulation settings to individual learning curves. Adaptive simulations that change depending on a trainee's performance would offer a more effective learning environment, therefore lowering cognitive load and increasing retention. Furthermore underlined should be cost–benefit studies evaluating the financial feasibility of hybrid models among several sectors. Research should look at how fairly affordable hybrid models could be utilized in environments with restricted resources. The future of training is finally in an integrated approach combining the best of both worlds' high-fidelity simulations for cognitive development with actual physical conditioning. By filling in the present void between virtual and real training settings, hybrid models will revolutionize how we equip people for high-stakes events in sports, healthcare, or other critical industries.

Acknowledgements The authors would like to thank the Ministry of Higher Education (MOHE) Malaysia and Universiti Malaysia Pahang Al-Sultan Abdullah for the financial support provided under the Fundamental Research Grant Scheme Grant No: FRGS/1/2024/TK10/UMP/02/2 (UMPSA Internal Reference: RDU240103) in completing this work.

References

Abbasi M, Asadoola Y, Ebrahimi H, Bahonar E, Dabirian Z, Esmaeili S, Mahdizadeh A, Sadeghi M (2024) Comparison of mannequin-based simulation training method with virtual training method on nursing students' learning cardiopulmonary resuscitation: a controlled randomized parallel trial. Iran J Nurs Midwifery Res 29(4):438–445. https://doi.org/10.4103/ijnmr.ijnmr_222_23

Armenia S, Thangamathesvaran L, Caine A, King N, Kunac A, Merchant A (2018) The role of high-fidelity team-based simulation in acute care settings: a systematic review. Surg J 04(03):e136–e151. https://doi.org/10.1055/s-0038-1667315

Bhat M, Mijar M, Dsouza L, Monteiro B, Kamath T (2023) Role of high-fidelity simulation in obstetrics ultrasound training: a fresh perspective in the COVID era. Muller J Med Sci Res 14(1):28–34. https://doi.org/10.4103/mjmsr.mjmsr_79_22

Brown WJ, Tortorella RAW (2020) Hybrid medical simulation—a systematic literature review. Smart Learn Environ 7(1). https://doi.org/10.1186/s40561-020-00127-6

Choi S, Shim H, Lee JH (2023) Efficiency analysis by training sequence of high-fidelity simulation-based neonatal resuscitation program (NRP). PLoS ONE 18(2):e0281515. https://doi.org/10.1371/journal.pone.0281515

Fortes LS, Almeida SS, Praça GM, Nascimento-Júnior JR, Lima-Junior D, Barbosa BT, Ferreira ME (2021) Virtual reality promotes greater improvements than video-stimulation screen on perceptual-cognitive skills in young soccer athletes. Hum Mov Sci 79:102856. https://doi.org/10.1016/j.humov.2021.102856

Guerrero JG, Rosales NS, Castro GMT (2022) Impact of high-fidelity simulation exposure of nursing students with their objective structured clinical examination: a quasi-experimental study. Nurs Open 10(2):765–772. https://doi.org/10.1002/nop2.1343

Liu C, Liu Y, Ni K (2023) Motion capture method for college basketball training based on AR/VR technology. Appl Math Nonlinear Sci 9(1). https://doi.org/10.2478/amns.2023.2.00298

Meyers L, Mahoney B, Schaffernocker T, Way D, Winfield S, Uribe A, Mavarez-Martinez A, Palettas M, Lipps J (2020) The effect of supplemental high fidelity simulation training in medical students. BMC Med Educ 20(1). https://doi.org/10.1186/s12909-020-02322-y

Moliterno NV, Paravidino VB, Robaina JR, Lima-Setta F, Da Cunha AJLA, Prata-Barbosa A, De Magalhães-Barbosa MC (2024) High-fidelity simulation versus case-based discussion for training undergraduate medical students in pediatric emergencies: a quasi-experimental study. J Pediat 100(4):422–429. https://doi.org/10.1016/j.jped.2024.03.007

Polak E, Kulasa J, VencesBrito A, Castro MA, Fernandes O (2016) Motion analysis systems as optimization training tools in combat sports and martial arts. Rev Artes Marcial Asiát 10(2):105. https://doi.org/10.18002/rama.v10i2.1687

Renganayagalu SK, Mallam S, Nazir S, Ernstsen J, Haavardtun P (2019) Impact of simulation fidelity on student self-efficacy and perceived skill development in maritime training. TransNav Int J Mar Navig Saf Sea Transp 13(3):663–669. https://doi.org/10.12716/1001.13.03.25

Richlan F, Weiß M, Kastner P, Braid J (2023) Virtual training, real effects: a narrative review on sports performance enhancement through interventions in virtual reality. Front Psychol 14. https://doi.org/10.3389/fpsyg.2023.1240790

Wong FMF, Wong DCN (2023) A modified guideline for high-fidelity patient simulation to improve student satisfaction and self-confidence in learning: a mixed study. Nurs Rep 13(3):1030–1039. https://doi.org/10.3390/nursrep13030090

Development of a Sensor-Based Tool for Assisting Beginners in Table Tennis: A Case Study on Forehand Grip Consistency

Syukran Hakim Norazam, Muhamad Ridzuan Radin Muhamad Amin, Mohd Amir Shahlan Mohd Aspar, Muhammad Nur Farhan Saniman, and Abdul Nasir Abd Ghafar

Abstract This study presents the development of a sensor-based training tool designed to assist beginner table tennis players in maintaining a consistent forehand grip and swing technique. The system integrates an MPU6050 motion sensor with an ESP32 microcontroller to monitor racket angle and acceleration in real time. Players receive immediate feedback through visual (LED) and auditory (buzzer) signals, allowing for real-time posture and grip corrections. The system was tested in controlled experimental conditions, analyzing stroke performance across multiple trials. The results indicated that proper strokes exhibited well-defined acceleration and rotational motion patterns, whereas inconsistent strokes showed irregular fluctuations and instability across all axes. Comparative data analysis highlighted that stable bat angles and smooth angular velocity trends were key indicators of well-executed strokes. This system provides an objective and data-driven approach to skill development, offering a structured method for players to refine their technique. Future enhancements will focus on expanding the system's capabilities to analyze additional stroke types and incorporating AI-driven adaptive feedback for personalized training.

Keywords Sensor-based feedback · Table tennis training · Forehand stroke analysis · Sports technology

S. H. Norazam · M. R. R. M. Amin (✉) · M. A. S. M. Aspar · A. N. A. Ghafar
Universiti Malaysia Pahang Al-Sultan Abdullah, Pekan, Pahang, Malaysia
e-mail: muhamadridzuan@umpsa.edu.my

M. N. F. Saniman
University of Kuala Lumpur Malaysia France Institute, Bandar Baru Bangi, Selangor, Malaysia

M. H. A. Hassan et al. (eds.), *Proceedings of the 3rd International Conference on Innovation and Technology in Sports*, Lecture Notes in Bioengineering,
https://doi.org/10.1007/978-981-95-0804-4_46

1 Introduction

1.1 Background

Table tennis is a widely recognized sport that requires a combination of speed, precision, and strategic execution. The forehand stroke is one of the most fundamental techniques in the game, as it enables players to generate controlled power and spin (Ma et al. 2024). Mastering this stroke requires a stable grip, proper swing mechanics, and consistent execution. However, beginner players often struggle with maintaining an optimal racket angle and stroke consistency, which directly affects their ability to control the ball and execute accurate shots (Jamil and Yusoff 2016).

Traditional coaching methods predominantly rely on visual observation and verbal feedback from coaches, which can be subjective and inconsistent (Smith and Brown 2020). While experienced players can self-correct based on prior training, beginners lack the necessary motor coordination and kinesthetic awareness to recognize their mistakes in real-time (Wang et al. 2019). According to research on sensor-based training in sports, incorporating wearable motion-tracking devices has significantly improved athletes' ability to refine their technique by providing quantitative performance metrics rather than subjective evaluation (Johnson and Lee 2020).

In table tennis, limited studies have explored the use of real-time feedback systems for stroke correction (Kim et al. 2019). Existing training aids, such as robotic ball feeders and video-based analysis, primarily focus on improving shot accuracy rather than refining biomechanical execution (Zhang and Liu 2020). This research aims to bridge this gap by developing a sensor-based training tool that provides instant corrective feedback based on stroke dynamics. By integrating an MPU6050 motion sensor with an ESP32 microcontroller, the system captures grip stability, angular movement, and stroke acceleration, enabling players to adjust their technique dynamically (Gupta et al. 2020). This data-driven approach offers an objective and repeatable method for assessing and improving forehand stroke mechanics, reducing reliance on conventional coaching techniques (Amin et al. 2024).

1.2 Problem Statement

Beginner table tennis players frequently experience difficulties in maintaining a consistent forehand grip and executing a controlled swing, leading to ineffective stroke mechanics (Patel and Shah 2020). This inconsistency can negatively impact shot accuracy, energy transfer efficiency, and overall skill progression (Chen et al. 2020). In traditional coaching setups, stroke assessment is predominantly qualitative, relying on a coach's expertise to observe and correct mistakes. However, such methods are inherently subjective, and beginners often fail to receive instant feedback that allows for immediate adjustments (Tanaka et al. Sept. 2020).

Moreover, most existing training tools focus primarily on shot accuracy rather than stroke execution analysis (Karumdin et al. 2024). There is a lack of real-time biomechanical feedback systems that quantitatively assess grip stability, swing path, and racket angle variations (Li and Wang 2020). Without access to such insights, beginner players may develop poor muscle memory, reinforcing incorrect techniques over time (Kumar et al. 2021). This research aims to overcome these challenges by designing a wearable motion-tracking system that provides instant feedback on stroke mechanics, allowing players to correct their form dynamically during training sessions (Park and Choi 2021).

1.3 Objective

The main objective of this study is to develop a sensor-based training system that enhances the learning process for beginner table tennis players. Specifically, the study aims:

1. To design and implement a motion sensor-based feedback tool capable of evaluating grip stability and swing consistency in real-time.
2. To develop a real-time feedback mechanism using LED and buzzer signals to provide instant corrective guidance during forehand stroke execution.
3. To analyze and compare stroke execution across different players, focusing on racket angle variations, acceleration trends, and rotational stability, with the goal of identifying key biomechanical indicators of an effective stroke.

1.4 Scope of Study

This study is specifically focused on beginner table tennis players, aiming to improve their forehand stroke consistency through an objective, motion-based feedback system. The developed tool is designed to monitor grip stability and swing dynamics using an MPU6050 sensor while providing instant feedback through LED and buzzer signals. The system captures key biomechanical stroke parameters, including racket angles, acceleration variations, and rotational stability, which are crucial for refining techniques.

At this stage, the system is limited to analyzing forehand strokes only, as backhand techniques, footwork mechanics, and spin analysis are not considered within the scope of this research. Future enhancements may include advanced tracking algorithms to monitor multiple stroke types and integrate AI-driven adaptive training feedback, further personalizing the training experience.

2 Methodology

2.1 System Architecture

The proposed system is designed to provide real-time feedback to beginner table tennis players by analyzing their forehand stroke mechanics. The system consists of two core components: a motion-sensing module and a feedback module. The motion-sensing module comprises an MPU6050 inertial measurement unit (IMU), which captures angular velocity and acceleration data, and an ESP32 microcontroller, which processes the sensor data. The feedback module consists of LED indicators and a buzzer, which provide immediate alerts when deviations from the optimal stroke mechanics are detected.

The workflow of the system is as follows:

1. The MPU6050 sensor, mounted on the player's racket, continuously captures movement data, including grip stability, stroke angle, and acceleration.
2. The ESP32 microcontroller processes this data and compares it against predefined thresholds for correct stroke mechanics, which were established based on data collected from experienced players.
3. If the player's stroke deviates from these optimal thresholds, the red LED and buzzer are triggered to indicate an incorrect motion. If the stroke aligns with the correct parameters, a green LED provides positive reinforcement.
4. The collected data is stored for further performance analysis, allowing players to review and refine their techniques over multiple training sessions.

By integrating sensor technology and real-time feedback mechanisms, the system provides objective performance evaluation rather than relying on subjective assessments from coaches.

2.2 Hardware Development

The hardware system consists of three key components:

1. MPU6050 Sensor—A 6-axis motion sensor that detects angular rotation and linear acceleration of the racket, capturing stroke motion in three dimensions.
2. ESP32 Microcontroller—A low-power, high-speed microcontroller that processes sensor data and controls feedback signals.
3. Feedback Indicators (LEDs and Buzzer)—Red and green LEDs indicate incorrect and correct strokes, respectively, while a buzzer provides auditory feedback.

The MPU6050 sensor is securely mounted onto the racket handle, ensuring minimal interference with player movement. The ESP32 microcontroller is attached to a wristband worn by the player, enabling wireless data transmission and reducing discomfort. The system is powered by a 5 V rechargeable battery, allowing for prolonged training sessions.

2.3 *Software Development*

The software for the system was developed using the Arduino IDE, which was compatible with the ESP32 microcontroller. The software's main function was to process the data from the MPU6050 sensor and determine whether the player's grip and swing fell within the acceptable range for a correct forehand stroke.

The program was structured into several key functions:

1. Sensor Data Acquisition: The MPU6050 sensor provided real-time data on the racket's angle and acceleration, which was processed by the ESP32.
2. Data Processing: Thresholds for acceptable racket angle and swing acceleration were pre-defined based on data collected from experienced players. These thresholds were used to compare the player's movements during each stroke.
3. Feedback Control: If the player's grip or swing deviated from the predefined thresholds, the system triggered the red LED and buzzer to indicate incorrect technique. If the player maintained the correct grip and swing, the green LED was activated.

2.4 *Flowchart*

The process flow of the system is outlined in the flowchart shown in Fig. 1. The sequence begins with the player preparing to hit the ball, during which the MPU6050 sensor continuously monitors the racket's angle and acceleration. The system processes the sensor data in real-time to detect whether the player is holding the racket correctly and swinging at an appropriate speed.

If an incorrect grip or swing is detected, the system activates the red LED and buzzer to alert the player. The player is then prompted to adjust their technique. If the system detects that the player is performing the stroke correctly, the green LED lights up, providing positive feedback.

2.5 *Experimental Setup*

2.5.1 Participants

The experimental study involved two categories of players: experienced players and beginner players. The experienced players provided a baseline dataset for ideal stroke

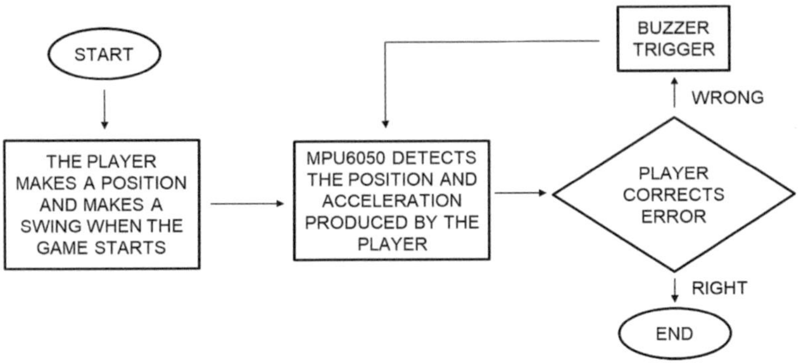

Fig. 1 Flowchart illustrating the operation of the sensor-based feedback system for grip and swing correction in table tennis training

mechanics, while the beginner players served as the target group for evaluating the effectiveness of the sensor-based training tool. Each participant used the sensor-equipped racket and wore a wrist-mounted ESP32 microcontroller during trials.

2.5.2 Experimental Procedure

The experiment was structured into three main phases:

A. **Baseline Measurement**

- Experienced players performed 50 forehand strokes to establish reference values for stroke consistency, grip stability, and angular motion patterns.
- Data from this group was used to define optimal threshold values for correct stroke mechanics, serving as a benchmark for comparison.

B. **System-Assisted Training for Beginners**

- Beginner players performed 50 forehand strokes while wearing the sensor-based system.
- The MPU6050 motion sensor continuously recorded bat angle (pitch and roll), stroke acceleration, and rotational movement.
- The system provided real-time feedback via LED and buzzer signals, enabling players to dynamically adjust their technique.

C. **Post-Practice Performance Analysis**

- The recorded data was analyzed to assess improvements in stroke stability, grip consistency, and acceleration control.
- Pre-training and post-training stroke variations were compared to evaluate the effectiveness of the real-time feedback mechanism.

2.5.3 Data Collection Processing

The MPU6050 motion sensor was configured to sample data at 100 Hz, ensuring high-resolution motion tracking. The ESP32 microcontroller processed and logged key performance metrics in real time, including:

- Stroke Speed (Angular Velocity): Captured using gyroscope data to analyze fluidity and control.
- Lateral and Vertical Stability: Analyzed through accelerometer data on the X-, Y-, and Z-axes to identify inconsistencies in stroke motion.
- All motion data was stored in a structured dataset for subsequent performance analysis and comparison.

2.5.4 Performance Evaluation Metrics

To validate the system's effectiveness, a comparative analysis was conducted between experienced and beginner players:

1. Good Strokes (Experienced Players): Displayed smooth acceleration profiles, minimal angular deviations, and stable rotational motion.
2. Poor Strokes (Beginner Players): Showed erratic fluctuations, inconsistent angular velocity, and unstable racket positioning.

The statistical evaluation of the collected data confirmed the system's capability to distinguish between well-executed and improper strokes. This validation reinforces its potential as an objective, sensor-driven training tool for table tennis players.

3 Results and Discussion

3.1 Overview

The experimental study aimed to evaluate the effectiveness of the sensor-based feedback system in improving forehand stroke consistency among beginner table tennis players. The results were analyzed by comparing stroke data collected from experienced players and beginners, focusing on key parameters such as bat angle stability, stroke speed, and lateral control.

Through the analysis of accelerometer and gyroscope readings, this section discusses how the system influences stroke execution, highlights observed trends and evaluates the potential impact of real-time feedback mechanisms on training effectiveness (Fig. 2).

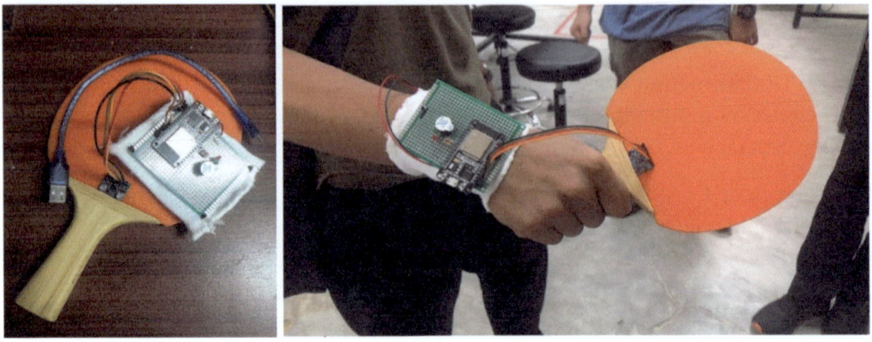

Fig. 2 Experimental setup of the sensor-based table tennis training system

3.2 Stroke Speed and Acceleration Analysis

The gyroscope readings, shown in Figs. 3, 4, and 5, were analyzed to evaluate the stroke speed and acceleration trends of both beginner and experienced players. The data revealed that experienced players exhibited smooth acceleration and deceleration phases, reinforcing well-coordinated and controlled stroke execution. Their gyroscope readings displayed periodic oscillations, which indicate predictable and stable stroke control.

Conversely, beginner players exhibited erratic fluctuations in stroke speed, particularly in the initial acceleration phase, as seen in Fig. 3 (Gyroscope X-Axis Data). Instead of a controlled motion, the data revealed abrupt changes in angular velocity,

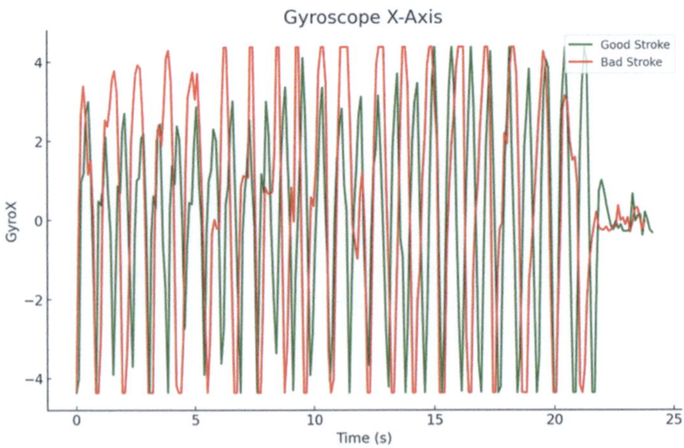

Fig. 3 Gyroscope X-axis data: stroke speed and rotational stability comparison between experienced and beginner players

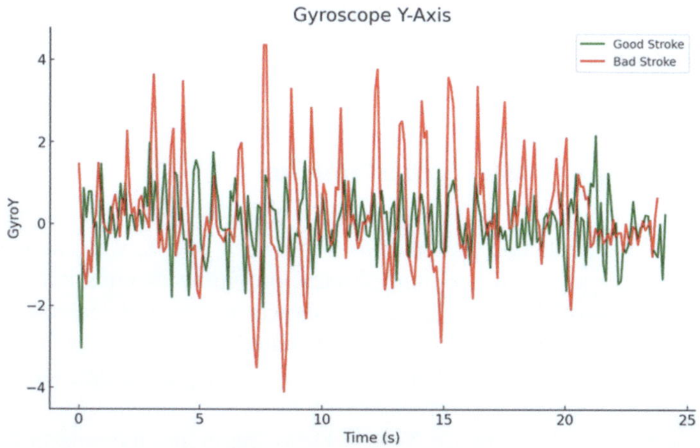

Fig. 4 Gyroscope Y-axis data: stroke control and rotational motion trends for experienced and beginner players

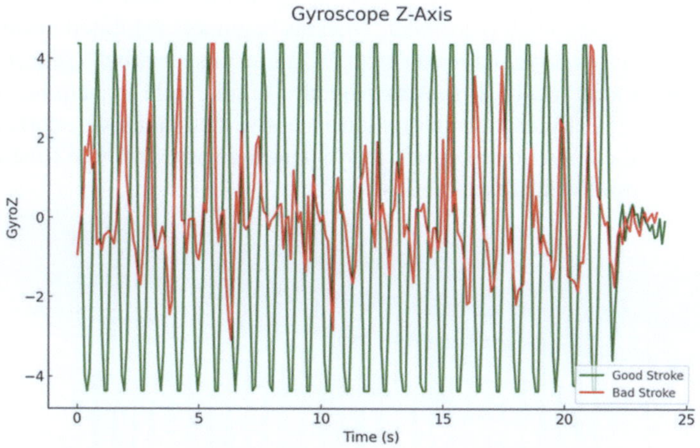

Fig. 5 Gyroscope Z-axis data: rotational movement variability between experienced and beginner players

reflecting inconsistent energy transfer and lack of follow-through mechanics. Similarly, Fig. 4 (Gyroscope Y-Axis Data) demonstrates how beginners struggled with rotational stability, often applying excessive or insufficient force in their swings. This inconsistency was further emphasized in Fig. 5 (Gyroscope Z-Axis Data), where irregular rotational movements were detected, affecting overall stroke efficiency.

The sensor-based feedback system addressed this issue by providing real-time alerts whenever irregular velocity patterns were detected, allowing players to adjust their motion dynamically. Over multiple trials, the ability to receive instant feedback helped beginners develop more stable acceleration trends, improving their overall stroke performance.

3.3 Lateral and Vertical Stroke Stability

The accelerometer readings, presented in Figs. 6, 7, and 8, were used to assess stroke stability across the X-, Y-, and Z-axes, providing insights into lateral and vertical stroke control. The X-axis motion, shown in Fig. 6 (Acceleration X-Axis Data), represents lateral movement and demonstrated that experienced players maintained a stable trajectory, while beginners displayed frequent erratic shifts, leading to inconsistent ball contact.

Similarly, Fig. 7 (Acceleration Y-Axis Data) highlights the side-to-side instability in beginner strokes, confirming that players had difficulty maintaining stroke alignment. Additionally, Fig. 8 (Acceleration Z-Axis Data) illustrates how vertical force application varied significantly between experienced and beginner players. The data indicates that experienced players applied controlled and periodic upward force, ensuring optimal ball placement, whereas beginners exhibited unstable force application, often resulting in weaker strokes and inconsistent ball contact.

The sensor-based feedback system played a crucial role in helping beginners refine their technique by alerting them when excessive deviations were detected. Over multiple trials, the reduction in erratic movement indicated that real-time feedback contributed to improved stroke stability.

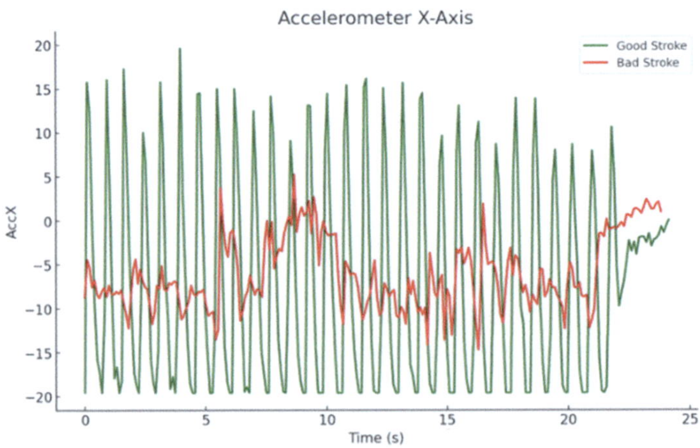

Fig. 6 Acceleration X-axis data: lateral stroke stability comparison between experienced and beginner players

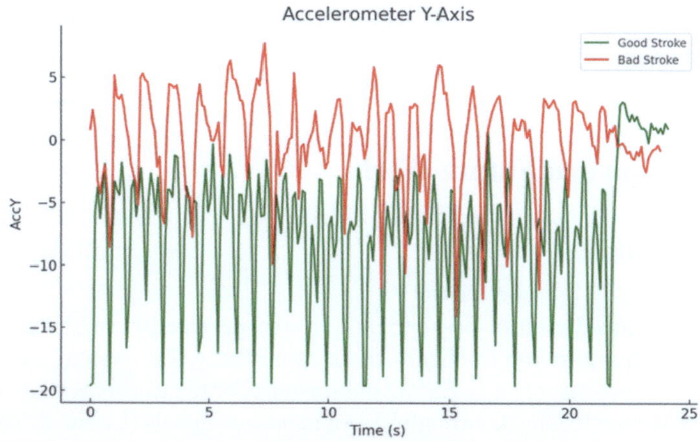

Fig. 7 Acceleration Y-axis data: side-to-side stroke control and consistency analysis

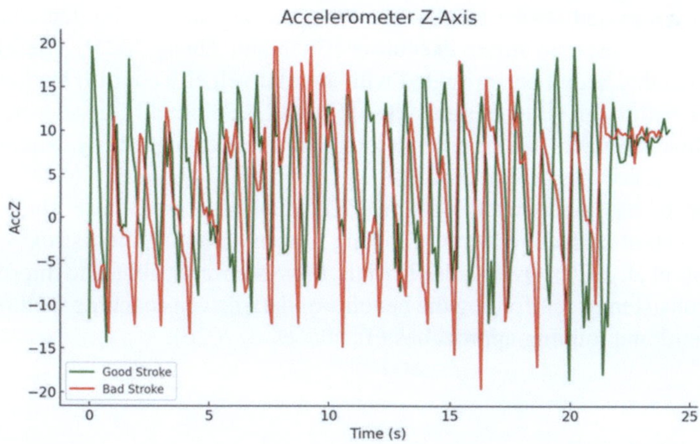

Fig. 8 Acceleration Z-axis data: vertical motion stability and stroke execution differences

3.4 Comparison with Traditional Training

When compared with traditional coaching methods, the sensor-based system provided certain advantages. Traditional coaching relies heavily on visual observation and verbal feedback from a coach, which can be subjective and may not always be immediate. In contrast, the sensor-based system provided instant and objective feedback, allowing players to adjust their technique in real time. This immediate feedback is crucial for beginners, as it prevents the formation of bad habits that can hinder long-term progress.

In addition, the system allowed players to practice independently without the need for constant supervision from a coach. This opens up opportunities for players who may not have regular access to professional coaching to improve their skills through self-guided practice.

4 Conclusion and Future Work

4.1 Conclusion

This study introduced a sensor-based training system designed to assist beginner table tennis players in improving forehand stroke consistency. The system integrated an MPU6050 motion sensor and an ESP32 microcontroller to measure stroke dynamics, providing real-time feedback through LED and buzzer signals (Lee et al. 2021). The experimental evaluation focused on stroke speed, acceleration trends, and stability, comparing performance between experienced and beginner players.

The results demonstrated that experienced players maintained stable acceleration trends and controlled stroke mechanics, whereas beginners exhibited erratic fluctuations and inconsistent stroke execution (Chen and Zhang 2021). The gyroscope data revealed that beginners struggled with angular velocity control, leading to poor stroke follow-through (Bagaber and Yusoff 2018). Additionally, accelerometer data indicated that beginners had significant lateral and vertical instability, which affected ball placement accuracy (Kim et al. 2019).

The sensor-based feedback system effectively addressed these challenges by providing instant corrective alerts, allowing players to adjust their strokes dynamically (Gupta et al. 2020). Over multiple trials, the system contributed to improvements in stroke consistency, reinforcing the benefits of data-driven coaching methodologies over conventional training approaches (Tanaka et al. 2020).

4.2 Limitation

Despite the system's effectiveness in enhancing forehand stroke mechanics, several limitations were identified. The current implementation only supports forehand stroke analysis, with no functionality for backhand strokes, spin tracking, or footwork evaluation (Karumdin et al. 2024). Additionally, the predefined stroke thresholds used for feedback were based on data from experienced players, which may not be universally applicable to all beginner playing styles (Li and Wang 2020).

Another limitation is the lack of personalized adaptation in the feedback mechanism. While the system provides alerts based on fixed thresholds, it does not yet have the capability to adapt dynamically to a player's improvement over time (Kumar et al. 2021).

4.3 Future Work

While the sensor-based training system has demonstrated its effectiveness in improving forehand stroke consistency, several key areas can be explored to enhance its functionality further. One of the most practical advancements is the integration of an AI-driven adaptive feedback mechanism. Currently, the system relies on fixed stroke thresholds derived from experienced players, which may not accommodate individual variations in playing styles. By incorporating machine learning algorithms, the system could analyze a player's progress over multiple training sessions and dynamically adjust feedback based on their learning curve. This would allow for personalized coaching, helping players refine their strokes in a way that aligns with their unique biomechanics (Lee et al. 2021).

Another crucial improvement is the expansion of the system to monitor additional stroke techniques, such as backhand strokes. The current implementation focuses solely on forehand strokes, limiting its application to a broader range of training scenarios (Chen and Zhang 2021). Enhancing the system to analyze both forehand and backhand mechanics would provide a more comprehensive training tool for table tennis players, enabling them to develop a well-rounded skill set.

Additionally, developing an interactive dashboard for real-time performance tracking would significantly improve usability. While the current system provides real-time feedback, it does not offer a structured way for players to review their stroke performance over time (Jamil and Yusoff 2016). A visual analytics dashboard displaying key parameters such as stroke stability, acceleration trends, and consistency scores would allow players and coaches to assess improvements more effectively and make informed adjustments to training strategies (Wang et al. 2019).

By implementing these enhancements, the sensor-based training system could evolve into an intelligent, data-driven coaching tool that offers real-time, personalized training feedback while providing valuable insights into player performance. These improvements would make the system more adaptive, user-friendly, and applicable to a wider range of training needs in table tennis (Amin et al. 2024).

References

Amin MRRM, Nasir A, Muhammad Sidik MH, Jamaludin ASB, Rosli AM (2024) Applying convolutional neural networks (CNNs): a machine learning method in modern sports. Lect Notes Bioeng 405–415

Bagaber SA, Yusoff AR (2018) Sustainable optimization of dry turning of stainless steel based on energy consumption and machining cost. Procedia CIRP 77:397–400. https://doi.org/10.1016/j.procir.2018.08.300

Chen M, Zhang L (2021) Real-time motion analysis in table tennis training using wearable technology. IEEE Trans Industr Electron 68(4):3210–3218

Chen Y et al (2020) A real-time coaching system for table tennis using wearable sensors. IEEE Trans Neural Syst Rehabil Eng 28(5):1150–1159

Gupta R et al (2020) Performance evaluation of table tennis players using wearable technology. IEEE Trans Instrum Meas 69(6):3421–3430
Jamil N, Yusoff AR (2016) Electromagnetic actuator for determining frequency response functions of dynamic modal testing on milling tool. Measurement 82:355–366. https://doi.org/10.1016/j.measurement.2016.01.016
Johnson M, Lee P (2020) Analyzing table tennis strokes with inertial measurement units. IEEE Access 8:123456–123465
Karumdin N, Mazlan MS, Saniman MNF, Amin MRRM (2024) Influence of injection molding manufacturing process on sports equipment durability. Lect Bioeng 431–441
Kim S et al (2019) Motion analysis of table tennis players using wearable sensors and machine learning. IEEE Trans Industr Inf 15(1):507–516
Kumar D et al (2021) Smart wearable system for real-time monitoring of table tennis players. IEEE Sens J 21(2):1234–1242
Lee J et al (2021) Wearable sensor system for quantitative assessment of table tennis skills. IEEE Trans Instrum Meas 70:1–9
Li F, Wang J (2020) Analysis of table tennis stroke techniques using wearable inertial sensors. IEEE Trans Hum Mach Syst 50(3):195–204
Ma D et al (2024) avaTTAR: table tennis stroke training with on-body and detached visualization in augmented reality. arXiv:2407.15373. https://arxiv.org/abs/2407.15373
Park S, Choi H (2021) Evaluation of table tennis player performance using IMU-based motion capture. IEEE Access 9:45678–45687
Patel A, Shah N (2020) Sensor-based analysis of table tennis techniques. IEEE Trans Comput Soc Syst 7(4):1012–1020
Smith J, Brown A (2020) Development of a wearable sensor system for table tennis skill assessment. IEEE Sens J 20(5):2500–2508
Tanaka K et al (2020) Development of a table tennis training support system using IMU sensors. IEEE Trans Biomed Eng 67(9):2542–2550
Wang L et al (2019) Real-time feedback system for table tennis training using IMU sensors. IEEE Trans Hum Mach Syst 49(5):485–495
Zhang H, Liu Y (2020) Design of a smart racket for table tennis training. IEEE Internet Things J 7(8):7505–7513

A Study on Muscle Activities During Stoop Lifting Movement

Nurhidayah Omar, Munawwarah Solihah Muhammad Isa, Mohammad Shahril Salim, Saidatul Ardeenawati Awang, and Suhizaz Sudin

Abstract Stoop lifting is used in many sports for functional purposes, especially when quick lifts or short bursts of strength are required such as in rugby, American football, and track and field. The purpose of this study was to observe the activity of the back muscles during stoop lifting. Seven healthy male subjects (age = 23 years; body height = 173 cm; body mass = 70.6 kg) participated in the study. All subjects needs to lift two different weights of weight plate. The muscle activities at eight different locations were measured to investigate muscle effort during stoop lifting. EMG for left and right side of the erector spinae, latisimus dorsi, external oblique and internal oblique were recorded and analyzed. The results showed that average RMS values of the back muscles increases as the weight of the load increases. In the case of repetitive movements which normally practices in sports, the study suggests a proper training techniques which emphasis on core stability, postural awareness, and gradual progression as this lifting method is risky on the lumbar spine health.

Keywords EMG · Root Mean Square (RMS) · Biomechanics

1 Introduction

Stoop lifting is a common lifting technique which have been used in various sports. Stoop lifting refers to a movement where an individual lifts an object while maintaining a bent posture at the waist or hips, rather than using a deep squat position. This technique, often contrasted with proper lifting techniques that involve the hips and knees in a deeper squat, has been widely discussed in sports and exercise science due to its potential implications on performance, injury risk, and biomechanics.

N. Omar (✉) · M. S. M. Isa
Institute of Engineering Mathematics, Universiti Malaysia Perlis, Arau, Perlis, Malaysia
e-mail: hidayahomar@unimap.edu.my

N. Omar · M. S. Salim · S. A. Awang · S. Sudin
Sport Engineering Research Centre, Universiti Malaysia Perlis, Arau, Perlis, Malaysia

M. H. A. Hassan et al. (eds.), *Proceedings of the 3rd International Conference on Innovation and Technology in Sports*, Lecture Notes in Bioengineering,
https://doi.org/10.1007/978-981-95-0804-4_47

Despite the potential injury risks, stoop lifting is still used in many sports for functional purposes, especially in disciplines requiring quick lifts or short bursts of strength. In sports like rugby, football, and track and field, athletes often use the stoop lift when they need to lift an object quickly or when there is a lack of time to perform a deeper squat or deadlift. Athletes involved in these sports often prioritize speed and efficiency over the biomechanics of the lift, and the stoop lift is preferred due to its quicker execution (Smitha et al. 2017; Fossmo and Tillaar 2022). In strongman competitions and crossfit, athletes are frequently trained to perform the stoop lift for lifting objects such as stones, logs, or other irregularly shaped objects. Athletes in these sports often perform the stoop lift due to the irregular shapes and shifting weight distributions, which necessitate less reliance on knee flexion and more on the back and posterior chain (Winwood 2011; Lee Bell et al. 2023). Stoop lifting can be used effectively in training for power and explosiveness as it demands quick and forceful activation of the posterior chain muscles (Crawley et al. 2024). The purpose of this study is to observe muscle activities during stoop lifting.

2 Methodology

Data collection was carried out in the Biomechanics laboratory at Faculty of Electronic Engineering and Technology (FKTEN), University Malaysia Perlis (UniMAP). Seven male students were participated in this study. The subjects had a mean age, height, weight of 23 years, 173 cm and 70.6 kg respectively. The subjects are healthy with no history of back disorders.

Muscle activities were collected using eight Trigno Wireless EMG System (Trigno™ Wireless System, Delsys Inc., Natick, MA, USA) and the Trigno parallel bar surface sensors (27 × 37 × 13 mm, Trigno™ Wireless, Delsys Inc.). EMG data collection and sensor configuration is initiated through Trigno software called EMG works Acquisition (Version 1.0.0.0, Delsys Inc.) installed on PC. The Trigno base station is connected to the PC via the USB port for data acquiring (Fig. 1).

The Adhesive Skin Interface is attached to the sensor for the sensor to be easily attached to the skin. By following the Delsys Trigno Wireless Biofeedback System user's guide, to attach the EMG sensors, the wireless surface EMGs bipolar electrodes are put oriented with the muscle fibers by aligning the sensors arrow with the direction of the underlying muscle fibers. Electromyographic activity was collected through the use of eight surface electromyography (EMG) bipolar electrodes. All eight sensors were placed at the skin of the back muscles on the left and right sides of erector spinae, latissimus dorsi, external oblique and internal oblique following the guideline as in Sarkar and Sarkar (2018).

All analyses were evaluated and processed using EMG works Analysis software (Version 1, Delsys Inc.).

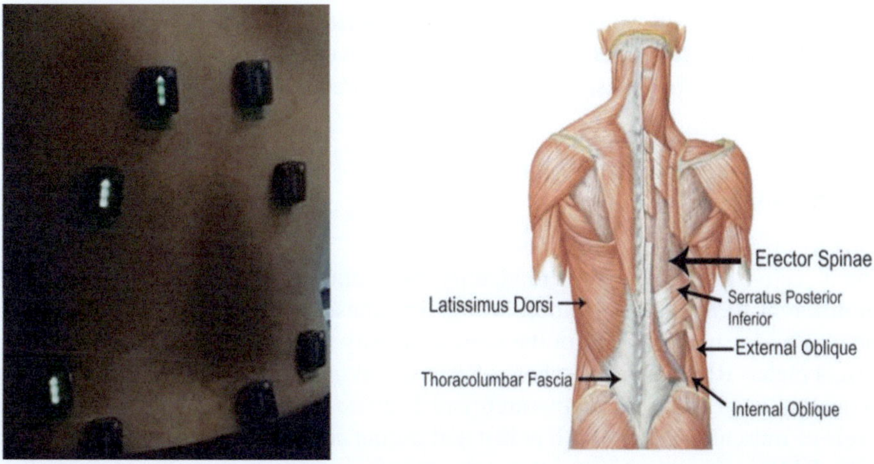

Fig. 1 Attachment of EMG sensors on the upper body

3 Results

The average root mean square (RMS) when stoop lifting for 3 kg and 5 kg is shown in Fig. 2.

There is a clear difference in RMS of muscle activity. The EMG analysis and RMS values demonstrated increases in signal amplitude as the weight increases. Muscle

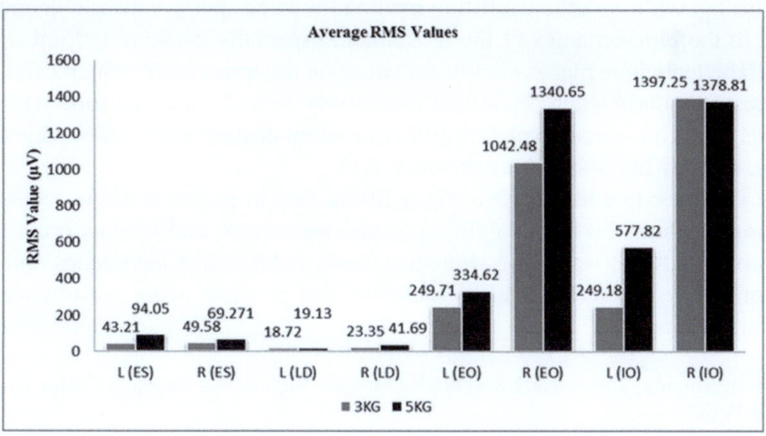

Fig. 2 Average root mean square (RMS) for lifting different weight plates in different back muscles. The bar graph illustrated the average RMS of the left and right erector spinae, latissimus dorsi, external oblique and internal oblique for all seven subjects for each weight plate. The grey area represented the RMS for lifting 3 kg weight plate while black region represented the RMS for lifting 5 kg weight plate

with the highest RMS value was internal oblique (IO) followed by external oblique (EO). Meanwhile, latissimus dorsi (LD) had the lowest RMS values where both right and left muscle were lowest in RMS when compared to other back muscles.

4 Discussion

The objective of this study was to identify the relationship of back muscle reactions on different weights during stoop lifting. The muscle pattern was evaluated using the root mean square (RMS) of the muscle activity for both lifting of 3 kg and 5 kg weight plates. RMS value was chosen because it is one of the parameters that is most frequently used in scientific research and because is more accurately depicts the levels of muscular activity both at rest and during contraction (Fukuda et al. 2010). Also, RMS is important to point out the muscle activation strength (Mohd Salleh et al. 2020).

Based on the results, lifting 5 kg weight plate produced higher RMS values than 3 kg. Past studies have reported the correlation of the muscle activities and the loads (Subramaniyam et al. 2019). Thus, as the weight increases, the muscle coactivity also increases (Davis and Marras 2000).

5 Conclusion

Stoop lifting, while an efficient lifting method for some sports, carries inherent risks related to the biomechanics of the movement, especially concerning lumbar spine health. The technique places significant stress on the lower back, which can lead to injuries, particularly when performed repetitively or with improper form. However, in sports requiring quick, explosive lifting or when dealing with awkwardly shaped objects, stoop lifting remains a necessary skill.

The key to safe and effective stoop lifting lies in proper training techniques, including emphasis on core stability, postural awareness, and gradual progression. Additionally, alternative lifting strategies, such as hip hinge techniques and squat variations, may reduce the load on the spine and promote better performance and injury prevention.

Acknowledgements This research was funded by Sport Engineering Research Centre, Universiti Malaysia Perlis.

References

Crawley K, Adams KJ, De Beliso M, Lawrence MM (2024) Effect of extreme volume-load differences for a single unilateral exercise during in-season resistance training on measures of bilateral strength, power, and speed in collegiate American football players. J Strength Cond Res 38(1):80–89

Davis KG, Marras WS (2000) Assessment of the relationship between box weight and trunk kinematics: does a reduction in box weight necessarily correspond to a decrease in spinal loading? Hum Factors 42(2):195–208

Fossmo JE, van den Tillaar R (2022) The effects of different relative loads in weight training on acceleration and acceleration from flying starts. Sports 10(148)

Fukuda TY, Echeimberg JO, Pompeu JE, Lucareli PRG, Garbelotti S, Gimenes RO (2010) Root mean square value of the electromyographic signal in the isometric torque of the quadriceps, hamstrings and brachial biceps muscles in female subjects. J Appl Res 10(1):32–39

Lee Bell L, Strafford BW, Coleman M, Korakakis PA, Nolan D (2023) Integrating deloading into strength and physique sports training programmes: an international delphi consensus approach. Sports Medi 9(87)

Mohd Salleh NA, Noor Sazali MSAB, Che Zakaria NA, Mustafah NM, Azaman A (2020) Electromyography analysis of lower extremity muscles during squat and stoop movement. IOP Conf Ser Mater Sci Eng 834(1)

Sarkar A, Sarkar D (2018) Early low back pain caused by bad posture and weak back and abdominal muscles. Int J Sci Res 7(7):8–14

Smitha TB, Hébert-Losier K, McClymont D (2017) An examination of the jump-and-lift factors influencing the time to reach peak catch height during a Rugby Union lineout. J Sports Sci

Subramaniyam M, Kim SE, Park SJ, Min SN (2019) Trunk muscle activities during lifting in upright and stooped postures. Int J Eng Res Technol 12(2):245–254

Winwood PW, Keogh JWL, Harris NK (2011) The strength and conditioning practices of strongman competitors. J Strength Cond Res 25(11):3118–3128

The Association Between Pain Characteristics, Psychological Aspects, Functional Disability and Postural Sway in Individuals with Nonspecific Chronic Low Back Pain: The Preliminary Study

Rahmat Adnan, Sarah Syafieqah Anuwar, Dzulhafiz Zikrie Dzolkapli, Siti Nurul Ezzah Che Man, Hammad Haziq Ismail, Muhammad Shaleh Mazuki, Motohide Arayama, Norasrudin Sulaiman, Hashbullah Ismail, Mazuin Mohd Razalli, and Shariman Ismadi Ismail

Abstract Pain characteristics, psychological and disability have been reported presence in Chronic Low Back Pain (CLBP). Alteration of postural balance and control have been found in individuals with CLBP. Currently, there are limited study to establish the relationship between pain characteristics, psychological, disability and postural instability in CLBP. Thus, this study aims to explore potential association between psychosocial factors and postural sway in CLBP. A total of 19 male volunteers with background of nonspecific CLBP were recruited. All participants completed in answering the self-reported questionnaire related to pain characteristics (Symptom duration, Current and Worst Numeric Rating Scale (NRS), psychological (Fear Avoidance Belief for Physical Activity (FABQ), Pain Coping Inventory Adaptive (PCIA), Pain Coping Inventory Maladaptive (PCIMA), Tampa Scale Kinesiophobia (TSK) and disability (Rolland Morris Disability Questionnaire (RMDQ), Oswestry Disability Index (ODI), Foot Ankle Disability Index (FADI)) prior to the experimental of postural balance test (Total Distance (TD) and Ellipse (EA)) measured using a plantar pressure measurement system. Result showed there are

R. Adnan (✉) · S. S. Anuwar · D. Z. Dzolkapli · S. N. E. C. Man · H. H. Ismail · M. S. Mazuki · N. Sulaiman · H. Ismail · S. I. Ismail
Faculty of Sports Science and Recreation, Universiti Teknologi MARA Shah Alam, Shah Alam, Selangor, Malaysia
e-mail: rahmatadnan@uitm.edu.my

M. M. Razalli
Department of Radiology, Hospital Al-Sultan Abdullah, UiTM Puncak Alam, Puncak Alam, Selangor, Malaysia

H. Ismail
Malaysia Exercise and Sports Science Association (MESSA), Kuala Lumpur, Malaysia

M. Arayama
DreamGP Inc, Osaka, Japan

M. H. A. Hassan et al. (eds.), *Proceedings of the 3rd International Conference on Innovation and Technology in Sports*, Lecture Notes in Bioengineering,
https://doi.org/10.1007/978-981-95-0804-4_48

significant positive association between duration of symptoms and PCIA ($r = -0.540$, $p = 0.025$) and PCIMA ($r = -0.570$, $p = 0.017$) in CLBP. However, there is only significant negative association between duration of symptoms and TD ($r = -0.533$, $p = 0.023$) in CLBP. No associations were found between duration of symptoms, FABQPA, TSK, RMDQ, ODI, FADI and EA ($p > 0.05$). A significant positive association between current NRS and FABQPA ($r = 0.531$, $p = 0.023$), and significant negative association between current NRS, TD ($r = -0.628$, $p = 0.009$) and EA ($r = -0.605$, $p = 0.013$) in CLBP. No association were found between current NRS, PCIA, PCIMA, TSK, RMDQ, ODI and FADI ($p > 0.05$). Furthermore, significant positive association between worst NRS, FABQPA ($r = 0.520$, $p = 0.043$), RMDQ ($r = 0.565$, $p = 0.023$), and negative association worst NRS and EA ($r = -0.604$, $p = 0.013$) in CLBP. No association were found between worst NRS, PCIA, PCIMA, TSK, ODI, FADI and TD ($p > 0.05$). Regarding psychological aspects, a significant negative association between FABQPA and FADI ($r = -0.584$, $p = 0.014$), positive association between PCIMA and TD ($r = 0.491$, $p = 0.045$) in CLBP. No association between FABQPA, PCIA, PCIMA, TSK, RMDQ, ODI, and EA were found. Lastly, regarding functional disability there are significant negative association between RMDQ ($r = -0.852$, $p = 0.001$), ODI ($r = -0.610$, $p = 0.009$) and FADI in CLBP. No association found between RMDQ, ODI, TD and EA ($p > 0.05$). This study has demonstrated the magnitude of relationship within pain characteristics, psychological, disability and postural sway in CLBP.

Keywords Psychosocial · Chronification · Proprioception · Low back pain · Postural sway

1 Introduction

Low back pain (LBP) is one of the most prevalent musculoskeletal disorders with huge impact on socioeconomic problems. About 80% of the complaints cannot been recognized to specific disease or anatomic abnormality, thus have been classified as nonspecific LBP (Bardin et al. 2017). Patients with LBP have different contributors to their condition and marked with differences during their LBP experience. Up to 10% of acute LBP complaints will develop into chronic LBP (CLBP) (Balagué et al. 2012; Hoogen et al. 1998).

Individuals with CLBP exhibit greater postural instability and control which further lead to postural sway (Ruhe et al. 2011). Alteration postural control mechanisms such as deterioration of the proprioceptive information are believed to be affected in CLBP patients (Casadio et al. 2005). Reduces the proprioception accuracy resulting in an imprecise estimation of center of mass position, hence compensating the center of pressure shift. In the study by Koch and Hansel, CLBP is often associated with restricted trunk movement by loss input lumbar proprioception but shift other sensory such as altering the stability surfaces or visual feedback (Koch and Hänsel 2019).

Several factors influence on the progression to chronicity in LBP such as longer duration of symptoms, psychological distress, higher disability level, psychological factors like patients attitude and belief, pain and movement related fear, pain coping and depression are important in CLBP (Adnan et al. 2020). Fear avoidance model first coined by Vlaeyen and Linton on a vicious cycle ensues negative cognitive behavior precede to avoidance behavior and end up with disability and depression in CLBP (Vlaeyen et al. 2016).

To gain further insight into relationships between pain characteristics, psychological, functional disability and postural control in CLBP. Our aim was to know what magnitude of association reveals between pain characteristics, psychological, functional disability and postural sway in CLBP patients.

2 Methodology

2.1 Participants

A total of 19 male participants were selected for this study. Non-specific CLBP is defined as pain in the whole lumbar region that has no clear anatomical cause and no recognizable medical history that could explain the complaints (Airaksinen et al. 2006). The LBP complaints had to be of such severity that the participant had consulted a physical therapist (PT) or medical doctor (MD) regarding the complaints. Individuals with a history of spinal surgery, spinal traumata, spinal deformities or severe orthopedic, respiratory, inflammatory, metabolic, neurologic or cardiovascular diseases were not eligible for study participation. The patients with CLBP had to suffer from LBP $\geq$ 3 months to be included in the study.

2.2 Research Procedure

This observational study took place at the Human Performance Laboratory Faculty Sports Science and Recreation UiTM Shah Alam and was approved by University ethical committee (FERC/11/2024/UG/MR/0443). All participants provided signed informed consent prior to the commencement of the study. During the experimental session participants were asked to fill out self-reported questionnaires related to pain intensity, psychological state, functional disability related to LBP and ankle. Subsequently, postural sway was assessed with plantar pressure measurement system during bipedal standing (Fig. 1).

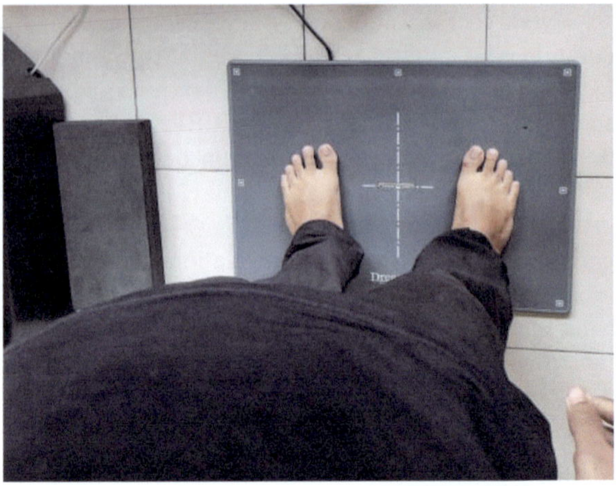

Fig. 1 The starting position of the bipedal standing test. Each subject was instructed to remain stationary with barefoot for 40 s during the trial

2.3 Pain Characteristics Outcome

Pain characteristics studied included the symptom duration refers to the first onset of LBP (in months) was recorded (Airaksinen et al. 2006). The NRS were used to examine current LBP intensity, and worst LBP intensity ever experienced. The NRS ranged from 0 to 10, with 0 representing "no pain" and 10 representing "unbearable pain" (Downie et al. 1978; Jensen et al. 1994; Price et al. 1994).

2.4 Psychological Related Outcome

Psychological aspects were assessed through validated and reliable self-report questionnaires. This study used fear avoidance belief questionnaire (FABQ) based on 16 questions. The present study was conducted on young adults, hence only physical activity subscale is used. A 7-point Likert scale, where 0 corresponded to "completely disagree", and 6 denoted "completely agree". The total scores spanned from 0 to 24, with higher scores meaning heightened fear-avoidance beliefs. FABQ known to be valid and reliable and responsive in CLBP (Osuka 2024). To measure cognitive and behaviour coping strategies, this study used pain coping inventory (PCI). PCI. Contains 33 questions consists of 6 domains of cognitive and behavioural strategies dealing with chronic. The first 3 domain that higher score refer as transformation, distraction and reducing demand know as adaptive strategies. The last 3 domain such as retreating, worrying and resting is dimension of maladaptive strategies. All questions rated according to a 4—point likert scale ranging from 1 (hardly ever) to 4 (very often) in the frequency when dealing with the pain (Kraaimaat and Evers 2003). The Tampa Scale Kinesiophobia (TSK) which consists of 17 items, was used to assess fear of movement. Tampa Scale for Kinesiophobia scores range from 17 to 68, with scores >37 indicating high fear of movement (Roelofs et al. 2004).

2.5 *Functional Disability*

Functional characteristics were measured using the Rolland Morris Disability Questionnaire (RMDQ), which evaluated how LBP impacts the functioning of daily physical activities thus providing an insight into the degree of LBP-related disability (Roland and Morris 1983). The RMDQ contains 24 items of which a total score ranging between 0 (no disability) and 24 (severe disability) is calculated, a with higher score indicating a higher degrees of LBP-related disability (Stevens et al. 2016). The RMDQ has a good internal consistency, test–retest reliability, and construct validity (Brouwer et al. 2004; Smeets et al. 2011; Turner et al. 2003; Beurskens et al. 1995). The Oswestry Disability Index (ODI) was used to measure LBP-related disability. 31 This questionnaire examines the disabling effects of LBP on the following 10 domains: pain intensity, personal care, lifting, walking, sitting, standing, sleeping, sexual function, social interaction, and traveling. Domain scores were retrieved, summed, and multiplied by 2 to calculate the percentage of LBP-related disability. The highest percentages refer to more severe degrees of disability. The ODI is known to be valid, reliable, and responsive in LBP patients (Fairbank and Pynsent 2000).

2.6 *Postural Balance*

The participants postural sway was assessed using a plantar pressure measurement system. All participants were instructed to complete trial of a 40-s quasi-static stance with their eyes open on a plantar pressure measurement system at a frequency of 300 Hz (Materialise, Belgium). Participants were instructed to stand still in a barefoot bipedal stance with an upright posture, arms hanging loosely by their sides, and their gaze directed forward. The stance width was selected individually (Heß et al. 2021). Data on the total travel distance (TD) of the center of pressure (COP) and the ellipse area (EA) of the COP movement measured from the foot's plantar pressure measurement system were recorded for postural balance analysis (Fig. 1).

2.7 Data Analysis

Statistical analyses were performed using IBM SPSS Statistics 22 (SPP Inc., Chicago, IL, USA). The baseline descriptive statistics for age, height weight, symptom duration, current NRS of pain intensity, Worst NRS of intensity, FABQPA, PCI, TSK, RMDQ, ODI, FADI, TD and EA were reported. Descriptive (mean, median, standard deviation (SD), and range (minimum to maximum)) were calculated and normality of the data was assessed using the Shapiro–Wilk's test and visual assessment of the histograms and boxplots. To quantify the bivariate relationship between variables, Pearson for parametric data and spearman for nonparametric correlation were calculated with $p < 0.05$.

3 Results

3.1 Participants Demographics

Table 1 reported the participants demographics profile and all their baseline related assessments. A total of 19 male participants with mean age of 22.3 ± 1.3 years and the mean of duration complaint of 31.89 ± 33.3 month from LBP. The fear avoidance belief from physical activity mean was 11.9 ± 5.2. The total mean score of pain PCIA was 32.9 ± 5.1 and PCIMA was 51.2 ± 10.4. For the functional disability, RMDQ mean score was 7.6 ± 4.3, the ODI mean score was 15.3 ± 10.6 and FADI mean score was 91.3 ± 23.1. Lastly, the mean score of TD was 55.8 ± 54.6 and mean score of EA was 8.2 ± 16.0.

3.2 Bivariate Correlation Between Pain Characteristics, Psychological Factors, Functional Disability and Postural Sway

Table 2 presents the correlations coefficients between variables. For pain characteristics, there are significant positive associations between duration of symptoms and PCIA ($r = -0.540$, $p = 0.025$) and PCIMA ($r = -0.570$, $p = 0.017$) in CLBP. However, there is only a significant negative association between duration of symptoms and TD ($r = -0.533$, $p = 0.023$) in CLBP. No associations were found between duration of symptoms, FABQPA, TSK, RMDQ, ODI, FADI and EA ($p > 0.05$). A significant positive association between current NRS and FABQPA ($r = 0.531$, $p = 0.023$), and significant negative association between current NRS, TD ($r = -0.628$, $p = 0.009$) and EA ($r = -0.605$, $p = 0.013$) in CLBP. No association were found between current NRS, PCIA, PCIMA, TSK, RMDQ, ODI and FADI ($p > 0.05$).

Table 1 Participants Demographic

	Mean	Standard deviation
Personal		
Age (years)	22.6	1.3
Height (cm)	170.2	4.8
Weight (Kg)	67.8	11.5
Pain characteristics		
Duration of symptoms (month)	31.89	33.3
Current NRS	4.7	1.9
Worst NRS	6	1.2
Psychological		
FABQPA	11.9	5.2
PCIA	32.9	5.1
PCIMA	51.2	10.4
Functional disability		
RMDQ	7.6	4.3
ODI	15.3	10.6
FADI	91.3	23.1
Postural sway		
TD	55.8	54.6
EA	8.2	16.0

cm = centimeters, Kg = kilogram, NRS = numeric rating scale, FABQPA = fear avoidance belief questionnaire from physical activity, PCIA = pain coping inventory adaptive, PCIMA = pain coping inventory maladaptive, RMDQ = Rolland Morris disability questionnaire, ODI = oswestry disability index, FADI = foot ankle disability index, TD = total distance, EA = ellipse area

Furthermore, significant positive association between worst NRS, FABQPA (r = 0.520, $p = 0.043$), RMDQ (r = 0.565, $p = 0.023$), and negative association worst NRS and EA (r = −0.604, $p = 0.013$) in CLBP. No association were found between worst NRS, PCIA, PCIMA, TSK, ODI, FADI and TD ($p > 0.05$). Regarding psychological aspects, a significant negative association between FABQPA and FADI (r = −0.584, $p = 0.014$), positive association between PCIMA and TD (r = 0.491, $p = 0.045$) in CLBP. No association between FABQPA, PCIA, PCIMA, TSK, RMDQ, ODI, and EA were found. Lastly, regarding functional disability there are significant negative association between RMDQ (r = −0.852, $p = 0.001$), ODI (r = −0.610, $p = 0.009$) and FADI in CLBP. No association found between RMDQ, ODI, TD and EA in CLBP ($p > 0.05$).

Table 2 Bivariate correlation between pain characteristics, psychological factors, functional disability and postural sway

	Duration of symptoms	Current NRS	Worst NRS	FABQPA	PCIA	PCIMA	TSK	RMDQ	ODI	FADI	TD	EA
Duration of symptoms (month)	1.00			– 0.135, 0.594	**r = – 0.540, 0.025***	**r = – 0.570, 0.017***	0.140, 0.604	0.317, 0.232	– 0.059, 0.823	– 0.177, 0.469	**r = – 0.533, 0.023***	– 0.357, 0.146
Current NRS		1.00		**r = 0.531, 0.023***	– 0.068, 0.801	0.010, 0.969	0.240, 0.370	0.456, 0.076	0.101, 0.710	– 0.425, 0.101	**r = – 0.628, 0.009***	**r = – 0.605, 0.013***
Worst NRS			1.00	**r = 0.520, 0.043***	– 0.080, 0.769	0.060, 0.826	– 0.141, 0.601	**r = 0.565, 0.023***	0.472, 0.065	– 0.351, 0.183	– 0.324, 0.22	**r = – 0.604, 0.013***
FABQPA				1.00				0.561, 0.024*	0.420, 0.065	**r = – 0.584, 0.014***	– 0.070, 0.782	– 0.122, 0.629
PCIA					1.00			0.147, 0.588	0.432, 0.083	– 0.399, 0.112	0.440, 0.077	0.221, 0.394
PCIMA						1.00		0.310, 0.242	0.474, 0.055	– 0.453, 0.068	**r = 0.491, 0.045***	0.290, 0.259
TSK							1.00	0.226, 0.406	– 0.124, 0.649	– 0.332, 0.210	– 0.484, 0.057	– 0.345, 0.190
RMDQ								1.00		**r = – 0.852, 0.001****	– 0.312, 0.239	– 0.377, 0.150
ODI									1.00	**r = – 0.610, 0.009****	0.187, 0.472	– 0.078, 0.765

(continued)

Table 2 (continued)

	Duration of symptoms	Current NRS	Worst NRS	FABQPA	PCIA	PCIMA	TSK	RMDQ	ODI	FADI	TD	EA
FADI										1.00	− 0.039, 0.881	0.080, 0.759
TD											1.00	

* Correlation is significant at the 0.05 level (2-tailed)

** Correlation is significant at the 0.01 level (2-tailed)

cm = centimeters, Kg = kilogram, NRS = numeric rating scale, FABQPA = fear avoidance belief questionnaire from physical activity, PCIA = pain coping inventory adaptive, PCIMA = pain coping inventory maladaptive, RMDQ = Rolland Morris disability questionnaire, ODI = oswestry disability index, FADI = foot ankle disability index, TD = total distance, EA = ellipse area

4 Discussions

This study examined the relationships between pain characteristics, psychological, functional disability and postural sway in CLBP patients. It was found that pain characteristics such as longer symptom duration, higher current LBP intensity and worst LBP intensity is associated with lower postural sway in CLBP. More interestingly, longer symptom duration of LBP associated with both adaptive and maladaptive coping pain strategies in CLBP patients. Patients experience with increased current LBP and worst LBP intensity is associated with a higher fear avoidance from physical activity. Furthermore, a worst LBP experience associated with higher LBP disability in CLBP patients. Regarding psychological aspects, patients embrace with higher fear avoidance belief from physical activity is associated with highly ankle disability in CLBP. Moreover, patients with higher score of maladaptive coping strategies is associated with higher postural sway in CLBP. Finally, patients with higher disability associated with higher ankle disability in CLBP.

Our study revealed patients with longer symptom duration of CLBP, higher current LBP intensity and worst LBP intensity associated with lower postural sway. This finding however in contrast with our earlier hypotheses that prolonged symptoms, higher pain intensity and worst pain intensity of CLBP could be associated with greater postural sway. First, it seems evident that besides sway strategy, postural bracing could be another way of CLBP patients' strategies to live with persistent pain. Therefore, adaptation to the persistent pain could explain why CLBP patients might have better postural bracing strategies. This study outcome was in line with other past studies that have investigated the changes of postural sway among low back pain patients were inconsistent and majority did not find increased postural sway in low back pain patients (Mazaheri et al. 2013). This inconsistency could be related to the duration of the pain experienced and analysing the relationship of LBP and postural sway could be more practical when the patients is categorized based on the symptom's duration. Further research is needed to confirm this claim. Moreover, CLBP patients who have longer symptom duration is also associated with lower pain coping strategies. This should suggest that prolonged CLBP symptoms might give positive effect to live with persistent pain.

Current LBP intensity and worst LBP intensity is associated with higher fear avoidance belief and worst LBP intensity associated with higher disability in CLBP patients. This important finding appear that higher current LBP intensity and worst LBP intensity have direct impact on fear avoidance belief from physical activity and disability related LBP.

This study was the first to establish maladaptive strategies associated with impair postural balance in CLBP. Hence, overcoming the maladaptive coping strategies might improve postural sway in CLBP. Future research should explore the role of coping pain in reducing alteration of postural sway in CLBP.

Our finding shows that LBP-disability and fear avoidance belief is associated with foot-disability in CLBP. It seem the impairment of LBP may not only present in lumbosacral regions but also widespread to distal region such as foot/ankle function Some previous studies showed foot mechanics influence low back function, for example foot eversion is associated with anterior pelvic tilt (Pinto et al. 2008; Betsch et al. 2011), lateral forefoot associated with paraspinal muscle activity (Bird et al. 2003) and low foot arch reduce shock absorption at low back muscle (Ogon et al. 1999). These might be explained the theory of compensation strategies between LBP and foot/ankle function in CLBP. This claim however requires further investigation to support this theory.

5 Conclusions

This preliminary study suggests pain characteristics, psychological factors, functional disability and postural control is associated in CLBP patients. Continuing with the current work should replicate these results and pursue exploration of the potential complex relation between pain-psychological and underlying mechanisms in chronification of LBP.

Acknowledgements Special thanks to all the participants in this research.

References

Adnan R et al (2020) Differences in psychological factors, disability and fatigue according to the grade of chronification in non-specific low back pain patients: a cross-sectional study. J Back Musculoskelet Rehabil 33(6):919–930

Airaksinen O et al (2006) Chapter 4 European guidelines for the management of chronic nonspecific low back pain. Eur Spine J 15:s192–s300

Balagué F et al (2012) Non-specific low back pain. Lancet 379(9814):482–491

Bardin LD, King P, Maher CG (2017) Diagnostic triage for low back pain: a practical approach for primary care. Med J Aust 206(6):268–273

Betsch M et al (2011) Influence of foot positions on the spine and pelvis. Arthritis Care Res 63(12):1758–1765

Beurskens AJ, Köke A, Knipschild P (1995) Measuring the functional status of patients with low back pain. Assessment of the quality of four disease-specific questionnaires. Spine 20(9):1017–1028

Bird AR, Bendrups AP, Payne CB (2003) The effect of foot wedging on electromyographic activity in the erector spinae and gluteus medius muscles during walking. Gait Posture 18(2):81–91

Brouwer S et al (2004) Reliability and stability of the Roland Morris disability questionnaire: intra class correlation and limits of agreement. Disabil Rehabil 26(3):162–165

Casadio M, Morasso PG, Sanguineti V (2005) Direct measurement of ankle stiffness during quiet standing: implications for control modelling and clinical application. Gait Posture 21(4):410–424

Downie W et al (1978) Studies with pain rating scales. Ann Rheum Dis 37(4):378–381

Fairbank JC, Pynsent PB (2000) The Oswestry disability index. Spine 25(22):2940–2953

Heß T et al (2021) Postural performance and plantar cutaneous vibration perception in patients with idiopathic normal pressure hydrocephalus. Heliyon **7**(1)
Jensen MP, Turner JA, Romano JM (1994) What is the maximum number of levels needed in pain intensity measurement? Pain 58(3):387–392
Koch C, Hänsel F (2019) Non-specific low back pain and postural control during quiet standing—a systematic review. Front Psychol 10:586
Kraaimaat FW, Evers AW (2003) Pain-coping strategies in chronic pain patients: psychometric characteristics of the pain-coping inventory (PCI). Int J Behav Med 10:343–363
Mazaheri M et al (2013) Low back pain and postural sway during quiet standing with and without sensory manipulation: a systematic review. Gait Posture 37(1):12–22
Ogon M et al (1999) Does arch height affect impact loading at the lower back level in running? Foot Ankle Int 20(4):263–266
Osuka S et al (2024) Fear-avoidance beliefs associated with non-specific chronic low back pain in college athletes. J Pain Res 2024:285–292
Pinto RZ et al (2008) Bilateral and unilateral increases in calcaneal eversion affect pelvic alignment in standing position. Man Ther 13(6):513–519
Price DD et al (1994) A comparison of pain measurement characteristics of mechanical visual analogue and simple numerical rating scales. Pain 56(2):217–226
Roelofs J et al (2004) The Tampa scale for Kinesiophobia: further examination of psychometric properties in patients with chronic low back pain and fibromyalgia. Eur J Pain 8(5):495–502
Roland M, Morris R (1983) A study of the natural history of back pain: part I: development of a reliable and sensitive measure of disability in low-back pain. Spine 8(2):141–144
Ruhe A, Fejer R, Walker B (2011) Center of pressure excursion as a measure of balance performance in patients with non-specific low back pain compared to healthy controls: a systematic review of the literature. Eur Spine J 20:358–368
Smeets R et al (2011) Measures of function in low back pain/disorders: low back pain rating scale (LBPRS), Oswestry disability index (ODI), progressive isoinertial lifting evaluation (PILE), Quebec back pain disability scale (QBPDS), and Roland-Morris disability questionnaire (RDQ). Arthritis Care Res 63(S11):S158–S173
Stevens ML, Lin CC, Maher CG (2016) The roland morris disability questionnaire. J Physiother 62(2):116
Turner JA et al (2003) Comparison of the Roland-Morris disability questionnaire and generic health status measures: a population-based study of workers' compensation back injury claimants. Spine 28(10):1061–1067
van den Hoogen HJ et al (1998) On the course of low back pain in general practice: a one year follow up study. Ann Rheum Dis 57(1):13–19
Vlaeyen JW, Crombez G, Linton SJ (2016) The fear-avoidance model of pain. Pain 157(8):1588–1589

MIX
Papier aus verantwortungsvollen Quellen
Paper from responsible sources
FSC® C105338

If you have any concerns about our products,
you can contact us on
ProductSafety@springernature.com

In case Publisher is established outside the EU,
the EU authorized representative is:
Springer Nature Customer Service Center GmbH
Europaplatz 3, 69115 Heidelberg, Germany

Printed by Libri Plureos GmbH
in Hamburg, Germany